FUNGAL BIOREMEDIATION
Fundamentals and Applications

T0203828

Editors

Araceli Tomasini
Departamento de Biotecnología
Universidad Autónoma Metropolitana-Iztapalapa
Ciudad de Mexico
Mexico

H. Hugo León-Santiesteban
Departamento de Energía
Universidad Autónoma Metropolitana-Azcapotzalco
Ciudad de Mexico
Mexico

CRC Press
Taylor & Francis Group
Boca Raton London New York

CRC Press is an imprint of the
Taylor & Francis Group, an **informa** business
A SCIENCE PUBLISHERS BOOK

Cover illustrations reproduced by kind courtesy of Arturo Ruiz Lara and Diana Cuervo Maya

CRC Press
Taylor & Francis Group
6000 Broken Sound Parkway NW, Suite 300
Boca Raton, FL 33487-2742

First issued in paperback 2021

Library of Congress Cataloging-in-Publication Data

Names: Tomasini, Araceli, editor. | Leon-Santiesteban, H. Hugo (Hector Hugo), editor.
Title: Fungal bioremediation : fundamentals and applications / editors, Araceli Tomasini (Departamento de Biotecnologia, Universidad Autonoma Metropolitana-Iztapalapa, Ciudad de Mexico, Mexico), H. Hugo Leâon-Santiesteban (Departamento de Energia, Universidad Autonoma Metropolitana-Azcapotzalco, Ciudad de Mexico, Mexico).
Description: Boca Raton, FL : CRC Press, 2019. | "A science publishers book." | Includes bibliographical references and index.
Identifiers: LCCN 2018057729 | ISBN 9781138636408 (hardback)
Subjects: LCSH: Fungal remediation. | Bioremediation. | Factory and trade waste--Biodegradation.
Classification: LCC TD192.72 .F8425 2019 | DDC 628.5--dc23
LC record available at https://lccn.loc.gov/2018057729

Visit the Taylor & Francis Web site at
http://www.taylorandfrancis.com

and the CRC Press Web site at
http://www.crcpress.com

Preface

The large number of people that inhabit our planet and all their activities have caused severe environmental pollution problems and disruption of the ecological balance. The air, ground, and water are polluted at different levels depending on the geographic zone. Some of these pollutants are compounds of natural origin or xenobiotics. Xenobiotics are human-made compounds for a specific use. Many researchers have dedicated themselves to studying and understanding the biodegradation of the pollutants, with the goal of proposing processes for their mitigation. These processes may be biological and physicochemical; and each of them has advantages and disadvantages. One of the main problems that prevent institutions and businesses from applying these processes is that the cost is very high without giving anything in return economically. Biological processes, also known as bioremediation, are less expensive to apply than the physicochemical, since they involve the use of living organisms, such as plants, bacteria, fungi, and/or algae.

We know that nature has her own ways of eliminating organic waste. One of the main actors involved in these processes are the filamentous fungi. However, the large number and diversity of compounds that are disposed of into the environment has overcome the natural capacity of these organisms.

In this book, studies in fungal bioremediation conducted by experts are mentioned. Studies of fungi bioremediation began since the second half of the last century. And there have been many advancements in different topics, such as the isolation of xenobiotic-degrading fungi, degradation mechanisms, enzyme involved in degradation, molecular studies, proteomic studies, microscopy studies, and application technologies.

This book shows the newest advancements in the field of fungal bioremediation, and is mainly addressed to grad students, as well as to professionals and researchers. The book has three parts: fundamentals, applications, and technologies and tools. Chapters were written by leading researchers who are doing work on fungi and bioremediation. In the first part, the characteristics of filamentous fungi and their role in the environmental remediation are briefly described. The main enzymes that participate in the processes of degradation of lignin and toxic compounds are also described. Cutting edge knowledge on fungal enzymes involved in the degradation of toxic compounds, their production, their activity, and their overexpression are written in this section.

Each chapter in the second part of the book deals with the degradation and/or removal of different toxic compounds. Examples of degradation of toxic xenobiotics are mentioned, which currently cause serious pollution problems, including emerging pollutants. The degradation of lignin or hydrocarbons are not mentioned, because both topics are mentioned in all referenced books of fungal bioremediation.

The reader can get information about state-of-the-art advancements and knowledge in the field of remediation by fungi of chlorophenols, dyes, heavy metals, emerging pollutants, radioactive compounds, as well as pollutants present in the air. The chapters mention the most used fungi, the detoxification mechanisms of the compounds, and the most relevant research conducted till date.

In the third part of the book, tools and cutting-edge technologies used in the study of the removal and degradation of pollutants by fungi are presented. The first chapter of this part mentions how proteomics has helped to understand some of the degradation mechanisms done by fungi. The second one describes genomics as a very important tool in the understanding of the fungal metabolism, and technological advancements, such as the use of different bioreactors for the culture of fungi. The last chapter describes a tool which is not mentioned in other books about these topics, but which is just as important: microscopy. Topics ranging from the most basic techniques to the forefront of microscopy are discussed.

We want to thank all contributing authors that put in many hours to the creation of this book in an enthusiastic and professional manner. Thank you to all the authors that helped us achieve this project in the best way possible. Finally, we want to thank our families for their patience, support, and help, which were fundamental for the creation of this book.

Araceli Tomasini
H. Hugo León-Santiesteban
August 2018

Contents

Preface   iii

PART I: FUNDAMENTALS

1. **The Role of the Filamentous Fungi in Bioremediation**   3
   Araceli Tomasini and H. Hugo León-Santiesteban

2. **Fungal Peroxidases Mediated Bioremediation of Industrial Pollutants**   22
   Misha Ali, Qayyum Husain and Hassan Mubarak Ishqi

3. **Phenoloxidases of Fungi and Bioremediation**   62
   Montiel-González Alba Mónica and Marcial Quino Jaime

PART II: APPLICATIONS

4. **Chlorophenols Removal by Fungi**   93
   H. Hugo León-Santiesteban and Araceli Tomasini

5. **Azo Dye Decoloration by Fungi**   125
   M. R. Vernekar, J. S. Gokhale and S. S. Lele

6. **Fungal Processes of Interaction with Chromium**   173
   P. Romo-Rodríguez and J. F. Gutiérrez-Corona

7. **Removal of Emerging Pollutants by Fungi: Elimination of Antibiotics**   186
   Carlos E. Rodríguez-Rodríguez, Juan Carlos Cambronero-Heinrichs,
   Wilson Beita-Sandí and J. Esteban Durán

8. **Fungi and Remediation of Radionuclide Pollution**   238
   John Dighton

9. **Removal of Gaseous Pollutants from Air by Fungi**   264
   Alberto Vergara-Fernández, Felipe Scott, Patricio Moreno-Casas and
   Sergio Revah

PART III: USEFUL TOOLS

10. **A Learning Journey On Toxico-Proteomics:**   287
    The Neglected Role of Filamentous Fungi in
    the Environmental Mitigation of Pentachlorophenol
    Celso Martins, Isabel Martins, Tiago Martins, Adélia Varela and
    Cristina Silva Pereira

11. **Bioreactors and other Technologies used in Fungal Bioprocesses**   319
    Avinash V. Karpe, Rohan M. Shah, Vijay Dhamale, Snehal Jadhav,
    David J. Beale and *Enzo A. Palombo*

12. **Electron Microscopy Techniques Applied to Bioremediation and**   354
    Biodeterioration Studies with Moulds: State of the Art and
    Future Perspectives
    Antoni Solé, Maria Àngels Calvo, María José Lora and
    Alejandro Sánchez-Chardi

Index   387

PART I
FUNDAMENTALS

CHAPTER-1

The Role of the Filamentous Fungi in Bioremediation

Araceli Tomasini[1,]* and *H. Hugo León-Santiesteban*[2]

1. Introduction

Environmental pollutants are a serious problem worldwide. They damage the ecosystems and all forms of life. Examples abound, such as climate change, diseases, and in some cases, mutagenic alterations that can lead to the death of organisms, just to name a few.

Pollution problems are increasing due to the rising human population and their anthropogenic activities. Compounds that cause pollution of the environment are increasing in quantity and diversity because of the lifestyle of this century that constantly demands new products and evolving technology. There are many toxic compounds used as pesticides, biocides, or as energy sources. Others are not toxic, but their accumulation in the environment is so high, that they can also cause pollution.

Given this problem, many studies have aimed at diminishing the toxic compounds present in the environment, either in the soil, water, or air. Physical, chemical, and biological processes have been proposed to remove the toxic compounds. All of them present advantages and disadvantages and their application depends on the type and concentration of pollutants and the site where they are present: soil, water, or air.

Biological processes involve organisms or biological activity to degrade or remove the pollutants; this is known as bioremediation. This chapter will describe the fundaments of bioremediation by fungi, deal with how they can remove toxic compounds, and provide some examples.

[1] Departamento de Biotecnología, Universidad Autónoma Metropolitana-Iztapalapa, C.P. 09340, Ciudad de Mexico, Mexico.
[2] Departamento de Energía, Universidad Autónoma Metropolitana-Azcapotzalco, C.P. 02200, Ciudad de Mexico, Mexico.
* Corresponding author: atc@xanum.uam.mx

2. Bioremediation

Many definitions are given for bioremediation. One of the most accepted is: "bioremediation is a process to clean a polluted site using organisms as plants, algae, bacteria, and fungi" (Vidali 2001). Another definition used by many authors is the following: "bioremediation is a biological process to degrade or remove environmental contaminants from a polluted site, water, soil, and air, using organisms or biological activity as plants, algae, bacteria, fungi, and enzymes".

The principal advantage of bioremediation compared to physical and chemical processes is the cost. In general, it is lower than the physical and chemical processes, besides that it is a process that does not produce more contaminants. The disadvantages are that the bioremediation process takes more time than physical or chemical processes. The second one is that the site to be cleaned must have the right conditions for the growth and development of organisms that carry out the processes of removal or degradation of the contaminants.

Bioremediation includes two mechanisms to diminish the polluted compounds, through degradation involving the chemical modification of the molecules, and removal, which is a sorption process of the polluting molecules. Removal of toxic compounds is accomplished through a sorption process, the compounds are absorbed by a substrate; in this case, a biological substrate, biomass, or agriculture waste that can be used wet or dry. Biodegradation involves a chemical change of the molecule to produce other compounds, generally less toxic, and the live organisms or their enzymatic system are responsible for the compound's degradation.

Bioremediation involves different organisms. When plants are responsible for the remediation it is called phytoremediation. This process is one of the most economic ones, but it takes more time to be accomplished.

Bacteria, fungi, and algae can also be responsible for bioremediation; the efficiency depends on the type of compounds, the initial concentration of the toxic compounds, the site to be cleaned, and the microorganism to be used. In many cases, high efficiency is obtained using a microbial consortium containing a wide diversity of microorganisms, including different types of bacteria, fungi, and microalgae.

This chapter is focused on studying the role of filamentous fungi in bioremediation processes; the role of other organisms, such as plants, bacteria, and algae are not discussed, but it is important to know that bioremediation could also be using some of these organisms.

3. Filamentous Fungi

Fungi are eukaryotic and heterotrophic organisms that include unicellular and pluricellular fungi. The unicellular are yeasts, and the pluricellular are represented by all types of filamentous fungi, called as such due to their way of growth. Yeast are anaerobes and aerobes, and filamentous fungi are exclusively aerobes. This chapter will deal only with filamentous fungi.

They develop a tubular structure named hypha, the hyphae form the mycelium, fungi can reproduce by sexual or asexual spores, and by vegetative means from a mycelium segment. There are many types of filamentous fungi; they exist in terrestrial and aquatic habitats.

These fungi can grow as saprotrophs, which obtain their nutrients from dead organisms. They can grow also in symbiosis, which means a common life between two organisms including parasitism and mutualism. Parasitism is when only one of the organisms benefits and mutualism when both organisms are benefited; for example, mycorrhizae (mutualism with roots of plants) or lichens (mutualisms with algae) (Carlile et al. 2001).

They can grow in soil producing many hyphae that spread profusely all over the ground; despite fungi being microscopic, hyphae of *Armillaria ostoyae* can reach a spread rate of 1 m year^{-1} (Peet et al. 1996). *Armillaria bulbosa*, a filamentous fungus type, has been reported as one of the largest and oldest-living organisms (Smith et al. 1992).

Some fungi are pathogenic agents of plants and they are responsible for the loss of agricultural products, which are an important food source. However, many other filamentous fungi are used to produce secondary metabolites, such as antibiotics, immunosuppressors, anticholesterolemics, etc. (Barrios-Gonzalez et al. 2003). An important role to note is that fungi are the primary organisms responsible for degrading organic matter in nature.

Filamentous fungi grow in soil, and many of them degrade organic compounds present in dead organisms, both animal and vegetal. Organic matter is converted into small molecules that can be used by other organisms as a source of energy, carbon, and nitrogen. The ability of fungi to degrade organic compounds has been exceeded by the large number of organic compounds present in the soil due to the increase of human populations in our planet. This ability has also been affected by the presence of toxic compounds, such as xenobiotics. These last compounds are not present in nature, but they are produced by humans for specific uses, such as pesticides, biocides, or are the result of waste or byproducts from industrial processes.

Bioremediation by fungi is an alternative process to remove, degrade, or render harmless toxic compounds using natural biological activity. The fungi belonging to the basidiomycetes were the first assayed to degrade toxic compounds, as it was observed that this class of fungi grows on fragments of trees that are lying on the ground in forests, causing their rotting. Depending on the type of rotting that they cause on wood, fungi are called white-, brown-, or bland-rot fungi. These fungi have a common characteristic, the production of an enzymatic system able to degrade lignin. For this reason, they are also named ligninolytic fungi. *Phanerochatete chrysosporium* was one of the first white-rot fungi studied, showing its ability to degrade lignin, toxic compounds, and its relationship with the enzymatic system (Bumpus et al. 1985).

From the 1980s onwards, many fungi have been studied and found, which can be used in environmental bioremediation. It has been shown that ligninolytic fungi could degrade diverse types of toxic compounds, such as hydrocarbons, chlorophenols, polychlorinated biphenyls, etc., because of their enzymatic system.

Basidiomycetes, such as *Phanerochaete, Trametes, Pleurotus, Lentinus*, among others, can degrade toxic compounds (Table 1.1). Many of them have been isolated from natural sources, mainly from the soil or wastewater contaminated with toxic compounds (Singh 2006, Fackler et al. 2007, Harms et al. 2011, Seigle-Murandi et al. 1993, Lee et al. 2014, Mineki et al. 2015).

Table 1.1 Filamentous fungi used in bioremediation.

Class	Fungi	Use	References
Basidiomycete	*Phanerochaete chrysosporium*	Pentachlorophenol and anthracene degradation	Mileski et al. (1988), Mohammadi et al. (2010)
	P. chrysosporium	Dichlorophenol degradation	Huang et al. (2017)
	P. chrysosporium	Blue 4 dye sorption	Bayramoğlu et al. (2006)
	Phanerochaete velutina	Wood decomposing	Darrah and Fricker (2014)
	Phanerochaete sordida	Dioxin degradation	Sato et al. (2003)
	Bjerkandera adusta	Pentachlorophenol degradation	Rubilar et al. (2007)
	Trametes versicolor	Dye decolorization BTEX* oxidation	Libra et al. (2003), Aranda et al. (2010)
	Trametes modesta	Trinitrotoluene degradation	Nyanhongo et al. (2006)
	T. versicolor	Sulfonamide degradation	Rodríguez-Rodríguez et al. (2012)
	Pleurotus ostreatus	Olive cake degradation	Saavedra et al. (2006)
	Pleurotus pulmonarius	Aromatic hydrocarbons	D'Annibale et al. (2005)
	P. ostreatus	Ni, Cr(VI), Zn, Cd sorption	Javaid et al. (2011)
	P. ostreatus	Naphthalen sulphonic acid polymers degradation	Palli et al. (2016)
	B. adusta	Naphthalene sulphonic acid polymers degradation	Palli et al. (2016)
	Lentinula edodoes	Phenol degradation	Ranjini and Padmavathi (2012)
	Irpex lacteus	Soil bioremediation	Novotný et al. (2000)
	Panus tigrinus	Chlorophenol degradation	Leontievsky et al. (2002)
	Anthracophyllum discolor	Pentachlorophenol degradation	Rubilar et al. (2007), Cea et al. (2010)
	Geotrichum sp.	Azo dyes biotransformation	Máximo et al. (2003)
	Gloeophyllum striatum	Fluorophenol degradation	Kramer et al. (2004)
	Daldinia concentrica	Dibutylphthalate degradation	Lee et al. (2004)
	Agaricus augustus	Tribromophenol degradation	Donoso et al. (2008)
	Agaricus bisporus	PAH and Pb removal	García-Delgado et al. (2015)
	Funalia trogii	Dyes decolorization	Özsoy et al. (2005)
	Phlebia tremellosa	Lignin degradation	Fackler et al. (2007)
	Oxyporus latemarginatus	Lignin degradation	Fackler et al. (2007)

	Lentinus tigrinus	Phenol degradation	Kadimaliev et al. (2011)
	Eisenia fetida	Olive cake transformation	Saavedra et al. (2006)
	T. versiclor	Naproxen degradation	Marco-Urrea et al. (2010)
	Pycnoporus sanguineus	Metal sorption	Zulfadhly et al. (2001)
	P. sanguineus	Cu(II) sorption	Yahaya et al. (2009)
Hyphomycete aquatic	*Clavariopsis aquatica*	Nonylphenol degradation	Junghanns et al. (2005)
Ascomycete	*Aspergillus* spp.	Banana waste degradation	Shah et al. (2005)
	Asperigllus niger var tubingensis	Cr(VI) Sorption/reduction	Coreño-Alonso et al. (2009), (2014)
	Asperigllus niger	Pentachlorophenol sorption	Mathialagan and Viraraghavan (2009)
	Aspergillus awamori	Phenol, catechol, dichlorophenol, dimethoxyphenol degradation	Stoilova et al. (2006)
	Aspergillus fumigatus	Anthracene degradation	Ye et al. (2011)
	A. fumigatus	Cd, Cu, Ni, Pb and Zn accumulation	Dey et al. (2016)
	Aspergillus spp.	Cr and Cd bioaccumulation	Zafar et al. (2007)
	Aspergillus oruzae	Anthraquinone bisorption	Zhang et al. (2015)
	Phylosticta spp.	Banana waste degradation	Shah et al. (2005)
	Trichoderma virgatum	Pentachlorophenol degradation	Cserjesi and Johnson (1972)
	Gloeophylhum striatum	Dichlorophenol and pentachlorophenol degradation	Fahr et al. (1999)
	Trichoderma longibrachiatum	Pentachlorophenol degradation	Carvalho et al. (2009)
	Trichoderma spp.	Soil bioremediation	Harman et al. (2004), Ezzi and Lynch (2005)
	Trichoderma spp.	Cyanide degradation	Ezzi and Lynch (2005)
	Trichoderma harzianum	Soil bioremediation	Matsubara et al. (2006)

Table 1.1 contd. ...

...Table 1.1 contd.

Class	Fungi	Use	References
Ascomycete	*Fusarium* spp.	Cyanide degradation	Ezzi and Lynch (2005)
	Fusarium solani	DDT degradation	Mitra et al. (2001)
	Fusarium oxysporum	Phthalate degradation	Kim et al. (2003)
	Fusarium spp.	Zn sorption	Velmurugan et al. (2010)
	F. solani	Anthracene and benzl[a]anthracene degradation	Wu et al. (2010)
	Penicillium janczewskii	Pentachlorophenol degradation	Carvalho et al. (2009)
	Penicillium restictum	Black 5 dye sorption	Iscen et al. (2007)
	Cordyceps sinensis	Dioxane degradation	Nakamiya et al. (2005)
	Talaromyces helicus	Biphenyl degradation	Romero et al. (2005)
	Byssochlamys nivea	Pentachlorophenol degradation	Bosso et al. (2015)
	Scopulariopsis brumptii	Pentachlorophenol degradation	Bosso et al. (2015)
Zygomycete	*zygomycetes*	Vanillic acid transformation	Seigle-Murandi et al. (1992)
	Cunninghamella echinulata	Pentachlorophenol degradation	Seigle-Murandi et al. (1993)
	Rhizopus oryzae	Pentachlorophenol degradation	León-Santiesteban et al. (2014), (2016)
	Rhizopus spp.	Cr and Cd bio-accumulation	Zafar et al. (2007)
	Rhizopus nigricans	Cr(VI) sorption	Bai and Abraham (2003)
	Rhizopus sp.	Arsenic sorption	Jaiswal et al. (2018)
	Amylomyces rouxii	Pentachlorophenol degradation	Montiel et al. (2004), Marcial et al. (2006)
	Mucor rouxii	Pb, Ni, Cd and Zn sorption	Yan and Viraraghavan (2003)
	Mucor plumbeus	Pentachlorophenol degradation	Carvalho et al. (2011), (2013)
	Mucor ramosissimus	Pentachlorophenol degradation	Szewczyk et al. (2003), Szewczyk and Długoński (2009)
	Mucor hiemalis	Ni(II) sorption	Shroff and Vaidya (2011)
	Mucor hiemalis	Cr(VI)	Tewari et al. (2005)
	Mucor sp.	Cd and Pb sorption and accumulation	Deng et al. (2011)

*benzene, toluene, ethyl benzene, and xylene

4. Biosorption

Removal of toxic compounds through sorption by fungal biomass is one of the bioremediation processes. The compounds are not modified, they are only removed or eliminated from the polluted site and concentrated into a support, and this is called sorption. If the support is biological, the process is called biosorption.

Sorption is a physicochemical process where the molecules from a fluid phase, also called sorbate, are adhered to a surface of a solid, the sorbent. Adsorption processes are useful tools to remove chemicals from polluted sites, adsorption technologies can use inorganic and organic sorbents and they may be physical or chemical. In physical sorption, the involved forces are weak, such as van der Waals interactions, and electrostatic interactions formed by gradient-dipole or -quadruple interactions. Chemical sorption is due to the formation of a chemical bond between the sorbate and the sorbent. These interaction forces are stronger and more specific than the forces in physical sorption.

The main advantage of biosorption is that sorbents are cheap and biodegradable; they include agricultural wastes, such as wood chips, sugar cane bagasse, wheat shell, microbial biomass from bacteria, fungi, yeasts, and microalgae, etc. Fungal biomass has been studied to remove toxic compounds from polluted water. It was applied first to the removal of heavy metals, because fungal biomass is efficient to remove these ions. Biosorption occurs when the metal is fixed to the inactive or non-viable biomass, wet or dry.

Biosorption can occur in different parts of the cell, mainly in the cell wall by mechanisms, such as ionic exchange, chelation, or formation of a complex. Fungal biomass waste from the fermentation industries could be used as sorbent (Kuyucak and Volesky 1988, Volesky 1990). Biosorption of toxic compounds or heavy metals by fungal biomass is due mainly to the cell wall of fungi, which contains between 80 and 90% of polysaccharides, mainly chitin, glucan, mannan, galactosan, and chitosan, in some cases, cellulose. The rest, 10–20%, is constituted by proteins, lipids, polyphosphates, and inorganic ions that constitute the cement of the cell wall. The specific composition depends on the fungus and the identity and quantity of the different functional groups of the fungal biomass. Sorption ability of fungi is related to the functional groups present in the cell wall.

Metals can be removed also by living biomass involving metabolic processes; in this case the process is named bioaccumulation. Removal of heavy metals by bioaccumulation and biosorption by fungi has been studied for approximately 40 years. Filamentous fungi, such as *Aspergillus* sp., *Rhizopus* sp., *Mucor rouxii*, *Penicillium* sp., *Trametes* sp., among others, have been widely used to remove heavy metals, including As, Cr, Ni, Pb, Cd, Zn, Hg, Al, etc.

Mechanisms used by fungi to remove metals from the environment have been reported (Tsezos et al. 1997, Coreño-Alonso et al. 2014, Awasthi et al. 2017). Biosorption of metal ions is due to the physicochemical interactions between the functional groups of the fungal cell wall and the metal ions by mechanisms, such as physical sorption, ion exchange, and chemical sorption, such as chelation, complexation, and coordination. This sorption can be reversible, which means the metal can be desorbed from the biomass. Metal bioaccumulation by fungi involves cell metabolism; the metal must pass through the cell membrane into the cytoplasm and this action has been related to different defense mechanisms of fungi against toxicity caused by metals. In some cases, it can be metal precipitation or extracellular bioaccumulation, involving both mentioned mechanisms (Sağ 2001, Dhankhar and Hooda 2011). The characteristics of fungal biomass have allowed adsorbing many types of toxic compounds besides metals.

The fungal biosorption approach is an alternative technique to remove different types of pollutants, such as phenolic compounds and dyes (Crini 2006, Kumar and Min 2011). The type of biomass and their previous treatment, type of pollutant, initial concentration of the compound, biomass concentration, pH, and temperature influence the process.

Fungal biomass is washed with water and it can be used either wet or dry. In some cases, the biomass is pre-treated to modify the functional groups of the fungal biomass, it can also be used immobilized, in both cases, the modification is made to increase the sorption ability. pH plays an important role because the solubility and ionization of toxic compounds depend largely on the pH of the phase where the toxicants are in the environment, mainly if the pollutants are in aqueous phase (Aksu 2005, León-Santiestebán et al. 2011, Kumar and Min 2011).

Fungal biosorption has been studied for many years; however, its application is limited mainly because more basic knowledge is needed on aspects such as physicochemical conditions, effect of parameters, and because the technology has not been fully developed yet.

5. Biodegradation

The degradation process is the chemical modification of the toxic compounds into less toxic or innocuous molecules. The mineralization process refers to the complete degradation of the toxic compound until CO_2 and water are obtained. However, mineralization is not always achieved, the most common is a partial degradation, in this case, some metabolites derived from the toxic compounds are produced. Generally, these metabolites are less toxic or completely innocuous; however, in a few cases the metabolites produced could be even more toxic than the original.

Fungi can degrade a vast variety of toxic compounds due to their enzymatic system. White rot-fungi degrade lignin, the most abundant resistant biopolymer in nature, have shown the ability to degrade toxic and xenobiotic compounds, using the same mechanisms that they use to degrade lignin. *Phanerochaete chrysosporiun* is one of the first and most studied fungus to degrade toxic compounds, its study began in the 1970s and, since then, many studies on the ability to degrade toxicants with this fungus have been reported (Mileski et al. 1988, Bogan and Lamar 1995, Laugero et al. 1997, Yadav et al. 1995). Table 1.1 shows some of the class and type of fungi used in bioremediation. These fungi have the characteristic to produce oxidase enzymes, which are involved in the degradation of the recalcitrant and xenobiotic toxicants, such as polycyclic aromatic compounds (PAHs), chlorophenols, polychlorinated biphenyls, synthetic polymers, dyes, pesticides, benzene, toluene, ethyl benzene, and xylene (BETX), chlorinated monoaromatics, halogenated organic compounds, and others. These compounds can be degraded by different fungi, mainly basidiomycetes, such as the white-rot fungi species that include *Phanerochaete, Bjerkandera, Trametes, Pleurotus, Phlebia,* and *Ceriporipsis* (Cameron et al. 2000), and brown-rot fungi, for example *Lentinus, Lentinulla,* and *Gloeophyllum.* Fungi different from basidiomycetes have also been studied regarding their ability to degrade toxic compounds.

Degradation of polycyclic aromatic hydrocarbons (PAHs) by fungi has been reported by many authors. Most studies showed that fungi degrade this group of toxicants by co-metabolism, which means fungi cannot use the PHAs as the sole source of carbon, as accomplished by bacteria (Ghosal et al. 2016). Ligninolytic and non-ligninolytic fungi can degrade PHAs, however ligninolytic fungi have been studied more and present some advantages.

The metabolic pathway of degradation of PHAs by non-ligninolytic fungus is different from the degradation pathway used by ligninolytic fungi. Degradation of PHAs by non-ligninolytic fungi, involving cytochrome P450, can produce dihydrodiols and epoxides that are more toxic than the original PHAs. Ligninolytic fungi degrade PHAs by peroxidases and phenoloxidases producing quinones, i.e., less toxic molecules (Hammel 1995, Tortella et al. 2005, Cerniglia and Sutherland 2010, Treu and Falandysz 2017). Co-metabolic anthracene and pyrene degradation by *A. fumigatus* and *Pseudotrametes giboosas,* respectively, was reported by Ye et al. (2011) and Wen et al. (2011), the last authors indicated that the degradation rate is high in co-metabolic conditions.

Romero et al. (2005) reported biotransformation of biphenyls by an ascomycete of the genus *Talaromyces,* finding that its oxidation produces dehydroxylate derivatives and their glucosyl conjugates. Pentachlorophenol degradation by zygomycetes, such as *Mucor, Amylomyces,* and *Rhizopus* has been reported by Carvalho et al. (2011), Montiel et al. (2004), and León-Santiesteban et al. (2014). Table 1.1 depicts other examples of toxicants' degradation by different filamentous fungi.

There are diverse mechanisms of toxic compound degradation by fungi, for example, dehalogenation, which could be through oxidative or reductive reactions, methylation, hydroxylation, conjugation, breaking of the aromatic rings, and polymerization are the most common reactions. The mechanism to degrade a toxicant by fungi depends on the fungi, the characteristic of toxic compounds, the conditions of the reaction, the composition of the culture media. For example, if fungi are grown in nitrogen-deficient cultures, or with glucose or another carbon source, the mechanism to degrade a specific toxicant would be different. León-Santiesteban et al. (2014) reported that in PCP degradation by *R. oryzae* ENHE, nine metabolites were found when the fungus grew with glucose-ammonium sulfate as source of carbon and nitrogen. When the fungus grew in glutamic acid-sodium nitrate as source of carbon and nitrogen, only three metabolites from PCP degradation were found.

Szewczyl et al. (2018) propose a biodegradation pathway of the herbicide ametryn, a herbicide from the triazines group, by *Metarhizium brunneum.* This fungus is an entomopathogen fungus tested to infect insects. These authors showed the ability to degrade ametryn and reported the metabolites produced from the degradation process.

As mentioned before, a great diversity of toxicants has been degraded by fungi, and in some cases, the mechanisms of degradation have been elucidated. We mention only some examples, there is much information about the degradation by fungi of toxic compounds including pesticides, PHAs, halogenated compounds, and emergent pollutants, such as pharmaceutical and personal care products (Table 1.1). Authors have focused the study on toxic degradation by fungi using different tools, such as fungi isolation, biochemistry, molecular biology, and analytical chemistry in order to understand and improve fungal bioremediation. Regardless to the focus of the study, the goal is to propose efficient processes of toxicants' degradation by fungi.

Currently, the enzymatic, redox reactions, and degradation pathways are known, as well as the metabolites produced from some toxic compounds degradation, such as PHAs, chlorophenols, azo dyes, by fungi. However, the degradation pathways by fungi of many recalcitrant compounds, such as emergent pollutants, are not yet known, just as the role of enzymes involved in degradation is not understood in many cases.

Molecular biology studies of fungi have been helping to understand and increase the ability of these microorganisms to degrade toxicants (Bogan et al. 1996, Huang et al. 2018, Szewczyk et al. 2018). Many studies about genetic improvement of fungal strains for degradation of toxic compounds have been reported (Mitra et al. 2001, Gao et al. 2010, Tripathi et al. 2013). For instance, Syed et al. (2010) reported the identification and functional characterization of P450 monooxygenase involved in the oxidation of PAHs by *P. chrysosporium* using a genome-to-function strategy. Molecular technologies have allowed better understanding of the role of *Ascomycota* fungi in the transformation of toxicants (Aranda 2016).

In addition, some other applications have been proposed to remove toxic compounds from polluted sites. Bioaugmentation or biostimulation is referred to increase the presence of fungi responsible for or involved in degradation processes of some compounds *in situ*. The above mentioned can be made first by isolating fungi from the site, then culturing them in the laboratory and, finally, inoculating the polluted soil with the cultured fungi. Lladó et al. (2013) reported the use of two fungi, *Trametes versocilor* and *Lentinus tigrinus*, for PAH removal from contaminated soil. The authors reported that PAH fractions remain in the soil contaminated by creosote after 180 days and that the biostimulation of the soil increases degradation rates. Degradation of anthracene by bioaugmentation of *T. versicolor* in a composting process was reported by Sayara et al. (2011).

While the use and the ability of fungi to degrade toxic compounds have been studied extensively, more studies are still needed for the scaling-up of the process. *In situ* bioremediation is still the objective to achieve.

6. Enzymes

The chemical modification of the toxicant is carried out by enzymatic reactions, involving oxidoreductase enzymes. This class of enzymes catalyzes redox reactions with transfer of hydrogen and oxygen atoms or electrons among molecules. Oxidoreductases include dehydrogenases, oxidases, peroxidases, and oxygenases. Table 1.2 shows the mechanisms of actions of these enzymes.

Peroxidases and phenoloxidases are the main enzymes produced by fungi. Peroxidases are hemoproteins and the hydrogen peroxide is necessary as an electron acceptor. Phenoloxidases are enzymes containing copper that catalyze the oxidation of phenolic compounds without cofactor, using molecular oxygen for the reaction.

Table 1.2 Mechanisms of action of oxidoreductase enzymes.

Enzyme	Mechanism of action
Oxygenases	Transfer oxygen atoms from diatomic oxygen
Oxidases	Transfer electrons to diatomic oxygen
Peroxidases	Transfer electrons to hydrogen peroxide
Dehydrogenases	Transfer hydride ions

Some of these enzymes were found in the culture broth of *P. chrysosporium*, and it was shown that their activity is dependent on H_2O_2 (Forney et al. 1982, Paszczyński et al. 1985, Hatakka 1994), and they were named lignin peroxidases by Tien and Kirk (1988). It was shown that the enzymatic system produced by *P. chysosporium* and other basidiomycetes can degrade lignin; these enzymes are called ligninolytic. Lignin peroxidase and manganese peroxidase use hydrogen peroxide produced by glyoxal oxidase or aryl alcohol oxidase produced by the same fungi. Laccases use the four copper atoms to catalyze the electron reduction of molecular oxygen to water (Kudanga et al. 2011, Sinsabaugh 2010). The main feature of these enzymes is that they are non-specific; thus, they can degrade diverse substrates, including toxic compounds. Many authors have studied fungal ligninolytic enzymes and their activity and ability to degrade recalcitrant compounds, such as BTX, polyaromatic hydrocarbons (PHA), polychlorinated biphenyls, etc.

The most studied fungal enzymes to degrade toxic compounds are lignin and manganese peroxidases, as well as the phenoloxidases, laccase, and tyrosinase. These enzymes are mainly produced and excreted by basidiomycetes, such as *P. chrysosoporium*, *Trametes*, *Pleurotus*, *Pycnoporus*, etc. Other fungi, such as *Ascomycetes* that include *Aspergillus*, *Penicillium*, *Trichoderma*, and *Zygomycetes* such as *Mucor*, *Rhizopus*, as well as *Amylomyces*, also produce peroxidases and phenoloxidases (Durán and Esposito 2000, Montiel et al. 2004). Other enzymes, such as versatile peroxidases, reductive dehalogenases, cytocrome P450, monooxygenases, and peroxygenases, produced by fungi are involved in the degradation process of toxicants (Hofrichter et al. 2010, Subramanian and Yadav 2009, Reddy and Gold 1999, McErlean et al. 2006, Syed et al. 2010). *P. chrysosporium*, *P. ostreatus*, and *B. adusta* showed ability to degrade PHAs and the enzymes produced by these fungi involved in the degradation pathways have been reported (Kadri et al. 2017).

Molecular studies aimed at increasing the enzymatic activity produced by fungi have also been published. There are works reporting the regulation and expression of peroxidases and phenoloxidases, heterologous expression, transcriptome and proteomic studies of fungi producing these enzymes (Kellner and Vandenbol 2010, Piscitelli et al. 2010, Manavalan et al. 2011, Huang et al. 2011, Huang et al. 2017, Carvalho et al. 2013, Janusz et al. 2013, Montiel-González et al. 2009).

Immobilization of enzymes or of the enzyme producing microorganism used in biodegradation processes has been proposed with the advantage that it could be used several times and the degradation process would be environmentally friendly (León-

Santiestebán et al. 2011, Voběrková et al. 2018). Immobilization of enzyme could be accomplished on inert supports such as nylon fibers or on organic ones such as sugar cane bagasse (León-Santiestebán et al. 2011, Mohammadi et al. 2010). Yang et al. (2017) reported the degradation of antibiotics by immobilized laccase at pH 7.0 in 48 hours, removing approximately 60% of tetracycline and oxytetracycline.

Enzymes play a key role in the degradation processes and have been widely studied; the next two chapters deal with these fungal enzymes.

Conclusions

The role of fungi as natural degraders of organic material is very important in the environment, although their capacity has been exceeded by the amount and the diversity of compounds. The interest of using fungi to degrade toxicants has increased, due to the current pollution problem and to the low cost that their use may imply.

To understand how fungi can degrade toxicants and increase their ability, physiological, biochemical, and molecular studies, including genomics, transcriptomics, and proteomics, have been made. All this effort has generated a very broad knowledge of fungi, revealing their potential in bioremediation processes to eliminate hazardous compounds.

However, for a practical application, bioremediation *in situ*, more studies are needed to know the technology that could be used. Bioaugmentation is not an efficient technology, so far, because of the competition among native microorganisms. Another drawback is the control of the environmental conditions required for fungi growth. The addition of enzymes to the site is complicated, since the enzymatic activity depends on parameters, such as pH, humidity, and temperature, which are not possible to control *in situ*. There are some studies reporting the scaling-up from laboratory to field applications; however, it is necessary to study this subject more to propose potential soil bioremediation processes using fungi.

Studies and financial investment should be directed to implement technologies applied *in situ*, which will help and increase the ability of fungi to degrade hazardous compounds, for the bioremediation of polluted environments. It should be emphasized that although bioremediation is a very slow process compared to chemical processes, it is not expensive and is friendly for the environment.

References

Aksu, Z. 2005. Application of biosorption for the removal of organic pollutants: a review. *Process Biochemistry* 40 (3): 997–1026. doi: https://doi.org/10.1016/j.procbio.2004.04.008.

Aranda, E. 2016. Promising approaches towards biotransformation of polycyclic aromatic hydrocarbons with *Ascomycota* fungi. *Current Opinion in Biotechnology* 38: 1–8. doi: https://doi.org/10.1016/j.copbio.2015.12.002.

Aranda, E., E. Marco-Urrea, G. Caminal, M. E. Arias, I. García-Romera and F. Guillén. 2010. Advanced oxidation of benzene, toluene, ethylbenzene and xylene isomers (BTEX) by *Trametes versicolor*. *Journal of Hazardous Materials* 181 (1): 181–186. doi: https://doi.org/10.1016/j.jhazmat.2010.04.114.

Awasthi, A. K., A. K. Pandey and J. Khan. 2017. Biosorption an innovative tool for bioremediation of metal-contaminated municipal solid waste leachate: optimization and mechanisms exploration.

International Journal of Environmental Science and Technology 14 (4): 729–742. doi: 10.1007/s13762-016-1173-2.

Bai, R. S. and T. E. Abraham. 2003. Studies on chromium(VI) adsorption–desorption using immobilized fungal biomass. *Bioresource Technology* 87 (1): 17–26. doi: https://doi.org/10.1016/S0960-8524(02)00222-5.

Barrios-Gonzalez, J., F. Fernández and A. Tomasini. 2003. Microbial secondary metabolites production and strain improvement. *Indian Journal of Biotechnology* 2 (3): 322–333.

Bayramoğlu, G., G. Celik and M. Y. Arica. 2006. Biosorption of Reactive Blue 4 dye by native and treated fungus *Phanerocheate chrysosporium*: Batch and continuous flow system studies. *Journal of Hazardous Materials* 137 (3): 1689–1697. doi: 10.1016/j.jhazmat.2006.05.005.

Bogan, B. W. and R. T. Lamar. 1995. One-electron oxidation in the degradation of creosote polycyclic aromatic hydrocarbons by *Phanerochaete chrysosporium*. *Applied and Environmental Microbiology* 61 (7): 2631–2635.

Bogan, B. W., B. Schoenike, R. T. Lamar and D. Cullen. 1996. Manganese peroxidase mRNA and enzyme activity levels during bioremediation of polycyclic aromatic hydrocarbon-contaminated soil with *Phanerochaete chrysosporium*. *Applied and Environmental Microbiology* 62 (7): 2381–2386.

Bosso, L., R. Scelza, A. Testa, G. Cristinzio and M. A. Rao. 2015. Depletion of Pentachlorophenol contamination in an agricultural soil treated with *Byssochlamys nivea*, *Scopulariopsis brumptii* and urban waste compost: A laboratory microcosm study. *Water, Air, & Soil Pollution* 226 (6): 183. doi: 10.1007/s11270-015-2436-0.

Bumpus, J. A., M. Tien, D. Wright and S. D. Aust. 1985. Oxidation of persistent environmental pollutants by a white rot fungus. *Science* 228 (4706): 1434–1436.

Cameron, M. D., S. Timofeevski and S. D. Aust. 2000. Enzymology of *Phanerochaete chrysosporium* with respect to the degradation of recalcitrant compounds and xenobiotics. *Applied Microbiology and Biotechnology* 54 (6): 751–758. doi: 10.1007/s002530000459.

Carlile, M. J., S. C. Watkinson and G. W. Gooday. 2001. *The Fungi*. Edited by Michael J. Carlile, Sarah C. Watkinson and Graham W. Gooday. 2 ed. London: Academic Press.

Carvalho, M. B., I. Martins, M. C. Leitão, H. Garcia, C. Rodrigues, V. San Romão, I. McLellan, A. Hursthouse and C. Silva Pereira. 2009. Screening pentachlorophenol degradation ability by environmental fungal strains belonging to the phyla Ascomycota and Zygomycota. *Journal of Industrial Microbiology & Biotechnology* 36 (10): 1249–1256. doi: 10.1007/s10295-009-0603-2.

Carvalho, M. B., S. Tavares, J. Medeiros, O. Núñez, H. Gallart-Ayala, M. C. Leitão, M. T. Galceran, A. Hursthouse and C. Silva Pereira. 2011. Degradation pathway of pentachlorophenol by *Mucor plumbeus* involves phase II conjugation and oxidation–reduction reactions. *Journal of Hazardous Materials* 198: 133–142. doi: https://doi.org/10.1016/j.jhazmat.2011.10.021.

Carvalho, M. B., I. Martins, J. Medeiros, S. Tavares, S. Planchon, J. Renaut, O. Núñez, H. Gallart-Ayala, M. T. Galceran, A. Hursthouse and C. Silva Pereira. 2013. The response of *Mucor plumbeus* to pentachlorophenol: A toxicoproteomics study. *Journal of Proteomics* 78: 159–171. doi: https://doi.org/10.1016/j.jprot.2012.11.006.

Cea, M., M. Jorquera, O. Rubilar, H. Langer, G. Tortella and M. C. Diez. 2010. Bioremediation of soil contaminated with pentachlorophenol by *Anthracophyllum discolor* and its effect on soil microbial community. *Journal of Hazardous Materials* 181 (1): 315–323. doi: https://doi.org/10.1016/j.jhazmat.2010.05.013.

Cerniglia, C. E. and J. B. Sutherland. 2010. Degradation of polycyclic aromatic hydrocarbons by fungi. In: *Handbook of Hydrocarbon and Lipid Microbiology*, edited by Kenneth N. Timmis, pp. 2079–2110. Berlin, Heidelberg: Springer Berlin Heidelberg.

Coreño-Alonso, A., F. J. Acevedo-Aguilar, G. E. Reyna-López, A. Tomasini, F. J. Fernández, K. Wrobel, K. Wrobel and J. F. Gutiérrez-Corona. 2009. Cr(VI) reduction by an *Aspergillus tubingensis* strain: Role of carboxylic acids and implications for natural attenuation and biotreatment of Cr(VI) contamination. *Chemosphere* 76 (1): 43–47. doi: https://doi.org/10.1016/j.chemosphere.2009.02.031.

Coreño-Alonso, A., A. Solé, E. Diestra, I. Esteve, J. F. Gutiérrez-Corona, G. E. Reyna López, F. J. Fernández and A. Tomasini. 2014. Mechanisms of interaction of chromium with *Aspergillus niger* var tubingensis strain Ed8. *Bioresource Technology* 158: 188–192. doi: https://doi.org/10.1016/j.biortech.2014.02.036.

Crini, G. 2006. Non-conventional low-cost adsorbents for dye removal: A review. *Bioresource Technology* 97 (9): 1061–1085. doi: https://doi.org/10.1016/j.biortech.2005.05.001.

Cserjesi, A. J. and E. L. Johnson. 1972. Methylation of pentachlorophenol by *Trichoderma virgatum*. *Canadian Journal of Microbiology* 18 (1): 45–49. doi: 10.1139/m72-007.

D'Annibale, A., M. Ricci, V. Leonardi, D. Quaratino, E. Mincione and M. Petruccioli. 2005. Degradation of aromatic hydrocarbons by white-rot fungi in a historically contaminated soil. *Biotechnology and Bioengineering* 90 (6): 723–731. doi: doi:10.1002/bit.20461.

Darrah, P. R. and M. D. Fricker. 2014. Foraging by a wood-decomposing fungus is ecologically adaptive. *Environmental Microbiology* 16 (1): 118–129. doi: doi:10.1111/1462-2920.12216.

Deng, Z., L. Cao, H. Huang, X. Jiang, W. Wang, Y. Shi and R. Zhang. 2011. Characterization of Cd- and Pb-resistant fungal endophyte *Mucor* sp. CBRF59 isolated from rapes (*Brassica chinensis*) in a metal-contaminated soil. *Journal of Hazardous Materials* 185 (2): 717–724. doi: https://doi.org/10.1016/j.jhazmat.2010.09.078.

Dey, P., D. Gola, A. Mishra, A. Malik, P. Kumar, D. K. Singh, N. Patel, M. von Bergen and N. Jehmlich. 2016. Comparative performance evaluation of multi-metal resistant fungal strains for simultaneous removal of multiple hazardous metals. *Journal of Hazardous Materials* 318: 679–685. doi: https://doi.org/10.1016/j.jhazmat.2016.07.025.

Dhankhar, R. and A. Hooda. 2011. Fungal biosorption—an alternative to meet the challenges of heavy metal pollution in aqueous solutions. *Environmental Technology* 32 (5): 467–491. doi: 10.1080/09593330.2011.572922.

Donoso, C., J. Becerra, M. Martínez, N. Garrido and M. Silva. 2008. Degradative ability of 2, 4, 6-tribromophenol by saprophytic fungi *Trametes versicolor* and *Agaricus augustus* isolated from chilean forestry. *World Journal of Microbiology and Biotechnology* 24 (7): 961–968. doi: 10.1007/s11274-007-9559-4.

Durán, N. and E. Esposito. 2000. Potential applications of oxidative enzymes and phenoloxidase-like compounds in wastewater and soil treatment: a review. *Applied Catalysis B: Environmental* 28 (2): 83–99. doi: https://doi.org/10.1016/S0926-3373(00)00168-5.

Ezzi, M. I. and J. M. Lynch. 2005. Biodegradation of cyanide by *Trichoderma* spp. and *Fusarium* spp. *Enzyme and Microbial Technology* 36 (7): 849–854. doi: https://doi.org/10.1016/j.enzmictec.2004.03.030.

Fackler, K., M. Schmutzer, L. Manoch, M. Schwanninger, B. Hinterstoisser, T. Ters, K. Messner and C. Gradinger. 2007. Evaluation of the selectivity of white rot isolates using near infrared spectroscopic techniques. *Enzyme and Microbial Technology* 41 (6): 881–887. doi: https://doi.org/10.1016/j.enzmictec.2007.07.016.

Fahr, K., H. G. Wetzstein, R. Grey and D. Schlosser. 1999. Degradation of 2, 4-dichlorophenol and pentachlorophenol by two brown rot fungi. *FEMS Microbiology Letters* 175 (1): 127–132. doi: doi:10.1111/j.1574-6968.1999.tb13611.x.

Forney, L. J., C. A. Reddy, M. Tien and S. D. Aust. 1982. The involvement of hydroxyl radical derived from hydrogen peroxide in lignin degradation by the white rot fungus *Phanerochaete chrysosporium*. *Journal of Biological Chemistry* 257 (19): 11455–62.

Gao, D., L. Du, J. Yang, W. M. Wu and H. Liang. 2010. A critical review of the application of white rot fungus to environmental pollution control. *Critical Reviews in Biotechnology* 30 (1): 70–77. doi: 10.3109/07388550903427272.

García-Delgado, C., F. Yunta and E. Eymar. 2015. Bioremediation of multi-polluted soil by spent mushroom (*Agaricus bisporus*) substrate: Polycyclic aromatic hydrocarbons degradation and Pb availability. *Journal of Hazardous Materials* 300: 281–288. doi: https://doi.org/10.1016/j.jhazmat.2015.07.008.

Ghosal, D., S. Ghosh, T. K. Dutta and Y. Ahn. 2016. Current state of knowledge in microbial degradation of polycyclic aromatic hydrocarbons (PAHs): A review. *Frontiers in Microbiology* 7 (1369). doi: 10.3389/fmicb.2016.01369.

Hammel, K. E. 1995. Mechanisms for polycyclic aromatic hydrocarbon degradation by ligninolytic fungi. *Environmental Health Perspectives* 103 (Suppl 5): 41–43.

Harman, G. E., M. Lorito and J. M. Lynch. 2004. Uses of *Trichoderma* spp. to alleviate or remediate soil and water pollution. *Advances in Applied Microbiology* 56: 313–330. doi: https://doi.org/10.1016/S0065-2164(04)56010-0.

Harms, H., D. Schlosser and L. Y. Wick. 2011. Untapped potential: exploiting fungi in bioremediation of hazardous chemicals. *Nature Reviews Microbiology* 9: 177–192. doi: 10.1038/nrmicro2519. https://www.nature.com/articles/nrmicro2519#supplementary-information.

Hatakka, A. 1994. Lignin-modifying enzymes from selected white-rot fungi: production and role from in lignin degradation. *FEMS Microbiology Reviews* 13 (2–3): 125–135. doi: 10.1111/j.1574-6976.1994.tb00039.x.

Hofrichter, M., R. Ullrich, M. J. Pecyna, C. Liers and T. Lundell. 2010. New and classic families of secreted fungal heme peroxidases. *Applied Microbiology and Biotechnology* 87 (3): 871–897. doi: 10.1007/s00253-010-2633-0.

Huang, B., W. Lin, P. C. K. Cheung and J. Wu. 2011. Differential proteomic analysis of temperature-induced autolysis in mycelium of *Pleurotus tuber-regium*. *Current Microbiology* 62 (4): 1160–1167. doi: 10.1007/s00284-010-9838-4.

Huang, Z., G. Chen, G. Zeng, Z. Guo, K. He, L. Hu, J. Wu, L. Zhang, Y. Zhu and Z. Song. 2017. Toxicity mechanisms and synergies of silver nanoparticles in 2, 4-dichlorophenol degradation by *Phanerochaete chrysosporium*. *Journal of Hazardous Materials* 321: 37–46. doi: https://doi.org/10.1016/j.jhazmat.2016.08.075.

Huang, Z., Z. Zeng, A. Chen, G. Zeng, R. Xiao, P. Xu, K. He, Z. Song, L. Hu, M. Peng, T. Huang and G. Chen. 2018. Differential behaviors of silver nanoparticles and silver ions towards cysteine: Bioremediation and toxicity to *Phanerochaete chrysosporium*. *Chemosphere* 203: 199–208. doi: 10.1016/j.chemosphere.2018.03.144.

Iscen, C. F., I. Kiran and S. Ilhan. 2007. Biosorption of Reactive Black 5 dye by *Penicillium restrictum*: The kinetic study. *Journal of Hazardous Materials* 143 (1): 335–340. doi: https://doi.org/10.1016/j.jhazmat.2006.09.028.

Jaiswal, V., S. Saxena, I. Kaur, P. Dubey, S. Nand, M. Naseem, S. B. Singh, P. K. Srivastava and S. K. Barik. 2018. Application of four novel fungal strains to remove arsenic from contaminated water in batch and column modes. *Journal of Hazardous Materials* 356: 98–107. doi: https://doi.org/10.1016/j.jhazmat.2018.04.053.

Janusz, G., K. H. Kucharzyk, A. Pawlik, M. Staszczak and A. J. Paszczynski. 2013. Fungal laccase, manganese peroxidase and lignin peroxidase: Gene expression and regulation. *Enzyme and Microbial Technology* 52 (1): 1–12. doi: https://doi.org/10.1016/j.enzmictec.2012.10.003.

Javaid, A., R. Bajwa, U. Shafique and J. Anwar. 2011. Removal of heavy metals by adsorption on *Pleurotus ostreatus*. *Biomass and Bioenergy* 35 (5): 1675–1682. doi: https://doi.org/10.1016/j.biombioe.2010.12.035.

Junghanns, C., M. Moeder, G. Krauss, C. Martin and D. Schlosser. 2005. Degradation of the xenoestrogen nonylphenol by aquatic fungi and their laccases. *Microbiology* 151 (1): 45–57. doi:10.1099/mic.0.27431-0.

Kadimaliev, D. A., V. V. Revin, N. A. Atykyan, O. S. Nadezhina and A. A. Parshin. 2011. The role of laccase and peroxidase of *Lentinus* (*Panus*) *tigrinus* fungus in biodegradation of high phenol concentrations in liquid medium. *Applied Biochemistry and Microbiology* 47 (1): 66–71. doi: 10.1134/s0003683811010066.

Kadri, T., T. Rouissi, S. K. Brar, M. Cledon, S. Sarma and M. Verma. 2017. Biodegradation of polycyclic aromatic hydrocarbons (PAHs) by fungal enzymes: A review. *Journal of Environmental Sciences* 51: 52–74. doi: https://doi.org/10.1016/j.jes.2016.08.023.

Kellner, H. and M. Vandenbol. 2010. Fungi unearthed: Transcripts encoding lignocellulolytic and chitinolytic enzymes in forest soil. *PLOS ONE* 5 (6): e10971. doi: 10.1371/journal.pone.0010971.

Kim, Y. H., J. Lee and S. H. Moon. 2003. Degradation of an endocrine disrupting chemical, DEHP [di-(2-ethylhexyl)-phthalate], by *Fusarium oxysporum* f. sp. pisi cutinase. *Applied Microbiology and Biotechnology* 63 (1): 75–80. doi: 10.1007/s00253-003-1332-5.

Kramer, C., G. Kreisel, K. Fahr, J. Käßbohrer and D. Schlosser. 2004. Degradation of 2-fluorophenol by the brown-rot fungus *Gloeophyllum striatum*: evidence for the involvement of extracellular Fenton chemistry. *Applied Microbiology and Biotechnology* 64 (3): 387–395. doi: 10.1007/s00253-003-1445-x.

Kudanga, T., G. S. Nyanhongo, G. M. Guebitz and S. Burton. 2011. Potential applications of laccase-mediated coupling and grafting reactions: A review. *Enzyme and Microbial Technology* 48 (3): 195–208. doi: https://doi.org/10.1016/j.enzmictec.2010.11.007.

Kumar, N. S. and K. Min. 2011. Phenolic compounds biosorption onto *Schizophyllum commune* fungus: FTIR analysis, kinetics and adsorption isotherms modeling. *Chemical Engineering Journal* 168 (2): 562–571. doi: https://doi.org/10.1016/j.cej.2011.01.023.

Kuyucak, N. and B. Volesky. 1988. Biosorbents for recovery of metals from industrial solutions. *Biotechnology Letters* 10 (2): 137–142. doi: 10.1007/bf01024641.

Laugero, C., J. C. Sigoillot, S. Moukha, M. Asther and C. Mougin. 1997. Comparison of static and agitated immobilized cultures of *Phanerochaete chrysosporium* for the degradation of pentachlorophenol and its metabolite pentachloroanisole. *Canadian Journal of Microbiology* 43 (4): 378–383. doi: 10.1139/m97-052.

Lee, H., Y. Jang, Y. S. Choi, M. J. Kim, J. Lee, H. Lee, J. H. Hong, Y. M. Lee, G. H. Kim and J. J. Kim. 2014. Biotechnological procedures to select white rot fungi for the degradation of PAHs. *Journal of Microbiological Methods* 97: 56–62. doi: https://doi.org/10.1016/j.mimet.2013.12.007.

Lee, S. M., B. W. Koo, S. S. Lee, M. K. Kim, D. H. Choi, E. J. Hong, E. B. Jeung and I. G. Choi. 2004. Biodegradation of dibutylphthalate by white rot fungi and evaluation on its estrogenic activity. *Enzyme and Microbial Technology* 35 (5): 417–423. doi: https://doi.org/10.1016/j.enzmictec.2004.06.001.

León-Santiestebán, H., M. Meraz, K. Wrobel and A. Tomasini. 2011. Pentachlorophenol sorption in nylon fiber and removal by immobilized *Rhizopus oryzae* ENHE. *Journal of Hazardous Materials* 190 (1): 707–712. doi: https://doi.org/10.1016/j.jhazmat.2011.03.101.

León-Santiesteban, H. H., K. Wrobel, L. A. García, S. Revah and A. Tomasini. 2014. Pentachlorophenol sorption by *Rhizopus oryzae* ENHE: pH and temperature effects. *Water, Air, & Soil Pollution* 225 (5): 1947. doi: 10.1007/s11270-014-1947-4.

León-Santiesteban, H. H., K. Wrobel, S. Revah and A. Tomasini. 2016. Pentachlorophenol removal by *Rhizopus oryzae* CDBB-H-1877 using sorption and degradation mechanisms. *Journal of Chemical Technology & Biotechnology* 91 (1): 65–71. doi: doi:10.1002/jctb.4566.

Leontievsky, A., N. Myasoedova, L. Golovleva, M. Sedarati and C. Evans. 2002. Adaptation of the white-rot basidiomycete *Panus tigrinus* for transformation of high concentrations of chlorophenols. *Applied Microbiology and Biotechnology* 59 (4): 599–604. doi: 10.1007/s00253-002-1037-1.

Libra, J. A., M. Borchert and S. Banit. 2003. Competition strategies for the decolorization of a textile-reactive dye with the white-rot fungi *Trametes versicolor* under non-sterile conditions. *Biotechnology and Bioengineering* 82 (6): 736–744. doi: doi:10.1002/bit.10623.

Lladó, S., S. Covino, A. M. Solanas, M. Viñas, M. Petruccioli and A. D'annibale. 2013. Comparative assessment of bioremediation approaches to highly recalcitrant PAH degradation in a real industrial polluted soil. *Journal of Hazardous Materials* 248–249: 407–414. doi: https://doi.org/10.1016/j.jhazmat.2013.01.020.

Manavalan, A., S. S. Adav and S. K. Sze. 2011. iTRAQ-based quantitative secretome analysis of *Phanerochaete chrysosporium*. *Journal of Proteomics* 75 (2): 642–654. doi: https://doi.org/10.1016/j.jprot.2011.09.001.

Marcial, J., J. Barrios-González and A. Tomasini. 2006. Effect of medium composition on pentachlorophenol removal by *Amylomyces rouxii* in solid-state culture. *Process Biochemistry* 41 (2): 496–500. doi: https://doi.org/10.1016/j.procbio.2005.07.010.

Marco-Urrea, E., M. Pérez-Trujillo, P. Blánquez, T. Vicent and G. Caminal. 2010. Biodegradation of the analgesic naproxen by *Trametes versicolor* and identification of intermediates using HPLC-DAD-MS and NMR. *Bioresource Technology* 101 (7): 2159–2166. doi: https://doi.org/10.1016/j.biortech.2009.11.019.

Mathialagan, T. and T. Viraraghavan. 2009. Biosorption of pentachlorophenol from aqueous solutions by a fungal biomass. *Bioresource Technology* 100 (2): 549–558. doi: https://doi.org/10.1016/j.biortech.2008.06.054.

Matsubara, M., J. M. Lynch and F. A. A. M. De Leij. 2006. A simple screening procedure for selecting fungi with potential for use in the bioremediation of contaminated land. *Enzyme and Microbial Technology* 39 (7): 1365–1372. doi: https://doi.org/10.1016/j.enzmictec.2005.04.025.

Máximo, C., M. T. P. Amorim and M. Costa-Ferreira. 2003. Biotransformation of industrial reactive azo dyes by *Geotrichum* sp. CCMI 1019. *Enzyme and Microbial Technology* 32 (1): 145–151. doi: https://doi.org/10.1016/S0141-0229(02)00281-8.

McErlean, C., R. Marchant and I. M. Banat. 2006. An evaluation of soil colonisation potential of selected fungi and their production of ligninolytic enzymes for use in soil bioremediation applications. *Antonie van Leeuwenhoek* 90 (2): 147–158. doi: 10.1007/s10482-006-9069-7.

Mileski, G. J., J. A. Bumpus, M. A. Jurek and S. D. Aust. 1988. Biodegradation of pentachlorophenol by the white rot fungus *Phanerochaete chrysosporium*. *Applied and Environmental Microbiology* 54 (12): 2885–2889.

Mineki, S., K. Suzuki, K. Iwata, D. Nakajima and S. Goto. 2015. Degradation of polyaromatic hydrocarbons by fungi isolated from soil in Japan. *Polycyclic Aromatic Compounds* 35 (1): 120–128. doi: 10.1080/10406638.2014.937007.

Mitra, J., P. K. Mukherjee, S. P. Kale and N. B. K. Murthy. 2001. Bioremediation of DDT in soil by genetically improved strains of soil fungus *Fusarium solani*. *Biodegradation* 12 (4): 235–245. doi: 10.1023/a:1013117406216.

Mohammadi, A., M. Enayatzadeh and B. Nasernejad. 2009. Enzymatic degradation of anthracene by the white rot fungus *Phanerochaete chrysosporium* immobilized on sugarcane bagasse. *Journal of Hazardous Materials* 2010 Apr 15; 176 (1-3): 1128 Enayatzadeh. doi: https://doi.org/10.1016/j.jhazmat.2008.03.132.

Montiel-González, A. M., F. J. Fernández, N. Keer and A. Tomasini. 2009. Increased PCP removal by *Amylomyces rouxii* transformants with heterologous *Phanerochaete chrysosporium* peroxidases supplementing their natural degradative pathway. *Applied Microbiology and Biotechnology* 84 (2): 335–340. doi: 10.1007/s00253-009-1981-0.

Montiel, A. M., F. J. Fernández, J. Marcial, J. Soriano, J. Barrios-González and A. Tomasini. 2004. A fungal phenoloxidase (tyrosinase) involved in pentachlorophenol degradation. *Biotechnology Letters* 26 (17): 1353–1357. doi: 10.1023/b:bile.0000045632.36401.86.

Nakamiya, K., S. Hashimoto, H. Ito, J. S. Edmonds and M. Morita. 2005. Degradation of 1, 4-dioxane and cyclic ethers by an isolated fungus. *Applied and Environmental Microbiology* 71 (3): 1254–1258. doi: 10.1128/AEM.71.3.1254-1258.2005.

Novotný, Č., P. Erbanová, T. Cajthaml, N. Rothschild, C. Dosoretz and V. Šašek. 2000. *Irpex lacteus*, a white rot fungus applicable to water and soil bioremediation. *Applied Microbiology and Biotechnology* 54 (6): 850–853. doi: 10.1007/s002530000432.

Nyanhongo, G. S., S. R. Couto and G. M. Guebitz. 2006. Coupling of 2, 4, 6-trinitrotoluene (TNT) metabolites onto humic monomers by a new laccase from *Trametes modesta*. *Chemosphere* 64 (3): 359–370. doi: https://doi.org/10.1016/j.chemosphere.2005.12.034.

Özsoy, H. D., A. Ünyayar and M. A. Mazmancı. 2005. Decolourisation of reactive textile dyes Drimarene Blue X3LR and Remazol Brilliant Blue R by *Funalia trogii* ATCC 200800. *Biodegradation* 16 (3): 195–204. doi: 10.1007/s10532-004-0627-2.

Palli, L., A. Gullotto, S. Tilli, D. Caniani, R. Gori and A. Scozzafava. 2016. Biodegradation of 2-naphthalensulfonic acid polymers by white-rot fungi: Scale-up into non-sterile packed bed bioreactors. *Chemosphere* 164: 120–127. doi: https://doi.org/10.1016/j.chemosphere.2016.08.071.

Paszczyński, A., V. B. Huynh and R. Crawford. 1985. Enzymatic activities of an extracellular, manganese-dependent peroxidase from *Phanerochaete chrysosporium*. *FEMS Microbiology Letters* 29 (1–2): 37–41. doi: 10.1111/j.1574-6968.1985.tb00831.x.

Peet, F. G., D. J. Morrison and K. W. Pellow. 1996. Rate of spread of *Armillaria ostoyae* in two Douglas-fir plantations in the southern interior of British Columbia. *Canadian Journal of Forest Research* 26 (1): 148–151. doi: 10.1139/x26-016.

Piscitelli, A., C. Pezzella, P. Giardina, V. Faraco and G. Sannia. 2010. Heterologous laccase production and its role in industrial applications. *Bioengineered Bugs* 1 (4): 252–262. doi: 10.4161/bbug.1.4.11438.

Ranjini, R. and T. Padmavathi. 2012. Differential phenol tolerance and degradation by *Lentinula edodes* cultured in differing levels of carbon and nitrogen. *Asiatic Journal of Biotechnology Resources* 3 (10): 1457–1471.

Reddy, G. V. B. and M. H. Gold. 1999. A two-component tetrachlorohydroquinone reductive dehalogenase system from the lignin-degrading basidiomycete *Phanerochaete chrysosporium*. *Biochemical and Biophysical Research Communications* 257 (3): 901–905. doi: https://doi.org/10.1006/bbrc.1999.0561.

Rodríguez-Rodríguez, C. E., M. J. García-Galán, P. Blánquez, M. S. Díaz-Cruz, D. Barceló, G. Caminal and T. Vicent. 2012. Continuous degradation of a mixture of sulfonamides by *Trametes versicolor* and identification of metabolites from sulfapyridine and sulfathiazole. *Journal of Hazardous Materials* 213-214: 347–354. doi: https://doi.org/10.1016/j.jhazmat.2012.02.008.

Romero, M. C., E. Hammer, R. Hanschke, A. M. Arambarri and F. Schauer. 2005. Biotransformation of biphenyl by the filamentous fungus *Talaromyces helicus*. *World Journal of Microbiology and Biotechnology* 21 (2): 101–106. doi: 10.1007/s11274-004-2779-y.

Rubilar, O., G. Feijoo, C. Diez, T. A. Lu-Chau, M. T. Moreira and J. M. Lema. 2007. Biodegradation of pentachlorophenol in soil slurry cultures by *Bjerkandera adusta* and *Anthracophyllum discolor*. *Industrial & Engineering Chemistry Research* 46 (21): 6744–6751. doi: 10.1021/ie061678b.

Saavedra, M., E. Benitez, C. Cifuentes and R. Nogales. 2006. Enzyme activities and chemical changes in wet olive cake after treatment with *Pleurotus ostreatus* or *Eisenia fetida*. *Biodegradation* 17 (1): 93–102. doi: 10.1007/s10532-005-4216-9.

Sağ, Y. 2001. Biosorption of heavy metals by fungal biomass and modeling of fungal biosorption: a review. *Separation and Purification Methods* 30 (1): 1–48. doi: 10.1081/SPM-100102984.

Sato, A., T. Watanabe, Y. Watanabe and R. Kurane. 2003. Enhancement of biodegradation of 2, 7-dichlorodibenzo-p-dioxin by addition of fungal culture filtrate. *World Journal of Microbiology and Biotechnology* 19 (4): 439–441. doi: 10.1023/a:1023917628659.

Sayara, T., E. Borràs, G. Caminal, M. Sarrà and A. Sánchez. 2011. Bioremediation of PAHs-contaminated soil through composting: Influence of bioaugmentation and biostimulation on contaminant biodegradation. *International Biodeterioration & Biodegradation* 65 (6): 859–865. doi: https://doi.org/10.1016/j.ibiod.2011.05.006.

Seigle-Murandi, F., P. Guiraud, R. Steiman and J. L. Benoit-Guyod. 1992. Phenoloxidase production and vanillic acid metabolism by Zygomycetes. *Microbiologica* 15 (2): 157–165.

Seigle-Murandi, F., R. Steiman, J. L. Benoit-Guyod and P. Guiraud. 1993. Fungal degradation of pentachlorophenol by micromycetes. *Journal of Biotechnology* 30 (1): 27–35. doi: https://doi.org/10.1016/0168-1656(93)90024-H.

Shah, M. P., G. V. Reddy, R. Banerjee, P. Ravindra Babu and I. L. Kothari. 2005. Microbial degradation of banana waste under solid state bioprocessing using two lignocellulolytic fungi (*Phylosticta* spp. MPS-001 and *Aspergillus* spp. MPS-002). *Process Biochemistry* 40 (1): 445–451. doi: https://doi.org/10.1016/j.procbio.2004.01.020.

Shroff, K. A. and V. K. Vaidya. 2011. Kinetics and equilibrium studies on biosorption of nickel from aqueous solution by dead fungal biomass of *Mucor hiemalis*. *Chemical Engineering Journal* 171 (3): 1234–1245. doi: https://doi.org/10.1016/j.cej.2011.05.034.

Singh, H. 2006. *Mycoremediation: Fungal Bioremediation*. Edited by H. Singh. Hoboken, New Jersey: John Wiley & Sons, Inc.

Sinsabaugh, R. L. 2010. Phenol oxidase, peroxidase and organic matter dynamics of soil. *Soil Biology and Biochemistry* 42 (3): 391–404. doi: https://doi.org/10.1016/j.soilbio.2009.10.014.

Smith, M. L., J. N. Bruhn and J. B. Anderson. 1992. The fungus *Armillaria bulbosa* is among the largest and oldest living organisms. *Nature* 356: 428. doi: 10.1038/356428a0.

Stoilova, I., A. Krastanov, V. Stanchev, D. Daniel, M. Gerginova and Z. Alexieva. 2006. Biodegradation of high amounts of phenol, catechol, 2, 4-dichlorophenol and 2, 6-dimethoxyphenol by *Aspergillus awamori* cells. *Enzyme and Microbial Technology* 39 (5): 1036–1041. doi: https://doi.org/10.1016/j.enzmictec.2006.02.006.

Subramanian, V. and J. S. Yadav. 2009. Role of P450 Monooxygenases in the degradation of the endocrine-disrupting chemical nonylphenol by the white rot fungus *Phanerochaete chrysosporium*. *Applied and Environmental Microbiology* 75 (17): 5570–5580. doi: 10.1128/AEM.02942-08.

Syed, K., H. Doddapaneni, V. Subramanian, Y. W. Lam and J. S. Yadav. 2010. Genome-to-function characterization of novel fungal P450 monooxygenases oxidizing polycyclic aromatic hydrocarbons (PAHs). *Biochemical and Biophysical Research Communications* 399 (4): 492–497. doi: https://doi.org/10.1016/j.bbrc.2010.07.094.

Szewczyk, R., P. Bernat, K. Milczarek and J. Dlugoński. 2003. Application of microscopic fungi isolated from polluted industrial areas for polycyclic aromatic hydrocarbons and pentachlorophenol reduction. *Biodegradation* 14 (1): 1–8. doi: 10.1023/a:1023522828660.

Szewczyk, R. and J. Długoński. 2009. Pentachlorophenol and spent engine oil degradation by *Mucor ramosissimus*. *International Biodeterioration & Biodegradation* 63 (2): 123–129. doi: https://doi.org/10.1016/j.ibiod.2008.08.001.

Szewczyk, R., A. Kuśmierska and P. Bernat. 2018. Ametryn removal by *Metarhizium brunneum*: Biodegradation pathway proposal and metabolic background revealed. *Chemosphere* 190: 174–183. doi: https://doi.org/10.1016/j.chemosphere.2017.10.011.

Tewari, N., P. Vasudevan and B. K. Guha. 2005. Study on biosorption of Cr(VI) by *Mucor hiemalis*. *Biochemical Engineering Journal* 23 (2): 185–192. doi: https://doi.org/10.1016/j.bej.2005.01.011.

Tien, M. and T. K. Kirk. 1988. Lignin peroxidase of *Phanerochaete chrysosporium*. In: *Methods in Enzymology*, 238–249. Academic Press.

Tortella, G. R., M. C. Diez and N. Durán. 2005. Fungal diversity and use in decomposition of environmental pollutants. *Critical Reviews in Microbiology* 31 (4): 197–212. doi: 10.1080/10408410500304066.

Treu, R. and J. Falandysz. 2017. Mycoremediation of hydrocarbons with basidiomycetes—a review. *Journal of Environmental Science and Health, Part B* 52 (3): 148–155. doi: 10.1080/03601234.2017.1261536.

Tripathi, P., P. Singh, A. Mishra, P. Chauhan, S. Dwivedi, R. T. Bais and R. D. Tripathi. 2013. *Trichoderma*: A potential bioremediator for environmental clean up. *Clean Technologies and Environmental Policy* 15. doi: 10.1007/s10098-012-0553-7.

Tsezos, M., Z. Georgousis and E. Remoundaki. 1997. Mechanism of aluminum interference on uranium biosorption by *Rhizopus arrhizus*. *Biotechnology and Bioengineering* 55: 16–27. doi: 10.1002/(SICI)1097-0290(19970705)55:1<16::AID-BIT3>3.0.CO;2-#.

Velmurugan, P., J. Shim, Y. You, S. Choi, S. Kamala-Kannan, K.-J. Lee, H. J. Kim and B.-T. Oh. 2010. Removal of zinc by live, dead, and dried biomass of *Fusarium* spp. isolated from the abandoned-metal mine in South Korea and its perspective of producing nanocrystals. *Journal of Hazardous Materials* 182 (1): 317–324. doi: https://doi.org/10.1016/j.jhazmat.2010.06.032.

Vidali, M. 2001. Bioremediation. An overview. *Pure and Applied Chemistry* 73 (7): 1163. doi: 10.1351/pac200173071163.

Voběrková, S., V. Solčány, M. Vršanská and V. Adam. 2018. Immobilization of ligninolytic enzymes from white-rot fungi in cross-linked aggregates. *Chemosphere* 202: 694–707. doi: https://doi.org/10.1016/j.chemosphere.2018.03.088.

Volesky, B. 1990. *Biosorption of Heavy Metals*. Edited by Bohumil Volesky: CRC Press USA.

Wen, J., D. Gao, B. Zhang and H. Liang. 2011. Co-metabolic degradation of pyrene by indigenous white-rot fungus *Pseudotrametes gibbosa* from the northeast China. *International Biodeterioration & Biodegradation* 65 (4): 600–604. doi: https://doi.org/10.1016/j.ibiod.2011.03.003.

Wu, Y. R., Z. H. Luo and L. L. P. Vrijmoed. 2010. Biodegradation of anthracene and benz[a]anthracene by two *Fusarium solani* strains isolated from mangrove sediments. *Bioresource Technology* 101 (24): 9666–9672. doi: https://doi.org/10.1016/j.biortech.2010.07.049.

Yadav, J. S., J. F. Quensen, J. M. Tiedje and C. A. Reddy. 1995. Degradation of polychlorinated biphenyl mixtures (Aroclors 1242, 1254, and 1260) by the white rot fungus *Phanerochaete chrysosporium* as evidenced by congener-specific analysis. *Applied and Environmental Microbiology* 61 (7): 2560–2565.

Yahaya, Y. A., M. M. Don and S. Bhatia. 2009. Biosorption of copper (II) onto immobilized cells of *Pycnoporus sanguineus* from aqueous solution: Equilibrium and kinetic studies. *Journal of Hazardous Materials* 161 (1): 189–195. doi: https://doi.org/10.1016/j.jhazmat.2008.03.104.

Yan, G. and T. Viraraghavan. 2003. Heavy-metal removal from aqueous solution by fungus *Mucor rouxii*. *Water Research* 37 (18): 4486–4496. doi: https://doi.org/10.1016/S0043-1354(03)00409-3.

Yang, J., Y. Lin, X. Yang, T. B. Ng, X. Ye and J. Lin. 2017. Degradation of tetracycline by immobilized laccase and the proposed transformation pathway. *Journal of Hazardous Materials* 322: 525–531. doi: https://doi.org/10.1016/j.jhazmat.2016.10.019.

Ye, J. S., H. Yin, J. Qiang, H. Peng, H. M. Qin, N. Zhang and B. Y. He. 2011. Biodegradation of anthracene by *Aspergillus fumigatus*. *Journal of Hazardous Materials* 185 (1): 174–181. doi: https://doi.org/10.1016/j.jhazmat.2010.09.015.

Zafar, S., F. Aqil and I. Ahmad. 2007. Metal tolerance and biosorption potential of filamentous fungi isolated from metal contaminated agricultural soil. *Bioresource Technology* 98 (13): 2557–2561. doi: https://doi.org/10.1016/j.biortech.2006.09.051.

Zhang, Z., D. Shi, H. Ding, H. Zheng and H. Chen. 2015. Biosorption characteristics of 1, 8-dihydroxy anthraquinone onto *Aspergillus oryzae* CGMCC5992 biomass. *International Journal of Environmental Science and Technology* 12 (10): 3351–3362. doi: 10.1007/s13762-015-0762-9.

Zulfadhly, Z., M. D. Mashitah and S. Bhatia. 2001. Heavy metals removal in fixed-bed column by the macro fungus *Pycnoporus sanguineus*. *Environmental Pollution* 112 (3): 463–470. doi: https://doi.org/10.1016/S0269-7491(00)00136-6.

CHAPTER-2

Fungal Peroxidases Mediated Bioremediation of Industrial Pollutants

Misha Ali, Qayyum Husain* and Hassan Mubarak Ishqi

1. Introduction

The use of chemicals in many aspects of our daily lives, industrial processes, and agricultural practices results in the deliberate or accidental spillage of potentially toxic chemicals into the environment. A major portion of the pollutants is constituted by aromatic compounds released by various industrial processes including petroleum refining, dye manufacturing, textile, leather, plastic and resin synthesis, coking and coal conversion, chemical and pharmaceutical industries, foundries, and pulp and paper plants (Husain and Ulber 2011). There are several classes of aromatic compounds that have been selected as priority pollutants by the United States Environmental Protection Agency due to their negative impact on the environment and aquatic and human life. Some of these classes include polycyclic aromatic hydrocarbons (PAHs), pentachlorophenol (PCP), polychlorinated biphenyls (PCBs), phenolic compounds, dyes, pesticides, explosives, and pharmaceuticals, etc. These aromatic pollutants are persistent/recalcitrant in the environment, and some of them are known carcinogenic and/or mutagenic agents (Husain et al. 2009, Yu et al. 2016). Therefore, the removal of such compounds from the environment is necessary. Many conventional methods which have been employed for the treatment of environmental pollutants include chemical, physical and biological methods, or a combination of these methods (Silva et al. 2013).

Conventional chemical and physical procedures which have been employed for the removal of aromatic pollutants include electrochemical oxidation, membrane

Department of Biochemistry, Faculty of Life Sciences, Aligarh Muslim University, Aligarh-202002, India.
* Corresponding author: qayyumbiochem@gmail.com

filtration, coagulation, sorption, ion exchange, flocculation, electrolysis, adsorption, advanced oxidation processes, bleaching, chlorination, ozonation, Fenton oxidation, chemical reduction, and photocatalytic oxidation (Hao et al. 2000, Husain 2006). Although these methods are effective in the clearance of toxic compounds from the environment, most of them suffer from some drawbacks, such as partial removal, applicability to a limited concentration range, high cost, and the formation of toxic by-products (Ashraf and Husain 2010). These methods of remediation and degradation of organic compounds are actually outdated due to some unresolved problems. Recently, the biological techniques based on microbial degradation of organopollutants have attracted much attention of environmentalists for the treatment of wastewater. Several researchers have shown partial or complete biodegradation of such chemicals by pure and mixed cultures of bacteria, fungi, and algae (Husain et al. 2009). The treatment of aromatic pollutants using microbiological tools has some of its own demerits, such as limited mobility and survival of cells in the soil, high cost of microbial culture production, alternative carbon sources, completeness of the indigenous populations, and metabolic inhibition (Steevensz et al. 2014). Numerous kinds of organisms have been used for the complete degradation of toxic compounds, but much success has not yet been achieved. Aromatic compounds are degraded by secondary metabolic pathways of bacteria, fungi, and algae, which require appropriate growth conditions aided by additional loads of chemicals (Robinson et al. 2001, Rabinovich et al. 2004). Moreover, the degradation of aromatic compounds is not constant with time, as the expression of enzymes involved in degradation depends on the growth phase of the organisms which is in turn influenced by inhibitors present in wastewater (Wesenberg et al. 2003, Kurnik et al. 2015).

In an attempt to overcome the problems associated with the traditional biological and chemical waste treatment procedures, recent researches have been focused on the applications of enzymes in such fields (Husain and Jan 2000, Rao et al. 2014). Due to a variety of chemical transformations catalyzed by enzymes, the enzyme based technology brought the focus of biotechnologists towards this novel approach. Cheap and more readily available enzymes are produced through better isolation and purification procedures due to recent advances in enzyme based technology. These can be applied in many remediation processes to target specific recalcitrant pollutants present in wastewaters (Xu and Salmon 2008). Enzymes isolated from microorganisms have been often preferred over intact organisms due to easy handling and storage, greater specificity, higher stability, and better optimization (Satar et al. 2012, Husain and Husain 2012). Moreover, unlike chemical catalysts, the enzymatic systems have the potential of carrying out complex chemical transformation under mild environmental conditions with high reaction velocity and efficiency (Husain and Husain 2007, Michniewicz et al. 2008). Due to high specificity for individual compounds, enzymatic processes have been developed specifically to target selected aromatic pollutants, which cannot be treated effectively using traditional methods (Alarcón-Payán et al. 2017).

The enzymatic treatment has an edge over the conventional methods of pollutants treatment because this technique can be applied—at very low to high concentrations of contaminants over the broad range of pH, temperature, and ionic strength, to persistent and recalcitrant materials, and in the absence of shock loading effects. It reduces sludge volume and delays associated with the acclimatization of biomass (Husain and Husain 2012, Goncalves et al. 2015). Among various enzymes being actively pursued for environmental purposes, oxidoreductases have attracted more attention. These are a large group of enzymes catalyzing oxidation/reduction reactions. Oxidoreductases catalyze the electron transfer from one substrate to another, where the oxidized substrate is referred to as electron donor in contrast to the reduced substrate which is electron acceptor (Husain 2017a,b). The potential application of oxidoreductases from bacteria, fungi, and plants is increasingly being demonstrated, and among these enzymes are peroxidases, laccases, and polyphenol oxidases (Durán et al. 2002).

Peroxidases are suitable catalysts for the bioremediation, as they exhibit broad substrate specificity and high redox potential (up to 1.3 V). The bioremediation mediated by these enzymes relies on the oxidation of toxic aromatic compounds into less toxic derivatives and also removes them from the polluted water, e.g., phenols are being oxidized and polymerized by such enzymes into insoluble products which can be easily removed from the wastewater (Ayala and Torres 2016). Peroxidases [donor: H_2O_2 oxidoreductases] are predominantly heme proteins which utilize H_2O_2 or organic hydroperoxides as co-substrate to oxidize a variety of organic and inorganic substrates (Fatima and Husain 2008). They are found in all five kingdoms of life. The "Non-animal heme-peroxidases" include intracellular peroxidases of bacterial and eukaryotic origin (Class I, e.g., yeast cytochrome c peroxidase), secreted fungal peroxidases (Class II, e.g., lignin, manganese, and versatile peroxidases), and classic secretory plant peroxidases (Class III, e.g., horseradish, turnip, and bitter gourd peroxidases) (Battistuzzi et al. 2010, Hofrichter et al. 2010).

2. Fungal Peroxidases

2.1 Ligninolytic peroxidases

Lignocellulose is the most abundant recalcitrant renewable material available in nature. Lignin is an amorphous, complex and three-dimensional aromatic polymer. These polymers have compact structures that are insoluble in water, as well as difficult to penetrate by microbes or enzymes (Sánchez 2009). The rigidity of lignin can be accounted to the intermonomeric linkages that comprise of many kinds of C–C and C–O bonds, with the β-aryl ether linkage being the most significant and these linkages are not hydrolysable. A conclusion from the above items is summarized as follows: (i) polymeric lignin degradation requires extracellular enzymes and/or small molecular weight mediators or factors, such as radicals, (ii) the lignin degrading system must be unspecific, and (iii) the enzymes must be oxidative, not hydrolytic (Hatakka and Hammel 2011, Satar et al. 2012). Fungi are the only organisms known that are able to completely mineralize lignocellulose by using their ligninolytic enzymes system. One of the major components of the ligninolytic system is the peroxidase enzyme family, which is one of the major components of the ligninolytic system and is comprised of several members. Peroxidases are oxidoreductases that utilize H_2O_2 to catalyze oxidative reactions of diverse substrates. The enzymatically generated H_2O_2 oxidizes lignin polymer in a reaction catalyzed by ligninolytic peroxidases, and this process is described as a key reaction in the process of enzymatic combustion (Ruiz-Dueñas and Martínez 2009).

The main lignin modifying enzymes which play a central role in lignin modification are lignin peroxidase (LiP), manganese peroxidase (MnP), and versatile peroxidase (VP). These peroxidases oxidize specific components of the lignin structure and may act in synergy if they are produced by the same organism. While VP has the capability of oxidizing both phenolic and non-phenolic structures, MnP oxidizes only phenolic structures of lignin, and LiP targets the non-phenolic components (Falade et al. 2017). These extracellular lignin degrading enzymes belong to Class II heme-peroxidases. Ligninolytic enzymes have very high reducing specificity for the substrates, because ligninolytic fungi grow on a variety of lignocellulosic substrates, and produce ligninases, which have the ability to mineralize distinct structures of lignin (Knop et al. 2015). The low specificity of many fungal enzymes enables the organisms producing them to co-metabolize structurally distinct compounds belonging to different pollutant classes, e.g., benzene, toluene, ethylbenzene and xylenes (BTEX) compounds, organochlorines (chloroaliphatics, chlorolignols, chlorophenols, polychlorinated biphenyls, and polychlorinated dibenzo-p-dioxins (PCDD)), 2, 4, 6-trinitrotoluene (TNT), PAHs, pesticides, synthetic dyes, and synthetic polymers (Harms et al. 2011). Interestingly, the structurally diverging representatives of a particular pollutant class (such as various low- and high-molecular-mass PAHs and different PCDD congeners) can be degraded by the same fungal organism, even in the mixture (Nakamiya et al. 2005, Cerniglia and Sutherland 2010).

2.2 *Non-ligninolytic peroxidases*

Other fungal peroxidases which belong to Class II heme-peroxidases are non-ligninolytic *Arthromyces ramosus* peroxidase (ARP) and *Coprinus cinereus* peroxidase (CIP). Because of the catalytic similarity of these enzymes to horse radish peroxidase (HRP), their application in the removal of phenolic compounds from wastewater has also drawn great attention. It was exhibited that various phenolic compounds could be removed from their aqueous solutions using ARP or CIP (Ikehata et al. 2004).

The other peroxidases that are less studied, but have shown their potential in the degradation of organopollutants, are heme-thiolate peroxidases (HTPs) and dye decolorization peroxidases (DyPs). They have considerably divergent protein sequences and unique catalytic properties, which does not allow their classification into the class II peroxidase family, but justifies their affiliation to separate (super) families of heme peroxidases (Sugano 2009, Lundell et al. 2010). Two main enzymes from the HTPs family are *Caldariomyces fumago* chloroperoxidase (CfuCPO) and aromatic peroxygenases (APOs). It is evident that the main activity of CPO is the oxidation of chloride (Cl^-) into hypochlorous acid at pH < 5, which in turn acts as a strong oxidant and can chlorinate organic compounds. However, it is still unclear to which end, this reaction is catalyzed. Initially, it had been thought that CfuCPO was involved in the synthesis of chlorinated metabolites, such as caldariomycin (Hofrichter et al. 2010), but later this assumption was questioned (van Pée 2001).

Recently, CfuCPO and also non-heme chloroperoxidases are rather believed to be involved in antimicrobial antagonism, since the hypochlorite formed has strong biocidal activities (Bengtson et al. 2009). The natural functions of APOs are still not known. Due to their versatility, they could be involved in the non-specific oxidation and detoxification of microbial metabolites or plant ingredients (e.g., methoxylated phytoalexins) and also in the conversion of lignin-derived compounds, *o*-demethylation of mono- and oligomeric lignin fragments (Hofrichter et al. 2010).

HTPs catalyze a wide range of reactions, including oxidations of various aliphatic and aromatic compounds (Ullrich and Hofrichter 2005, Gutiérrez et al. 2011). DyP-type peroxidases are catalytically highly stable and versatile (Pühse et al. 2009). The physiological role of DyPs is also still not clear, but it has been suggested to be a part of the catalytic toxic aromatic products, or as a defense mechanism against oxidative stress. They possess a high redox potential (1.1–1.2 V) so that numerous substrates can serve as electron donors (Liers et al. 2013, Liers et al. 2014). DyP-type peroxidases are therefore able to oxidize a wide spectrum of complex dyes, in particular, xenobiotic anthraquinone derivatives, typical peroxidase substrates, such as ABTS (2,2'-azino-bis(3-ethylthiazoline-6-sulfonate)) and phenols (Sugano 2009). Figure 2.1 illustrates all the fungal peroxidases involved in bioremediation of different kinds of pollutants.

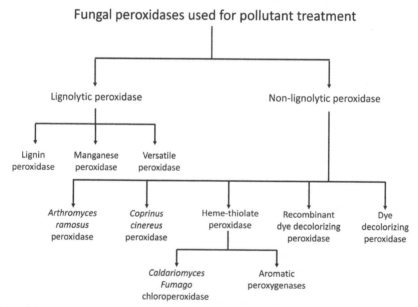

Figure 2.1 Demonstrates fungal peroxidases employed for bioremediation.

3. Main Fungi Producing Peroxidases

The organisms predominantly responsible for lignocellulose degradation are fungi, which are classified by the type of degradation. They are, namely, white-rot fungi (WRF), brown-rot fungi (BRF), and soft-rot fungi (SRF). WRF break down lignin in wood, leaving lighter colored (white) cellulose behind. Some of these fungi break down both lignin and cellulose by producing powerful extracellular oxidative and hydrolytic enzymes (Manavalan et al. 2015). WRF is a heterogeneous group of fungi that usually belong to basidiomycetes, although there are ascomycetous fungi that cause pseudo-white-rot, also designated as soft-rot type II, such as fungi belong to the family Xylariaceae (Liers et al. 2006). Basidiomycetous WRF and some related litter-decomposing fungi are the only organisms known which are capable of efficiently mineralizing lignin. BRF also belong to basidiomycetes that degrade wood in a manner which yields brown, shrunken specimens that typically exhibit a pattern of cubical cracks and easily disintegrate upon handling. They degrade wood polysaccharides, cellulose, and hemicellulose, by producing hydrolytic enzymes cellulases and hemicellulases, while they catalyze partial modification of lignin. SRF from ascomycetes secretes cellulase that breaks down cellulose from their hyphae, leading to the formation of microscopic cavities inside the wood and sometimes to a discoloration and cracking pattern resembling that of BRF (Manavalan et al. 2015). Xylariaceous ascomycetes from genera, such as *Daldinia*, *Hypoxylon*, and *Xylaria* are grouped with the SRFs, they cause a typical erosive soft rot. Compared to basidiomycetous fungi, the knowledge about lignocellulose degradation by ascomycetes is limited, and very little is known about its mechanism of lignin degradation (Hatakka and Hammel 2011). Hence, it can be concluded that the main fungi responsible for lignin degradation is WRF.

LiP, MnP, VP, and laccase are the major lignin degrading enzymes from WRF. LiP and MnP were discovered in the mid-1980s and were purified from extracellular culture medium of a basidiomycete *Phanerochaete chrysosporium. P. chrysosporium* is certainly the most studied fungus for ligninolytic peroxidase production. Soon after the discovery of *P. chrysosporium* LiP and MnP, the production of these enzymes by many different WRF were reported, such as *Trametes* (*Coriolus* or *Polyporus*) *versicolor, Chrysonila sitophila,* and *Phlebia radiata* among many others. Besides *P. chrysosporium*, LiP activity was found in the culture filtrates of *Chrysosporium pruinosum* and *Coriolus* (*Trametes*) *versicolor* (Lundell et al. 2010, Liu et al. 2012, Pinto et al. 2012).

Pleurotus species is another class of extensively studied WRF and these fungi have also been exploited for the production of ligninolytic peroxidases. The ligninolytic peroxidases produced by one of *Pleurotus* spp., *P. eryngii* had remarkably high catalytic activities as compared to the conventionally produced LiP and MnP by *P. chrysosporium* (Gómez-Toribio et al. 2001). Papinutti et al. (2003) discussed the production of MnP and laccase by a WRF *Fomes sclerodermeus BAFC 2752*. Some other distinguished fungi discovered to produce either LiP or MnP are marine ascomycetes *Sordaria fimicola* and *Halosarpheia ratnagiriensis*; MnP and laccase, a marine basidiomycete *Flavodon flavus*; LiP, MnP, and laccase (Raghukumar 2000), litter-decomposing basidiomycetes *Agrocybe praecox*, *Collybia dryophila*, and *Stropharia coronilla*, MnP (Steffen et al. 2002), and yeasts *Candida tropicalis* and *Debaryomyces polymorphus*, MnP (Yang et al. 2003).

DyP was isolated for the first time from WRF, *Bjerkandera adusta* (Kim et al. 1995). It was also reported that DyP was produced by jelly fungus *Auricularia auricula-judae*, *Exidia glandulosa*, and *Mycena epipterygia* (Liers et al. 2010, Liers et al. 2013, Linde et al. 2014). DyPs were also reported from *Irpex lacteus*, a white rot basidiomycete and WRF *Pleurotus sapidus* (Salvachúa et al. 2013, Lauber et al. 2017). One of the HTP, i.e., chloroperoxidase, was isolated from the ascomycete *Leptoxyphium fumago*. A second HTP described as an aromatic peroxygenase (APO) has been produced from agaric basidiomycetes, *Agrocybe aegerita* (Ullrich et al. 2004, Gutiérrez et al. 2011). APO of somewhat different catalytic properties has been isolated from ink cap mushroom *Coprinus radians* (Anh et al. 2007). Gröbe et al. (2011) described isolation of APO from agaric fungus *Marasmius rotula*. ARP and CIP were mainly produced by the imperfect fungus (deuteromycete) *Arthromyces ramosus* obtained from soil, and from an inky cap basidiomycete *Coprinus cinereus*, respectively.

4. Characteristics of Fungal Peroxidases

4.1 Lignin peroxidase (LiP)

LiP is a heme-containing glycoprotein oxidoreductase which was first found and reported 25 years back in white-rot basidiomycete *P. chrysosporium* as lignin degrading fungi. However, many other fungi have also been known to produce LiP (Miki et al. 2010). Most of the LiPs used these days in different bioremediation processes and lignin degradation has been produced by fungus *P. chrysosporium* where this enzyme is secreted during secondary metabolism as a result of nitrogen limitation (Martínez et al. 2009, Hofrichter et al. 2010). The catalytic cycle involves initial oxidation of the enzyme by H_2O_2 to form a two-electron deficient intermediate named as compound I. The compound I consists of a Fe(IV) oxoferryl centre and a porphyrin-based cation radical. It further oxidizes substrates by one electron and forms compound II, which is a more reduced Fe(IV) oxoferryl intermediate (Breen and Singleton 1999).

LiP is catalytically the most prominent fungal peroxidase in class II. It has the ability to oxidize directly several phenolic and non-phenolic substrates with calculated ionization potential up to 9.0 eV (ten Have and Teunissen 2001). It completely oxidizes methylated lignin and lignin model compounds in addition to several polyaromatic hydrocarbons, PAHs. Among the different oxidation reactions catalyzed by LiP are the cleavage of Cα–Cβ and aryl Cα bond, demethylation, and aromatic ring opening. Veratryl alcohol (VA), a secondary metabolite, has been the focus of many studies. VA is a rich substrate for LiP and increases the oxidation of otherwise weak or terminal LiP substrates (Ollikka et al. 1993). VA is playing three important roles. Firstly, VA behaves as a mediator in electron-transfer reactions.

Secondly, it is a good substrate for compound II and for that reason; it is required for completing the catalytic cycle of LiP via oxidation of terminal substrates. Thirdly, H_2O_2-dependent inactivation of LiP by reducing compound II back to native LiP is prevented by VA. Moreover, if the inactive LiP compound III is established, the intermediate VA+ is able to reduce LiP compound III back to its original form (ten Have and Teunissen 2001). The oxidation site in LiP is confined to catalytically active tryptophan (Trp171 in *P. chrysosporium*) on the enzyme surface for the high-redox potential compound like VA. This site is very important as it has been demonstrated by the mutagenesis studies. It is a substrate-intermediate-protein radical center upon LiP catalysis and initiation of the long-range electron transfer pathway leading to the heme (Smith et al. 2009).

4.2 *Manganese peroxidase (MnP)*

Another heme-containing glycoprotein oxidoreductase which preferentially uses Mn^{2+} as the electron donor is MnP. This enzyme has been produced by only a few basidiomycetous fungi and there is no record of its isolation from bacterium, yeast, mold, or mycorrhiza-forming basidiomycete (Hofrichter 2002). It was first identified in WRF *P. chrysosporium* along with LiP. The first phase of the catalytic cycle involving 2 electron oxidation of enzyme by H_2O_2 to compound I and subsequent reduction to compound II by oxidation of the suitable substrate (numerous monomeric and dimeric phenols which also includes phenolic lignin model compounds) or Mn^{2+}, is analogous to that of other peroxidases. However, unlike other peroxidases, compound II strictly requires Mn^{2+} as an electron donor (Kuan et al. 1993).

MnP is one of the most dominating enzymes of the class II peroxidases. Thus, the MnP encoding genes are much more widely prevalent compared to LiP in agaric (cap forming) and polyploid as well as corticoid fungi, all of which belong to Agaricomycotina subphylum (Morgenstern et al. 2008). Recent phylogenetic analysis has revealed that there were at least two pre-MnP lineages that separated early in the lignin-degrading fungi. This led to the formation of two groups of MnPs, namely, the short hybrid MnP variants or group A, and the characteristic long MnP enzymes or group B. The former are evolutionarily more relevant to VPs and LiP than to classical MnPs (Lundell et al. 2010). The MnP is around 20 to 30 amino acid long protein (longer than LiP or VP) having a 5th cysteine disulfide bond with a Mn^{2+} binding site near the C-terminus (Sundaramoorthy et al. 2005). MnP is exceedingly particular for oxidation of Mn^{2+} ions in an acidic environment (Gold et al. 2000). This is due to a solvent exposed and conserved Mn^{2+} binding site neighborhood with either of the 2 heme propionates. It was first observed in the crystal structure of *P. chrysosporium* MnP1, consisting of three acidic amino-acid residues (two Glu, E35 and E39; and one Asp, D179). These amino acids are essential for the Mn^{2+} ion hexa-coordination, thus enabling the rapid electron transfer to the ferryl iron and heme (Sundaramoorthy et al. 2005).

The catalytic cycle of MnP completes when two Mn^{2+} ions are oxidized and converted into two Mn^{3+} ions. These ions form chelate complexes with anions of dicarboxylic acids such as oxalate or malonate (produced by the WRF in copious amount) and are dispersed out of the binding site (Gold et al. 2000). The scope of the oxidation capability of MnP is clear in its products, the chelated Mn^{3+} ions, which act as diffusible mediators for charge transfer and then attack phenolic compounds containing bigger biopolymers, such as natural and synthetic lignin, milled wood, humic substances, and various xenobiotic compounds (Hofrichter et al. 2001, Husain et al. 2009). The cloned and enzymatically described long, classical, and highly conserved MnPs have only been described for the efficiently wood-lignin-degrading fungi of the basidiomycetous families *Corticiaceae* or *Polyporaceae*, such as *Ceriporiopsis subvermispora, Dichomitus squalens, P. chrysosporium, Phanerochaete sordida, Phlebia radiata,* and *Phlebia* spp. (isolates Nf-b19 and MG-60), thus indicating the importance of this particular peroxidase for the development of typical white-rot decay of wood (Lundell et al. 2010).

4.3 Versatile peroxidase (VP)

VP is the third largest lignin-degrading and high redox potential peroxidase of class II fungal peroxidases. It was first described in the plant roots- and wood colonizing white-rot basidiomycete *Pleurotus eryngii* (Pérez-Boada et al. 2005) and is also found in *Bjerkandera* spp. (Moreira et al. 2005). It is a hybrid-type heme containing enzyme that combines the catalytic functions of both MnP and LiP. They can oxidize Mn^{2+} as well as high redox-potential compounds, such as Reactive Black 5. Physical and catalytic properties have shown that the multiple catalytic properties are due to co-existence of different catalytic sites on the same enzyme. In VP, the three specific MnP amino acid residues involved in the Mn^{2+} binding and specific LiP amino acid exposed tryptophan residue (Trp-164) is found (Ruiz-Dueñas et al. 2009).

The catalytic cycle of the enzyme combines activities resonant of both MnP and LiP. At the lignin-peroxidase site, the enzyme is capable of Mn^{2+}-independent oxidation of various organic compounds, including high redox potential non-phenolic aromatics and dyes following the catalytic cycle in a similar fashion as that of LiP, as described earlier. Due to steric hindrance, long-range electron transfer from a protein radical at the surface of the enzyme (which would act as the substrate oxidizer) to the heme cofactor, has been suggested to explain oxidation of large compounds, such as the large aromatic substrates, lignin polymer, or complex redox mediators (Ruiz-Dueñas et al. 2009). At the manganese oxidation site, Mn^{2+} is oxidized to Mn^{3+} employing the mechanism identical to the MnP reaction mechanism. The generated Mn^{3+} acts as a diffusible oxidizer for free phenols and phenolic lignin and has also been reported to initiate lipid peroxidation reactions, which may be implicated in the biodegradation of recalcitrant materials (Kapich et al. 2005). The *Bjerkandera* spp. peroxidase is apparently another VP with the conserved Trp and Mn^{2+}-binding site residues found in the primary structure (Moreira et al. 2005). Moreover, in *P. radiata*, two divergent MnPs have been described, the first of the classical-long MnP type and the second of a short-VP-type comprising the conserved Mn^{2+}-binding site but lacking the exposed tryptophan (Hildén et al. 2005). Due to this, the latter MnP is unable to oxidize veratryl alcohol but shares with VP the Mn^{2+}-independent ability to oxidize small dye phenolic and amine compounds.

4.4 *Arthromyces ramosus peroxidase (ARP) and coprinus cinereus peroxidase (CIP)*

Arthromyces ramosus peroxidase (ARP) and *Coprinus cinereus* peroxidase (CIP) of class II fungal peroxidases with no ligninolytic activities were discovered in the mid-1980s. ARP and CIP were found in an imperfect fungus (Deuteromycete) *Arthromyces ramosus* and inky cap basidiomycete *Coprinus cinereus*, respectively (Morita et al. 1988). *Coprinus macrorhizus* peroxidase (CMP) is another group of fungal peroxidase obtained from *Coprinus macrorhizus*. However, *C. macrorhizus* is recognized as the same species as *C. cinereus*, and these two *Coprinus* peroxidases were suggested to be identical. Certainly, the analyses of three-dimensional molecular structures, amino acid sequences, and catalytic activities of these fungal peroxidases; CIP, CMP, and ARP have illustrated that the enzymes are identical only with a minor change in glycosylation, which did not influence the reaction rate of the enzyme (Sawai-Hatanaka et al. 1995).

These three peroxidases readily oxidize phenols and smaller dye molecules similar to the class III horse radish peroxidase (HRP). However, the 3-D structure and heme packing with conserved helices, two domains with bound Ca^{2+} ions, N-glycosylation sites, and the conserved proximal and distal residues near the heme resembled other class II peroxidases (Houborg et al. 2003). The distal side substrate channel of CIP/ARP is more open and accessible to phenolic and aromatic reducing compounds (Kunishima et al. 1994, Petersen et al. 1994). There is a 40–50% amino-acid identity of CIP with LiPs, VPs, and MnPs, and the size (340 amino acids in the mature protein) is almost the same. Instead of very similar catalytic activities of ARP and CIP to a classical plant peroxidase, HRP, the similarity in overall amino acid sequences among the genes encoding these plant and fungal enzymes is less. Although ARP and CIP amino acid sequences have shown better agreement with those of LiP and MnP compared to those of HRP, but no ligninolytic activity is exhibited by these enzymes. Some significant structural differences revealed by X-ray crystallography at the active sites of the two fungal peroxidases have demonstrated the reason for the differences in substrate specificity (Kunishima et al. 1994, Petersen et al. 1994).

4.5 *Dye decolorization peroxidases (DyP)*

DyPs are a new superfamily of heme-containing glycoproteins that oxidize typical peroxidase substrates and also additionally oxidize anthraquinone-type dyes and their derivatives (Sugano et al. 2009, Sugano 2009). Furthermore, some DyPs are able to oxidize aromatic sulphides (van Bloois et al. 2010), non-phenolic β-O-4 lignin model compound (Liers et al. 2010), as well as cleave β-carotene and other carotenoids (Zelena et al. 2009). The enzyme was first identified in the fungus *Geotrichum candidum* Dec 1, which was later assigned to the phytopathogenic basidiomycete *Thanatephorus cucumeris* that was a misidentified strain (Strittmatter et al. 2013). The present correctly identified strain for DyP secretion is *Bjerkandera adusta* (Ruiz-Dueñas et al. 2011). Afterwards, it was also isolated from bacteria (Hofrichter et al. 2010).

DyP is a glycosylated heme protein (17% sugars) with a Soret band at 406 nm, a pI of 3.8, and a molecular mass of 60 kDa. The sequence alignments and structural comparison of DyPs with representative members of all classes of the plant, bacterial, and fungal peroxidases demonstrated that DyPs do not belong to either of these families (Sugano et al. 2007, Zubieta et al. 2007). These are structurally different from all other peroxidases, however, these enzymes showed little sequence similarity (0.5–5%) to classic fungal peroxidases and lack the typical heme-binding region, which was conserved in plant peroxidases (Faraco et al. 2007, Sugano 2009). A characteristic feature of all DyPs investigated so far is their ability to oxidize synthetic high-redox potential anthraquinone dyes, which were recalcitrant to oxidize by other peroxidases. This catalytic feature made DyPs a potential tool to target pollutants (Syoda et al. 2006). However, the mechanism of the catalytic cycle of DyP and the cleavage of dyes are not fully understood yet, but there are interesting indications that includes, besides typical peroxidative reactions, the hydrolytic fission of the anthraquinone backbone (Sugano et al. 2009). The latter notion is supported by the identification of phthalic acid as a cleavage product of the recalcitrant dye Reactive Blue 5. The spectroscopic characteristics of DyP in the presence of an equivalent of peroxide showed a similarity to those of a typical peroxidase compound I but the formation of compound II has not been noticed so far (Sugano et al. 2007). DyPs differ from other heme peroxidases in the process of compound I formation, since an aspartate replaces the distal histidine that is usually responsible for H_2O_2 binding. In this respect, DyPs resemble heme-thiolate peroxidases, HTPs (CfuCPO, APOs), which contain identical amino acid glutamate at the respective position (Pecyna et al. 2009).

The high stability of DyPs is another significant property that attracted it for applications under rather harsh environmental conditions. Pühse et al. (2009) reported that purified dimeric MsP1 (DyP-type peroxidase of *M. scorodonius*) was markedly more stable at high temperatures (up to 70°C) and pressures (up to 2,500 bar). DyPs also appeared quite stable under acidic conditions. DyP isoform of *A. auricula-judae* exhibited very high stability in a buffer of pH 2.5 even after 2 hours of incubation. The secretion of this DyP by *A. auricula-judae* under conditions close to nature, i.e., in wood cultures and their ability to oxidize non-phenolic lignin model compounds (indicative for a ligninolytic activity) may throw some light on the possible physiological function of these enzymes (Liers et al. 2010). According to a method of Ayala et al. (2007), which is based on the oxidation of different *p*-substituted phenols, there are indications that the redox potential of DyPs range between those of high-redox potential peroxidases (i.e., ~ 1.5 V for LiP and VP) as well as low-redox potential peroxidases, i.e., ~ 0.8–1.0 V for plant peroxidases.

4.6 *Heme-thiolate peroxidases (HTPs)*

The most interesting catalytic activity of secreted HTPs is the transfer of peroxide-oxygen (from H_2O_2 or R-OOH) to substrate molecules, and has been described as a peroxygenase activity. The most relevant enzymes in this category include CfuCPO (EC 1.11.1.10) and APOs (EC 1.11.2.1).

4.6.1 Caldariomyces fumago chloroperoxidase (CfuCPO)

CfuCPO is a halogenating enzyme and it does not fit into the class II of heme peroxidases, due to the difference in amino-acid sequences and type of reactions they catalyze (Passardi et al. 2007). This enzyme also illustrates remarkable structural differences compared to all other non-animal peroxidases, since it bears a cysteine-ligated heme, while the other peroxidases have a histidine as axial heme ligand, heme-imidazole type (Liu and Wang 2007). It has been the only HTP characterized on the protein level for almost 50 years, until similar enzymes were discovered in several agaric basidiomycetes (Omura 2005, Hofrichter and Ullrich 2006, 2011).

The main activity of CPO is to oxidize chloride (Cl^-) into hypochlorous acid at low pH (< 5). Thus, it appears as a strong oxidant that catalyzes chlorination of organic compounds. The first part of the cycle including peroxide binding and formation of compound 0, heterolytic O–O fission, and compound I formation is similar to other heme peroxidases (even if a Glu residue is involved in H_2O_2 binding instead of a His). However, the subsequent oxidation of halide by compound I via two electron oxidation to form compound X, a hypothetic Fe(III) hypohalite intermediate, differs from the typical cycle of heme peroxidases (Kühnel et al. 2006). This unstable compound X is proposed to rapidly decay into the native ferric enzyme and hypohalous acid that oxidizes electrophilic substrates via halogenation. Compound II is not formed in this cycle; however, CfuCPO can act as a "normal" peroxidase in the absence of halides (Hofrichter et al. 2010).

It catalyzes the oxidation of phenols and anilines via a peroxidative mechanism in the absence of halides. These reactions include also the oxidation of highly chlorinated derivatives; e.g., PCP, which were earlier reported to be partially dehalogenated while being polymerized into insoluble polymeric materials. Contrasting to simple phenol oxidation, the conversion of PCP by CfuCPO was one to two magnitudes faster than the reactions of HRP, LiP, or VP (Longoria et al. 2008). These one-electron oxidations/H abstractions follow the typical cycle of heme peroxidases involving compound I and II formation, and may occur in distance to the active site heme at the protein surface. CfuCPO is also able to catalyze oxygen transfers like it epoxidizes double bonds of alkenes and cycloalkenes as well as hydroxylates benzylic carbon via a peroxygenase mechanism (oxygenation mechanism/two-electron oxidation) that resembles the peroxide "shunt" pathway of P450s (Osborne et al. 2007, Manoj and Hager 2008). However, oxygen transfers to less activated molecules, such as aromatic rings or n-alkanes cannot be catalyzed by CfuCPO (Ullrich and Hofrichter 2007).

The reasons for describing CfuCPO as a "heme peroxidase cytochrome P450 functional hybrid" or a "Janus enzyme" were the different catalytic properties of CfuCPO, i.e., one electron oxidations as well as two-electron oxidations coupled with halide oxidation or oxygen transfer (Manoj and Hager 2008). It is important to remember that CfuCPO does not share any sequence homology with P450s. In contrast, molecular architecture (arrangement of α-helices in the tertiary structure) of both enzyme types was reported to show some similarity. A problem that limited the practical utility of CfuCPO is its inherent activity to dismutate parts of the added H_2O_2. This leads to the formation of superoxide ($O_2^{\cdot-}$) and in turn, to heme bleaching and enzyme inactivation (Park and Clark 2006). To overcome this and other shortcomings and also to improve the stability and performance of CfuCPO, specific reaction systems and ingenious catalytic devices have been developed.

4.6.2 Aromatic peroxygenases (APOs)

The second group of HTPs was discovered in the wood and litter-dwelling black poplar mushroom *A. aegerita*, synonym *A. cylindracea*. This later turned out to be a true peroxygenase that efficiently transfers oxygen from peroxides to various organic substrates including heterocyclic, aromatic, and aliphatic compounds (Hofrichter and Ullrich 2011). Afterward, similar enzymatic activities were reported in the cultures of other agaric basidiomycetes, for example, *Coprinopsis verticillata*, *Coprinellus radians*, and *Agrocybe alnetorum* (Anh et al. 2007).

Depending on the reaction conditions and the particular substrate, APOs catalyze various reactions, including one-electron abstractions, peroxygenations, and brominations. However, it catalyzes no chlorinations. APO enzyme needs merely H_2O_2 as co-substrate to function in all these cases. Toluene and naphthalene were the first substrates found to undergo epoxidation and hydroxylation during the progress of APO catalyzed reaction. The toluene molecule is hydroxylated both at the methyl group and at the aromatic ring, resulting in the formation of mixtures of *p*- and *o*-cresol, methylhydroquinone, as well as benzyl alcohol, benzoic acid, and benzaldehyde (Ullrich and Hofrichter 2005). Various APOs regioselectively epoxidize naphthalene into naphthalene 1, 2-oxide that spontaneously hydrolyze into 1- and 2-naphthol (Anh et al. 2007, Kluge et al. 2009). Other polyaromatic compounds, such as anthracene, phenanthrene, fluorine, and pyrene, as well as dibenzofuran were also found to be the subject of APO-catalyzed oxygenation resulting in mixtures of mono- and polyhydroxylated products (Aranda et al. 2010). There is ostensibly a limitation in molecule size concerning PAH oxygenation and bulky compounds, such as the 5-ring system of perylene which cannot be oxidized by APO.

Kinne et al. (2009b) have reported the presence of "etherase" activity in APO. The cleavage of ether bond is initiated by the hydroxylation of one of the ether oxygens adjacent carbon atoms. The subsequent unstable hemiacetal spontaneously hydrolyzes into phenol/alcohol and aldehyde products. Ether cleavage was found to take place between aliphatic and aromatic molecule parts (1,4-dimethoxybenzene) but also in alicyclic and aliphatic ethers; dioxane, tertrahydrofuran, di-isopropyl ether. Besides oxygen transfers, APOs also catalyze one electron oxidation of phenolic substrates; peroxidative activity. Aromatic peroxygenation and phenol oxidation can consecutively take place when the former reaction introduces an OH-group into the substrate, which in turn, serves as a target for the peroxidative activity. In such cases, the radical scavenger ascorbic acid prevents further oxidation and polymerization of phenolic intermediates by reducing phenoxyl radicals formed (Aranda et al. 2008, Kinne et al. 2009a, Kinne et al. 2009b).

Although no spectroscopic data on reactive APO intermediates is present, still a catalytic cycle can be proposed, which combines elements of that of heme peroxidases (e.g., CfuCPO) with the peroxide "shunt" pathway of P450 monooxygenases (Makris et al. 2005, Ullrich and Hofrichter 2007, Hofrichter and Ullrich 2011). Compound I in the peroxygenase mode reacts directly with the substrate by transferring ferryl oxygen to the latter while receiving two electrons. For oxygen insertion into ethers, intramolecular isotope effects were observed with asymmetrically deuterium-labeled [O-CD3]1,4-dimethoxybenzene that indicated an H-abstraction rebound mechanism (Kinne et al. 2009b). Contrary to this, the oxygenation of aromatic rings, e.g., naphthalene by APO may involve the oxidation of one of the π-bonds to form an epoxide intermediate instead of direct insertion of oxygen into one of the aromatic C–H bonds (Kluge et al. 2009).

The ability to transfer peroxide-borne oxygen to diverse substrates is the most significant activity of these enzymes in environmental pollution control. Some earlier workers have also reported the transfer of peroxide oxygen to recalcitrant aromatics, heterocycles, and ethers (Ullrich et al. 2004, Ullrich and Hofrichter 2005). Of great environmental importance is the oxidative dehalogenation of highly chlorinated derivatives (e.g., PCP) into insoluble polymeric materials. This occurs about one to two magnitudes faster than the reactions of LiP, HRP, or VP (Longoria et al. 2008). Furthermore, CfuCPO and APOs can catalyze epoxidation reactions by adding oxygen to double bonds of unsaturated compounds, such as alkenes and cycloalkenes, while APOs can also epoxidize and hydroxylate aromatic rings (Hofrichter et al. 2010).

5. Role of Fungal Peroxidases in Pollutants Degradation

5.1 *Polycyclic aromatic hydrocarbons (PAHs)*

PAHs are also known as polyarenes, and are a major class of chemicals with two or more fused aromatic rings in linear, angular, or clustered arrangements. PAHs with less than six aromatic rings are frequently denominated as small PAHs, and those containing more than this number are often referred to as large PAHs (Haritash and Kaushik 2009, Karim and Husain 2010, Acevedo et al. 2011). These PAHs occur as colorless, white/pale yellow solids with low solubility in water, high melting and boiling points, and lower vapor pressure. As the molecular weight of PAHs increases, their solubility in water decreases; melting and boiling point increases, and vapor pressure decreases (Zhang et al. 2016).

Recent findings showed that extracellular peroxidases of fungi are responsible for the initial oxidation of PAHs (Li et al. 2014, Zhang et al. 2015). Fungal LiPs are known to oxidize a number of PAHs directly, while fungal MnPs co-oxidize them indirectly through enzyme-mediated lignin peroxidation. Table 2.1 summarizes the PAHs degraded by fungal peroxidases.

Ligninolytic enzymes undergo a one electron radical oxidation, producing aryl cation radicals from contaminants and subsequent generation of quinones (Cerniglia and Sutherland 2010). Purified forms of LiP and MnP similarly were able to oxidize anthracene, pyrene, fluorene, and benzo[a]pyrene to the corresponding quinones. Hammel et al. (1986) demonstrated that LiP produced by *P. chrysosporium* catalyzes the degradation of certain PAHs with IP < 7.55 eV. Therefore, H_2O_2-oxidized states of LiP are more oxidizing than the analogous states of standard peroxidases. Studies on pyrene as a substrate showed that pyrene-1,6-dione and pyrene-1,8-dione are the

Table 2.1 Various PAHs degraded by fungal peroxidases.

Enzyme	Source	PAHs	References
LiP	*P. chrysosporium*	B[a]P	Haemmerli et al. (1986)
MnP	*Nematoloma frowardii*	Phenanthrene, pyrene, fluoranthene, anthracene, & B[a]P	Sack et al. (1997)
MnP-LiP hybrid	*Bjerkandera adusta*	Anthracene, pyrene & B[a]P	Wang et al. (2003)
MnP	*Irpex lacteus*	Phenanthrene, anthracene, fluoranthene, & pyrene	Baborová et al. (2006)
MnP	*Anthracophyllum discolor*	Pyrene & anthracene	Acevedo et al. (2010)
APO	*Agrocybe aegerita*	2-Methylnaphthalene, 1-methylnaphthalene, dibenzofuran, fluorene, phenanthrene, anthracene & pyrene	Aranda et al. (2010)
MnP-Tra-48424	*Trametes* sp. 48424	Fluorene, fluoranthene, pyrene, phenanthrene & anthracene	Zhang et al. (2016)

principle oxidation products. Gas chromatography/mass spectrometry analysis of LiP-catalyzed pyrene oxidation done in the presence of H_2O_2 revealed that the quinone oxygens come from water. The one-electron oxidative mechanism of LiP is relevant to lignin and lignin-related substructures, as well as certain PAH and heteroaromatic contaminants. Purified LiP from *P. chrysosporium* had been shown to attack benzo[a]pyrene using one-electron abstractions, causing unstable benzo[a]pyrene radicals which undergo further spontaneous reactions to hydroxylated metabolites and many benzo[a]pyrenequinines. The oxidation rate is ameliorated more than 14 times in the presence of VA, and most of the enzyme activity was retained during the benzo[a]pyreneoxidation (Haemmerli et al. 1986). Benzo[a]pyrene-1,6-, 3,6-, and 6,12-quinones were detected as the products of benzo[a]pyreneoxidation by LiP from *P. chrysosporium*. At the same time with the appearance of oxidation products, LiP was inactivated. Similar to all peroxidases, LiP is inhibited by the presence of H_2O_2; the addition of VA to the reaction mixture has demonstrated the stabilization of enzyme (Valderrama et al. 2002).

Vazquez-Duhalt et al. (1994) utilized LiP from *P. chrysosporium* to investigate the oxidation of anthracene, 1-, 2-, and 9-methylanthracenes, acenaphthene, fluoranthene, pyrene, carbazole, and dibenzothiophene. Among the studied compounds, LiP was able to oxidize compounds with IP < 8 eV. The greatest specific activity of PAHs oxidation was shown when pHs were between 3.5 and 4.0. The reaction products involve hydroxyl and keto groups. Anthraquinone is the major product of anthracene oxidation by LiP obtained from *P. chrysosporium* (Field et al. 1996). The degradation of PAHs by MnP crude preparation of *Nematoloma frowardii* was demonstrated for a mixture of eight different PAHs, as well as the five individual PAHs phenanthrene, pyrene, fluoranthene, anthracene, and benzo[a]pyrene (B[a]P).

The oxidation of PAHs was enhanced by the addition of a mediator substance, glutathione, which is able to form reactive thiyl radicals (Sack et al. 1997). Günther et al. (1998) demonstrated the induction effect of reduced glutathione (GSH) on MnP and they found an enhancement in its oxidative strength. Consequently, anthracene was fully reduced and 60% of pyrene was degraded after only 24 hours. PAHs anthracene and its methyl derivatives, pyrene, and benzo[a]pyrene with ionization potentials of 7.43 eV or lower were oxidized by a purified Mn-LiP hybrid isoenzyme obtained from *B. adusta* in the presence and absence of manganous ions. The rate of PAH oxidation was decreased by the presence of Mn. The PAH metabolites of the Mn-independent reaction were identified as the corresponding quinones (Wang et al. 2003). Baborová et al. (2006) demonstrated the degradation of four PAHs, i.e., phenanthrene, anthracene, fluoranthene, and pyrene by purified MnP from *Irpex lacteus*. Major degradation products of anthracene were also identified, and it was found that MnP was able to cleave the aromatic ring of the PAH. Anthracene was degraded to phthalic acid and was found to have the highest degradation rate, followed by dibenzothiophene and then pyrene (Eibes et al. 2006).

Aranda et al. (2010) reported the hydroxylation of aromatic rings of 2-methylnaphthalene, 1-methylnaphthalene, dibenzofuran, fluorene, phenanthrene, anthracene, and pyrene catalyzed by APO from *Agrocybe aegerita*. MnP produced by *Anthracophyllum discolor* the degraded pyrene (> 86%) and anthracene (> 65%) alone or in the mixture. It also degraded fluoranthene and phenanthrene, but less effectively, < 15.2% and < 8.6%, respectively (Acevedo et al. 2010). This enzyme catalyzed oxidation of PAHs resulted in the formation of respective quinones. Anthrone, an expected intermediate was produced during degradation of anthracene by MnP, followed by the production of 9,10-anthraquinone. Anthraquinone has earlier been shown as the typical oxidation product in *in vitro* reactions of peroxidases. The high hydrophobicity of PAHs significantly inhibits their degradation in liquid media but MnP degrades anthracene, dibenzothiophene, and pyrene in the presence of acetone (36%, v/v), which is a water miscible organic solvent. Different PAHs (10 mg L^{-1}) fluorene, fluoranthene, pyrene, phenanthrene, and anthracene were completely degraded by MnP-Tra-48424 from *Trametes* sp. 48424 after 24 hours of incubation. MnP-Tra-48424 was highly efficient in the degradation of anthracene and removed around 80% of anthracene within 3 hours of treatment (Zhang et al. 2016).

5.2 Dyes

In spite of lack of recent data on worldwide dye production, annual production of over 700,000 tones has been often reported in the literature (Liu et al. 2014). In the textile industry, after the dyeing process, up to 50% of the dyes are lost and about 10–15% of them are discharged in the effluents (Langhals 2004, Husain 2010a). Color is generally the first contaminant in these effluents. A very small quantity of dye in water (10–20 mg L^{-1}) is highly noticeable. The presence of these dyes in water can affect its transparency and also sunlight penetration. The environmental discharge of these dyes has become a major concern as they are not only particularly problematic for color pollution in water bodies (Husain 2006), but also because of their metabolites may work as mutagenic or carcinogenic agents. In view of strict legislation and regulations, it is necessary to treat these industrial dye-laden effluents before they are discharged into the environment. The ligninolytic enzymes have wide applications that are currently used in the removal of dyes discharged from industrial effluents (Alam et al. 2009).

It is well known that LiP decolorizes azo dyes either directly or indirectly in the presence of a mediator such as VA, whereas MnP appeared to play only a minor role in decolorization, even in the presence of manganese ion (Young and Yu 1997). However, unlike the MnP from *P. chrysosporium*, Mn-oxidizing peroxidases isolated from *B. adusta* and *Pleurotus eryngii* regardless of the presence of manganese ion were reported to decolorize multiple azo dyes (Camarero et al. 1999). Interestingly, these enzymes exhibit both MnP-like and LiP-like characteristics and were able to oxidize Mn^{2+} to Mn^{3+} at around pH 5 while also oxidizing aromatic compounds such as VA, a typical substrate of LiP at around pH 3, irrespective of the presence of Mn^{2+} (Ruiz-Dueñas et al. 2001). Based on the mechanism of action these enzymes were thus termed MnP–LiP hybrid peroxidases, VPs, or manganese-independent peroxidases. The high redox potential (1.1–1.2 V) of DyP-type peroxidases allows it to oxidize a wide spectrum of complex dyes, in particular, xenobiotic anthraquinone derivatives (Liers et al. 2014). Table 2.2 summarizes dyes degraded by fungal peroxidases.

Ferreira et al. (2000) reported the degradation of Methylene Blue and Azure B by LiP from *P. chrysosporium*. The study of reaction products showed that N-demethylation reactions were involved in the degradation of such dyes. The role of MnP in dye decolorization of polymeric dye, Poly R478 was assayed with isoenzyme PCH4 from *P. chrysosporium* and isoenzyme BOS2 from *Bjerkandera* sp. BOS55. The assays were conducted with semi-continuous addition of H_2O_2. The two pure MnP have different catalytic properties because MnP from *Bjerkandera* produced a certain decolorization in the absence of Mn^{+2}, while very little effect was observed in the case of MnP from *P. chrysosporium*. The enzymatic treatment triggered not only the destruction of the chromophoric groups, but there was also a noticeable breakdown of the chemical structure of the dye. Experiments conducted with pure enzymes proved that MnP was the main enzyme which is responsible for the dye decolorization (Moreira et al. 2001). An azo-reactive dye, reactive red 120 was successfully decolorized by purified MnP of *P. sordida* in the presence of Mn(II) and Tween 80. The involvement of lipid peroxidation during decolorization with MnP was considered. The decolorization of the dye did not occur under anaerobic conditions. This suggested that dye decolorization by MnP was influenced by dissolved oxygen (Harazono et al. 2003).

In vitro decolorization of Reactive Brilliant Red K-2BP, an industrial azo dye by crude LiP and MnP was evaluated and compared for their degradation characteristics. Decolorization by LiP can be enhanced to the greatest extent (89%) with higher addition of H_2O_2 and VA. In contrast, decolorization by MnP was optimized only with a suitable dose of H_2O_2 (0.1 mM) and decreased by the addition of Mn^{2+}. At high dye concentrations, there is a decline in decolorization; LiP was able to decolorize a dye concentration of 60 mg L^{-1} and below to no less than 85%, and MnP of 10 mg L^{-1} to a maximum of 71%. Also, the combined decolorization of dye by LiP

Table 2.2 Various dyes degraded by fungal peroxidases.

Enzyme	Source	Dyes	References
LiP	*P. chrysosporium*	Methylene blue & azure B	Ferreira et al. (2000)
MnP-PCH4	*P. chrysosporium*	Poly R478	Moreira et al. (2001)
MnP-BOS2	*Bjerkandera* sp. BOS55	Poly R478	Moreira et al. (2001)
MnP	*P. sordida*	Reactive red 120	Harazono et al. (2003)
VP and DyP	*Thanatephorus cucumeris* Dec 1	Reactive blue 5	Sugano et al. (2006)
rDyP	*Aspergillus oryzae*	RBBR	Shakeri and Shoda (2007)
MnP	*Ischnoderma resinosum*	Reactive black 5, reactive blue 19, reactive red 22 & reactive yellow 15	Kokol et al. (2007)
MnP	*Schizophyllum* sp. F17	Orange G & orange IV	Yao et al. (2013)
CD2-MnP	*Irpex lacteus* CD2	RBBR, direct red 5B, remazol brilliant violet 5R, indigo carmine & methyl green	Qin et al. (2014)
MnP	*Pleurotus ostreatus* MTCC 142	Tartrazine, trypan blue & ponceau S	Arunkumar and Sheik Abdulla (2015)
CPO	*Caldariomyces fumago*	Crystal violet & alizarin red	Liu et al. (2014)
VP	*Bjerkandera adusta*	Phthalocyanines & azo dyes	Baratto et al. (2015)
MnP-Tra-48424	*Trametes* sp. 48424	RBBR, remazol brilliant violet 5R, indigo carmine & methyl green	Zhang et al. (2016)
MnP	*Irpex lacteus* F17	Malachite green	Yang et al. (2016)

and MnP was somewhat lower than that obtained by LiP alone. It is suggested that the optimization of the H_2O_2 supply was mainly responsible for a high efficiency in continuous dye degradation by crude LiP and MnP (Yu et al. 2006). Coapplication of VP and DyP from *Thanatephorus cucumeris* Dec 1 was able to completely decolorize Reactive Blue 5, *in vitro*. This process of decolorization proceeded sequentially. DyP decolorized Reactive Blue 5 to light red-brown compounds, and then VP decolorized these colored intermediates to colorless. Following stretched reactions, the absorbance corresponding to the conjugated double bond from phenyl (250–300 nm) decreased, which illustrated the degradation of aromatic rings (Sugano et al. 2006). Another anthraquinone dye, Remazol Brilliant Blue R (RBBR) was degraded by crude recombinant DyP (rDyP) generated by *Aspergillus oryzae*. It was observed that in a batch process, equimolar batch addition of H_2O_2, and RBBR yielded complete decolorization of RBBR by rDyP, with a turnover capacity of 4.75. In stepwise fed-batch addition of H_2O_2, the turnover capacity increased to 5.76, and by further increasing dye concentration, it reached 14.3. Upon addition of H_2O_2 in continuous fed-batch to minimize rDyP inactivation and adding 1.6 mM dye in stepwise fed-batch mode, the turnover capacity increased to 20.4. At this turnover capacity, 1 L of crude rDyP solution containing 5,000 U could decolorize up to 102 g RBBR in 650 minutes (Shakeri and Shoda 2007).

Kokol et al. (2007) reported the decolorization of four reactive dyes, i.e., Reactive Black 5, Reactive Blue 19, Reactive Red 22, and Reactive Yellow 15 by purified MnP from *Ischnoderma resinosum*. The efficiency of fungal LiP and HRP was compared for decolorization of Methylene Blue and its demethylated derivatives (Ferreira-Leitão et al. 2007). It was shown that both enzymes were able to oxidize methylene blue and its derivatives. However, it was stated that LiP was more effective as its oxidation potential was almost double that of HRP. Ferreira-Leitao et al. (2007) described that LiP would be more suitable for degradation of phenothiazine dyes and decolourization of wastewater. Additionally, a LiP produced from sewage sludge treatment plant was reported to exhibit potential for textile effluent treatment and dye decolorization (Alam et al. 2009). Yao et al. (2013) employed crude MnP from *Schizophyllum* sp. F17 for the decolorization of three structurally different azo dyes. Under optimized conditions, after 20 minutes of reaction at pH 4 and 35°C, the decolorization of Orange G and Orange IV by MnP, at an initial dye concentration of 50 mg L^{-1}, was 57 and 76%, respectively. However, only 8% decolorization of Congo Red was observed under the same experimental conditions. A rapid decolorization of azo dyes with a low MnP activity of 24 U L^{-1} was revealed through this enzymatic reaction system. A novel MnP named CD2-MnP was purified and characterized from *I. lacteus* CD2. This novel MnP had a strong capability for tolerating different metal ions such as Ca^{2+}, Co^{2+}, Cd^{2+}, Ni^{2+}, Mg^{2+}, and Zn^{2+} as well as organic solvents such as methanol, dimethyl sulfoxide, ethanol, ethylene glycol, butanediol, isopropyl alcohol, and glycerin. The diverse types of dyes including the anthraquinone dye (RBBR), azo dye (Direct Red 5B, Remazol Brilliant Violet 5R), indigo dye (Indigo Carmine), and triphenylmethane dye (methyl green), as well as simulated textile wastewater could be efficiently decolorized by CD2-MnP. CD2-MnP also decolorized different dyes in the presence of metal ions and organic solvents (Qin et al. 2014).

MnP purified from mutant *Pleurotus ostreatus* MTCC 142 was exploited for decolorization of structurally different azo dyes. Among the selected azo dyes, Tartrazine, Trypan Blue, and Ponceau S were decolorized rapidly by the purified enzyme extracts under optimized condition and it decolorized 56, 59, and 78% of these dyes, respectively (Arunkumar and Sheik Abdulla 2015). Liu et al. (2014) demonstrated an efficient and rapid enzymatic decolorization of triphenylmethane dyes (Crystal Violet) and anthraquinone (Alizarin Red) by CPO from *Caldariomyces fumago*. At mild conditions and with an enzyme concentration below ppm level, the decolorization efficiency of Crystal Violet was 97.68%, and that of alizarin red reached to 98.23%, both within 7 minutes. The analysis of products through UV–VIS and HPLC–MS indicated that the chromophoric groups were destructed and the dye molecules were broken-down into small products. Baratto et al. (2015) elucidated the mechanisms of industrial dye transformation by purified VP from *B. adusta*. The enzyme was able to decolorize different classes of dyes, including phthalocyanines and azo, but cannot transform any of the investigated anthraquinones. The enzymatic decolorization products showed the cleavage of the azo bond in azo dyes and the total disruption of the phthalocyaninic ring in phthalocyanine dyes by EPR and MS analyses. RBBR, Remazol Brilliant Violet 5R, Indigo Carmine, and Methyl Green were efficiently decolorized by MnP-Tra-48424 purified from *Trametes* sp. 48424. The enzyme also decolorized Indigo Carmine and Methyl Green combined with metal ions and organic solvents. The decolorization capability of MnP-Tra-48424 was not inhibited by selected metal ions and organic solvents (Zhang et al. 2016). Malachite Green, a recalcitrant, carcinogenic, and mutagenic triphenylmethane dye, was decolorized and detoxified using crude MnP prepared from *Irpex lacteus* F17, a WRF. 96 % of 200 mg L^{-1} of Methyl Green was decolorized when 66.32 U L^{-1} of MnP was added for 1 hour, under optimal conditions. Degradation products identified by UV–VIS and UPLC analysis revealed that the dye was degraded via two different routes, either by N-demethylation of Methyl Green or the oxidative cleavage of the C–C double bond in this dye (Yang et al. 2016).

5.3 *Endocrine disrupting compounds (EDCs)*

There had been accumulation of a large number of scientific data regarding the hormone-like effects of anthropogenic chemicals in the environment over the last 50 years (Husain and Qayyum 2013). Hormonal disruptions in populations at the receiving end of the ecosystems are generated by the release of human hormones into environmental matrices. The ability of these endocrine disrupting compounds (EDCs) to act as a hormone agonist or antagonist and also their capability to disturb the synthesis of endogenous hormones or hormone receptors are believed to be the reasons of their effects. These EDCs can disrupt the secretion, synthesis, binding, transport, action, and elimination of the endogenous hormones that are accountable for maintaining homeostasis, development, reproduction, and integrity in living organisms and their offspring (Husain and Qayyum 2013). Owing to their high toxicity, bioactivity, ubiquitous nature and persistence, removal of these substances from the environment is of the most importance. A promising way to get rid of these compounds is the application of isolated fungal enzymes for the biodegradation. Recent findings from *in vitro* experiments using purified enzymes confirmed that ligninolytic enzymes were able to extensively degrade EDCs (Tanaka and Taniguchi 2003). Table 2.3 summarizes EDCs degraded by fungal peroxidases.

MnP from *P. ostreatus* metabolized bisphenol A (BPA) *in vitro* into phenol, 4-isopropenylphenol, 4-isopropylphenol, and hexestrol. The BPA degradation products were expected to be formed by the one-electron oxidation of the substrate (Hirano et al. 2000). Tsutsumi et al. (2001) reported the complete removal of estrogenic activities of BPA and nonylphenol by MnP from *P. chrysosporium* ME-446 after 12 hours of treatment. Estrogenic activities of the steroidal hormones 17β-estradiol and ethinylestradiol were also completely removed by the same enzyme after 8 hours-treatment (Suzuki et al. 2003). Another natural steroidal hormone estrone was treated by MnP from *P. sordida* YK-624 and it was observed that estrogenic activity was completely removed after 2 hours of incubation (Tamagawa et al. 2006). Tamagawa et al. (2007) demonstrated the complete removal of estrogenic activity of 4-*t*-octylphenol by MnP from *P. sordida* YK-624 after 2 hours of incubation. LiP from *P. chrysosporium* effectively transformed and eliminated 17β-estradiol in the presence of VA. The combination of two effects by which VA boosted the transformation and removal of 17β-estradiol are mitigating LiP inactivation and modifying the enzyme

Table 2.3 Various EDCs degraded by fungal peroxidases.

Enzyme	Source	EDCs	References
MnP	*Pleurotus ostreatus*	BPA	Hirano et al. (2000)
MnP	*P. chrysosporium* ME-446	BPA & nonylphenol	Tsutsumi et al. (2001)
MnP	*P. chrysosporium* ME-446	17 β-estradiol & ethinylestradiol	Suzuki et al. (2003)
MnP	*P. sordida* YK-624	Estrone	Tamagawa et al. (2006)
MnP	*P. sordida* YK-624	4-tert-octylphenol	Tamagawa et al. (2007)
LiP	*P. chrysosporium*	17 β-estradiol	Mao et al. (2010b)
YK-LiP1	*P. sordida* YK-624	BPA, *p*–t-octylphenol, 17 β-estradiol, estrone & ethinylestradiol	Wang et al. (2012)
CPO	*Caldariomyces fumago*	BPA	Dong et al. (2016)
MnP	*Ganoderma lucidum* IBL-05	Nonylphenol & triclosan	Bilal et al. (2017)

catalytic kinetics (Mao et al. 2010a). In another study the authors revealed that a total of six products were formed after the treatment of 17β-estradiol by LiP from *P. chrysosporium* (Mao et al. 2010b). The major products appeared to be oligomers resulting from 17β-estradiol coupling. These products possibly formed colloidal solids in water that can be removed via ultrafiltration or settled during ultracentrifugation. It was demonstrated that EDCs, BPA, p-t-octylphenol, 17β-estradiol, estrone, and ethinylestradiol were removed by LiP from WRF *P. sordida* YK-624 (YK-LiP1) more efficiently than LiP from *P. chrysosporium* (Wang et al. 2012). After a 24 hour-treatment, p-t-octylphenol and BPA disappeared almost completely in the reaction mixture containing YK-LiP1. BPA (150 mg L^{-1}) was completely degraded by CPO (μg mL^{-1}) within 7 minutes at room temperature. HPLC-MS analysis of degradation products revealed multiple steps in degradation pathway. Ecotoxicity assessment of degradation products with *Chlorella pyrenoidosa* reported a decrease in toxicity compared to the parent compound (Dong et al. 2016). Furthermore, 5,50-bis(1,1,3,3-tetramethylbutyl)-[1,10-biphenyl]-2,20-diol and 5,50-bis-[1-(4-hydroxy-phenyl)-1-methyl-ethyl]-biphenyl-2,20-diol were recognized as the main metabolite from *p*-t-octylphenolor BPA, respectively. Over 80% of nonylphenol and triclosan were degraded by cross-linked enzyme aggregates of MnP from *Ganoderma lucidum* IBL-05 in a packed bed reactor system (Bilal et al. 2017).

5.4 Pharmaceuticals

Till the last decade pharmaceuticals were discharged into the environment without caring about their hazardous effects. Halling-Sørensen et al. (1998) reviewed the work on the knowledge of the exposures, fates, and effects of pharmaceuticals on the live plants, aquatic, and terrestrial animals. In recent years, this issue of pharmaceuticals release in our environment attracted increasing attention and emerged as an important research topic (Sarmah et al. 2006). In 1999, only twenty-four kinds of pharmaceuticals were reportedly found in the aquatic environment in the USA (Daughton and Ternes 1999). Reports in the recent years from either developed or developing countries exhibited those pharmaceuticals and their metabolites occur

Table 2.4 Various pharmaceuticals degraded by fungal peroxidases.

Enzyme	Source	Pharmaceuticals	References
LiP & MnP	*P. chrysosporium*	TC and OTC	Wen et al. (2009)
MnP	*P. chrysosporium* ME-446	Triclosan	Inoue et al. (2010)
LiP	*P. chrysosporium*	Carbamazepine & diclofenac	Zhang and Geißen (2010)
VP	*Bjerkandera adusta*	Citalopram hydrobromide, fluoxetine hydrochloride, diclofenac, naproxen, sulfamethoxazole & carbamazepine	Eibes et al. (2011)
CPO	*Caldarimyces fumago*	Diclofenac & naproxen	Li et al. (2017)

in a wide range of environmental samples, including groundwater, surface water, as well as drinking water (Moldovan 2006, Duong et al. 2008). Currently, there is great interest in lignin-degrading WRF and their ligninolytic enzymes ability, such as MnP and LiP, to degrade pharmaceuticals (Marco-Urrea et al. 2010). Table 2.4 illustrates the pharmaceuticals degraded by fungal peroxidases.

Wen et al. (2009) showed that tetracycline (TC) and oxytetracycline (OTC) can be strongly degraded by LiP. At 40 U L^{-1} of the enzyme activity and 50 mg L^{-1} of TC and OTC, the degradation of TC and OTC reached about 95% in 5 minutes. The pH and temperature dependent degradation of TC and OTC by LiP was greatly enhanced by increasing the concentrations of VA and initial H_2O_2. The optimum conditions for degradation were determined as 37°C, pH 4.2, 2 mM VA, 0.4 mM H_2O_2. Also, the MnP from *P. chrysosporium*, was found to be highly efficient in the degradation of TC and OTC. 72.5% of 50 mg L^{-1} of TC was degraded in the presence of 40 U L^{-1} of MnP, while 84.3% of 50 mg L^{-1} of OTC was degraded with the same amount of the catalyst added after 4 hours (Wen et al. 2009). Triclosan 28.95 mg L^{-1}, an antimicrobial and preservative agent, was effectively reduced up to 94% in 30 minutes after treatment with 0.5 nkat mL^{-1} MnP from *P. chrysosporium* ME-446 and was completely removed after 60 minutes (Inoue et al. 2010). Carbamazepine and diclofenac was *in vitro* degraded by crude LiP from *P. chrysosporium*. Complete degradation of diclofenac at pH 3.0–4.5 and 3–24 ppm H_2O_2 by LiP was observed, while the degradation efficiency of carbamazepine was mostly below 10%. The degradation efficiency of carbamazepine was not enhanced even by the addition of VA and the increased temperature (Zhang and Geißen 2010). Eibes et al. (2011) demonstrated the potential degradation of a series of pharmaceuticals including antidepressives (citalopram hydrobromide and fluoxetine hydrochloride), anti-inflammatory drugs (diclofenac and naproxen), antibiotics (sulfamethoxazole), antiepileptics (carbamazepine), and estrogen hormones (estrone, 17β-estradiol, 17α-ethinylestradiol) by VP from ligninolytic fungus *B. adusta*. Estrogens and diclofenac were completely eliminated with a very low VP activity (10 U L^{-1}) and only within 5–25 minutes. Very low or undetectable removal yields were observed for fluoxetine (lower than 10%), citalopram (up to 18%), and carbamazepine (not degraded). High degradation (80%) was achieved for sulfamethoxazole and naproxen. Diclofenac and naproxen were completely converted by CPO into the products of lower eco-toxicity. Complete conversion was achieved in only 9 and 7 minutes for diclofenac and naproxen respectively, with 0.1 mmol L^{-1} H_2O_2 and nanomolar enzyme concentration at pH 3.0 (Li et al. 2017).

5.5 Pesticides

Pesticides are the only chemicals that are intentionally made to be toxic and introduced into the environment directly. These can be swallowed, absorbed through the skin, or inhaled (Davila-Vazquez et al. 2005). The different kinds of pesticides, such as herbicides, fungicides, and insecticides have been in use for decades without any check. This resulted in a strong contamination of air, water, foods, and also in the development of pesticide resistant organisms. Aggravation of this serious problem during the last decade resulted in high risks to human health (Coelho-Moreira et al. 2013). The highly non-specific and non-stereo-selective ligninolytic enzymes are capable of degrading a wide range of recalcitrant compounds that exhibit structural similarity to lignin, such as pesticides. This occurs through the production of free radicals using H_2O_2 and molecular oxygen (Husain and Ulber 2011, Karim et al. 2011, Reddy and Mathew 2001). Table 2.5 summarizes the pesticides degraded by fungal peroxidases.

Hirai et al. (2004) described that methoxychlor level was decreased by about 65% after a 24-hour treatment of MnP-Tween 80. Methoxychlor was converted into methoxychlor olefin and 4,40-dimethoxybenzophenone by MnP-Tween 80 treatment. These results indicate that MnP from WRF can catalyze the oxidative dechlorination of methoxychlor. Moreover, a metabolite methoxychlor olefin was also degraded by MnP-Tween 80 treatment. Dichlorophen, bromoxynil, and PCP were transformed by VP *Bjerkandera adusta* UAMH 8258 in the presence and absence of Mn(II). The activity was found to be higher in the absence of Mn(II) at pH 3 than in the presence of Mn(II) at pH 4 for all the pesticides transformed. Gas chromatography–mass spectrometry analysis reaction products showed the presence of 2,3,5,6-tetrachloroquinone as one of the products from pentachlorophenol oxidation, whereas the main product from dichlorophen was 4-chlorophenol-2,2'-methylenequinone. From the peroxidase oxidation of bromoxynil, various polymers were obtained. It was noticed in all cases that the dehalogenation was catalyzed by VP (Davila-Vazquez et al. 2005). Glyphosate was completely degraded by MnP from *Nematoloma frowardii* in the presence of $MnSO_4$ and Tween 80, with or without H_2O_2 (Pizzul et al. 2009).

Table 2.5 Various pesticides degraded by fungal peroxidases.

Enzyme	Source	Pesticides	References
MnP	WRF	Methoxychlor	Hirai et al. (2004)
VP	*Bjerkandera adusta* UAMH 8258	Dichlorophen, bromoxynil & PCP	Davila-Vazquez et al. (2005)
MnP	*Nematoloma frowardii*	Glyphosate	Pizzul et al. (2009)

5.6 *Phenols*

Several industrial and agricultural activities are known to pollute the wastewater by discharging varying kinds of phenols and their derivatives. Phenolic compounds and their derivatives, e.g., catechols, are detrimental to living organisms even at low concentrations and are thus considered priority pollutants (Akhtar and Husain 2006, Ashraf and Husain 2010). Both the ligninolytic and non-ligninolytic peroxidases are being involved in the removal of phenolic compounds. Ikehata and Buchanan (2002) reported the degradation of aqueous phenol by the cultivated CIP in the presence and in the absence of additives, including polyethylene glycol (PEG) and chitosan. The phenol transformation efficiency catalyzed by CIP was enhanced on addition of chitosan and PEG by the factor of 1.3 and 1.5, respectively. Aqueous phenol was also effectively removed by ARP in the presence of high molecular weight PEG as a protective additive (Villalobos and Buchanan 2002). LiP from *P. chrysosporium* detoxified the toxic halogenated phenols by their oxidization and subsequent polymerization. Upon oxidation of the target substrates, analysis of gel permeation–HPLC demonstrated the formation of dimers, trimers, and tetramers. Effective detoxification of the aqueous phase accompanies the polymerization during oxidation of 2,4-dibromophenol, the degree of which correlated with the degree of oxidation and polymerization. The inclusion of redox mediators significantly improved the operational stability of LiP. This resulted in improved oxidation and more swift reaction rates (Ward et al. 2003). Table 2.6 represents phenols degraded by fungal peroxidases.

Murphy (2007) reported the oxidation of fluorophenol by CPO from *Caldariomyces fumago*. LiP from *P. chrysosporium* was able to remove four catechols (1,2-dihydroxybenzene), namely catechol, 4-chlorocatechol, 4,5-dichlorocatechol, and 4-methylcatechol distinctive pollutants in wastewater derived from paper and oil industries. LiP (1 µM) after 1 hour removes 2 mM catecholic substrates in the following order: 4,5-dichlorocatechol (95%) > 4-chlorocatechol (90%) > catechol (55%) > 4-methylcatechol (43%). Insoluble products were formed in all reactions except for 4-methylcatechol (Cohen et al. 2009). Xu et al. (2017) described the

Table 2.6 Various phenols degraded by fungal peroxidases.

Enzyme	Source	Phenols	References
CIP	*Coprinus cinereus*	Phenol	Ikehata and Buchanan (2002)
ARP	*Arthromyces ramosus*	Phenol	Villalobos and Buchanan (2002)
LiP	*P. chrysosporium*	Halogenated phenols	Ward et al. (2003)
CPO	*Caldariomyces fumago*	Fluorophenol	Murphy (2007)
LiP	*P. chrysosporium*	Catechol, 4-chlorocatechol, 4,5-dichlorocatechol & 4-methylcatechol	Cohen et al. (2009)
rGluMnP1	*Ganoderma lucidum*	Phenol	Xu et al. (2017)

degradation of phenol by recombinant MnP (rGluMnP1) from *Ganoderma lucidum*, expressed and produced in *Pichia pastoris*. In the experiments, phenol and its principal degradation products, such as resorcinol, hydroquinone, pyrocatechol, benzoquinone, were detected successfully.

6. Immobilized fungal peroxidases in bioremediation

The use of soluble enzymes has some inherent limitations in industrial and environmental applications, such as non-reusability, instability towards extreme pH, elevated temperatures, and organic solvents, product inhibition, and marginal life span (Husain 2010b). To overcome such limitations, the immobilization strategies appear to be the most obvious solution. Immobilization protects enzymes from denaturation and helps to retain them in biochemical reactors to further catalyze the subsequent feed and offer more economical exploitation of biocatalysts in industry, waste treatment, and in the development of bioprocess monitoring devices, such as the biosensor. Immobilization helps in maintaining homogeneity of enzymes in the reaction media since it avoids aggregation of enzyme particles. It is an efficient way to prevent inactivation and extend the enzyme's life span. Moreover, a proper immobilization technique may permit not only reusability of the biocatalyst but also increase structural rigidity of the enzyme and thus improves pH, heat, and organic solvent tolerance (Satar and Husain 2009). Many inorganic and organic materials have been used as carriers (enzyme supports) for the immobilization of enzymes, e.g., glass (silica) beads, natural and synthesized polymers, ion exchange resins, magnetic particles, metal oxides, and membranes (Husain and Ulber 2011, Kim et al. 2012). Recently nano-materials have exhibited advantages over traditional bulk materials owing to their miniature size, large surface area, and high enzyme loading capacity. Immobilization on nano sized structures also reduces the limitation of substrate diffusion. Moreover, the physical characteristics of nanoparticles, such as enhanced diffusion and particle mobility, can impact the inherent catalytic activity of attached enzymes (Ansari and Husain 2012, Husain 2017c, d, 2018).

There are five different kinds of methods employed for the immobilization of enzymes. These are adsorption, covalent attachment, entrapment, micro encapsulation, and chemical aggregation. All these methods have their own merits and demerits and have been successfully considered for the immobilization of enzymes of varying classes (Husain 2010b, 2019).

Purified MnP from *L. edodes* was immobilized on azlactone-functional copolymer functionalized with ethylenediamine through its carboxyl groups via covalent attachments using 2-ethoxy-1-ethoxycarbonyl-1,2-dihydroquinoline as a crosslinking agent. The immobilized MnP was used in a two-reactor system, producing chelated Mn(III) in one reactor and subsequently oxidizing chlorophenols in second reactor (Grabski et al. 1998). The significant decolorization of azo dyes in the static and shaky situation by gelatin-immobilized MnP from WRF was investigated. The immobilized enzyme retained its full activity after two repeated uses in batch processes (Cheng et al. 2007). Shakeri and Shoda (2008) demonstrated the immobilization of rDyP produced from *Aspergillus oryzae* using silica-based mesoporous materials, AlSBA-15 and FSM-16. rDyP immobilized on FSM-16 decolorized eight sequential batches of an anthraquinone dye, RBBR in repeated batch decolorization process at pH 4.0.

Co-immobilization of LiP from *P. chrysosporium* and glucose oxidase (GOD) on nanoporous gold has been described earlier by Qiu et al. (2009). High LiP activity was sustained by *in situ* release of H_2O_2 by a co-immobilized glucose oxidase. This co-immobilization system was demonstrated to be very effective for LiP-mediated decolorization of fuchsine, rhodamine B, and pyrogallol red. Natural nanomaterial, i.e., nanoclay has been utilized for the adsorption of MnP from *Anthracophyllum discolour* and found to be highly efficient in the removal of PAHs (Acevedo et al. 2010). rDyP produced from *A. oryzae* was immobilized in synthesized silica-based mesocellular foam (average pore size 25 nm) and used for decolorization of the RBBR, an anthraquinone dye. In repeated dye-decolorization tests, 20 batches of RBBR were decolorized by the mesocellular foam-immobilized rDyP (Shakeri and Shoda 2010). Recalcitrant azo dyes; Reactive Black 5, Direct Violet 51, Bismark Brown R, and Ponceau Xylidine were decolorized by *P. chrysosporium* MnP immobilized into Ca-alginate beads in successive batch cultures (Enayatzamir et al. 2010). Hu et al. (2013) employed of LiP from *P. chrysosporium* BKM-F-1767 covalently immobilized on mesoporous silica for the decolorization of acid orange II. The dye was decolorized more than 50% after 5 reuses. MnP from *Ganoderma lucidum* entrapped in sol-gel matrix of tetramethoxysilane and proplytrimethoxysilane was used for the decolorization of different textile effluents. After a 4 hour reaction time, the industrial effluents were decolorized to different extents with a maximum of 99.2%. The maximally decolorized effluent was further analyzed for formaldehyde and nitroamines and results showed that toxicity parameters were below the permissible limits (Iqbal and Asgher 2013). Combined cross-linked enzyme aggregates were prepared from VP and GOD. VP from *B. adusta* was first insolubilized in the form of cross-linked enzyme aggregates (CLEAs) with 67% recovery of initial enzyme activity, while co-aggregation of VP with GOD from *Aspergillus niger* led to an increased activity recovery of 89%. Combined CLEAs were able to remove endocrine disruptors effectively in a batch process (Taboada-Puig et al. 2011). In another study the three oxidative enzymes laccase from *T. versicolor*, VP from *B. adusta*, and GOD from *A. niger* were concomitantly cross-linked after aggregation, thus, making a combined CLEAs that was versatile and involved in an enzymatic cascade reaction to transform a wide range of pharmaceutically active compounds (Touahar et al. 2014).

Purified MnP isolated from *Ganoderma lucidum* IBL-05 was immobilized onto polyvinyl alcohol-alginate beads and its potential was investigated for the decolorization and detoxification of new class of reactive dyes and textile wastewater. The immobilized MnP exhibited high decolorization efficiency for Sandal reactive dyes (78.14–92.29%) and textile wastewater (61–80%). Polvinyl alcohol-alginate bound MnP retained more than 60% of dye decolorization activity after six repeated uses. The cytotoxicity of treated dyes was remarkably reduced (Bilal and Asgher 2015). Bilal et al. (2016a) used MnP from *P. chrysosporium* immobilized on glutaraldehyde activated chitosan beads for the remediation of dyes present in textile effluents. Maximum color removal of 97.31% was recorded and up to 82.40%, 78.30%, and 91.7% reductions in COD, TOC, and BOD were observed, respectively. The cytotoxicity of bio-treated effluents reduced significantly and 38.46%, 43.47%, and 41.83% *Allium cepa* root length, root count, and mitotic index were increased, respectively. Brine shrimp nauplii death was reduced up to 63.64%. The chitosan immobilized MnP retained 60% activity after 10 repeated decolorization batches. The decolorization of three textile reactive dyes; Reactive Red 195A, Reactive Yellow 145A, and Reactive Blue 21 by MnP from *Ganoderma lucidum* IBL-05 entrapped into agar-agar gel has been investigated. The studies concluded that the toxicity of dyes aqueous solutions was significantly reduced after treatment (Bilal et al. 2016b). MnP from *Ganoderma lucidum* IBL-05 was immobilized using hydrophobic sol-gel matrix of tetramethoxysilane and propyltrimethoxysilane, and the immobilized enzyme retained nearly 93% dye decolorization efficiency from textile effluent. In a recent study, chitosan nanoparticles have also been utilized for the encapsulation of VP in order to remove phenolic compounds from wastewater (Alarcón-Payán et al. 2017). Table 2.7 lists immobilized fungal peroxidases used for the remediation of pollutants.

Conclusion

Majority of fungal peroxidases are produced by basidiomycete while some ligninolytic and non-ligninolytic peroxidases are also produced by ascomycete. The exception to this is ARP which is exclusively produced by deuteromycete. Fungal peroxidases have great potential for the degradation and removal of a different class of environmental pollutants (Fig. 2.2). Both the ligninolytic and non-ligninolytic peroxidases have unique catalytic mechanisms for the degradation of their respective organopollutants. Ligninolytic LiP, MnP, and VP can degrade different class of pollutants, while non-ligninolytic ARP, CIP, HTPs, and DyPs mainly target the phenols and dyes. DyPs have a high redox potential and therefore can degrade phenols and dyes with complex structure. For the large-scale application of fungal peroxidases, high production is required which cannot be met by their host organisms. Therefore, a well suited heterologous expression system and optimized cultivation/fermentation

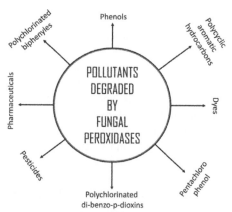

Figure 2.2 Illustrates different classes of pollutants degraded by fungal peroxidases.

Table 2.7 Immobilized fungal peroxidases for bioremediation.

Enzyme	Source	Support for immobilization	Treated pollutant	References
MnP	*Lentinula edodes*	Ethylenediamine functionalized azlactone-functional copolymer	Chlorophenols	Grabski et al. (1998)
MnP	WRF	Gelatin	Azo dyes	Cheng et al. (2007)
rDyP	*Aspergillus oryzae*	Silica-based mesoporous material (FSM-16)	RBBR	Shakeri and Shoda (2008)
LiP	*P. chrysosporium*	Nanoporous gold	Fuchsine, rhodamine B & pyrogallol red	Qiu et al. (2009)
MnP	*Anthracophyllum discolour*	Nanoclay	PAHs	Acevedo et al., (2010)
rDyP	*Aspergillus oryzae*	Silica-based mesocellular foam	RBBR	Shakeri and Shoda (2010)
MnP	*P. chrysosporium*	Ca-alginate beads	Reactive black 5, direct violet 51, bismark brown R & ponceau xylidine	Enayatzamir et al. (2010)
VP	*Bjerkandera adusta*	CLEAs	Endocrine disruptors	Taboada-Puig et al. (2011)
LiP	*P. chrysosporium* BKM-F-1767	Mesoporous silica	Acid orange II	Hu et al. (2013)
MnP	*Ganoderma lucidum*	Sol-gel matrix of tetramethoxysilane & proplytrimethoxysilane	Textile effluent	Iqbal and Asgher (2013)
VP	*Bjerkandera adusta*	CLEAs	Pharmaceutically active compounds	Touahar et al. (2014)
MnP	*Ganoderma lucidum* IBL-05	Polyvinyl alcohol-alginate beads	New class of reactive dyes & textile effluents	Bilal and Asgher (2015)
MnP	*P. chrysosporium*	Chitosan beads	Textile effluent	Bilal et al. (2016a)
MnP	*Ganoderma lucidum* IBL-05	Agar-agar gel	Reactive red 195A, reactive yellow 145A & reactive blue 21	Bilal et al. (2016b)
VP	*Bjerkandera adusta*	Chitosan nanoparticles	Phenolic compounds	Alarcón-Payán et al. (2017)

conditions are required for the high-titre production of these enzymes. Also, to make these enzymes more efficient in wastewater treatment, protein engineering is requisite to enhance their stability against pH, temperature, metal ions, organic solvents, and other harsh environmental conditions. Such engineered fungal peroxidases can prove to be a "novel green catalysts" for the wastewater treatment.

References

Acevedo, F., L. Pizzul, M. d.P. Castillo, M. E. González, M. Cea, L. Gianfreda and M. C. Diez. 2010. Degradation of polycyclic aromatic hydrocarbons by free and nanoclay-immobilized manganese peroxidase from *Anthracophyllum discolor*. *Chemosphere* 80 (3): 271–278. doi: https://doi. org/10.1016/j.chemosphere.2010.04.022.

Acevedo, F., L. Pizzul, M. d. P. Castillo, R. Cuevas and M. C. Diez. 2011. Degradation of polycyclic aromatic hydrocarbons by the Chilean white-rot fungus *Anthracophyllum discolor*. *Journal of Hazardous Materials* 185 (1): 212–219. doi: https://doi.org/10.1016/j.jhazmat.2010.09.020.

Akhtar, S. and Q. Husain. 2006. Potential applications of immobilized bitter gourd (*Momordica charantia*) peroxidase in the removal of phenols from polluted water. *Chemosphere* 65 (7): 1228–1235. doi: https://doi.org/10.1016/j.chemosphere.2006.04.049.

Alam, M. Z., M. F. Mansor and K. C. A. Jalal. 2009. Optimization of decolorization of methylene blue by lignin peroxidase enzyme produced from sewage sludge with *Phanerocheate chrysosporium*. *Journal of Hazardous Materials* 162 (2): 708–715. doi: https://doi.org/10.1016/j.jhazmat.2008.05.085.

Alarcón-Payán, D. A., R. D. Koyani and R. Vazquez-Duhalt. 2017. Chitosan-based biocatalytic nanoparticles for pollutant removal from wastewater. *Enzyme and Microbial Technology* 100: 71–78. doi: https://doi.org/10.1016/j.enzmictec.2017.02.008.

Anh, D. H., R. Ullrich, D. Benndorf, A. Svatoś, A. Muck and M. Hofrichter. 2007. The coprophilous mushroom *Coprinus radians* secretes a haloperoxidase that catalyzes aromatic peroxygenation. *Applied and Environmental Microbiology* 73 (17): 5477–5485. doi: 10.1128/aem.00026-07.

Ansari, S. A. and Q. Husain. 2012. Potential applications of enzymes immobilized on/in nano materials: A review. *Biotechnology Advances* 30 (3): 512–523. doi: https://doi.org/10.1016/j. biotechadv.2011.09.005.

Aranda, E., M. Kinne, M. Kluge, R. Ullrich and M. Hofrichter. 2008. Conversion of dibenzothiophene by the mushrooms *Agrocybe aegerita* and *Coprinellus radians* and their extracellular peroxygenases. *Applied Microbiology and Biotechnology* 82 (6): 1057. doi: 10.1007/s00253-008-1778-6.

Aranda, E., R. Ullrich and M. Hofrichter. 2010. Conversion of polycyclic aromatic hydrocarbons, methyl naphthalenes and dibenzofuran by two fungal peroxygenases. *Biodegradation* 21 (2): 267–281. doi: 10.1007/s10532-009-9299-2.

Arunkumar, M. and S. H. Sheik Abdulla. 2015. Hyper-production of manganese peroxidase by mutant *Pleurotus ostreatus* MTCC 142 and its applications in biodegradation of textile azo dyes. *Desalination and Water Treatment* 56 (2): 509–520. doi: 10.1080/19443994.2014.937766.

Ashraf, H. and Q. Husain. 2010. Use of DEAD cellulose adsorbed and crosslinked white radish (*Raphanus sativus*) peroxidase for the removal of α-naphthol in batch and continuous process. *International Biodeterioration & Biodegradation* 64 (1): 27–31. doi: https://doi.org/10.1016/j.ibiod.2009.10.003.

Ayala, M., R. Roman and R. Vazquez-Duhalt. 2007. A catalytic approach to estimate the redox potential of heme-peroxidases. *Biochemical and Biophysical Research Communications* 357 (3): 804–808. doi: https://doi.org/10.1016/j.bbrc.2007.04.020.

Ayala, M. and E. Torres. 2016. Peroxidases as potential industrial biocatalysts. In: *Heme Peroxidases*, edited by E. Raven and B. Dunford, pp. 309–333. Europe: The Royal Society of Chemistry.

Baborová, P., M. Möder, P. Baldrian, K. Cajthamlová and T. Cajthaml. 2006. Purification of a new manganese peroxidase of the white-rot fungus *Irpex lacteus*, and degradation of polycyclic aromatic hydrocarbons by the enzyme. *Research in Microbiology* 157 (3): 248–253. doi: https://doi. org/10.1016/j.resmic.2005.09.001.

Baratto, M. C., K. Juarez-Moreno, R. Pogni, R. Basosi and R. Vazquez-Duhalt. 2015. EPR and LC-MS studies on the mechanism of industrial dye decolorization by versatile peroxidase from *Bjerkandera adusta*. *Environmental Science and Pollution Research* 22 (11): 8683–8692. doi: 10.1007/s11356-014-4051-9.

Battistuzzi, G., M. Bellei, C. A. Bortolotti and M. Sola. 2010. Redox properties of heme peroxidases. *Archives of Biochemistry and Biophysics* 500 (1): 21–36. doi: https://doi.org/10.1016/j. abb.2010.03.002.

Bengtson, P., D. Bastviken, W. De Boer and G. Öberg. 2009. Possible role of reactive chlorine in microbial antagonism and organic matter chlorination in terrestrial environments. *Environmental Microbiology* 11 (6): 1330–1339. doi: doi:10.1111/j.1462-2920.2009.01915.x.

Bilal, M. and M. Asgher. 2015. Sandal reactive dyes decolorization and cytotoxicity reduction using manganese peroxidase immobilized onto polyvinyl alcohol-alginate beads. *Chemistry Central Journal* 9: 47. doi: 10.1186/s13065-015-0125-0.

Bilal, M., M. Asgher, M. Iqbal, H. Hu and X. Zhang. 2016a. Chitosan beads immobilized manganese peroxidase catalytic potential for detoxification and decolorization of textile effluent. *International Journal of Biological Macromolecules* 89: 181–189. doi: https://doi.org/10.1016/j. ijbiomac.2016.04.075.

Bilal, M., M. Asgher, H. M. N. Iqbal, H. Hu and X. Zhang. 2017. Bio-based degradation of emerging endocrine-disrupting and dye-based pollutants using cross-linked enzyme aggregates. *Environmental Science and Pollution Research* 24 (8): 7035–7041. doi: 10.1007/s11356-017-8369-y.

Bilal, M., M. Asgher, M. Shahid and H. N. Bhatti. 2016b. Characteristic features and dye degrading capability of agar-agar gel immobilized manganese peroxidase. *International Journal of Biological Macromolecules* 86: 728–740. doi: https://doi.org/10.1016/j.ijbiomac.2016.02.014.

Breen, A. and F. L. Singleton. 1999. Fungi in lignocellulose breakdown and biopulping. *Current Opinion in Biotechnology* 10 (3): 252–258. doi: https://doi.org/10.1016/S0958-1669(99)80044-5.

Camarero, S., S. Sarkar, F. J. Ruiz-Dueñas, M. J. Martínez and Á. T. Martínez. 1999. Description of a versatile peroxidase involved in the natural degradation of lignin that has both manganese peroxidase and lignin peroxidase substrate interaction sites. *Journal of Biological Chemistry* 274 (15): 10324–10330. doi: 10.1074/jbc.274.15.10324.

Cerniglia, C. E. and J. B. Sutherland. 2010. Degradation of polycyclic aromatic hydrocarbons by fungi. In: *Handbook of Hydrocarbon and Lipid Microbiology*, edited by Kenneth N. Timmis, pp. 2079–2110. Berlin, Heidelberg: Springer Berlin Heidelberg.

Coelho-Moreira, J. d. S., A. Bracht, A. C. d. S. d. Souza, R. F. Oliveira, A. B. d. Sá-Nakanishi, C. G. M. d. Souza and R. M. Peralta. 2013. Degradation of diuron by *Phanerochaete chrysosporium*: Role of ligninolytic enzymes and cytochrome P450. *BioMed Research International* 2013: 9. doi: 10.1155/2013/251354.

Cohen, S., P. A. Belinky, Y. Hadar and C. G. Dosoretz. 2009. Characterization of catechol derivative removal by lignin peroxidase in aqueous mixture. *Bioresource Technology* 100 (7): 2247–2253. doi: https://doi.org/10.1016/j.biortech.2008.11.007.

Cheng, X. B., R. Jia, P. S. Li, Q. Zhu, S. Q. Tu and W. Z. Tang. 2007. Studies on the properties and co-immobilization of manganese peroxidase. *Chinese Journal of Biotechnology* 23 (1): 90–96. doi: https://doi.org/10.1016/S1872-2075(07)60006-5.

Daughton, C. G. and T. A. Ternes. 1999. Pharmaceuticals and personal care products in the environment: agents of subtle change? *Environmental Health Perspectives* 107 (Suppl 6): 907–938.

Davila-Vazquez, G., R. Tinoco, M. A. Pickard and R. Vazquez-Duhalt. 2005. Transformation of halogenated pesticides by versatile peroxidase from *Bjerkandera adusta*. *Enzyme and Microbial Technology* 36 (2): 223–231. doi: https://doi.org/10.1016/j.enzmictec.2004.07.015.

Dong, X., H. Li, Y. Jiang, H. Mu, S. Li and Q. Zhai. 2016. Rapid and efficient degradation of bisphenol A by chloroperoxidase from *Caldariomyces fumago*: product analysis and ecotoxicity evaluation of the degraded solution. *Biotechnology Letters* 38 (9): 1483–1491. doi: 10.1007/s10529-016-2137-9.

Duong, H. A., N. H. Pham, H. T. Nguyen, T. T. Hoang, H. V. Pham, V. C. Pham, M. Berg, W. Giger and A. C. Alder. 2008. Occurrence, fate and antibiotic resistance of fluoroquinolone antibacterials in hospital wastewaters in Hanoi, Vietnam. *Chemosphere* 72 (6): 968–973. doi: https://doi. org/10.1016/j.chemosphere.2008.03.009.

Durán, N., M. A. Rosa, A. D'Annibale and L. Gianfreda. 2002. Applications of laccases and tyrosinases (phenoloxidases) immobilized on different supports: a review. *Enzyme and Microbial Technology* 31 (7): 907–931. doi: https://doi.org/10.1016/S0141-0229(02)00214-4.

Eibes, G., T. Cajthaml, M. T. Moreira, G. Feijoo and J. M. Lema. 2006. Enzymatic degradation of anthracene, dibenzothiophene and pyrene by manganese peroxidase in media containing acetone. *Chemosphere* 64 (3): 408–414. doi: https://doi.org/10.1016/j.chemosphere.2005.11.075.

Eibes, G., G. Debernardi, G. Feijoo, M. T. Moreira and J. M. Lema. 2011. Oxidation of pharmaceutically active compounds by a ligninolytic fungal peroxidase. *Biodegradation* 22 (3): 539–550. doi: 10.1007/s10532-010-9426-0.

Enayatzamir, K., H. A. Alikhani, B. Yakhchali, F. Tabandeh and S. Rodríguez-Couto. 2010. Decolouration of azo dyes by *Phanerochaete chrysosporium* immobilised into alginate beads. *Environmental Science and Pollution Research* 17 (1): 145–153. doi: 10.1007/s11356-009-0109-5.

Falade, A. O., U. U. Nwodo, B. C. Iweriebor, E. Green, L. V. Mabinya and A. I. Okoh. 2017. Lignin peroxidase functionalities and prospective applications. *Microbiology Open* 6 (1): e00394. doi: 10.1002/mbo3.394.

Faraco, V., A. Piscitelli, G. Sannia and P. Giardina. 2007. Identification of a new member of the dye-decolorizing peroxidase family from *Pleurotus ostreatus*. *World Journal of Microbiology and Biotechnology* 23 (6): 889–893. doi: 10.1007/s11274-006-9303-5.

Fatima, A. and Q. Husain. 2008. Purification and characterization of a novel peroxidase from bitter gourd (*Momordica charantia*). *Protein & Peptide Letters* 15 (4): 377–384. doi: http://dx.doi.org/10.2174/092986608784246452.

Ferreira-Leitão, V. S., M. E. A. de Carvalho and E. P. S. Bon. 2007. Lignin peroxidase efficiency for methylene blue decolouration: Comparison to reported methods. *Dyes and Pigments* 74 (1): 230–236. doi: https://doi.org/10.1016/j.dyepig.2006.02.002.

Ferreira, V. S., D. B. Magalhães, S. H. Kling, J. G. Da Silva and E. P. S. Bon. 2000. N-demethylation of methylene blue by lignin peroxidase from *Phanerochaete chrysosporium*. *Applied Biochemistry and Biotechnology* 84 (1): 255–265. doi: 10.1385/abab:84-86:1-9:255.

Field, J. A., R. H. Vledder, J. G. van Zelst and W. H. Rulkens. 1996. The tolerance of lignin peroxidase and manganese-dependent peroxidase to miscible solvents and the *in vitro* oxidation of anthracene in solvent: water mixtures. *Enzyme and Microbial Technology* 18 (4): 300–308. doi: https://doi.org/10.1016/0141-0229(95)00109-3.

Gold, M., H. Youngs and M. Gelpke. 2000. Manganese peroxidase. *Metal Ions in Biological Systems* 37: 559–586.

Gómez-Toribio, V., A. T. Martínez, M. J. Martínez and F. Guillén. 2001. Oxidation of hydroquinones by the versatile ligninolytic peroxidase from *Pleurotus eryngii*. *European Journal of Biochemistry* 268 (17): 4787–4793. doi: doi:10.1046/j.1432-1327.2001.02405.x.

Goncalves, I., C. Silva and A. Cavaco-Paulo. 2015. Ultrasound enhanced laccase applications. *Green Chemistry* 17 (3): 1362–1374. doi: 10.1039/c4gc02221a.

Grabski, A. C., H. J. Grimek and R. R. Burgess. 1998. Immobilization of manganese peroxidase from *Lentinula edodes* and its biocatalytic generation of MnIII-chelate as a chemical oxidant of chlorophenols. *Biotechnology and Bioengineering* 60 (2): 204–215. doi: doi:10.1002/(SICI)1097-0290(19981020)60:2<204::AID-BIT8>3.0.CO;2-R.

Gröbe, G., R. Ullrich, M. J. Pecyna, D. Kapturska, S. Friedrich, M. Hofrichter and K. Scheibner. 2011. High-yield production of aromatic peroxygenase by the agaric fungus *Marasmius rotula*. *AMB Express* 1 (1): 31. doi: 10.1186/2191-0855-1-31.

Günther, T., U. Sack, M. Hofrichter and M. Lätz. 1998. Oxidation of PAH and PAH-derivatives by fungal and plant oxidoreductases. *Journal of Basic Microbiology* 38 (2): 113–122. doi: doi:10.1002/(SICI)1521-4028(199805)38:2<113::AID-JOBM113>3.0.CO;2-D.

Gutiérrez, A., E. D. Babot, R. Ullrich, M. Hofrichter, A. T. Martínez and J. C. del Río. 2011. Regioselective oxygenation of fatty acids, fatty alcohols and other aliphatic compounds by a basidiomycete heme-thiolate peroxidase. *Archives of Biochemistry and Biophysics* 514 (1): 33–43. doi: https://doi.org/10.1016/j.abb.2011.08.001.

Haemmerli, S. D., M. S. Leisola, D. Sanglard and A. Fiechter. 1986. Oxidation of benzo(a)pyrene by extracellular ligninases of *Phanerochaete chrysosporium*. Veratryl alcohol and stability of ligninase. *Journal of Biological Chemistry* 261 (15): 6900–6093.

Halling-Sørensen, B., S. N. Nielsen, P. F. Lanzky, F. Ingerslev, H. C. Holten Lützhøft and S. E. Jørgensen. 1998. Occurrence, fate and effects of pharmaceutical substances in the environment—A review. *Chemosphere* 36 (2): 357–393. doi: https://doi.org/10.1016/S0045-6535(97)00354-8.

Hammel, K. E., B. Kalyanaraman and T. K. Kirk. 1986. Oxidation of polycyclic aromatic hydrocarbons and dibenzo[p]-dioxins by *Phanerochaete chrysosporium* ligninase. *Journal of Biological Chemistry* 261 (36): 16948–16952.

Hao, O. J., H. Kim and P. C. Chiang. 2000. Decolorization of wastewater. *Critical Reviews in Environmental Science and Technology* 30 (4): 449–505. doi: 10.1080/10643380091184237.

Harazono, K., Y. Watanabe and K. Nakamura. 2003. Decolorization of azo dye by the white-rot basidiomycete *Phanerochaete sordida* and by its manganese peroxidase. *Journal of Bioscience and Bioengineering* 95 (5): 455–459. doi: https://doi.org/10.1016/S1389-1723(03)80044-0.

Haritash, A. K. and C. P. Kaushik. 2009. Biodegradation aspects of polycyclic aromatic hydrocarbons (PAHs): A review. *Journal of Hazardous Materials* 169 (1): 1–15. doi: https://doi.org/10.1016/j.jhazmat.2009.03.137.

Harms, H., D. Schlosser and L. Y. Wick. 2011. Untapped potential: exploiting fungi in bioremediation of hazardous chemicals. *Nature Reviews Microbiology* 9: 177–192. doi: 10.1038/nrmicro2519.

Hatakka, A. and K. E. Hammel. 2011. Fungal biodegradation of lignocelluloses. In: *Industrial Applications*, edited by Martin Hofrichter, 319–340. Berlin, Heidelberg: Springer Berlin Heidelberg.

Hildén, K., A. T. Martinez, A. Hatakka and T. Lundell. 2005. The two manganese peroxidases Pr-MnP2 and Pr-MnP3 of *Phlebia radiata*, a lignin-degrading basidiomycete, are phylogenetically and structurally divergent. *Fungal Genetics and Biology* 42 (5): 403–419. doi: https://doi.org/10.1016/j.fgb.2005.01.008.

Hirai, H., S. Nakanishi and T. Nishida. 2004. Oxidative dechlorination of methoxychlor by ligninolytic enzymes from white-rot fungi. *Chemosphere* 55 (4): 641–645. doi: https://doi.org/10.1016/j.chemosphere.2003.11.035.

Hirano, T., Y. Honda, T. Watanabe and M. Kuwahara. 2000. Degradation of bisphenol a by the lignin-degrading enzyme, manganese peroxidase, produced by the white-rot basidiomycete, *Pleurotus ostreatus*. *Bioscience, Biotechnology, and Biochemistry* 64 (9): 1958–1962. doi: 10.1271/bbb.64.1958.

Hofrichter, M. 2002. Review: lignin conversion by manganese peroxidase (MnP). *Enzyme and Microbial Technology* 30 (4): 454–466. doi: https://doi.org/10.1016/S0141-0229(01)00528-2.

Hofrichter, M., T. Lundell and A. Hatakka. 2001. Conversion of milled pine wood by manganese peroxidase from *Phlebia radiata*. *Applied and Environmental Microbiology* 67 (10): 4588–4593. doi: 10.1128/aem.67.10.4588-4593.2001.

Hofrichter, M. and R. Ullrich. 2006. Heme-thiolate haloperoxidases: versatile biocatalysts with biotechnological and environmental significance. *Applied Microbiology and Biotechnology* 71 (3): 276. doi: 10.1007/s00253-006-0417-3.

Hofrichter, M. and R. Ullrich. 2011. New Trends in Fungal Biooxidation. In: *Industrial Applications*, edited by Martin Hofrichter, 425–449. Berlin, Heidelberg: Springer Berlin Heidelberg.

Hofrichter, M., R. Ullrich, M. J. Pecyna, C. Liers and T. Lundell. 2010. New and classic families of secreted fungal heme peroxidases. *Applied Microbiology and Biotechnology* 87 (3): 871–897. doi: 10.1007/s00253-010-2633-0.

Houborg, K., P. Harris, J. C. N. Poulsen, P. Schneider, A. Svendsen and S. Larsen. 2003. The structure of a mutant enzyme of *Coprinus cinereus* peroxidase provides an understanding of its increased thermostability. *Acta Crystallographica Section D* 59 (6): 997–1003. doi: doi:10.1107/S0907444903006784.

Hu, Z., L. Xu and X. Wen. 2013. Mesoporous silicas synthesis and application for lignin peroxidase immobilization by covalent binding method. *Journal of Environmental Sciences* 25 (1): 181–187. doi: https://doi.org/10.1016/S1001-0742(12)60008-4.

Husain, M. and Q. Husain. 2008. Applications of redox mediators in the treatment of organic pollutants by using oxidoreductive enzymes: A review. *Critical Reviews in Environmental Science and Technology* 38 (1): 1–42. doi: 10.1080/10643380701501213.

Husain, Q. 2006. Potential applications of the oxidoreductive Enzymes in the decolorization and detoxification of textile and other synthetic dyes from polluted water: A review. *Critical Reviews in Biotechnology* 26 (4): 201–221. doi: 10.1080/07388550600969936.

Husain, Q. 2010a. Peroxidase mediated decolorization and remediation of wastewater containing industrial dyes: a review. *Reviews in Environmental Science and Bio/Technology* 9 (2): 117–140. doi: 10.1007/s11157-009-9184-9.

Husain, Q. 2010b. β Galactosidases and their potential applications: a review. *Critical Reviews in Biotechnology* 30 (1): 41–62. doi: 10.3109/07388550903330497.

Husain, Q. 2017a. Biosensor applications of graphene-nanocomposites bound oxido reductive and hydrolytic enzymes. *Analytical Methods* 9 (48): 6734–6746. doi: 10.1039/c7ay02606d.

Husain, Q. 2017b. High yield immobilization and stabilization of oxidoreductases using magnetic nanosupports and their potential applications: An update. *Current Catalysis* 6 (3): 168–187. doi: http://dx.doi.org/10.2174/2211544706666170704141828.

Husain, Q. 2017c. Nanomaterials as novel supports for the immobilization of amylolytic enzymes and their applications: A review. *Biocatalysis* 3: 37–53. doi: 10.1515/boca-2017-0004.

Husain, Q. 2017d. Nanomaterials immobilized cellulolytic enzymes and their industrial applications: a literature review. *SM Biochemistry & Molecular Biology* 4 (3): 1029.

Husain, Q. 2018. Nanocarriers immobilized proteases and their industrial applications: An overview. *Journal of Nanoscience and Nanotechnology* 18 (1): 486–499.

Husain, Q. 2019. Remediation of phenolic compounds from polluted water by immobilized peroxidases. In: *Emerging and Eco-Friendly Approaches for Waste Management*, edited by Ram Naresh Bharagava and Pankaj Chowdhary, 329–358. Singapore: Springer Singapore.

Husain, Q. and M. Husain. 2012. Peroxidases as a potential tool for the decolorization and removal of synthetic dyes from polluted water. In: *Environmental Protection Strategies for Sustainable*

Development, edited by Abdul Malik and Elisabeth Grohmann, 453–498. Dordrecht: Springer Netherlands.

Husain, Q., M. Husain and Y. Kulshrestha. 2009. Remediation and treatment of organopollutants mediated by peroxidases: a review. *Critical Reviews in Biotechnology* 29 (2): 94–119. doi: 10.1080/07388550802685306.

Husain, Q. and U. Jan. 2000. Detoxification of phenols and aromatic amines from polluted wastewater by using phenol oxidases. *Journal of Scientific & Industrial Research* 59: 286–293.

Husain, Q. and S. Qayyum. 2013. Biological and enzymatic treatment of bisphenol A and other endocrine disrupting compounds: a review. *Critical Reviews in Biotechnology* 33 (3): 260–292. doi: 10.3109/07388551.2012.694409.

Husain, Q. and R. Ulber. 2011. Immobilized peroxidase as a valuable tool in the remediation of aromatic pollutants and xenobiotic compounds: A review. *Critical Reviews in Environmental Science and Technology* 41 (8): 770–804. doi: 10.1080/10643380903299491.

Ikehata, K. and I. D. Buchanan. 2002. Screening of coprinus species for the production of extracellular peroxidase and evaluation of the enzyme for the treatment of aqueous phenol. *Environmental Technology* 23 (12): 1355–1367. doi: 10.1080/09593332508618441.

Ikehata, K., I. D. Buchanan and D. W. Smith. 2004. Recent developments in the production of extracellular fungal peroxidases and laccases for waste treatment. *Journal of Environmental Engineering and Science* 3 (1): 1–19. doi: 10.1139/s03-077.

Inoue, Y., T. Hata, S. Kawai, H. Okamura and T. Nishida. 2010. Elimination and detoxification of triclosan by manganese peroxidase from white rot fungus. *Journal of Hazardous Materials* 180 (1): 764–767. doi: https://doi.org/10.1016/j.jhazmat.2010.04.024.

Iqbal, M. N. H. and M. Asgher. 2013. Decolorization applicability of sol–gel matrix immobilized manganese peroxidase produced from an indigenous white rot fungal strain *Ganoderma lucidum*. *BMC Biotechnology* 13 (1): 56. doi: 10.1186/1472-6750-13-56.

Kapich, A. N., K. T. Steffen, M. Hofrichter and A. Hatakka. 2005. Involvement of lipid peroxidation in the degradation of a non-phenolic lignin model compound by manganese peroxidase of the litter-decomposing fungus *Stropharia coronilla*. *Biochemical and Biophysical Research Communications* 330 (2): 371–377. doi: https://doi.org/10.1016/j.bbrc.2005.02.167.

Karim, Z. and Q. Husain. 2010. Removal of anthracene from model wastewater by immobilized peroxidase from *Momordica charantia* in batch process as well as in a continuous spiral-bed reactor. *Journal of Molecular Catalysis B: Enzymatic* 66 (3): 302–310. doi: https://doi.org/10.1016/j.molcatb.2010.06.007.

Karim, Z., Q. Husain, R. Adnan and N. Akhtar. 2011. Remediation of model wastewater polluted with methyl parathion by reverse micelle entrapped peroxidase. *Water Quality Research Journal of Canada* 46 (4): 345–354. doi: 10.2166/wqrjc.2012.020.

Kim, H. J., Y. Suma, S. H. Lee, J. A. Kim and H. S. Kim. 2012. Immobilization of horseradish peroxidase onto clay minerals using soil organic matter for phenol removal. *Journal of Molecular Catalysis B: Enzymatic* 83: 8–15. doi: https://doi.org/10.1016/j.molcatb.2012.06.012.

Kim, S. J., K. Ishikawa, M. Hirai and M. Shoda. 1995. Characteristics of a newly isolated fungus, *Geotrichum candidum* Dec 1, which decolorizes various dyes. *Journal of Fermentation and Bioengineering* 79 (6): 601–607. doi: https://doi.org/10.1016/0922-338X(95)94755-G.

Kinne, M., M. Poraj-Kobielska, E. Aranda, R. Ullrich, K. E. Hammel, S. Scheibner and M. Hofrichter. 2009a. Regioselective preparation of 5-hydroxypropranolol and 4′-hydroxydiclofenac with a fungal peroxygenase. *Bioorganic & Medicinal Chemistry Letters* 19 (11): 3085–3087. doi: https://doi.org/10.1016/j.bmcl.2009.04.015.

Kinne, M., M. Poraj-Kobielska, S. A. Ralph, R. Ullrich, M. Hofrichter and K. E. Hammel. 2009b. Oxidative cleavage of diverse ethers by an extracellular fungal peroxygenase. *The Journal of Biological Chemistry* 284 (43): 29343–29349. doi: 10.1074/jbc.M109.040857.

Kluge, M., R. Ullrich, C. Dolge, K. Scheibner and M. Hofrichter. 2009. Hydroxylation of naphthalene by aromatic peroxygenase from *Agrocybe aegerita* proceeds via oxygen transfer from H_2O_2 and intermediary epoxidation. *Applied Microbiology and Biotechnology* 81 (6): 1071–1076. doi: 10.1007/s00253-008-1704-y.

Knop, D., O. Yarden and Y. Hadar. 2015. The ligninolytic peroxidases in the genus *Pleurotus*: divergence in activities, expression, and potential applications. *Applied Microbiology and Biotechnology* 99 (3): 1025–1038. doi: 10.1007/s00253-014-6256-8.

Kokol, V., A. Doliška, I. Eichlerová, P. Baldrian and F. Nerud. 2007. Decolorization of textile dyes by whole cultures of *Ischnoderma resinosum* and by purified laccase and Mn-peroxidase. *Enzyme and Microbial Technology* 40 (7): 1673–1677. doi: https://doi.org/10.1016/j.enzmictec.2006.08.015.

Kuan, I. C., K. A. Johnson and M. Tien. 1993. Kinetic analysis of manganese peroxidase. The reaction with manganese complexes. *Journal of Biological Chemistry* 268 (27): 20064–20070.

Kühnel, K., W. Blankenfeldt, J. Terner and I. Schlichting. 2006. Crystal structures of chloroperoxidase with its bound substrates and complexed with formate, acetate, and nitrate. *Journal of Biological Chemistry* 281 (33): 23990–23998. doi: 10.1074/jbc.M603166200.

Kunishima, N., K. Fukuyama, H. Matsubara, H. Hatanaka, Y. Shibano and T. Amachi. 1994. Crystal structure of the fungal peroxidase from *Arthromyces ramosus* at 1·9 Å resolution: Structural comparisons with the lignin and cytochrome c peroxidases. *Journal of Molecular Biology* 235 (1): 331–344. doi: https://doi.org/10.1016/S0022-2836(05)80037-3.

Kurnik, K., K. Treder, M. Skorupa-Kłaput, A. Tretyn and J. Tyburski. 2015. Removal of phenol from synthetic and industrial wastewater by potato pulp peroxidases. *Water, Air, & Soil Pollution* 226 (8): 254. doi: 10.1007/s11270-015-2517-0.

Langhals, H. 2004. Color chemistry. Synthesis, properties and applications of organic dyes and pigments. In: *Angewandte Chemie International Edition*, ed Heinrich Zollinger. https://onlinelibrary.wiley.com/doi/abs/10.1002/anie.200385122.

Lauber, C., T. Schwarz, Q. K. Nguyen, P. Lorenz, G. Lochnit and H. Zorn. 2017. Identification, heterologous expression and characterization of a dye-decolorizing peroxidase of *Pleurotus sapidus*. *AMB Express* 7 (1): 164. doi: 10.1186/s13568-017-0463-5.

Li, X., Q. He, H. Li, X. Gao, M. Hu, S. Li, Q. Zhai, Y. Jiang and X. Wang. 2017. Bioconversion of non-steroidal anti-inflammatory drugs diclofenac and naproxen by chloroperoxidase. *Biochemical Engineering Journal* 120: 7–16. doi: https://doi.org/10.1016/j.bej.2016.12.018.

Li, X., Y. Wang, S. Wu, L. Qiu, L. Gu, J. Li, B. Zhang and W. Zhong. 2014. Peculiarities of metabolism of anthracene and pyrene by laccase-producing fungus *Pycnoporus sanguineus* H1. *Biotechnology and Applied Biochemistry* 61 (5): 549–554. doi: doi:10.1002/bab.1197.

Liers, C., E. Aranda, E. Strittmatter, K. Piontek, D. A. Plattner, H. Zorn, R. Ullrich and M. Hofrichter. 2014. Phenol oxidation by DyP-type peroxidases in comparison to fungal and plant peroxidases. *Journal of Molecular Catalysis B: Enzymatic* 103: 41–46. doi: https://doi.org/10.1016/j.molcatb.2013.09.025.

Liers, C., C. Bobeth, M. Pecyna, R. Ullrich and M. Hofrichter. 2010. DyP-like peroxidases of the jelly fungus *Auricularia auricula*-judae oxidize non phenolic lignin model compounds and high-redox potential dyes. *Applied Microbiology and Biotechnology* 85 (6): 1869–1879. doi: 10.1007/s00253-009-2173-7.

Liers, C., M. J. Pecyna, H. Kellner, A. Worrich, H. Zorn, K. T. Steffen, M. Hofrichter and R. Ullrich. 2013. Substrate oxidation by dye-decolorizing peroxidases (DyPs) from wood- and litter-degrading agaricomycetes compared to other fungal and plant heme-peroxidases. *Applied Microbiology and Biotechnology* 97 (13): 5839–5849. doi: 10.1007/s00253-012-4521-2.

Liers, C., R. Ullrich, K. T. Steffen, A. Hatakka and M. Hofrichter. 2006. Mineralization of 14C-labelled synthetic lignin and extracellular enzyme activities of the wood-colonizing ascomycetes *Xylaria hypoxylon* and *Xylaria polymorpha*. *Applied Microbiology and Biotechnology* 69 (5): 573–579. doi: 10.1007/s00253-005-0010-1.

Linde, D., C. Coscolín, C. Liers, M. Hofrichter, A. T. Martínez and F. J. Ruiz-Dueñas. 2014. Heterologous expression and physicochemical characterization of a fungal dye-decolorizing peroxidase from *Auricularia auricula*-judae. *Protein Expression and Purification* 103: 28–37. doi: https://doi.org/10.1016/j.pep.2014.08.007.

Liu, D., J. Gong, W. Dai, X. Kang, Z. Huang, H. M. Zhang, W. Liu, L. Liu, J. Ma, Z. Xia, Y. Chen, Y. Chen, D. Wang, P. Ni, A. Y. Guo and X. Xiong. 2012. The genome of *Ganderma lucidum* provide insights into triterpense biosynthesis and wood degradation. *PLOS ONE* 7 (5): e36146. doi: 10.1371/journal.pone.0036146.

Liu, J. Z. and M. Wang. 2007. Improvement of activity and stability of chloroperoxidase by chemical modification. *BMC Biotechnology* 7 (1): 23. doi: 10.1186/1472-6750-7-23.

Liu, L., J. Zhang, Y. Tan, Y. Jiang, M. Hu, S. Li and Q. Zhai. 2014. Rapid decolorization of anthraquinone and triphenylmethane dye using chloroperoxidase: Catalytic mechanism, analysis of products and degradation route. *Chemical Engineering Journal* 244: 9–18. doi: https://doi.org/10.1016/j.cej.2014.01.063.

Longoria, A., R. Tinoco and R. Vázquez-Duhalt. 2008. Chloroperoxidase-mediated transformation of highly halogenated monoaromatic compounds. *Chemosphere* 72 (3): 485–490. doi: https://doi.org/10.1016/j.chemosphere.2008.03.006.

Lundell, T. K., M. R. Mäkelä and K. Hildén. 2010. Lignin-modifying enzymes in filamentous basidiomycetes—ecological, functional and phylogenetic review. *Journal of Basic Microbiology* 50 (1): 5–20. doi: doi:10.1002/jobm.200900338.

Makris, T. M., I. Denisov, I. Schlichting and S. G. Sligar. 2005. Activation of molecular oxygen by cytochrome P450. In: *Cytochrome P450: Structure, Mechanism, and Biochemistry*, edited by Paul R. Ortiz de Montellano, 149–182. Boston, MA: Springer US.

Manavalan, T., A. Manavalan and K. Heese. 2015. Characterization of lignocellulolytic enzymes from white-rot fungi. *Current Microbiology* 70 (4): 485–498. doi: 10.1007/s00284-014-0743-0.

Manoj, K. M. and L. P. Hager. 2008. Chloroperoxidase, a janus enzyme. *Biochemistry* 47 (9): 2997–3003. doi: 10.1021/bi7022656.

Mao, L., J. Lu, S. Gao and Q. Huang. 2010a. Transformation of 17ß-estradiol mediated by lignin peroxidase: The role of veratryl alcohol. *Archives of Environmental Contamination and Toxicology* 59 (1): 13–19. doi: 10.1007/s00244-009-9448-y.

Mao, L., J. Lu, M. Habteselassie, Q. Luo, S. Gao, M. Cabrera and Q. Huang. 2010b. Ligninase-mediated removal of natural and synthetic estrogens from water: II. reactions of 17β-estradiol. *Environmental Science & Technology* 44 (7): 2599–2604. doi: 10.1021/es903058k.

Marco-Urrea, E., J. Radjenović, G. Caminal, M. Petrović, T. Vicent and D. Barceló. 2010. Oxidation of atenolol, propranolol, carbamazepine and clofibric acid by a biological fenton-like system mediated by the white-rot fungus *Trametes versicolor*. *Water Research* 44 (2): 521–532. doi: https://doi.org/10.1016/j.watres.2009.09.049.

Martínez, Á. T., F. J. Ruiz-Dueñas, M. J. Martínez, J. C. del Río and A. Gutiérrez. 2009. Enzymatic delignification of plant cell wall: from nature to mill. *Current Opinion in Biotechnology* 20 (3): 348–357. doi: https://doi.org/10.1016/j.copbio.2009.05.002.

Michniewicz, A., S. Ledakowicz, R. Ullrich and M. Hofrichter. 2008. Kinetics of the enzymatic decolorization of textile dyes by laccase from *Cerrena unicolor*. *Dyes and Pigments* 77 (2): 295–302. doi: https://doi.org/10.1016/j.dyepig.2007.05.015.

Miki, Y., H. Ichinose and H. Wariishi. 2010. Molecular characterization of lignin peroxidase from the white-rot basidiomycete *Trametes cervina*: a novel fungal peroxidase. *FEMS Microbiology Letters* 304 (1): 39–46. doi: 10.1111/j.1574-6968.2009.01880.x.

Moldovan, Z. 2006. Occurrences of pharmaceutical and personal care products as micropollutants in rivers from romania. *Chemosphere* 64 (11): 1808–1817. doi: https://doi.org/10.1016/j.chemosphere.2006.02.003.

Moreira, M. T., C. Palma, I. Mielgo, G. Feijoo and J. M. Lema. 2001. In vitro degradation of a polymeric dye (Poly R-478) by manganese peroxidase. *Biotechnology and Bioengineering* 75 (3): 362–368. doi: doi:10.1002/bit.10052.

Moreira, P. R., C. Duez, D. Dehareng, A. Antunes, E. Almeida-Vara, J. M. Frère, F. X. Malcata and J. C. Duarte. 2005. Molecular characterisation of a versatile peroxidase from a *Bjerkandera* strain. *Journal of Biotechnology* 118 (4): 339–352. doi: https://doi.org/10.1016/j.jbiotec.2005.05.014.

Morgenstern, I., S. Klopman and D. S. Hibbett. 2008. Molecular evolution and diversity of lignin degrading heme peroxidases in the agaricomycetes. *Journal of Molecular Evolution* 66 (3): 243–257. doi: 10.1007/s00239-008-9079-3.

Morita, Y., H. Yamashita, B. Mikami, H. Iwamoto, S. Aibara, M. Terada and J. Minami. 1988. Purification, crystallization, and characterization of peroxidase from *Coprinus cinereus*. *The Journal of Biochemistry* 103 (4): 693–699.

Murphy, C. D. 2007. Fluorophenol oxidation by a fungal chloroperoxidase. *Biotechnology Letters* 29 (1): 45–49. doi: 10.1007/s10529-006-9207-3.

Nakamiya, K., S. Hashimoto, H. Ito, J. S. Edmonds, A. Yasuhara and M. Morita. 2005. Degradation of dioxins by cyclic ether degrading fungus, *Cordyceps sinensis*. *FEMS Microbiology Letters* 248 (1): 17–22. doi: 10.1016/j.femsle.2005.05.013.

Ollikka, P., K. Alhonmäki, V. M. Leppänen, T. Glumoff, T. Raijola and I. Suominen. 1993. Decolorization of azo, triphenyl methane, heterocyclic, and polymeric dyes by lignin peroxidase iso enzymes from *Phanerochaete chrysosporium*. *Applied and Environmental Microbiology* 59 (12): 4010–4016.

Omura, T. 2005. Heme–thiolate proteins. *Biochemical and Biophysical Research Communications* 338 (1): 404–409. doi: https://doi.org/10.1016/j.bbrc.2005.08.267.

Osborne, R. L., M. K. Coggins, J. Terner and J. H. Dawson. 2007. *Caldariomyces fumago* chloroperoxidase catalyzes the oxidative dehalogenation of chlorophenols by a mechanism involving two one-electron steps. *Journal of the American Chemical Society* 129 (48): 14838–14839. doi: 10.1021/ja0746969.

Papinutti, V. L., L. A. Diorio and F. Forchiassin. 2003. Production of laccase and manganese peroxidase by *Fomes sclerodermeus* grown on wheat bran. *Journal of Industrial Microbiology and Biotechnology* 30 (3): 157–160. doi: 10.1007/s10295-003-0025-5.

Park, J. B. and D. S. Clark. 2006. Deactivation mechanisms of chloroperoxidase during biotransformations. *Biotechnology and Bioengineering* 93 (6): 1190–1195. doi:10.1002/bit.20825.

Passardi, F., G. Theiler, M. Zamocky, C. Cosio, N. Rouhier, F. Teixera, M. Margis-Pinheiro, V. Ioannidis, C. Penel, L. Falquet and C. Dunand. 2007. PeroxiBase: The peroxidase database. *Phytochemistry* 68 (12): 1605–1611. doi: https://doi.org/10.1016/j.phytochem.2007.04.005.

Pecyna, M. J., R. Ullrich, B. Bittner, A. Clemens, K. Scheibner, R. Schubert and M. Hofrichter. 2009. Molecular characterization of aromatic peroxygenase from *Agrocybe aegerita*. *Applied Microbiology and Biotechnology* 84 (5): 885–897. doi: 10.1007/s00253-009-2000-1.

Pérez-Boada, M., F. J. Ruiz-Dueñas, R. Pogni, R. Basosi, T. Choinowski, M. J. Martínez, K. Piontek and A. T. Martínez. 2005. Versatile peroxidase oxidation of high redox potential aromatic compounds: site-directed mutagenesis, Spectroscopic and crystallographic investigation of three long-range electron transfer pathways. *Journal of Molecular Biology* 354 (2): 385–402. doi: https://doi.org/10.1016/j.jmb.2005.09.047.

Petersen, J. F. W., A. Kadziola and S. Larsen. 1994. Three-dimensional structure of a recombinant peroxidase from *Coprinus cinereus* at 2.6 Å resolution. *FEBS Letters* 339 (3): 291–296. doi: https://doi.org/10.1016/0014-5793(94)80433-8.

Pinto, P. A., A. A. Dias, I. Fraga, G. Marques, M. A. M. Rodrigues, J. Colaço, A. Sampaio and R. M. F. Bezerra. 2012. Influence of ligninolytic enzymes on straw saccharification during fungal pretreatment. *Bioresource Technology* 111: 261–267. doi: https://doi.org/10.1016/j.biortech.2012.02.068.

Pizzul, L., M. d. P. Castillo and J. Stenström. 2009. Degradation of glyphosate and other pesticides by ligninolytic enzymes. *Biodegradation* 20 (6): 751. doi: 10.1007/s10532-009-9263-1.

Pühse, M., R. T. Szweda, Y. Ma, C. Jeworrek, R. Winter and H. Zorn. 2009. *Marasmius scorodonius* extracellular dimeric peroxidase—Exploring its temperature and pressure stability. *Biochimica et Biophysica Acta (BBA)—Proteins and Proteomics* 1794 (7): 1091–1098. doi: https://doi.org/10.1016/j.bbapap.2009.03.015.

Qin, X., J. Zhang, X. Zhang and Y. Yang. 2014. Induction, purification and characterization of a novel manganese peroxidase from *Irpex lacteus* CD2 and its application in the decolorization of different types of dye. *PLOS ONE* 9 (11): e113282. doi: 10.1371/journal.pone.0113282.

Qiu, H., Y. Li, G. Ji, G. Zhou, X. Huang, Y. Qu and P. Gao. 2009. Immobilization of lignin peroxidase on nanoporous gold: Enzymatic properties and *in situ* release of H2O2 by co-immobilized glucose oxidase. *Bioresource Technology* 100 (17): 3837–3842. doi: https://doi.org/10.1016/j.biortech.2009.03.016.

Rabinovich, M. L., A. V. Bolobova and L. G. Vasil'chenko. 2004. Fungal decomposition of natural aromatic structures and xenobiotics: A review. *Applied Biochemistry and Microbiology* 40 (1): 1–17. doi: 10.1023/b:abim.0000010343.73266.08.

Raghukumar, C. 2000. Fungi from marine habitats: an application in bioremediation. *Mycological Research* 104 (10): 1222–1226. doi: https://doi.org/10.1017/S095375620000294X.

Rao, M. A., R. Scelza, F. Acevedo, M. C. Diez and L. Gianfreda. 2014. Enzymes as useful tools for environmental purposes. *Chemosphere* 107: 145–162. doi: https://doi.org/10.1016/j.chemosphere.2013.12.059.

Reddy, C. A. and Z. Mathew. 2001. Bioremediation potential of white rot fungi. In: *Fungi in Bioremediation*, edited by G. M. Gadd, 52–78. Cambridge: Cambridge University Press.

Robinson, T., G. McMullan, R. Marchant and P. Nigam. 2001. Remediation of dyes in textile effluent: a critical review on current treatment technologies with a proposed alternative. *Bioresource Technology* 77 (3): 247–255. doi: https://doi.org/10.1016/S0960-8524(00)00080-8.

Ruiz-Dueñas, F. J., S. Camarero, M. Pérez-Boada, M. J. Martínez and A. T. Martínez. 2001. A new versatile peroxidase from *Pleurotus*. *Biochemical Society Transactions* 29 (2): 116–122. doi: 10.1042/bst0290116.

Ruiz-Dueñas, F. J., E. Fernández, M. J. Martínez and A. T. Martínez. 2011. *Pleurotus ostreatus* heme peroxidases: An *in silico* analysis from the genome sequence to the enzyme molecular structure. *Comptes Rendus Biologies* 334 (11): 795–805. doi: https://doi.org/10.1016/j.crvi.2011.06.004.

Ruiz-Dueñas, F. J. and Á. T. Martínez. 2009. Microbial degradation of lignin: how a bulky recalcitrant polymer is efficiently recycled in nature and how we can take advantage of this. *Microbial Biotechnology* 2 (2): 164–177. doi: doi:10.1111/j.1751-7915.2008.00078.x.

Ruiz-Dueñas, F. J., M. Morales, E. García, Y. Miki, M. J. Martínez and A. T. Martínez. 2009. Substrate oxidation sites in versatile peroxidase and other basidiomycete peroxidases. *Journal of Experimental Botany* 60 (2): 441–452. doi: 10.1093/jxb/ern261.

Sack, U., M. Hofrichter and W. Fritsche. 1997. Degradation of polycyclic aromatic hydrocarbons by manganese peroxidase of *Nematoloma frowardii*. *FEMS Microbiology Letters* 152 (2): 227–234. doi: https://doi.org/10.1016/S0378-1097(97)00202-4.

Salvachúa, D., A. Prieto, Á. T. Martínez and M. J. Martínez. 2013. Characterization of a novel dye-decolorizing peroxidase (DyP)-type enzyme from *Irpex lacteus* and its application in enzymatic hydrolysis of wheat straw. *Applied and Environmental Microbiology* 79 (14): 4316–4324. doi: 10.1128/aem.00699-13.

Sánchez, C. 2009. Lignocellulosic residues: Biodegradation and bioconversion by fungi. *Biotechnology Advances* 27 (2): 185–194. doi: https://doi.org/10.1016/j.biotechadv.2008.11.001.

Sarmah, A. K., M. T. Meyer and A. B. A. Boxall. 2006. A global perspective on the use, sales, exposure pathways, occurrence, fate and effects of veterinary antibiotics (VAs) in the environment. *Chemosphere* 65 (5): 725–759. doi: https://doi.org/10.1016/j.chemosphere.2006.03.026.

Satar, R. and Q. Husain. 2009. Use of bitter gourd (*Momordica charantia*) peroxidase together with redox mediators to decolorize disperse dyes. *Biotechnology and Bioprocess Engineering* 14 (2): 213–219. doi: 10.1007/s12257-008-0175-4.

Satar, R., A. B. A. Jerah and Q. Husain. 2012. Role of redox mediators in enhancing dye decolorization by using oxidoreductive enzymes. In: *Non-Conventional Textile Waste Water Treatment*, edited by Ahmed El Nemr, 123–160. NY, USA: Nova Science Publishers.

Sawai-Hatanaka, H., T. Ashikari, Y. Tanaka, Y. Asada, T. Nakayama, H. Minakata, N. Kunishima, K. Fukuyama, H. Yamada, Y. Shibano and T. Amachi. 1995. Cloning, sequencing, and heterologous expression of a gene coding for *Arthromyces ramosus* peroxidase. *Bioscience, Biotechnology, and Biochemistry* 59 (7): 1221–1228. doi: 10.1271/bbb.59.1221.

Shakeri, M. and M. Shoda. 2007. Change in turnover capacity of crude recombinant dye-decolorizing peroxidase (rDyP) in batch and fed-batch decolorization of remazol brilliant blue R. *Applied Microbiology and Biotechnology* 76 (4): 919–926. doi: 10.1007/s00253-007-1042-5.

Shakeri, M. and M. Shoda. 2008. Decolorization of an anthraquinone dye by the recombinant dye-decolorizing peroxidase (rDyP) immobilized on mesoporous materials. *Journal of Molecular Catalysis B: Enzymatic* 54 (1): 42–49. doi: https://doi.org/10.1016/j.molcatb.2007.12.009.

Shakeri, M. and M. Shoda. 2010. Efficient decolorization of an anthraquinone dye by recombinant dye-decolorizing peroxidase (rDyP) immobilized in silica-based mesocellular foam. *Journal of Molecular Catalysis B: Enzymatic* 62 (3): 277–281. doi: https://doi.org/10.1016/j.molcatb.2009.11.007.

Silva, M. C., J. A. Torres, L. R. Vasconcelos de Sá, P. M. B. Chagas, V. S. Ferreira-Leitão and A. D. Corrêa. 2013. The use of soybean peroxidase in the decolourization of remazol brilliant blue R and toxicological evaluation of its degradation products. *Journal of Molecular Catalysis B: Enzymatic* 89: 122–129. doi: https://doi.org/10.1016/j.molcatb.2013.01.004.

Smith, A. T., W. A. Doyle, P. Dorlet and A. Ivancich. 2009. Spectroscopic evidence for an engineered, catalytically active trp radical that creates the unique reactivity of lignin peroxidase. *Proceedings of the National Academy of Sciences* 106 (38): 16084–16089. doi: 10.1073/pnas.0904535106.

Steevensz, A., S. Madur, W. Feng, K. E. Taylor, J. K. Bewtra and N. Biswas. 2014. Crude soybean hull peroxidase treatment of phenol in synthetic and real wastewater: Enzyme economy enhanced by Triton X-100. *Enzyme and Microbial Technology* 55: 65–71. doi: https://doi.org/10.1016/j.enzmictec.2013.12.005.

Steffen, K. T., M. Hofrichter and A. Hatakka. 2002. Purification and characterization of manganese peroxidases from the litter-decomposing basidiomycetes *Agrocybe praecox* and *Stropharia coronilla*. *Enzyme and Microbial Technology* 30 (4): 550–555. doi: https://doi.org/10.1016/S0141-0229(01)00525-7.

Strittmatter, E., C. Liers, R. Ullrich, S. Wachter, M. Hofrichter, D. A. Plattner and K. Piontek. 2013. First crystal structure of a fungal high-redox potential dye-decolorizing peroxidase: substrate interaction sites and long-range electron transfer. *Journal of Biological Chemistry* 288 (6): 4095–4102. doi: 10.1074/jbc.M112.400176.

Sugano, Y. 2009. DyP-type peroxidases comprise a novel heme peroxidase family. *Cellular and Molecular Life Sciences* 66 (8): 1387–1403. doi: 10.1007/s00018-008-8651-8.

Sugano, Y., Y. Matsushima and M. Shoda. 2006. Complete decolorization of the anthraquinone dye reactive blue 5 by the concerted action of two peroxidases from *Thanatephorus cucumeris* dec 1. *Applied Microbiology and Biotechnology* 73 (4): 862–871. doi: 10.1007/s00253-006-0545-9.

Sugano, Y., Y. Matsushima, K. Tsuchiya, H. Aoki, M. Hirai and M. Shoda. 2009. Degradation pathway of an anthraquinone dye catalyzed by a unique peroxidase DyP from *Thanatephorus cucumeris* Dec 1. *Biodegradation* 20 (3): 433–440. doi: 10.1007/s10532-008-9234-y.

Sugano, Y., R. Muramatsu, A. Ichiyanagi, T. Sato and M. Shoda. 2007. DyP, a unique dye-decolorizing peroxidase, represents a novel heme peroxidase family: ASP171 REPLACES THE DISTAL HISTIDINE OF CLASSICAL PEROXIDASES. *Journal of Biological Chemistry* 282 (50): 36652–36658. doi: 10.1074/jbc.M706996200.

Sundaramoorthy, M., H. L. Youngs, M. H. Gold and T. L. Poulos. 2005. High-resolution crystal structure of manganese peroxidase: substrate and inhibitor complexes. *Biochemistry* 44 (17): 6463–6470. doi: 10.1021/bi047318e.

Suzuki, K., H. Hirai, H. Murata and T. Nishida. 2003. Removal of estrogenic activities of 17β-estradiol and ethinylestradiol by ligninolytic enzymes from white rot fungi. *Water Research* 37 (8): 1972–1975. doi: https://doi.org/10.1016/S0043-1354(02)00533-X.

Syoda, M., Y. Sugano and H. Kubota. 2006. Enzyme having decolorizing activity and method for decolorizing dyes by using the same. U.S. Patent 7, 041, 486.

Taboada-Puig, R., C. Junghanns, P. Demarche, M. T. Moreira, G. Feijoo, J. M. Lema and S. N. Agathos. 2011. Combined cross-linked enzyme aggregates from versatile peroxidase and glucose oxidase: Production, partial characterization and application for the elimination of endocrine disruptors. *Bioresource Technology* 102 (11): 6593–6599. doi: https://doi.org/10.1016/j.biortech.2011.03.018.

Touahar, I. E., L. Haroune, S. Ba, J. P. Bellenger and H. Cabana. 2014. Characterization of combined cross-linked enzyme aggregates from laccase, versatile peroxidase and glucose oxidase, and their utilization for the elimination of pharmaceuticals. *Science of the Total Environment* 481: 90–99. doi: https://doi.org/10.1016/j.scitotenv.2014.01.132.

Tamagawa, Y., H. Hirai, S. Kawai and T. Nishida. 2007. Removal of estrogenic activity of 4-tert-octylphenol by ligninolytic enzymes from white rot fungi. *Environmental Toxicology* 22 (3): 281–286. doi: doi:10.1002/tox.20258.

Tamagawa, Y., R. Yamaki, H. Hirai, S. Kawai and T. Nishida. 2006. Removal of estrogenic activity of natural steroidal hormone estrone by ligninolytic enzymes from white rot fungi. *Chemosphere* 65 (1): 97–101. doi: https://doi.org/10.1016/j.chemosphere.2006.02.031.

Tanaka, T. and M. Taniguchi. 2003. Treatment of phenolic endocrine-disrupting chemicals by lignin-degrading enzymes. In: *Wastewater treatment using enzymes*, edited by A. Sakurai. India: Research Signpost.

ten Have, R. and P. J. M. Teunissen. 2001. Oxidative mechanisms involved in lignin degradation by white-rot fungi. *Chemical Reviews* 101 (11): 3397–3414. doi: 10.1021/cr0001151.

Touahar, I. E., L. Haroune, S. Ba, J. P. Bellenger and H. Cabana. 2014. Characterization of combined cross-linked enzyme aggregates from laccase, versatile peroxidase and glucose oxidase, and their utilization for the elimination of pharmaceuticals. *Science of the Total Environment* 481: 90–99. doi: https://doi.org/10.1016/j.scitotenv.2014.01.132.

Tsutsumi, Y., T. Haneda and T. Nishida. 2001. Removal of estrogenic activities of bisphenol A and nonylphenol by oxidative enzymes from lignin-degrading basidiomycetes. *Chemosphere* 42 (3): 271–276. doi: https://doi.org/10.1016/S0045-6535(00)00081-3.

Ullrich, R. and M. Hofrichter. 2005. The haloperoxidase of the agaric fungus *Agrocybe aegerita* hydroxylates toluene and naphthalene. *FEBS Letters* 579 (27): 6247–6250. doi: https://doi.org/10.1016/j.febslet.2005.10.014.

Ullrich, R. and M. Hofrichter. 2007. Enzymatic hydroxylation of aromatic compounds. *Cellular and Molecular Life Sciences* 64 (3): 271–293. doi: 10.1007/s00018-007-6362-1.

Ullrich, R., J. Nüske, K. Scheibner, J. Spantzel and M. Hofrichter. 2004. Novel haloperoxidase from the agaric basidiomycete *Agrocybe aegerita* oxidizes aryl alcohols and aldehydes. *Applied and Environmental Microbiology* 70 (8): 4575–4581. doi: 10.1128/aem.70.8.4575-4581.2004.

Valderrama, B., M. Ayala and R. Vazquez-Duhalt. 2002. Suicide inactivation of peroxidases and the challenge of engineering more robust enzymes. *Chemistry & Biology* 9 (5): 555–565. doi: 10.1016/s1074-5521(02)00149-7.

van Bloois, E., D. E. Torres Pazmiño, R. T. Winter and M. W. Fraaije. 2010. A robust and extracellular heme-containing peroxidase from thermobifida fusca as prototype of a bacterial peroxidase superfamily. *Applied Microbiology and Biotechnology* 86 (5): 1419–1430. doi: 10.1007/s00253-009-2369-x.

van Pée, K. H. 2001. Microbial biosynthesis of halometabolites. *Archives of Microbiology* 175 (4): 250–258. doi: 10.1007/s002030100263.

Vazquez-Duhalt, R., D. W. S. Westlake and P. M. Fedorak. 1994. Lignin peroxidase oxidation of aromatic compounds in systems containing organic solvents. *Applied and Environmental Microbiology* 60 (2): 459–466.

Villalobos, D. A. and I. D. Buchanan. 2002. Removal of aqueous phenol by *Arthromyces ramosus* peroxidase. *Journal of Environmental Engineering and Science* 1 (1): 65–73. doi: 10.1139/s01-003.

Wang, J., N. Majima, H. Hirai and H. Kawagishi. 2012. Effective removal of endocrine-disrupting compounds by lignin peroxidase from the white-rot fungus *Phanerochaete sordida* YK-624. *Current Microbiology* 64 (3): 300–303. doi: 10.1007/s00284-011-0067-2.

Wang, Y., R. Vazquez-Duhalt and M. A. Pickard. 2003. Manganese–lignin peroxidase hybrid from *Bjerkandera adusta* oxidizes polycyclic aromatic hydrocarbons more actively in the absence of manganese. *Canadian Journal of Microbiology* 49 (11): 675–682. doi: 10.1139/w03-091.

Ward, G., Y. Hadar and C. G. Dosoretz. 2003. Lignin peroxidase-catalyzed polymerization and detoxification of toxic halogenated phenols. *Journal of Chemical Technology & Biotechnology* 78 (12): 1239–1245. doi: doi:10.1002/jctb.933.

Wen, X., Y. Jia and J. Li. 2009. Degradation of tetracycline and oxytetracycline by crude lignin peroxidase prepared from *Phanerochaete chrysosporium*—A white rot fungus. *Chemosphere* 75 (8): 1003–1007. doi: https://doi.org/10.1016/j.chemosphere.2009.01.052.

Wesenberg, D., I. Kyriakides and S. N. Agathos. 2003. White-rot fungi and their enzymes for the treatment of industrial dye effluents. *Biotechnology Advances* 22 (1): 161–187. doi: https://doi.org/10.1016/j.biotechadv.2003.08.011.

Xu, F. and S. Salmon. 2008. Potential applications of oxidoreductases for the re-oxidation of leuco vat or sulfur dyes in textile dyeing. *Engineering in Life Sciences* 8 (3): 331–337. doi: 10.1002/elsc.200700070.

Xu, H., M. Y. Guo, Y. H. Gao, X. H. Bai and X. W. Zhou. 2017. Expression and characteristics of manganese peroxidase from *Ganoderma lucidum* in pichia pastoris and its application in the degradation of four dyes and phenol. *BMC Biotechnology* 17: 19. doi: 10.1186/s12896-017-0338-5.

Yang, Q., M. Yang, K. Pritsch, A. Yediler, A. Hagn, M. Schloter and A. Kettrup. 2003. Decolorization of synthetic dyes and production of manganese-dependent peroxidase by new fungal isolates. *Biotechnology Letters* 25 (9): 709–713. doi: 10.1023/a:1023454513952.

Yang, X., J. Zheng, Y. Lu and R. Jia. 2016. Degradation and detoxification of the triphenylmethane dye malachite green catalyzed by crude manganese peroxidase from *Irpex lacteus* F17. *Environmental Science and Pollution Research* 23 (10): 9585–9597. doi: 10.1007/s11356-016-6164-9.

Yao, J., R. Jia, L. Zheng and B. Wang. 2013. Rapid decolorization of azo dyes by crude manganese peroxidase from *Schizophyllum* sp. F17 in solid-state fermentation. *Biotechnology and Bioprocess Engineering* 18 (5): 868–877. doi: 10.1007/s12257-013-0357-6.

Young, L. and J. Yu. 1997. Ligninase-catalysed decolorization of synthetic dyes. *Water Research* 31 (5): 1187–1193. doi: https://doi.org/10.1016/S0043-1354(96)00380-6.

Yu, G., X. Wen, R. Li and Y. Qian. 2006. *In vitro* degradation of a reactive azo dye by crude ligninolytic enzymes from non immersed liquid culture of *Phanerochaete chrysosporium*. *Process Biochemistry* 41 (9): 1987–1993. doi: https://doi.org/10.1016/j.procbio.2006.04.008.

Yu, S., X. Wang, Y. Ai, X. Tan, T. Hayat, W. Hu and X. Wang. 2016. Experimental and theoretical studies on competitive adsorption of aromatic compounds on reduced graphene oxides. *Journal of Materials Chemistry A* 4 (15): 5654–5662. doi: 10.1039/c6ta00890a.

Zelena, K., B. Hardebusch, B. Hülsdau, R. G. Berger and H. Zorn. 2009. Generation of norisoprenoid flavors from carotenoids by fungal peroxidases. *Journal of Agricultural and Food Chemistry* 57 (21): 9951–9955. doi: 10.1021/jf901438m.

Zhang, H., S. Zhang, F. He, X. Qin, X. Zhang and Y. Yang. 2016. Characterization of a manganese peroxidase from white-rot fungus *Trametes* sp. 48424 with strong ability of degrading different types of dyes and polycyclic aromatic hydrocarbons. *Journal of Hazardous Materials* 320: 265–277. doi: https://doi.org/10.1016/j.jhazmat.2016.07.065.

Zhang, S., Y. Ning, X. Zhang, Y. Zhao, X. Yang, K. Wu, S. Yang, G. La, X. Sun and X. Li. 2015. Contrasting characteristics of anthracene and pyrene degradation by wood rot fungus *Pycnoporus sanguineus* H1. *International Biodeterioration & Biodegradation* 105: 228–232. doi: https://doi.org/10.1016/j.ibiod.2015.09.012.

Zhang, Y. and S. U. Geißen. 2010. *In vitro* degradation of carbamazepine and diclofenac by crude lignin peroxidase. *Journal of Hazardous Materials* 176 (1): 1089–1092. doi: https://doi.org/10.1016/j.jhazmat.2009.10.133.

Zubieta, C., R. Joseph, K. S. Sri, D. McMullan, M. Kapoor, H. L. Axelrod, M. D. Miller, P. Abdubek, C. Acosta, T. Astakhova, D. Carlton, H. J. Chiu, T. Clayton, M. C. Deller, L. Duan, Y. Elias, M. A. Elsliger, J. Feuerhelm, S. K. Grzechnik, J. Hale, G. W. Han, L. Jaroszewski, K. K. Jin, H. E. Klock, M. W. Knuth, P. Kozbial, A. Kumar, D. Marciano, A. T. Morse, K. D. Murphy, E. Nigoghossian, L. Okach, S. Oommachen, R. Reyes, C. L. Rife, P. Schimmel, C. V. Trout, B. H. van den, D. Weekes, A. White, Q. Xu, K. O. Hodgson, J. Wooley, A. M. Deacon, A. Godzik, S. A. Lesley and I. A. Wilson. 2007. Identification and structural characterization of heme binding in a novel dye-decolorizing peroxidase, TyrA. *Proteins: Structure, Function, and Bioinformatics* 69 (2): 234–243. doi: doi:10.1002/prot.21673.

CHAPTER-3

Phenoloxidases of Fungi and Bioremediation

Montiel-González Alba Mónica[1],* and
Marcial Quino Jaime[2]

1. Introduction

In fungi, the production of phenoloxidase enzymes is a common and natural phenomenon that is present in several of their groups, such as the phylum Ascomycomicota and Basidiomycota (Bollag and Leonowicz 1984). Apparently, in these types of fungi, the enzymes are necessary for obtaining nutrients from organic matter in the environment and are actively involved in the carbon recycling process. For other fungi, these enzymes have been described as participants in the morphogenesis process, delignification, sporulation, pigment production, fruiting body formation, and pathogenesis of plants (Leontievsky et al. 2000, Tsai et al. 1999, Wösten and Wessels 2006). Several researchers have reported the actions of phenol oxidase enzymes produced by fungi; however, the most commonly studied phenol oxidases are laccase and tyrosinase, and both enzymes are characterized by the presence of Cu atoms at their catalytic sites. These enzymes show differences in physiological functions and can act on different phenolic compounds. In general, the fundamental functions of these enzymes are: (i) storage, transportation, and uptake of metal ions and oxygen, (ii) transfer of electrons, and (iii) catalysis (Solomon et al. 2014, Kanteev et al. 2015).

Fungi of the phylum Basidiomycota, specifically white rot fungi, have been considered important producers of laccase enzymes, because of this, they have great ability to degrade the lignocellulosic components of wood. These fungi have potential abilities in biotechnological processes, including environmental

---
[1] Centro de Investigación en Genética y Ambiente, Universidad Autónoma de Tlaxcala. Km. 10.5 aut. San Martín Texmelucan-Tlaxcala, San Felipe Ixtacuixtla, Tlax. C.P. 90120, Mexico.
[2] CONACYT-Instituto Nacional de Pediatría. Laboratorio de Bioquímica Genética, Secretaría de Salud. Mexico City, C.P. 04530, Mexico.
* Corresponding author: amonicamg@yahoo.com

bioremediation (Halaouli et al. 2006, Kües 2015, Martínková et al. 2016). Laccases are multicopper proteins that use molecular oxygen to oxidize various aromatic and nonaromatic compounds by a radical-catalyzed reaction mechanism; laccases use the available oxygen as the final electron acceptor, and there is no requirement for cofactors (Thurston 1994). Moreover, laccases have a broad substrate spectrum, which makes them capable of transforming xenobiotic compounds that are pollutants of the environment. Among the pollutants of major interest are dyes because of their relevance in textile, pharmaceutical, leather, cosmetics, and paper industries, as well as their impact on the health of living organisms when effluent wastes discharged into water bodies are found (Viswanath et al. 2014). As an option, to reduce the contamination by dyes, laccases have been tested during the transformation of textile dyes, and promising results were obtained. Laccases have been tested less frequently for the degradation of phenolic compounds, such as hydroxylated polychlorinated biphenyls, polychlorophenols, and polycyclic aromatic hydrocarbons, endocrine-disrupting chemicals, insecticides, fungicides, and herbicides. The success in the utilization of laccases in biodegradation processes is tangible; however, it depends on the standardization of several factors, such as the selection of laccase isoenzymes with major activity for this process, a task that is currently being researched actively in this field (Schückel et al. 2011). On the other hand, tyrosinases are other phenoloxidase enzymes that are different depending on the organism of origin. These metalloenzymes, according to their structure and spectroscopic properties, are considered a type-3 class, indicating that the metalloenzymes contain two copper atoms and have the ability to activate molecular oxygen (Solomon et al. 1992, Solomon et al. 1996, Kanteev et al. 2015). Tyrosinases are bifunctional enzymes by carrying out monophenol oxygenase and catechol oxidase reactions in subsequent ways (Sánchez-Ferrer et al. 1995, Decker et al. 2006). Although their reactions are known, the function of tyrosinase is uncertain; however, it has been described that, in prokaryotes, as well as in plants, fungi, arthropods, and mammals, tyrosinases are responsible for pigmentation, wound healing, radiation protection, and primary immune responses (Kanteev et al. 2015). Another role is the phenomenon of browning produced in edible fungi, such as *Agaricus bisporus* and *Lentinula edodes*, during their growth, collection, and storage, which is a major problem in the food industry (Jolivet et al. 1998, Seo et al. 2003). Therefore, a lesion, cut, or infection in the tissue of a fungus structure causes tyrosinase to react with phenolic substrates, giving rise to *o*-quinones. These quinones can bind to proteins and inactivate them, or they can be polymerized to give rise to melanins, which are molecules that darken or give a brown color to the affected area, making them inaccessible for certain pathogens (Martinez and Whitaker 1995, Ramsden and Riley 2014).

Given the relevance of phenoloxidases, this section describes some of the main characteristics of laccases and tyrosinases identified in fungi, as well as their application in the degradation of toxic compounds present in different environments.

2. Chemical Characteristics of Lacasse and Tyrosinase

Laccases (benzenediol: oxygen oxidoreductase; EC 1.10.3.2) are blue multicopper oxidase enzymes (MOCs). They are dimeric or tetrameric glycoproteins that have the ability to carry out oxidation of the substrate through of a four-electron reduction of molecular oxygen to water. They belong to a family of enzymes that contain at least four copper ions comprising a mononuclear blue type-1 or blue copper center and a trinuclear cluster at a 12–13 Å distance, formed by two type-3 copper ions and one type-2 copper ion (Messerschmidt and Huber 1990, Solomon et al. 1992). According to visible UV and electronic paramagnetic resonance (EPR) spectra, the four copper atoms show differing characteristics. Type-1 copper shows coordination with two histidines, one cysteine, and one methionine as ligands, and shows a strong electronic absorption in the visible region at 600 nm (ε = 5000 M^{-1} cm^{-1}), which is responsible for the blue color (Solomon et al. 1992). There are laccases that lack typical absorption at 600 nm; for example, the "white laccase" produced by *Pleurotus ostreatus* (Palmieri et al. 1997), which contain the particularity of an arrangement of 1 Cu, 1 Fe, 2 Zn atoms, and "yellow laccases" that are present in *Panus tigrinus* (Leontievsky et al. 1997), which contain copper but an altered oxidation state (Dwivedi et al. 2011). Type-2 copper has two histidines and water as ligands, undetectable absorption in the visible spectrum, EPR spectra as a low molecular weight copper complex, and is positioned close to type-3 copper. Type-3 copper has three histidines and a hydroxyl bridge that maintains the strong anti-ferromagnetic coupling between its atoms (Piontek et al. 2002) and shows an electron absorption at 330 nm (Fig. 3.1).

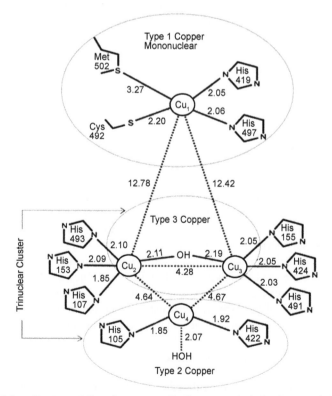

Figure 3.1 Schematic representation of cooper coordination centers, including interatomic distances among all relevant ligands (Taken from Dwivedi et al. (2011)).

The molecular structure and properties of the copper centers of laccases show some differences among laccases from diverse species, and this phenomenon can explain why these enzymes show varied redox potentials (Piontek et al. 2002), dividing them into low-redox potential and high-redox potential laccases. Laccases from basidiomycetes, especially white rot fungi, are high-redox potential laccases (Gutiérrez et al. 2006), whereas bacterial and plant laccases are examples of low-redox potential laccases (Mikolasch and Schauer 2009). Typical fungal laccases are proteins that have molecular weights from approximately 60 to 70 kDa; however, a variety of laccases isolated from ascomycetes show different molecular weights. These enzymes have a pH of approximately 4.0 in the range of pH 3 to 5 (Baldrian 2006). Laccases are stable in a range of 30°C to 50°C and are deactivated at temperatures above 60°C, as occurs with laccases of *Mauginiella* sp. (Palonen et al. 2003). Laccases can be monomeric, homo-oligomeric, and hetero-oligomeric proteins, containing two or three identical monomers, or a combination of them until there are three different polypeptides (Baldrian 2006). Various fungal species produce laccase isoenzymes, e.g., *Pleurotus ostreatus* (Palmeiri et al. 1993, Palmieri et al. 1997, Palmieri et al. 2000, Palmieri et al. 2003, Giardina et al. 1999, Sannia et al. 1986), *Marasmius quercophilus* (Farnet et al. 2000, Farnet et al. 2002, Farnet et al. 2004), and *P. chrysosporium* (Dittmer et al. 1997). Isoenzymes are enzymatic variants with different molecular weights that oxidize the same substrates but have different properties, and their emergence is explained due to the presence of multiple laccase genes in fungi (Chen et al. 2003), but also to the presence of copper (Palmieri et al. 2000, Palmieri et al. 2003) or physico-chemical changes. Relevant laccases in the bioremediation field are principally extracellular; these enzymes are proteins binding to glycans, which suggests that the glycans protect laccases from proteolytic degradation (Yoshitake et al. 1993). These glycoproteins generally have a glycosylation range from 10% to 25%; nevertheless, in some species such as *Coriolopsis fulvocinnerea*, *Pleurotus pulmonarius*, and *Botrytis cinerea*, the percentages of glycosylation are 32%, 44%, and from 49% to 84%, respectively, whereas in *Pleurotus eryngii*, laccases were found in 1% to 7% of bound sugar (Baldrian 2006). Fungal laccase properties depend on a great variety of molecular structures, such as the strain from which they came and the culture conditions in which the fungal enzyme producer is cultivated, including the inductor presence; these circumstances maintain the current studies and the search for new laccase enzymes.

Other enzymes of biotechnological interest due to its physiological implications and biochemical properties are tyrosinases, also known as polyphenol oxidase. Tyrosinases are copper-containing enzyme like laccases, but unlike these, tyrosinases perform two reactions: catalyze the ortho-hydroxylation of monophenol and the subsequent oxidation of the diphenolic product to the resulting quinone. Tyrosinase is widely distributed in microorganisms and animals in general. In vegetables, fruits, and mushrooms it is synthesized, being one of the key enzymes in the browning that occurs upon bruising or long-term storage; for this reason it is of agricultural importance.

In this sense, tyrosinases, depending on their biological origin, are quite significant in agriculture and different industries, such as in the cosmetic industry, for the development and/or screening of inhibitors of these enzymes. At the present time, fungal tyrosinase has become popular because it is readily available; even through genetic manipulation, these enzymes have been overexpressed by heterologous expression, which is useful for different applications, such as processing different toxic compounds, mainly phenols, present in different environments. For example, partially purified tyrosinase from the mushroom *Agaricus bisporus* is available commercially, and it has served as a widely used experimental model for the study of tyrosinase to determine its structural and catalytic characteristics. This enzyme is encoded by at least five genes in *A. bisporus* (http://genome.jgipsf.org/Agabi varbisH97 2/Agabi varbisH97 2.home.html.). The purification of these enzymes has shown multiple forms (Bouchilloux et al. 1963). Initially, Jolley et al. (1969) determined five different forms (I, Ia, II–IV), ranging from monomers to octamers. The authors suggested that the predominant enzyme was tetrameric and composed of four identical sub-units of 32 kDa. Later, other researchers identified the existence of the forms α, β, γ, and δ, and they were heteropolymers comprising two different polypeptidic subunits: a heavy chain (H) and a light chain (L) with sizes of 43 kDa and 13.4 kDa, respectively (Strothkamp et al. 1976, Bouchilloux et al. 1963). Additionally, two monomeric isoenzymes of *A. bisporus* (43 kDa) were obtained (Wichers et al. 1996). In recent years, a crystallographic structure of *A. bisporus* was obtained, and it was determined that it was tetrameric with a molecular weight of 120 kDa and composed of two subunits of ≈ 43 kDa (heavy subunit or H) and two subunits of ≈ 14 kDa (light subunit or L) (Inlow 2012). The identity, function, and origin of the L subunit are still unknown (Strothkamp et al. 1976, Mayer 2006, Schurink et al. 2007, Flurkey and Inlow 2008).

Tyrosinase contains three major domains: the N-terminal, C-terminal, and the central domain (Kanteev et al. 2015). Tyrosinases, which contain two copper atoms, use molecular oxygen to catalyze reactions (Solomon et al. 1992). The most important structural feature of these enzymes is the presence of two copper atoms (CuA and CuB) in their active center, each linked to three histidines and the whole to a single cysteine residue. In addition, a series of hydrophobic amino acids surrounds it with aromatic rings, which are also important in the activity of the enzyme (specifically, for the binding of the substrates). This arrangement has been conserved throughout evolution in all the enzymes of this type (Mayer 2006, Inlow 2012). Tyrosinases are enzymes characterized by catalyzing two reactions. The first is the monophenol oxygenase or cresolase activity (EC 1.14.18.1), which consists of hydroxylation in the ortho position of monophenols and oxidation of the intermediate *o*-diphenol to *o*-quinone. The other activity is diphenolase or catechol oxidase (EC 1.10.3.1), which consists of the oxidation of two molecules, an *o*-diphenol to two molecules of *o*-quinone, and the reduction of molecular oxygen, forming two water molecules (Sánchez-Ferrer et al. 1995). Monophenolase activity of tyrosinases is known to be the initial rate-determining reaction (Rodríguez-López et al. 1991, Robb 1984). In tyrosinase-catalyzed reactions, molecular oxygen is used as an electron acceptor and it is reduced to water.

The active site of tyrosinases can exist in three intermediate states: deoxy (Cu^I–Cu^I), oxy (Cu^{II}–O_2–Cu^{II}), and met (Cu^{II}–Cu^{II}), but this depends on the copper-ion valence and linking to molecular oxygen (Sánchez-Ferrer et al. 1995). Although the action mechanism of these enzymes is complex, it is thought to be allosteric and involves two oxygen-binding sites and aromatic compounds (Duckworth and Coleman 1970). All three forms participate in the catecholase cycle. In this cycle, an *o*-diphenol reduces Cu(II) from the "met" form to Cu(I), and then the "deoxy" form reacts with molecular oxygen, forming the "oxy" form. Each of the Cu(II) atoms of the "oxy" form then attaches to an oxygen atom of the *o*-diphenol hydroxyl groups, giving rise to the O_2-diphenol-tyrosinase complex. In the last step, the *o*-diphenol is oxidized to *o*-benzoquinone and the enzyme is reduced to the "met" form. To complete the cycle, another *o*-diphenol molecule binds to the "met" form, and oxidation of *o*-diphenol to *o*-quinone occurs, with the corresponding reduction of the "met" form to the "deoxy" form of the enzyme (Sánchez-Ferrer et al. 1995, Solomon et al. 1996). In the cresolase cycle, only the "deoxy" and "oxy" forms participate in the process. In this cycle, the "oxy" form reacts with a monophenolic substrate to form a ternary complex, which is reorganized, giving rise to a highly reactive intermediate and causes the hydroxylation of the substrate, forming an *o*-diphenol bound to the enzyme.

The tyrosinases of the mushroom *A. bisporus* and of the ascomycete *Neurospora crassa* are the most studied at the molecular, structural, and kinetic levels (Lerch 1983, Gerritsen et al. 1994, Sánchez-Ferrer et al. 1995, Fan and Flurkey 2004, Wichers et al. 2003). These enzymes have also been isolated from other fungi, such as *Aspergillus oryzae* (Fujita et al. 1995), *Trichoderma reseei* (Gasparetti et al. 2012), *Lentinus edodes* (Kanda et al. 1996), and *Pycnoporus sanguineus* (Halaouli et al. 2004). However, the function of tyrosinase remains unclear due to the characteristics of the enzymatic system, including its bivalent activity and the complexity of its extraction and purification.

Another frequent feature of fungal tyrosinases is that, in the native state, it exists as a latent enzyme until the enzyme is activated. Activation can be carried out *in vivo* by proteases as in *A. bisporus* and *Plutella xylostella* (Espín and Wichers 1999, Han et al. 2014). In addition, *in vitro* studies have demonstrated proteolytic activation (Laveda et al. 2001) with detergents, SDS, and polyamines (Laveda et al. 2000, Jiménez-Atiénzar et al. 1991).

3. Enzymatic Mechanisms of the Phenol Oxidases and Substrates

In some groups of fungi, such as in white rot, it is assumed that natural laccase enzymes allow them to access wood polysaccharides. Therefore, laccases are enzymes with specificity for phenolic subunits in lignin, and they degrade both β-1 and β-O-4 dimers via Cα-Cβ cleavage, Cα oxidation, and alkyl–aryl cleavage. In the redox reaction accomplished by laccases, the electrons of the substrate molecules are transferred to type 1 copper ion and, then, sent to the trinuclear cluster where the di-oxygen binds after the transfer, and the electrons are reduced to two molecules of water (Bento et al. 2006). Laccase operates as a battery and stores electrons from individual oxidation reactions to reduce molecular oxygen. Hence, the oxidation of four reducing substrate molecules is necessary for the complete reduction of molecular oxygen to water. When laccase oxidizes the substrate, free radicals are generated. This oxidation results in an oxygen-centered free radical, which may promote the conversion into a second enzyme-catalyzed reaction of a substrate, such as phenol to quinone. Alternatively, an amorphous insoluble melanin-like product could arise through its participation in polymerization reactions and/or non-enzymatic reactions, such as hydration or disproportionation (a reaction in which a substance is simultaneously oxidized and reduced, giving two different products) (Thurston 1994).

The laccase specificity for phenolic compounds allows one-electron oxidation of a wide range of aromatic compounds, such as polyphenolic compounds, methoxy-substituted phenols, and aromatic diamines, such as 2,6-dimethoxyphenol or guaiacol catechol or hydroquinone, as well as some inorganic compounds (iodine, $Mo(CN)_8^{4-}$, and $Fe(CN)_6^{4-}$). Syringaldazine [N, N0-bis (3,5-dimethoxy-4-hydroxybenzylidene hydrazine)] is often considered to be a specific laccase substrate (Harkin et al. 1974, Dwivedi et al. 2011, Viswanath et al. 2014). Laccase substrates have been grouped as three kinds of substituted compounds, i.e., *ortho*, *meta*, and *para*, which possess a single electron pair and are the best substrates for most laccases as *ortho*-substituted compounds (Dwivedi et al. 2011). However, laccases in the presence of mediators can oxidize and cleave non-phenolic substrates. Bourbonnais and Paice (1990) reported that laccases can catalyze the oxidation of non-phenolic benzyl alcohols in the presence of mediators, such as 2,2-azinobis [3-ethylbenzothiazoline-6-sulfonic acid] (ABTS), which is a diffusible electron carrier. The action of the laccase-ABTS couple is performed through carbon–hydrogen abstraction, resulting in a C–C bond cleavage in condensed lignin (Bourbonnais et al. 1995). Similarly, more than 100 mediator compounds have been found, which allow the oxidization of lignin or lignin model compounds through the selective oxidation of their benzylic hydroxyl groups (Eggert et al. 1996, Riva 2006). Many xenobiotic compounds show some substructures that are similar to those of lignin (Magan et al. 2010). Thus, they may function as a substrate for laccase enzymes allowing various species, especially of fungi, to survive in the presence of these compounds.

Tyrosinases, as was mentioned above, catalyze two types of coupled reactions involving molecular oxygen: monophenolase activity and diphenolase activity. The low substrate specificity and the high reactivity of the o-quinones generated by this enzyme determine their participation in physiological processes as diverse as lignin biosynthesis, sclerotization of the arthropod cuticle, and melanin biosynthesis (Sugumaran 2010, Solomon et al. 2014, Ramsden and Riley 2014, Abebe et al. 2010).

In fungi, the active site is located on the surface of the protein and the enzyme has a wide cavity to react as revealed by the structure of the enzyme of *A. bisporus* (Ismaya et al. 2011b, Ismaya et al. 2011a, Inlow 2012). These characteristics facilitate the access of different types of molecules and, consequently, the enzyme shows little substrate specificity (García-Molina et al. 2005, García-Molina et al. 2007, Garcia-Molina et al. 2012, Muñoz-Muñoz et al. 2008, Martínez Ortiz et al. 1988, Zawistowski et al. 1991, Rodríguez-López et al. 1991, Rodríguez-López et al. 1992). Other particularities of tyrosinase are that diphenolase activity is clearly detectable even from different biological sources, whereas the monophenolase activity, being the most specific and complex hydroxylation reaction, is not readily detectable and may generate erroneous results with respect to other enzymes, such as catechol oxidases

(EC 1.14.18.1). One of the specific substrates to determine monophenolase activity is tyrosine; however, its activity on this substrate is slow and complicated to quantify (Espin et al. 1995, Espín et al. 1996). It has also been suggested that the active site is located very close to the surface of the protein, thus bringing other molecules that are analogous to the specific substrates of the enzyme (monophenols and *o*-diphenols) and that interact easily with the active site. This characteristic causes tyrosinase to participate in other activities, such as ascorbate oxidase, oxidizing ascorbic acid to dehydroascorbic acid (Ros et al. 1993, Muñoz-Muñoz et al. 2009), catalase activity, decomposition of hydrogen peroxide (Yamazaki and Itoh 2003, García-Molina et al. 2005), oxidation of tetrahydropterins, oxidation of NADH, and oxidation of tetrahydrofolic acid (Garcia-Molina et al. 2010b, Garcia-Molina et al. 2010a, García-Molina et al. 2011). Due to this non-specificity, tyrosinases may also act on other types of non-natural synthetic substrates.

4. Fungi Phenoloxidases Production and Physiological Functions

Copper-containing polyphenol oxidase was discovered in the exudates of the Japanese lacquer tree *Rhus vernicifera* (Yoshida 1883); its presence in some fungi was reported by Bertrand (1896) and Laborde (1896), and its presence in basidiomycetes and ascomycetes was reported almost a century later (Bollag and Leonowicz 1984). Reports on laccase enzymes produced by plants, insects, fungi, and bacteria evidenced distinct molecular structures, properties, and physiological roles: lignin degradation (Leonowicz et al. 2001), fruiting body formation (Kües and Liu 2000, Wösten and Wessels 2006), pigment formation (Clutterbuck 1972, Tsai et al. 1999), etc., in fungi; lignification processes in plants (Sterjiades et al. 1992, Liu et al. 1994); cuticular sclerotization (Andersen et al. 1996, Kramer et al. 2001) in insects; participation in the melanization pathway and oxidation of toxic compounds in animals; and melanization processes in bacteria (Faure et al. 1994). Thus, laccase enzymes have the capability to oxidize a wide range of substrates. However, the laccase family could be considered as another type of enzyme. Hoegger et al. (2006) carried out a phylogenetic study in which they compared multicopper oxidase (MCO) sequences of fungi, plants, insects, and bacteria to redefine the classification of laccases. In laccases *sensu stricto*, represented by basidiomycetes and ascomycetes laccases, insect laccases, fungal pigment MCOs, fungal ferroxidases, ascorbate oxidases, plant laccase-like MCOs, and bilirubin oxidases, it has been argued that laccase enzymes have different properties, structures, and origins. MCO composition is variable in different species, the laccases *sensu stricto* sequences are grouped according to the taxonomic association of the corresponding species, and the fungi MCOs are grouped partially according to their lifestyle and are separated from the other MCO clusters (Hoegger et al. 2006). Some gene sequences encoding these enzymes indicate that the enzymes are "not typical laccases" in *Phanerochaete chrysosporium* and *Cryptococcus neoformans*. They have been characterized as having weak activity. They were grouped together with the ferroxidases sequences and not with laccases *sensu stricto*, effectively showing strong ferroxidase activity. In a different cluster, the fungal pigment MCOs are grouped and include enzymes required in conidial pigment biosynthesis (Clutterbuck 1972, Tsai et al. 1999), for example, the enzymes required for the 1,8-dihydroxynaphthalene biosynthesis pathway in *Aspergillus* species (Tsai et al. 1999). The problem is that these enzymes have been named laccases only because of their ability to oxidize typical laccase substrates, but they do not share chemical structures or gene sequences with other laccases. For example, considering the *P. chrysosporium* laccases, Larrondo et al. (2003) reported that these laccases are MCOs and are different from true laccases because of their particular differences. Later, it was confirmed that gene sequences for laccases have not been found in the complete genome sequence of this fungus (Martinez et al. 2004). Additionally, many species of fungi showed a combination of different MCO sequences simultaneously, and this could explain their participation in different fungal physiologies. Nevertheless, among some fungi laccase enzymes *sensu stricto* (basidiomycete) that have been characterized, the differences in expression patterns and biochemical properties of the enzymes have suggested that, despite belonging to this enzymatic group, they participate in different physiological roles (Klonowska et al. 2002), such as adaptation to high pH environments, participation in the lignin degradation pathway, and playing a cytoprotective role (Stoj and Kosman 2003), as well as participating during different phases of the lifecycle of fungi.

Laccases have been considered desirable enzyme sources for industrial and medical fields due to their specificity, reactivity, physicochemical, catalytic, and biological properties. However, their functional diversity has made it difficult to understand their physiological role and use in many processes, which has led to revising the evolution process and considering the importance of demands on oxidative enzymes that are dependent on ecological habitats (Hoegger et al. 2006). Therefore, laccase identification, characterization, and physiological role studies are important for the subsequent effective selection of laccases that are specific for toxic compound biodegradation processes. There have been reports on fungal species considered to be laccase producers since the 1970s. We can observe that the list of fungal species reported is short (Table 3.1), considering the number of fungal species estimated for the world, and the list of fungal species that are laccase producers and that have been tested for bioremediation processes is even shorter (Table 3.2).

Nearly half of the fungal laccases tested in bioremediation processes included in this chapter have been used for their action on dye compounds, because a great variety of them are present in wastewater of industries, such as textiles, food, pharmaceuticals, etc. In a few studies, fungal laccases have been tested on hydroxylated polychlorinated biphenyls, polychlorophenols, and polycyclic aromatic hydrocarbons, endocrine-disrupting chemicals, insecticides, fungicides, and herbicides.

Tyrosinase is the rate-limiting enzyme in the biosynthesis of melanin pigments (melanogenesis), starting from tyrosine as the precursor substrate to the formation of L-dopaquinone and L-dopachrome (Sánchez-Ferrer et al. 1995, Ramsden and Riley 2014). Melanins are heterogeneous polyphenolic polymers with colors ranging from yellow to black. In fungi, it has been established that melanin participates in the formation of reproductive organs, spore formation, virulence of pathogenic fungi, and tissue protection after damage. Fungal tyrosinases are cytoplasmic enzymes. This characteristic does not allow tyrosinases to contain a transit peptide, although

Table 3.1 Fungi species reported as laccase producers. The majority belong to basidiomycetes.

Ascomycetes			
Fungal species	**Reported by**	**Fungal species**	**Reported by**
Podospora anserine	Molitoris et al. (1972)	*Hortaea acidophila*	Tetsch et al. (2006)
Botrytis cinerea	Marc Dubernet et al. (1975)	*Myrioconium* sp.	Martin et al. (2007)
Penicillium cyclopium	Bartsch et al. (1979)	*Lamprospora wrightii*	Mueangtoom et al. (2010)
Aspergillus nidulans	Kurtz and Champe (1982)	*Aspergillus fumigatus*	Vivekanand et al. (2011)
Neurospora crassa	Tamaru and Inoue (1989)	*Chaetomium* sp. and	Qasemian et al. (2012)
Cryphonectria parasitica	Rigling and Van Alfen (1991)	*Xilogone sphaerospora*	
Myceliophthora thermophile	Berka et al. (1997)	*Colletotrichum orbiculare*	Lin et al. (2012)
Gaeumannomyces graminis	Edens et al. (1999)	*Aureobasidium pullulans*	Rich et al. (2013)
Melanocarpus albomyces	Kiiskinen et al. (2002)	*Aspergillus flavus*	Gomaa and Momtaz (2015)
Tricoderma atroviride and *Tricoderma harzianum*	Hölker et al. (2002)	*Leptosphaerulina* sp.	Copete et al. (2015)
Botryosphaeria sp.	Alves da Cunha et. al. (2003)	*Botrytis aclada*	Osipov et al. (2015)
Stachybotrys chartarum	Janssen et al. (2004)	*Talaromyces marneffei*	Sapmak et al. (2016)
Yarrowia lipolytica	Jolivalt et al. (2005)	*Trichoderma viride*	Divya and Sadasivan (2016)
Basidiomycetes			
Fungal species	**Reported by**	**Fungal species**	**Reported by**
Trametes (coriolus) versicolor	Archibald and Roy (1992)	*Phlebia radiata*	Mäkelä et al. (2006)
Agaricus bisporus	Perry et al. (1993)	*Cerrena maxima*	Lyashenko et al. (2006)
Ceriporiopsis subvermispora	Lobos et al. (1994)	*Coprinellus congregatus*	Kim et al. (2006)
Pleurotus ostreatus	Marzullo et al. (1995)	*Physisporinus rivulosus*	Hildén et al. (2007)
Rhizoctonia solani	Wahleithner et al. (1996)	*Polyporus brumalis*	Ryu et al. (2008)
Trametes villosa	Yaver and Golightly (1996)	*Fomitella fraxinea*	Park and Sang-Shin (2008)
Pycnoporus cinnabarius		*Steccherinum ochraceum*	Chernykh et al. (2008)
Pleurotus eringii	Eggert et al. (1996)	*Pycnoporus coccineus*	Uzan et al. (2010)
Coriolus versicolor	Muñoz et al. (1997)	*Grifola frondosa*	Nitheranont et al. (2011)
Dichomitus squalens	Mikuni and Morohoshi (2006)	*Polyporus grammocephalus*	Huang et al. (2010)
Rigidoporus lignosus	Pèrié et al. (1998)	*Agrocybe cylindracea*	Hu et al. (2011)
Ceriporiopsis subvermispora	Bonomo et al. (1998)	*Agaricus placomyces*	Sun et al. (2012)
Schizophylum commune	Karahanian et al. (1998)	*Cerrena unicolor*	Kucharzyk et al. (2012)
Lentinula edodes	Hatamoto et al. (1999)	*Coprinopsis* sp.	Qasemian et al. (2012)
Cyathus stercoreus	Zhao and Kwan (1999)	*Rigidoporus lignosus*	Cambria et al. (2012)
Marasmius quercophilus	Sethuraman et al. (1999)	*Hericium coralloides*	Zou et al. (2012)
Coriolus hirsutus	Farnet et al. (2002)	*Fomes durissimus*	Sahay et al. (2012)
Pleurotus sajor-caju	Gorbatova et al. (2000)	*Coprinus comatus*	Bao et al. (2013)
Phellinus ribis	Soden and Dobson (2001)	*Flammulina velutipes*	Otsuka Saito et al.
Trametes hirsuta and *Sclerotium rolfsii*	Min et al. (2001)	*Lentinus polychrous lév*	Khammuang and Sarnthima (2013)
Trametes trogii	Campos et al. (2001)	*Physisporinus rivulosus*	Hildén et al. (2013)
Coriolopsis rigida	Garzillo et al. (2001)	*Trametes pubescens*	Gonzalez et al. (2013)
Trametes modesta	Saparrat et al. (2002)	*Trametes polyzona*	Chairin et al. (2013)
Phanerochaete flavido-alba	Nyanhongo et al. (2002)	*Auricularia auricula-judae*	Fan et al. (2014)
Cyathus bullery	de la Rubia et al. (2002)		
Coprinus cinereus	Dhawan et al. (2002)	*Phytophthora capsici*	Feng and Li (2014)
Chantarellus cibarius, Tricholoma giganteum, Albatrella dispansus	Hoegger et al. (2004) Ng and Wang (2004)	*Agaricus brasiliensis*	Matsumoto-Akanuma et al. (2014)
Heterobasidion annosum		*Lentinus crinitus*	Valle et al. (2014)
		Heterobasidion annosum	Kuo et al. (2015)

Table 3.1 contd. ...

...Table 3.1 contd.

Basidiomycetes			
Fungal species	**Reported by**	**Fungal species**	**Reported by**
Trametes gallica	Asiegbu et al. (2004)	*Praxillus involutus*	Ellström et. al. (2015)
Tricoderma reesei	Dong and Zhang (2004)	*Tricholoma matsutake*	Xu et al. (2015)
Cryptococcus neoformans	Kiiskinen et al. (2002)	*Ceriporiopsis*	Vieira et al. (2015)
Panus tigrinus	Cadimaliev et al. (2005)	*subvermispora*	
Botryosphaeria rhodina	Rezende et al. (2005)	*Hypsizygus marmoreus*	Zhang et al. (2015)
Cyathus bulleri	Salony and Bisaria (2006)	*Agaricus blazei*	Valle et al. (2015)
Ganoderma lucidum	Wang and Ng (2006)	*Lepista nuda*	Zhu et al. (2016)
Pycnoporus sanguineus	Garcia et al. (2006)	*Ceriporiopsis*	Chmelová and Ondrejovič
		subvermispora	(2016)
		Pleurotus nebrodensis	Yuan et al. (2016)

Table 3.2 Fungi laccase producers tested in bioremediation processes.

Fungal species	**Compound**	**Action**	**Reported by**
Aspergillus niger	Bisphenol A, Nonylphenol	Degradation	Macellaro et al. (2014)
	Methylparaben, Butylparaben,	Degradation	Anastasi et al. (2009)
	Dimethyphthalate	Decolorization	Chmelová and Ondrejovič
Bjerkandera adusta	Pyrene		(2016)
Bjerkandera fumosa		Transformation	Leontievsky et al. (2000)
Ceriporiopsis subvermispora	Triphenylmethane dyes	Degradation	Wu and Nian (2014)
Coriolus versicolor	2,4,6-trichlorophenol	Degradation	Wong et al. (2012)
Fusarium solani	Anthracene	Decolorization	Hsu et al. (2012)
Lentinula edodes	Dyes and polyaromatic	Degradation	Anastasi et al. (2009)
	hydrocarbons	Bleaching	Schückel et al. (2011)
Lentinus sp.	Anthroquinone and azo dyes	Transformation	Leontievsky et al. (2000)
Lopharia spadicea	Pyrene	Decolorization	Faraco et al. (2009)
Marasmius sp.	Textile dyes	Degradation	Fountoulakis et al. (2002)
Panus tigrinus	2,4,6-trichlorophenol		Gayosso-Canales et al. (2012)
Phanerochaete chrysosporium	Textile industry dyes		Macellaro et al. (2014)
Pleurotus ostreatus	Phenolic compounds	Decolorization	Ulčnik et al. (2013)
	Polychlorinated biphenyls		Rodríguez Couto, et al. (2006)
	Bisphenol A, Nonylphenol		Zapata-Castillo et al. (2015)
	Methylparaben, Butylparaben,	Degradation	Balcázar-López et al. (2016)
	Dimethyphthalate	Degradation	Anastasi et al. (2009)
	Lindane and endosulfan	Dechlorination	Cabana et al. (2011)
Trametes hirsuta	Indigo Carmine, Bromophenol	Decolorization	Şaşmaz et al. (2011)
	Blue, Methyl Orange, Poly	Degradation	Zille et al. (2005)
	R-478	Bioremediation	Divya et al. (2014)
Trametes sanguineus	Benzo[α]pyrene, cresols,		
	phenanthrene, nitrophenols,		
	chlorophenols		
Trametes versicolor	Pyrene		
	Triclosan		
	Reactive Red 198, Rem Blue		
	RR, Dylon Navy 17, Rem Red		
	RR, and Rem Yellow RR		
Trametes villosa	Azo dyes		
Trichoderma viride	Phenol		

some have been reported in which transit peptides are associated with the cell wall, and these enzymes can be secreted, as in *Trametes reesei* (Selinheimo et al. 2009, Gasparetti et al. 2012) and *Pycnoporus sanguineus* (Duarte et al. 2012).

As mentioned previously, tyrosinase in fungi was initially characterized in *A. bisporus* (Wichers et al. 1996) due to the enzymatic darkening of the fungus during its development and post-conservation (Jolivet et al. 1998). Tyrosinase has also been reported in other fungi, such as *N. crassa* (Kupper et al. 1989), *L. edodes* (Kanda et al. 1996), *A. oryzae* (Nakamura et al. 2000), *Trametes* spp. (Tomšovský and Homolka 2004), *P. sanguineus* (Halaouli et al. 2004, Duarte et al. 2012), *T. reesei* (Selinheimo et al. 2009), and *Pholiota nameko* (Kawamura-Konishi et al. 2007), because this enzyme is of industrial interest.

Fungal tyrosinases currently have different applications, especially of biotechnological and environmental interest (Durán and Esposito 2000, Halaouli et al. 2006, Agarwal et al. 2016). As these enzymes have the ability to convert monophenols to diphenols, they allow the production of *ortho*-diphenol antioxidants, with beneficial properties as additives in food or for the design of pharmaceutical drugs. These enzymes have also been used in bioremediation processes of soils and water contaminated with phenolic compounds (Girelli et al. 2006).

5. Target Contaminants and Applications of the Phenoloxidases for Bioremediation

The above information allows visualization of the potential application of fungal laccases, especially from white rot fungi, to degrade a wide range of organic pollutants.

Vegetal cell wall constituents, such as lignin, have been recalcitrant compounds from an environmental point of view and their degradation is still the objective of many studies. These studies have shown that, despite fungi being good lignin degraders, the laccase enzyme causes minor changes to the lignin structure (Asina et al. 2016). In the case of other contaminants, a great variety of them are xenobiotics that have been created to sustain the current human lifestyle. The chemical nature of these compounds is variable. Among them are dyes, polycyclic aromatic hydrocarbons, pentachlorophenols, polychlorinated biphenyls, 1,1,1-trichloro-2,2-bis (4-chlorophenyl) ethane, benzene, toluene, ethylbenzene, xylene, and trinitrotoluene. Many studies have reported their environmental persistence and their toxic and mutagenic effects on living organisms and increased the interest in studying strategies for their degradation.

In this sense, several studies have shown the fungal ability to degrade or discolor a wide range of compounds. In some of these studies, it has not been clear which enzymes are responsible for this phenomenon (Rodríguez Couto et al. 2006, Barrasa et al. 2009, Divya et al. 2014, Anastasi et al. 2009). In other studies, it has been shown that laccases have the capacity to remove some phenolic toxins and aromatic amine contents from effluents (Niku-Paavola and Viikari 2000).

In the case of dyes, one of the priority fields that require intervention is degradation strategies. This intervention is due to the fact that dyes are compounds produced in large amounts (approximately 7×10^5 tons annually) and are used in several industries, such as the textile industry, which utilizes two-thirds of the commercialized dyes, pharmaceutical, foods, cosmetics, personal care products, etc., combined with the fact that there is a great variety, with almost 1,000,000 dyes, including reagents of various chemical compositions that are resistant to light, water, and other chemicals (Viswanath et al. 2014). Laccase has shown that it is capable of decolorizing the dye content in textile effluents. Zille et al. (2005) tested *Trametes villosa* laccase for azo dyes 3-(4-dimethylamino-1-phenylazo) benzenesulfonic acid and 3-(2-hydroxy-1-naphthylazo) benzenesulfonic acid degradation. The authors reported that laccase was capable of azo dye degradation; however, over long periods of oxidation, a polymerization reaction was observed. Degradation products became coupled among them or with the unreacted dye and retained the azo group in their molecules, showing some color after maximum decolorization. The presence of coupled compounds provided only azo dye in incomplete degradation, which limits the application of laccases in bioremediation processes. Faraco et al. (2009) tested the ability of free and immobilized laccase mixtures from *P. ostreatus* to decolorize dye in a wastewater model. Laccase enzymes in both forms showed the most affinity for a decolorizing acid dye wastewater model (Acid Blue 62, anthraquinonic type dye, and Acid Red 266, aniline mono-azo dye) achieving decolorization of 35% in 24 hours. In this study, the toxicity of the wastewater treated by fungal cultures and laccase enzymes was also tested. It was observed that the reduction of toxicity was better in the wastewater treated with a fungal culture compared to the reduction obtained with enzymatic treatment. The authors suggested the presence of other fungal components in the detoxification process *in vivo*. A laccase-related enzyme from *Marasmius* sp. showed the ability to bleach textile dyes with different chemical structures industrially. This laccase showed complete decolorization of Blue H-EGN 125, Lanaset Blue 5G, Levafix Blue E-RA, Procion Royal H-EXL, Remazol Brilliant Blue BB, Remazol Brilliant Blue R, and partial decolorization of Blue PN-3R, Crystal Violet, Remazol Bordo B, Remazol Brilliant Orange FR, and Remazol Golden Yellow RNL. Laccase was thus proposed for application in industrial bleaching processes (Schückel et al. 2011). The effects of laccase from *Lentinus* sp. on anthraquinone and azo dye degradation were tested by Hsu et al. (2012). This enzyme showed effective decolorization for Acid Blue 80, RBBR, and Acid Red 37 (97%, 88%, and 61% in 1 hour of reaction at room temperature). Additionally, the researchers confirmed the effect of treated dyes by laccase action on rice seeds and observed that the products obtained after treatment did not have a negative effect on these seeds. For this reason, the authors proposed the importance of bioremediation and detoxification of dyes in textiles. Laccase from *Ceriporiopsis subvermispora* was tested for triphenylmethane decolorization. This laccase was able to degrade Malachite Green (87.8%), Bromocresol Purple (71.6%), and Methyl Violet (68.1%) without a redox mediator, and Phenol Red (42%), Bromophenol Blue (49.8%), and Brilliant Blue R-250 (44.0%) with a redox mediator (Chmelová and Ondrejovič 2016). There are reports in which laccase enzymes and laccase-mediated systems have been tested for dye decolorization. Şaşmaz et al. (2011) suggested vanillin as a laccase mediator to increase their range of substrates, and they tested laccase enzymes from *Trametes versicolor* and a laccase-mediator system for dye decolorization under optimal conditions. Rem Blue RR and Dylon Navy 17 dyes were decolorized by laccase alone, but not Rem Yellow RR, RR 198, and Rem Red RR, whereas R198 and Rem Red RR were decolorized successfully by laccase and vanillin. The authors proposed laccases and laccase mediator systems as good options for the treatment of textile wastewaters.

On the other hand, the toxic compounds that derive from industrial activities include pesticides and contaminated air, soil, and water with phenolic compounds, hydrocarbons, endocrine disruptors, etc. (Shraddha et al. 2011). Recent studies have tested fungal abilities to degrade these kinds of compounds and have been successful. Ulčnik et al. (2013) reported the purified laccase effect on the degradation of organochlorine insecticides, such as lindane and endosulfan, commonly used on food crops or as wood preservatives, which are highly toxic to aquatic organisms, as well as to humans and most animal species, and are present in rivers and lakes. The authors reported that less than 30% of lindane was degraded, whereas no endosulfan removal occurred. They suggested that although several studies on fungal remediation of lindane and endosulfan have been performed, laccases are not the main enzymes in this process. In another case, contaminants called endocrine disrupting compounds (EDCs) were found to be derivative products of the cosmetic industry or natural chemicals, such as phytoestrogens. They are dangerous contaminants due to their antagonist effects on endogenous hormones, showing alteration of physiological functions, such as reproduction and development of different species, including humans. Recently, the use of phenoloxidases for EDC elimination from aqueous effluents was proposed. In this regard, Macellaro et al. (2014) tested the ability of four laccases to act on bisphenol A, nonylphenol, methylparaben, butylparaben, and dimethylphthalate degradation. The results showed that all laccases were capable of degrading EDCs. Bisphenol A substrate was oxidized under all conditions tested. The laccase from *P. ostreatus* was the best enzyme for bioremediation in these conditions, and they have been proposed as excellent candidates for the bioremediation process.

Polycyclic aromatic hydrocarbons (PAHs) are another type of contaminant compound. They are generated from the incomplete combustion or pyrolysis of organic materials, such as coal, oil, petroleum, gas, and wood. It is possible to find PAHs in water and sediment, and they are resistant to biodegradation, showing mutagenic and carcinogenic effects. Fungi constitute a potential source for PAH biodegradation. In the case of purifying laccases, laccases from *T. versicolor*, *C. hirsutus*, *P. ostreatus*, and *Coriolopsis gallica* have been tested in PAHs, such as acenaphthene, acenaphthylene, phenantrene, fluoranthene, pyrene, anthracene, benz[α]pyrene, fluoranthene, 9-methylanthracene, 2-methylanthracene, and biphenylene degradation. All tests occurred in the presence of mediators, especially 1-hydrobenzotriazole (HBT) and 2,2'-azino-bis-(3-ethylbenzothiazoline-6-sulfonic acid) (ABTS). In these studies, there were reported rates of degradation from 40% to 100% for different PAHs. It was also observed that yellow laccases are more effective than blue laccases, which are both from *P. ostreatus*. Despite this information, it has been suggested that enzymatic biodegradation of PAHs must be significantly reviewed because many factors need to be considered at the same time (Kadri et al. 2017).

Tyrosinases—although these enzymes have been used in less proportion than the laccases, several authors have reported that they can transform different phenols and/or aromatic compounds (Agarwal et al. 2016). Mainly, the phylum Basidiomycota includes some fungi with great potential for the elimination of toxic compounds present in wastewater or contaminated soils (Martínková et al. 2016). Some fungi that have been used for the isolation of tyrosinase, such as *N. crassa*, *A. muscaria*, *L. edodes*, *A. oryzae*, *P. mushroom*, *P. sanguineus*, *L. boryana*, and *A. bisporus*, this last one is the fungus with the most reported work and analysis of their produced tyrosinases. Nowadays, the contamination of the environment by petroleum products or chloro- and nitrophenols, polycyclic aromatic hydrocarbons, organic dyes generated in the wastewater of a number of industries, such as resin and plastics, textiles, dyes, iron and steel, and pulp and paper, as well as the use of pesticides, are a serious environmental concern. In this context, the application of tyrosinase for the removal of phenol and its derivatives has become a very important and effective method (Martínková et al. 2016, Agarwal et al. 2016), because this enzyme has the ability to use atmospheric oxygen as an electron acceptor; their broad substrate specificities for mono- and diphenols makes these enzymes attractive catalysts for environmental technologies. Besides, biosensors based on tyrosinases have been designed for measuring phenols, polyphenols, and pesticides. Tyrosinase electrodes are also applied to monitor phenolic and catechol compounds.

The analysis of tyrosinases produced by fungi has revealed that, when the enzyme comes from mushroom extracts (*A. bisporus*), it is effective with different phenols. Tyrosinases have the capacity to degrade phenol in concentrations of 100 to 1552 mg L^{-1} (Pigatto et al. 2013), polyphenols (1444 mg L^{-1}) from olive oil wastewater (Pigatto et al. 2013), *p*-cresol (270 and 540 mg L^{-1}) (Xu and Yang 2013, Chmátal et al. 2015), 4-chlorophenol (320) (Xu and Yang 2013), and bisphenol A (0.1 mg L^{-1}) present in spiked samples of river water (Kampmann et al. 2014). Tyrosinase isolated from *A. bisporus* (commercial) has also been analyzed on different compounds that have the ability to degrade phenol (5.6 mg L^{-1}), *p*-cresol (6.5 mg L^{-1}), catechol (6.6 mg L^{-1}), 4-methylcatechol (7.4 mg L^{-1}) (Girelli et al. 2006), and 1-, 2-naphthol (43 mg L^{-1}) (Kimura et al. 2012). Yamada et al. (2006) found degradation at different concentrations of the compounds when analyzing different *p*-alkylphenols from aqueous solutions by combined use of mushroom tyrosinase and chitosan beads: 4-n-octyl-, 4-isopropyl-, 4-n-heptyl, 4-n-hexyl, 4-n-nonyl-, 4-sec-butyl-, 4-n-pentyl, 4-tert-pentylphenol (10, 11, 19, 53, 68, 75, 82, and 82 mg L^{-1}, respectively). Commercial tyrosinase immobilized on polyacrylonitrile beads was able to degrade bisphenol A (228 mg L^{-1}), bisphenol B (242 mg L^{-1}), bisphenol F (200 mg L^{-1}), and tetrachlorobisphenol A (366 mg L^{-1}) (Nicolucci et al. 2011). One of the advantages of these enzymes is that they can degrade different compounds in a wide range of pH values and temperatures.

Some supports used for the immobilization of this enzyme are D-sorbitol cinnamic ester (Marín-Zamora et al. 2006), derivatized glass (Girelli et al. 2006), polyacrylonitrile (Nicolucci et al. 2011), silica aerogel (Sani et al. 2011), cross-linked chitosan-nanoclay beads (Dinçer et al. 2012), diatom biosilica (Bayramoglu et al. 2013), and cross-linking combined with encapsulation in alginate (Xu and Yang 2013), which have allowed for the detection of phenolic compounds in wastewaters as well as their degradation. Kim et al. (2008) reported the development of electrodes, using tyrosinase with gold nanoparticles, for the measurement of pesticides. Due to their complexity in purification processes, tyrosinases have been scarcely applied in processes of biodegradation in spite of acting on chlorophenols, diphenols, methylphenols, and naphtanols. This type of analysis has been performed with commercial tyrosinases purified from *A. bisporus* or from fungus fruiting bodies (Martínková et al. 2016).

Other species of fungi, such as *Amylomyces rouxii* and *Rhizopus oryzae* ENHE, have been shown to have the ability to degrade pentachlorophenol, a xenobiotic compound used as a pesticide (Montiel et al. 2004, León-Santiestebán et al. 2011).

Conclusions

Polyphenoloxidase (laccase and tyrosinase) enzymes are a group of diverse enzymes that were included here due to their capability to oxidize specific substrates, despite the fact that their structures and functions are different. These different structures and functions result in differential oxidation levels for a great variety of compounds they use as substrates, making it difficult to predict their contribution on degradation. Therefore, the main challenge for the enzymatic bioremediation field is the selection of the best laccases focused on a particular compound, as well as the establishment of physicochemical conditions of operation for the bioremediation process, use of mediators, etc.

Understanding enzymatic biodegradation, in the sense of decreasing the toxicity grades of intermediate compounds evaluated in biological models, is an important step for the design of large-scale processes and for the massive production of laccases and/or tyrosinases to achieve efficient effects on bioremediation, which is a significant task that needs to be addressed.

References

Abebe, A., D. Zheng, J. Evans and M. Sugumaran. 2010. Reexamination of the mechanisms of oxidative transformation of the insect cuticular sclerotizing precursor, 1,2-dehydro-N-acetyldopamine. *Insect Biochemistry and Molecular Biology* 40 (9): 650–659. doi: https://doi.org/10.1016/j.ibmb.2010.06.005.

Agarwal, P., R. Gupta and N. Agarwal. 2016. A review on enzymatic treatment of phenols in wastewater. *Journal of Biotechnology & Biomaterials* 6: 249. doi: 10.4172/2155-952X.1000249.

Alves da Cunha, M. A., A. M. Barbosa, E. C. Giese and R. F. H. Dekker. 2003. The effect of carbohydrate carbon sources on the production of constitutive and inducible laccases by *Botryosphaeria* sp. *Journal of Basic Microbiology* 43 (5): 385–392. doi: 10.1002/jobm.200310250.

Anastasi, A., T. Coppola, V. Prigione and G. C. Varese. 2009. Pyrene degradation and detoxification in soil by a consortium of basidiomycetes isolated from compost: Role of laccases and peroxidases. *Journal of Hazardous Materials* 165 (1): 1229–1233. doi: https://doi.org/10.1016/j.jhazmat.2008.10.032.

Andersen, S. O., M. G. Peter and P. Roepstorff. 1996. Cuticular sclerotization in insects. *Comparative Biochemistry and Physiology Part B: Biochemistry and Molecular Biology* 113 (4): 689–705. doi: https://doi.org/10.1016/0305-0491(95)02089-6.

Archibald, F. and B. Roy. 1992. Production of manganic chelates by laccase from the lignin-degrading fungus *Trametes* (*Coriolus*) *versicolor. Applied and Environmental Microbiology* 58 (5): 1496–1499.

Asiegbu, F. O., S. Abu, J. Stenlid and M. Johansson. 2004. Sequence polymorphism and molecular characterization of laccase genes of the conifer pathogen *Heterobasidion annosum. Mycological Research* 108 (2): 136–148. doi: https://doi.org/10.1017/S0953756203009183.

Asina, F., I. Brzonova, K. Voeller, E. Kozliak, A. Kubátová, B. Yao and Y. Ji. 2016. Biodegradation of lignin by fungi, bacteria and laccases. *Bioresource Technology* 220: 414–424. doi: https://doi.org/10.1016/j.biortech.2016.08.016.

Balcázar-López, E., L. H. Méndez-Lorenzo, R. A. Batista-García, U. Esquivel-Naranjo, M. Ayala, V. V. Kumar, O. Savary, H. Cabana, A. Herrera-Estrella and J. L. Folch-Mallol. 2016. Xenobiotic compounds degradation by heterologous expression of a *Trametes sanguineus* laccase in *Trichoderma atroviride*. *PLOS ONE* 11 (2): e0147997. doi: 10.1371/journal.pone.0147997.

Baldrian, P. 2006. Fungal laccases – occurrence and properties. *FEMS Microbiology Reviews* 30 (2): 215–242. doi: 10.1111/j.1574-4976.2005.00010.x.

Bao, S., Z. Teng and S. Ding. 2013. Heterologous expression and characterization of a novel laccase isoenzyme with dyes decolorization potential from *Coprinus comatus*. *Molecular Biology Reports* 40 (2): 1927–1936. doi: 10.1007/s11033-012-2249-9.

Barrasa, J. M., A. T. Martínez and M. J. Martínez. 2009. Isolation and selection of novel basidiomycetes for decolorization of recalcitrant dyes. *Folia Microbiologica* 54 (1): 59. doi: 10.1007/s12223-009-0009-6.

Bartsch, E., W. Lerbs and M. Luckner. 1979. Phenol oxidase activity and pigment synthesis in conidiospores of *Penicillium cyclopium*. *Zeitschrift für allgemeine Mikrobiologie* 19 (2): 75–82. doi: 10.1002/jobm.19790190202.

Bayramoglu, G., A. Akbulut and M. Yakup Arica. 2013. Immobilization of tyrosinase on modified diatom biosilica: Enzymatic removal of phenolic compounds from aqueous solution. *Journal of Hazardous Materials* 244-245:528-536. doi: https://doi.org/10.1016/j.jhazmat.2012.10.041.

Bento, I., M. A. Carrondo and P. F. Lindley. 2006. Reduction of dioxygen by enzymes containing copper. *JBIC Journal of Biological Inorganic Chemistry* 11 (5): 539–547. doi: 10.1007/s00775-006-0114-9.

Berka, R. M., P. Schneider, E. J. Golightly, S. H. Brown, M. Madden, K. M. Brown, T. Halkier, K. Mondorf and F. Xu. 1997. Characterization of the gene encoding an extracellular laccase of *Myceliophthora thermophila* and analysis of the recombinant enzyme expressed in *Aspergillus oryzae*. *Applied and Environmental Microbiology* 63 (8): 3151–7.

Bertrand, G. 1896. Sur la presence simultanee de la laccase et de la tyrosinase dans le suc de quelques champignons. *Comptes rendus hebdomadaires des seances de l'Academie des sciences* 123: 463–465.

Bollag, J.-M. and A. Leonowicz. 1984. Comparative studies of extracellular fungal laccases. *Applied and Environmental Microbiology* 48 (4): 849–854.

Bonomo, R. P., A. M. Boudet, R. Cozzolino, E. Rizzarelli, A. M. Santoro, R. Sterjiades and R. Zappalà. 1998. A comparative study of two isoforms of laccase secreted by the "white-rot" fungus *Rigidoporus lignosus*, exhibiting significant structural and functional differences. *Journal of Inorganic Biochemistry* 71 (3): 205–211. doi: https://doi.org/10.1016/S0162-0134(98)10057-0.

Bouchilloux, S., P. McMahill and H. S. Mason. 1963. The multiple forms of mushroom tyrosinase: Purification and molecular properties of the enzymes. *Journal of Biological Chemistry* 238 (5): 1699–1707.

Bourbonnais, R. and G. Paice Michael. 1990. Oxidation of non-phenolic substrates. *FEBS Letters* 267 (1): 99–102. doi: 10.1016/0014-5793(90)80298-w.

Bourbonnais, R., M. G. Paice, I. D. Reid, P. Lanthier and M. Yaguchi. 1995. Lignin oxidation by laccase isozymes from trametes versicolor and role of the mediator 2,2'-azinobis(3-ethylbenzthiazoline-6-sulfonate) in kraft lignin depolymerization. *Applied and Environmental Microbiology* 61 (5): 1876–80.

Cabana, H., A. Ahamed and R. Leduc. 2011. Conjugation of laccase from the white rot fungus *Trametes versicolor* to chitosan and its utilization for the elimination of triclosan. *Bioresource Technology* 102 (2): 1656–1662. doi: https://doi.org/10.1016/j.biortech.2010.09.080.

Cadimaliev, D. A., V. V. Revin, N. A. Atykyan and V. D. Samuilov. 2005. Extracellular oxidases of the lignin-degrading fungus *Panus tigrinus*. *Biochemistry (Moscow)* 70 (6): 703–707. doi: 10.1007/s10541-005-0171-7.

Cambria, M. T., D. Gullotto, S. Garavaglia and A. Cambria. 2012. *In silico* study of structural determinants modulating the redox potential of *Rigidoporus lignosus* and other fungal laccases. *Journal of Biomolecular Structure and Dynamics* 30 (1): 89–101. doi: 10.1080/07391102.2012.674275.

Campos, R., A. Kandelbauer, K. H. Robra, A. Cavaco-Paulo and G. M. Gübitz. 2001. Indigo degradation with purified laccases from *Trametes hirsuta* and *Sclerotium rolfsii*. *Journal of Biotechnology* 89 (2): 131–139. doi: https://doi.org/10.1016/S0168-1656(01)00303-0.

Clutterbuck, A. J. 1972. Absence of laccase from yellow-spored mutants of *Aspergillus nidulans*. *Microbiology* 70 (3): 423–435. doi: doi:10.1099/00221287-70-3-423.

Copete, L. S., X. Chanagá, J. Barriuso, M. F. López-Lucendo, M. J. Martínez and S. Camarero. 2015. Identification and characterization of laccase-type multicopper oxidases involved in dye-decolorization by the fungus *Leptosphaerulina* sp. *BMC Biotechnology* 15: 74. doi: 10.1186/s12896-015-0192-2.

Chairin, T., T. Nitheranont, A. Watanabe, Y. Asada, C. Khanongnuch and S. Lumyong. 2013. Purification and characterization of the extracellular laccase produced by *Trametes polyzona* WR710-1 under solid-state fermentation. *Journal of Basic Microbiology* 54 (1): 35–43. doi: 10.1002/jobm.201200456.

Chen, D. M., B. A. Bastias, A. F. S. Taylor and J. W. G. Cairney. 2003. Identification of laccase-like genes in ectomycorrhizal basidiomycetes and transcriptional regulation by nitrogen in *Piloderma byssinum*. *New Phytologist* 157 (3): 547–554. doi: doi:10.1046/j.1469-8137.2003.00687.x.

Chernykh, A., N. Myasoedova, M. Kolomytseva, M. Ferraroni, F. Briganti, A. Scozzafava and L. Golovleva. 2008. Laccase isoforms with unusual properties from the basidiomycete *Steccherinum ochraceum* strain 1833. *Journal of Applied Microbiology* 105 (6): 2065–2075. doi: 10.1111/j.1365-2672.2008.03924.x.

Chmátal, M., A. Veselá, A. Rinágelová, O. Kaplan and L. Martínková. 2015. Using the tyrosinase from Agaricus bisporus and the cyanide hydratase from *Aspergillus niger* K10 in the degradation of phenol and o-, m- and p-cresol and the free cyanide from coke plant wastewater. Proceedings of the 3rd International. Conference on Chemical Technology, Mikulov, Czech Republic.

Chmelová, D. and M. Ondrejovič. 2016. Purification and characterization of extracellular laccase produced by *Ceriporiopsis subvermispora* and decolorization of triphenylmethane dyes. *Journal of Basic Microbiology* 56 (11): 1173–1182. doi: doi:10.1002/jobm.201600152.

de la Rubia, T., E. Ruiz, J. Pérez and J. Martínez. 2002. Properties of a laccase produced by *Phanerochaete flavido-alba* induced by vanillin. *Archives of Microbiology* 179 (1): 70. doi: 10.1007/s00203-002-0501-8.

Decker, H., T. Schweikardt and F. Tuczek. 2006. the first crystal structure of tyrosinase: all questions answered? *Angewandte Chemie International Edition* 45 (28): 4546–4550. doi: doi:10.1002/anie.200601255.

Dhawan, S., R. Lal and R. C. Kuhad. 2002. Ethidium bromide stimulated hyper laccase production from bird's nest fungus *Cyathus bulleri*. *Letters in Applied Microbiology* 36 (1): 64–67. doi: 10.1046/j.1472-765X.2003.01267.x.

Dinçer, A., S. Becerik and T. Aydemir. 2012. Immobilization of tyrosinase on chitosan–clay composite beads. *International Journal of Biological Macromolecules* 50 (3): 815–820. doi: https://doi.org/10.1016/j.ijbiomac.2011.11.020.

Dittmer, J. K., N. J. Patel, S. W. Dhawale and S. S. Dhawale. 1997. Production of multiple laccase isoforms by *Phanerochaete chrysosporium* grown under nutrient sufficiency. *FEMS Microbiology Letters* 149 (1): 65–70. doi: https://doi.org/10.1016/S0378-1097(97)00055-4.

Divya, L. and C. Sadasivan. 2016. *Trichoderma viride* Laccase plays a crucial role in defense mechanism against antagonistic organisms. *Frontiers in Microbiology* 7: 741. doi: 10.3389/fmicb.2016.00741.

Divya, L. M., G. K. Prasanth and C. Sadasivan. 2014. Potential of the salt-tolerant laccase-producing strain *Trichoderma viride* Pers. NFCCI-2745 from an estuary in the bioremediation of phenol-polluted environments. *Journal of Basic Microbiology* 54 (6): 542–547. doi: doi:10.1002/jobm.201200394.

Dong, J. L. and Y. Z. Zhang. 2004. Purification and characterization of two laccase isoenzymes from a ligninolytic fungus *Trametes gallica*. *Preparative Biochemistry and Biotechnology* 34 (2): 179–194. doi: 10.1081/pb-120030876.

Duarte, L. T., J. B. Tiba, M. F. Santiago, T. A. Garcia and M. T. F. Bara. 2012. Production and characterization of tyrosinase activity in *Pycnoporus sanguineus* CCT-4518 crude extract. *Brazilian Journal of Microbiology* 43: 21–29.

Duckworth, H. W. and J. E. Coleman. 1970. Physicochemical and kinetic properties of *Mushroom Tyrosinase*. *Journal of Biological Chemistry* 245 (7): 1613–1625.

Durán, N. and E. Esposito. 2000. Potential applications of oxidative enzymes and phenoloxidase-like compounds in wastewater and soil treatment: a review. *Applied Catalysis B: Environmental* 28 (2): 83–99. doi: https://doi.org/10.1016/S0926-3373(00)00168-5.

Dwivedi, U. N., P. Singh, V. P. Pandey and A. Kumar. 2011. Structure–function relationship among bacterial, fungal and plant laccases. *Journal of Molecular Catalysis B: Enzymatic* 68 (2): 117–128. doi: https://doi.org/10.1016/j.molcatb.2010.11.002.

Edens, W. A., T. Q. Goins, D. Dooley and J. M. Henson. 1999. Purification and characterization of a secreted laccase of *Gaeumannomyces graminis* var. *tritici*. *Applied and Environmental Microbiology* 65 (7): 3071–3074.

Eggert, C., U. Temp, J. F. D. Dean and K.-E. L. Eriksson. 1996. A fungal metabolite mediates degradation of non-phenolic lignin structures and synthetic lignin by laccase. *FEBS Letters* 391 (1): 144–148. doi: https://doi.org/10.1016/0014-5793(96)00719-3.

Ellström, M., F. Shah, T. Johansson, D. Ahrén, P. Persson and A. Tunlid. 2015. The carbon starvation response of the ectomycorrhizal fungus *Paxillus involutus*. *FEMS Microbiology Ecology* 91 (4): fiv027. doi: 10.1093/femsec/fiv027.

Espin, J. C., M. Morales, R. Varon, J. Tudela and F. Garciacanovas. 1995. A continuous spectrophotometric method for determining the monophenolase and diphenolase activities of apple polyphenol oxidase. *Analytical Biochemistry* 231 (1): 237–246. doi: https://doi.org/10.1006/abio.1995.1526.

Espín, J. C., M. Morales, R. Varón, J. Tudela and F. García-Cánovas. 1996. Continuous spectrophotometric method for determining monophenolase and diphenolase activities of pear polyphenoloxidase. *Journal of Food Science* 61 (6): 1177–1182. doi: 10.1111/j.1365-2621.1996.tb10955.x.

Espín, J. C. and H. J. Wichers. 1999. Kinetics of activation of latent mushroom (*Agaricus bisporus*) tyrosinase by benzyl alcohol. *Journal of Agricultural and Food Chemistry* 47 (9): 3503–3508. doi: 10.1021/jf981334z.

Fan, X., Y. Zhou, Y. Xiao, Z. Xu and Y. Bian. 2014. Cloning, expression and phylogenetic analysis of a divergent laccase multigene family in *Auricularia auricula-judae*. *Microbiological Research* 169 (5): 453–462. doi: https://doi.org/10.1016/j.micres.2013.08.004.

Fan, Y. and W. H. Flurkey. 2004. Purification and characterization of tyrosinase from gill tissue of *Portabella* mushrooms. *Phytochemistry* 65 (6): 671–678. doi: https://doi.org/10.1016/j.phytochem.2004.01.008.

Faraco, V., C. Pezzella, A. Miele, P. Giardina and G. Sannia. 2009. Bio-remediation of colored industrial wastewaters by the white-rot fungi *Phanerochaete chrysosporium* and *Pleurotus ostreatus* and their enzymes. *Biodegradation* 20 (2): 209–220. doi: 10.1007/s10532-008-9214-2.

Farnet, A., S. Criquet, S. Tagger, G. Gil and J. Le Petit. 2000. Purification, partial characterization, and reactivity with aromatic compounds of two laccases from *Marasmius quercophilus* strain 17. *Canadian Journal of Microbiology* 46 (3): 189–94. doi: 10.1139/w99-138.

Farnet, A. M., S. Criquet, E. Pocachard, G. Gil and E. Ferre. 2002. Purification of a new isoform of laccase from a *Marasmius quercophilus* strain isolated from a cork oak litter (*Quercus suber* L.). *Mycologia* 94 (5): 735–740. doi: 10.2307/3761687.

Farnet, A.-M., S. Criquet, M. Cigna, G. Gil and E. Ferré. 2004. Purification of a laccase from *Marasmius quercophilus* induced with ferulic acid: reactivity towards natural and xenobiotic aromatic compounds. *Enzyme and Microbial Technology* 34 (6): 549–554. doi: https://doi.org/10.1016/j.enzmictec.2003.11.021.

Faure, D., M. L. Bouillant and R. Bally. 1994. Isolation of *Azospirillum lipoferum* 4T Tn5 mutants affected in melanization and laccase activity. *Applied and Environmental Microbiology* 60 (9): 3413–3415.

Feng, B. Z. and P. Li. 2014. Cloning, characterization and expression of a novel laccase gene *Pclac2* from *Phytophthora capsici*. *Brazilian Journal of Microbiology* 45 (1): 351–357. doi: 10.1590/s1517-83822014005000021.

Flurkey, W. H. and J. K. Inlow. 2008. Proteolytic processing of polyphenol oxidase from plants and fungi. *Journal of Inorganic Biochemistry* 102 (12): 2160–2170. doi: https://doi.org/10.1016/j.jinorgbio.2008.08.007.

Fountoulakis, M. S., S. N. Dokianakis, M. E. Kornaros, G. G. Aggelis and G. Lyberatos. 2002. Removal of phenolics in olive mill wastewaters using the white-rot fungus *Pleurotus ostreatus*. *Water Research* 36 (19): 4735–4744. doi: https://doi.org/10.1016/S0043-1354(02)00184-7.

Fujita, Y., Y. Uraga and E. Ichisima. 1995. Molecular cloning and nucleotide sequence of the protyrosinase gene, *melO*, from *Aspergillus oryzae* and expression of the gene in yeast cells. *Biochimica et Biophysica Acta (BBA)—Gene Structure and Expression* 1261 (1): 151–154. doi: https://doi.org/10.1016/0167-4781(95)00011-5.

García-Molina, F., M. J. Peñalver, L. G. Fenoll, J. N. Rodríguez-López, R. Varón, F. García-Cánovas, and J. Tudela. 2005. Kinetic study of monophenol and o-diphenol binding to oxytyrosinase. *Journal of Molecular Catalysis B: Enzymatic* 32 (5): 185–192. doi: https://doi.org/10.1016/j.molcatb.2004.12.005.

García-Molina, F., J. L. Muñoz, R. Varón, J. N. Rodríguez-López, F. García-Cánovas and J. Tudela. 2007. A Review on Spectrophotometric Methods for Measuring the Monophenolase and Diphenolase Activities of Tyrosinase. *Journal of Agricultural and Food Chemistry* 55 (24): 9739–9749. doi: 10.1021/jf0712301.

Garcia-Molina, F., J. L. Munoz-Munoz, M. Garcia-Molina, P. A. Garcia-Ruiz, J. Tudela, F. García-Cánovas and J. N. Rodriguez-Lopez. 2010a. Melanogenesis Inhibition Due to NADH. *Bioscienc,, Biotechnolog, and Biochemistry* 74 (9): 1777–1787. doi: 10.1271/bbb.90965.

Garcia-Molina, F., J. L. Munoz-Munoz, F. Martinez-Ortiz, J. Tudela, F. García-Cánovas and J. N. Rodriguez-Lopez. 2010b. Effects of tetrahydropterines on the generation of quinones catalyzed by tyrosinase. *Bioscienc, Biotechnolog, and Biochemistry* 74 (5): 1108–1109. doi: 10.1271/bbb.90924.

García-Molina, F., J. L. Muñoz-Muñoz, F. Martínez-Ortiz, P. A. García-Ruíz, J. Tudela, F. García-Cánovas and J. N. Rodríguez-López. 2011. Tetrahydrofolic acid is a potent suicide substrate of mushroom tyrosinase. *Journal of Agricultural and Food Chemistry* 59 (4): 1383–1391. doi: 10.1021/jf1035433.

Garcia-Molina, M. d. M., J. L. Muñoz-Muñoz, F. Garcia-Molina, P. A. García-Ruiz and F. Garcia-Canovas. 2012. Action of tyrosinase on ortho-substituted phenols: possible influence on browning and melanogenesis. *Journal of Agricultural and Food Chemistry* 60 (25): 6447–6453. doi: 10.1021/jf301238q.

Garcia, T. A., M. F. Santiago and C. J. Ulhoa. 2006. Properties of laccases produced by *Pycnoporus sanguineus* Induced by 2,5-xylidine. *Biotechnology Letters* 28 (9): 633–636. doi: 10.1007/s10529-006-0026-3.

Garzillo, A. M., M. C. Colao, V. Buonocore, R. Oliva, L. Falcigno, M. Saviano, A. Santoro, R. Zappala, R. Bonomo, C. Bianco, P. Giardina, G. Palmieri and G. Sannia. 2001. Structural and Kinetic Characterization of Native Laccases from *Pleurotus ostreatus, Rigidoporus lignosus,* and *Trametes trogii. Journal of Protein Chemistry* 20 (3): 191–201. doi: 10.1023/a:1010954812955.

Gasparetti, C., E. Nordlund, J. Jänis, J. Buchert, and K. Kruus. 2012. Extracellular tyrosinase from the fungus *Trichoderma reesei* shows product inhibition and different inhibition mechanism from the intracellular tyrosinase from *Agaricus bisporus. Biochimica et Biophysica Acta (BBA) - Proteins and Proteomics* 1824 (4): 598–607. doi: https://doi.org/10.1016/j.bbapap.2011.12.012.

Gayosso-Canales, M., R. Rodríguez-Vázquez, F. J. Esparza-García and R. M. Bermúdez-Cruz. 2012. PCBs stimulate laccase production and activity in *Pleurotus ostreatus* thus promoting their removal. *Folia Microbiologica* 57 (2): 149–158. doi: 10.1007/s12223-012-0106-9.

Gerritsen, Y. A. M., C. G. J. Chapelon, and H. J. Wichers. 1994. The low-isoelectric point tyrosinase of *Agaricus bisporus* may be a glycoprotein. *Phytochemistry* 35 (3): 573–577. doi: https://doi.org/10.1016/S0031-9422(00)90563-6.

Giardina, P., G. Palmieri, A. Scaloni, B. Fontanella, V. Faraco, G. Cennamo and G. Sannia. 1999. Protein and gene structure of a blue laccase from Pleurotus ostreatus1. *Biochemical Journal* 341 (Pt 3): 655–663.

Girelli, A. M., E. Mattei and A. Messina. 2006. Phenols removal by immobilized tyrosinase reactor in on-line high performance liquid chromatography. *Analytica Chimica Acta* 580 (2): 271–277. doi: https://doi.org/10.1016/j.aca.2006.07.088.

Gomaa, O. M., and O. A. Momtaz. 2015. Copper induction and differential expression of laccase in *Aspergillus flavus. Brazilian Journal of Microbiology* 46 (1): 285–292. doi: 10.1590/s1517-838246120120118.

Gonzalez, J. C., S. C. Medina, A. Rodriguez, J. F. Osma, C. J. Alméciga-Díaz and O. F. Sánchez. 2013. Production of *Trametes pubescens* Laccase under submerged and semi-solid culture conditions on agro-industrial wastes. *PLOS ONE* 8 (9): e73721. doi: 10.1371/journal.pone.0073721.

Gorbatova, O., E. V. Stepanova and O. V. Koroleva. 2000. Certain biochemical and physicochemical properties of the inducible form of extracellular laccase from basidiomycetes *Coriolus hirsutus. Prikladnaia Biokhimiia i Mikrobiologiia* 36 (3): 272–277.

Gutiérrez, A., J. C. del Río, D. Ibarra, J. Rencoret, J. Romero, M. Speranza, S. Camarero, M. J. Martínez and Á. T. Martínez. 2006. Enzymatic removal of free and conjugated sterols forming pitch deposits in environmentally sound bleaching of eucalypt paper pulp. *Environmental Science & Technology* 40 (10): 3416–3422. doi: 10.1021/es052547p.

Halaouli, S., M. Asther, K. Kruus, L. Guo, M. Hamdi, J. C. Sigoillot, M. Asther, and A. Lomascolo. 2004. Characterization of a new tyrosinase from *Pycnoporus* species with high potential for food technological applications. *Journal of Applied Microbiology* 98 (2): 332–343. doi: 10.1111/j.1365-2672.2004.02481.x.

82   *Fungal Bioremediation: Fundamentals and Applications*

Halaouli, S., M. Asther, J.-C. Sigoillot, M. Hamdi and A. Lomascolo. 2006. Fungal tyrosinases: new prospects in molecular characteristics, bioengineering and biotechnological applications. *Journal of Applied Microbiology* 100 (2): 219–232. doi: doi:10.1111/j.1365-2672.2006.02866.x.

Han, P., J. Fan, Y. Liu, A. G. S. Cuthbertson, S. Yan, B.-L. Qiu and S. Ren. 2014. RNAi-mediated knockdown of serine protease inhibitor genes increases the mortality of *Plutella xylostella* challenged by destruxin A. *PLOS ONE* 9 (5): e97863. doi: 10.1371/journal.pone.0097863.

Harkin, J. M., M. J. Larsen and J. R. Obst. 1974. Use of syringaldazine for detection of laccase in sporophores of wood rotting fungi. *Mycologia* 66 (3): 469–476. doi: 10.2307/3758490.

Hatamoto, O., H. Sekine, E. Nakano and K. Abe. 1999. Cloning and expression of a cDNA encoding the laccase from *Schizophyllum commune*. *Bioscienc,, Biotechnolog,, and Biochemistry* 63 (1): 58–64. doi: 10.1271/bbb.63.58.

Hildén, K., T. K. Hakala, P. Maijala, T. K. Lundell and A. Hatakka. 2007. Novel thermotolerant laccases produced by the white-rot fungus *Physisporinus rivulosus*. *Applied Microbiology and Biotechnology* 77 (2): 301–309. doi: 10.1007/s00253-007-1155-x.

Hildén, K., M. R. Mäkelä, T. Lundell, J. Kuuskeri, A. Chernykh, L. Golovleva, D. B. Archer and A. Hatakka. 2013. Heterologous expression and structural characterization of two low pH laccases from a biopulping white-rot fungus *Physisporinus rivulosus*. *Applied Microbiology and Biotechnology* 97 (4): 1589–1599. doi: 10.1007/s00253-012-4011-6.

Hoegger, P. J., M. Navarro-González, S. Kilaru, M. Hoffmann, E. D. Westbrook and U. Kües. 2004. The laccase gene family in *Coprinopsis cinerea* (*Coprinus cinereus*). *Current Genetics* 45 (1): 9–18. doi: 10.1007/s00294-003-0452-x.

Hoegger, P. J., S. Kilaru, T. Y. James, J. R. Thacker and U. Kües. 2006. Phylogenetic comparison and classification of laccase and related multicopper oxidase protein sequences. *The FEBS Journal* 273 (10): 2308–2326. doi: doi:10.1111/j.1742-4658.2006.05247.x.

Hölker, U., J. Dohse and M. Höfer. 2002. Extracellular laccases in ascomycetes *Trichoderma atroviride* and *Trichoderma harzianum*. *Folia Microbiologica* 47 (4): 423–427. doi: 10.1007/bf02818702.

Hsu, C.-A., T.-N. Wen, Y.-C. Su, Z.-B. Jiang, C.-W. Chen and L.-F. Shyur. 2012. Biological degradation of anthroquinone and azo dyes by a novel laccase from *Lentinus* sp. *Environmental Science & Technology* 46 (9): 5109–5117. doi: 10.1021/es2047014.

http://genome.jgipsf.org/Agabi varbisH97 2/Agabi varbisH97 2.home.html.

Hu, D. D., R. Y. Zhang, G. Q. Zhang, H. X. Wang and T. B. Ng. 2011. A laccase with antiproliferative activity against tumor cells from an edible mushroom, white common *Agrocybe cylindracea*. *Phytomedicine* 18 (5): 374–379. doi: https://doi.org/10.1016/j.phymed.2010.07.004.

Huang, S. J., Z. M. Liu, X. L. Huang, L. Q. Guo and J. F. Lin. 2010. Molecular cloning and characterization of a novel laccase gene from a white-rot fungus *Polyporus grammocephalus* TR16 and expression in *Pichia pastoris*. *Letters in Applied Microbiology* 52 (3): 290–297. doi: 10.1111/j.1472-765X.2010.02997.x.

Inlow, J. K. 2012. Homology models of four *Agaricus bisporus* tyrosinases. *International Journal of Biological Macromolecules* 50 (1): 283–293. doi: https://doi.org/10.1016/j.ijbiomac.2011.11.010.

Ismaya, W. T., H. J. Rozeboom, M. Schurink, C. G. Boeriu, H. Wichers and B. W. Dijkstra. 2011a. Crystallization and preliminary X-ray crystallographic analysis of tyrosinase from the mushroom *Agaricus bisporus*. *Acta Crystallographica Section F: Structural Biology and Crystallization Communications* 67 (Pt 5): 575–578. doi: 10.1107/s174430911100738x.

Ismaya, W. T., H. J. Rozeboom, A. Weijn, J. J. Mes, F. Fusetti, H. J. Wichers and B. W. Dijkstra. 2011b. Crystal structure of *Agaricus bisporus* mushroom tyrosinase: Identity of the tetramer subunits and interaction with tropolone. *Biochemistry* 50 (24): 5477–5486. doi: 10.1021/bi200395t.

Janssen, G. G., T. M. Baldwin, D. S. Winetzky, L. M. Tierney, H. Wang and C. J. Murray. 2004. Selective targeting of a laccase from *Stachybotrys chartarum* covalently linked to a carotenoid-binding peptide. *The Journal of Peptide Research* 64 (1): 10–24. doi: 10.1111/j.1399-3011.2004.00150.x.

Jiménez-Atiénzar, M., M. Angeles Pedreño and F. García-Carmona. 1991. Activation of polyphenol oxidase by polyamines. *Biochemistry International* 25 (5): 861–868.

Jolivet, S., N. Arpin, H. J. Wichers and G. Pellon. 1998. *Agaricus bisporus* browning: a review. *Mycological Research* 102 (12): 1459–1483. doi: https://doi.org/10.1017/S0953756298006248.

Jolivalt, C., C. Madzak, A. Brault, E. Caminade, C. Malosse and C. Mougin. 2005. Expression of laccase IIIb from the white-rot fungus Trametes versicolor in the yeast Yarrowia lipolytica for environmental applications. *Applied Microbiology and Biotechnology* 66(4): 450–6.

Jolley, R. L., D. A. Robb and H. S. Mason. 1969. The multiple forms of mushroom tyrosinase: association-dissociation phenomena. *Journal of Biological Chemistry* 244 (6): 1593–1599.

Kadri, T., T. Rouissi, S. Kaur Brar, M. Cledon, S. Sarma, and M. Verma. 2017. Biodegradation of polycyclic aromatic hydrocarbons (PAHs) by fungal enzymes: A review. *Journal of Environmental Sciences* 51: 52–74. doi: https://doi.org/10.1016/j.jes.2016.08.023.

Kampmann, M., S. Boll, J. Kossuch, J. Bielecki, S. Uhl, B. Kleiner and R. Wichmann. 2014. Efficient immobilization of mushroom tyrosinase utilizing whole cells from *Agaricus bisporus* and its application for degradation of bisphenol A. *Water Research* 57: 295–303. doi: https://doi.org/10.1016/j.watres.2014.03.054.

Kanda, K., T. Sato, S. Ishii, H. Enei and S.-i. Ejiri. 1996. Purification and Properties of Tyrosinase Isozymes from the Gill of *Lentinus edodes* Fruiting Body. *Bioscienc, Biotechnolog, and Biochemistry* 60 (8): 1273–1278. doi: 10.1271/bbb.60.1273.

Kanteev, M., M. Goldfeder and A. Fishman. 2015. Structure–function correlations in tyrosinases. *Protein Science* 24 (9): 1360–1369. doi: doi:10.1002/pro.2734.

Karahanian, E., G. Corsini, S. Lobos and R. Vicuña. 1998. Structure and expression of a laccase gene from the ligninolytic basidiomycete *Ceriporiopsis subvermispora*. *Biochimica et Biophysica Acta (BBA)—Gene Structure and Expression* 1443 (1): 65–74. doi: https://doi.org/10.1016/S0167-4781(98)00197-3.

Kawamura-Konishi, Y., M. Tsuji, S. Hatana, M. Asanuma, D. Kakuta, T. Kawano, E. B. Mukouyama, H. Goto and H. Suzuki. 2007. Purification, Characterization, and Molecular Cloning of Tyrosinase from *Pholiota nameko*. *Bioscienc,, Biotechnolog,, and Biochemistry* 71 (7): 1752–1760. doi: 10.1271/bbb.70171.

Khammuang, S. and R. Sarnthima. 2013. Decolorization of synthetic melanins by crude laccases of *Lentinus polychrous Lév*. *Folia Microbiologica* 58 (1): 1–7. doi: 10.1007/s12223-012-0151-4.

Kiiskinen, L. L., L. Viikari and K. Kruus. 2002. Purification and characterisation of a novel laccase from the ascomycete *Melanocarpus albomyces*. *Applied Microbiology and Biotechnology* 59 (2): 198–204. doi: 10.1007/s00253-002-1012-x.

Kim, G.-Y., J. Shim, M.-S. Kang and S.-H. Moon. 2008. Optimized coverage of gold nanoparticles at tyrosinase electrode for measurement of a pesticide in various water samples. *Journal of Hazardous Materials* 156 (1): 141–147. doi: https://doi.org/10.1016/j.jhazmat.2007.12.007.

Kim, S., Y. Leem, K. Kim and T. Choi Hyoung. 2006. Cloning of an acidic laccase gene (*clac2*) from *Coprinus congregatus* and its expression by external pH. *FEMS Microbiology Letters* 195 (2): 151–156. doi: 10.1111/j.1574-6968.2001.tb10513.x.

Kimura, Y., A. Kashiwada, K. Matsuda and K. Yamada. 2012. Use of chitosan for removal of naphthols through tyrosinase-catalyzed quinone oxidation. *Journal of Applied Polymer Science* 125 (S2): E42-E50. doi: 10.1002/app.36661.

Klonowska, A., C. Gaudin, A. Fournel, M. Asso, J. Le Petit, M. Giorgi and T. Tron. 2002. Characterization of a low redox potential laccase from the basidiomycete C30. *European Journal of Biochemistry* 269 (24): 6119–6125. doi: 10.1046/j.1432-1033.2002.03324.x.

Kramer, K. J., M. R. Kanost, T. L. Hopkins, H. Jiang, Y. C. Zhu, R. Xu, J. L. Kerwin and F. Turecek. 2001. Oxidative conjugation of catechols with proteins in insect skeletal systems. *Tetrahedron* 57 (2): 385–392. doi: https://doi.org/10.1016/S0040-4020(00)00949-2.

Kucharzyk, K. H., G. Janusz, I. Karczmarczyk and J. Rogalski. 2012. Chemical Modifications of Laccase from White-Rot Basidiomycete *Cerrena unicolor*. *Applied Biochemistry and Biotechnology* 168 (7): 1989–2003. doi: 10.1007/s12010-012-9912-4.

Kües, U. and Y. Liu. 2000. Fruiting body production in basidiomycetes. *Applied Microbiology and Biotechnology* 54 (2): 141–152. doi: 10.1007/s002530000396.

Kües, U. 2015. Fungal enzymes for environmental management. *Current Opinion in Biotechnology* 33: 268–278. doi: https://doi.org/10.1016/j.copbio.2015.03.006.

Kupper, U., D. M. Niedermann, G. Travaglini and K. Lerch. 1989. Isolation and characterization of the tyrosinase gene from *Neurospora crassa*. *Journal of Biological Chemistry* 264 (29): 17250–17258.

Kurtz, M. B. and S. P. Champe. 1982. Purification and characterization of the conidial laccase of *Aspergillus nidulans*. *Journal of Bacteriology* 151 (3): 1338–1345.

Laborde, J. 1896. Sur la casse des vins C R Hebd Seanes. *Academy Science* 123: 1074–1075.

Larrondo, L. F., L. Salas, F. Melo, R. Vicuña and D. Cullen. 2003. A Novel Extracellular Multicopper Oxidase from *Phanerochaete chrysosporium* with Ferroxidase Activity. *Applied and Environmental Microbiology* 69 (10): 6257–6263. doi: 10.1128/aem.69.10.6257-6263.2003.

Laveda, F., E. Núñez-Delicado, F. García-Carmona and A. Sánchez-Ferrer. 2000. Reversible sodium dodecyl sulfate activation of latent peach polyphenol oxidase by cyclodextrins. *Archives of Biochemistry and Biophysics* 379 (1): 1–6. doi: https://doi.org/10.1006/abbi.2000.1838.

Laveda, F., E. Núñez-Delicado, F. García-Carmona and A. Sánchez-Ferrer. 2001. Proteolytic activation of latent paraguaya Peach PPO. Characterization of monophenolase activity. *Journal of Agricultural and Food Chemistry* 49 (2): 1003–1008. doi: 10.1021/jf001010m.

León-Santiestebán, H., M. Meraz, K. Wrobel and A. Tomasini. 2011. Pentachlorophenol sorption in nylon fiber and removal by immobilized *Rhizopus oryzae* ENHE. *Journal of Hazardous Materials* 190 (1): 707–712. doi: https://doi.org/10.1016/j.jhazmat.2011.03.101.

Leonowicz, A., N. Cho, J. Luterek, A. Wilkolazka, M. Wojtas-Wasilewska, A. Matuszewska, M. Hofrichter, D. Wesenberg and J. Rogalski. 2001. Fungal laccase: properties and activity on lignin. *Journal of Basic Microbiology* 41 (34): 185–227. doi: 10.1002/1521-4028(200107)41:3/4<185::aid-jobm185>3.0.co;2-t.

Leontievsky, A., N. Myasoedova, N. Pozdnyakova and L. Golovleva. 1997. 'Yellow' laccase of *Panus tigrinus* oxidizes non-phenolic substrates without electron-transfer mediators. *FEBS Letters* 413 (3): 446–448. doi: https://doi.org/10.1016/S0014-5793(97)00953-8.

Leontievsky, A. A., N. M. Myasoedova, B. P. Baskunov, C. S. Evans and L. A. Golovleva. 2000. Transformation of 2,4,6-trichlorophenol by the white rot fungi *Panus tigrinus* and *Coriolus versicolor*. *Biodegradation* 11 (5): 331–340. doi: 10.1023/a:1011154209569.

Lerch, K. 1983. *Neurospora tyrosinase*: structural, spectroscopic and catalytic properties. *Molecular and Cellular Biochemistry* 52 (2): 125–138. doi: 10.1007/bf00224921.

Lin, S. Y., S. Okuda, K. Ikeda, T. Okuno and Y. Takano. 2012. LAC2 encoding a secreted laccase is involved in appressorial melanization and conidial pigmentation in *Colletotrichum orbiculare*. *Molecular Plant-Microbe Interactions* 25 (12): 1552–1561. doi: 10.1094/mpmi-05-12-0131-r.

Liu, L., F. D. Dean Jeffrey, E. Friedman William and L. Eriksson Karl-Erik. 1994. A laccase-like phenoloxidase is correlated with lignin biosynthesis in *Zinnia elegans* stem tissues. *The Plant Journal* 6 (2): 213–224. doi: 10.1046/j.1365-313X.1994.6020213.x.

Lobos, S., J. Larraín, L. Salas, D. Cullen and R. Vicuña. 1994. Isoenzymes of manganese-dependent peroxidase and laccase produced by the lignin-degrading basidiomycete *Ceriporiopsis subvermispora*. *Microbiology* 140 (10): 2691–2698. doi: doi:10.1099/00221287-140-10-2691.

Lyashenko, A. V., N. E. Zhukhlistova, A. G. Gabdoulkhakov, Y. N. Zhukova, W. Voelter, V. N. Zaitsev, I. Bento, E. V. Stepanova, G. S. Kachalova, O. g. V. Koroleva, E. A. Cherkashyn, V. I. Tishkov, V. S. Lamzin, K. Schirwitz, E. Y. Morgunova, C. Betzel, P. F. Lindley and A. b. M. Mikhailov. 2006. Purification, crystallization and preliminary X-ray study of the fungal laccase from *Cerrena maxima*. *Acta Crystallographica Section F: Structural Biology and Crystallization Communications* 62 (Pt 10): 954–957. doi: 10.1107/s1744309106036578.

Macellaro, G., C. Pezzella, P. Cicatiello, G. Sannia and A. Piscitelli. 2014. Fungal laccases degradation of endocrine disrupting compounds. *BioMed Research International* 2014: 8. doi: 10.1155/2014/614038.

Magan, N., S. Fragoeiro, and C. Bastos. 2010. Environmental factors and bioremediation of xenobiotics using white rot fungi. *Mycobiology* 38 (4): 238–248. doi: 10.4489/myco.2010.38.4.238.

Mäkelä, M. R., K. S. Hildén, T. K. Hakala, A. Hatakka and T. K. Lundell. 2006. Expression and molecular properties of a new laccase of the white rot fungus *Phlebia radiata* grown on wood. *Current Genetics* 50 (5): 323–333. doi: 10.1007/s00294-006-0090-1.

Marc Dubernet, M. M., P. Ribéreau Gayon and M. Prévot. 1975. Properties of the laccase of *Botrytis cinerea*. *C R Acad Sci Hebd Seances Acad Sci D* 280 (10): 1313–1316.

Marín-Zamora, M. E., F. Rojas-Melgarejo, F. García-Cánovas and P. A. García-Ruiz. 2006. Direct immobilization of tyrosinase enzyme from natural mushrooms (*Agaricus bisporus*) on d-sorbitol cinnamic ester. *Journal of Biotechnology* 126 (3): 295–303. doi: https://doi.org/10.1016/j.jbiotec.2006.04.024.

Martin, C., M. Pecyna, H. Kellner, N. Jehmlich, C. Junghanns, D. Benndorf, M. von Bergen and D. Schlosser. 2007. Purification and biochemical characterization of a laccase from the aquatic fungus *Myrioconium* sp. UHH 1-13-18-4 and molecular analysis of the laccase-encoding gene. *Applied Microbiology and Biotechnology* 77 (3): 613–624. doi: 10.1007/s00253-007-1207-2.

Martinez, D., L. F. Larrondo, N. Putnam, M. D. S. Gelpke, K. Huang, J. Chapman, K. G. Helfenbein, P. Ramaiya, J. C. Detter, F. Larimer, P. M. Coutinho, B. Henrissat, R. Berka, D. Cullen and D. Rokhsar. 2004. Genome sequence of the lignocellulose degrading fungus *Phanerochaete chrysosporium* strain RP78. *Nature Biotechnology* 22:695. doi: 10.1038/nbt967.

Martinez, M. V., and J. R. Whitaker. 1995. The biochemistry and control of enzymatic browning. *Trends in Food Science & Technology* 6 (6): 195–200. doi: https://doi.org/10.1016/S0924-2244(00)89054-8.

Martínez Ortiz, F., J. Tudela Serrano, J. N. Rodríguez López, R. Varón Castellanos, J. Lozano Teruel and F. García-Cánovas. 1988. Oxidation of 3,4-dihydroxymandelic acid catalyzed by tyrosinase. *Biochimica et Biophysica Acta (BBA)—Protein Structure and Molecular Enzymology* 957 (1): 158–163. doi: https://doi.org/10.1016/0167-4838(88)90169-0.

Martínková, L., M. Kotik, E. Marková and L. Homolka. 2016. Biodegradation of phenolic compounds by Basidiomycota and its phenol oxidases: A review. *Chemosphere* 149: 373–382. doi: https://doi.org/10.1016/j.chemosphere.2016.01.022.

Marzullo, L., R. Cannio, P. Giardina, M. T. Santini and G. Sannia. 1995. Veratryl Alcohol Oxidase from *Pleurotus ostreatus* Participates in Lignin Biodegradation and Prevents Polymerization of Laccase-oxidized Substrates. *Journal of Biological Chemistry* 270 (8): 3823–3827. doi: 10.1074/jbc.270.8.3823.

Matsumoto-Akanuma, A., S. Akanuma, M. Motoi, A. Yamagishi and N. Ohno. 2014. Cloning and Characterization of Laccase DNA from the Royal Sun Medicinal Mushroom, *Agaricus brasiliensis* (Higher Basidiomycetes). *International Journal of Medicinal Mushrooms* 16 (4): 375–393. doi: 10.1615/IntJMedMushrooms.v16.i4.80.

Mayer, A. M. 2006. Polyphenol oxidases in plants and fungi: Going places? A review. *Phytochemistry* 67 (21): 2318–2331. doi: https://doi.org/10.1016/j.phytochem.2006.08.006.

Messerschmidt, A. and R. Huber. 1990. The blue oxidases, ascorbate oxidase, laccase and ceruloplasmin Modelling and structural relationships. *European Journal of Biochemistry* 187 (2): 341–352. doi: 10.1111/j.1432-1033.1990.tb15311.x.

Mikolasch, A. and F. Schauer. 2009. Fungal laccases as tools for the synthesis of new hybrid molecules and biomaterials. *Applied Microbiology and Biotechnology* 82 (4): 605–624. doi: 10.1007/s00253-009-1869-z.

Mikuni, J. and N. Morohoshi. 2006. Cloning and sequencing of a second laccase gene from the white-rot fungus *Coriolus versicolor*. *FEMS Microbiology Letters* 155 (1): 79–84. doi: 10.1111/j.1574-6968.1997.tb12689.x.

Min, K.-L., Y.-H. Kim, Y. W. Kim, H. S. Jung and Y. C. Hah. 2001. Characterization of a Novel Laccase Produced by the Wood-Rotting Fungus *Phellinus ribis*. *Archives of Biochemistry and Biophysics* 392 (2): 279–286. doi: https://doi.org/10.1006/abbi.2001.2459.

Molitoris, H. P., J. F. L. Van Breemen, E. F. J. Van Bruggen and K. Esser. 1972. The phenoloxidases of the ascomycete *Podospora anserina* X. Electron microscopic studies on the structure of laccases I, II and III. *Biochimica et Biophysica Acta (BBA)—Protein Structure* 271 (2): 286–291. doi: https://doi.org/10.1016/0005-2795(72)90202-4.

Montiel, A. M., F. J. Fernández, J. Marcial, J. Soriano, J. Barrios-González and A. Tomasini. 2004. A fungal phenoloxidase (tyrosinase) involved in pentachlorophenol degradation. *Biotechnology Letters* 26 (17): 1353–1357. doi: 10.1023/b:bile.0000045632.36401.86.

Mueangtoom, K., R. Kittl, O. Mann, D. Haltrich and R. Ludwig. 2010. Low pH dye decolorization with ascomycete *Lamprospora wrightii* laccase. *Biotechnology Journal* 5 (8): 857–870. doi: 10.1002/biot.201000120.

Muñoz-Muñoz, Jose L., F. García-Molina, Pedro A. García-Ruiz, M. Molina-Alarcón, J. Tudela, F. García-Cánovas and Jose N. Rodríguez-López. 2008. Phenolic substrates and suicide inactivation of tyrosinase: kinetics and mechanism. *Biochemical Journal* 416 (3): 431.

Muñoz-Muñoz, J. L., F. Garcia-Molina, P. A. García-Ruiz, R. Varon, J. Tudela, F. García-Cánovas and J. N. Rodriguez-Lopez. 2009. Stereospecific inactivation of tyrosinase by l- and d-ascorbic acid. *Biochimica et Biophysica Acta (BBA)—Proteins and Proteomics* 1794 (2): 244–253. doi: https://doi.org/10.1016/j.bbapap.2008.10.002.

Muñoz, C., F. Guillén, A. T. Martínez and M. J. Martínez. 1997. Induction and Characterization of Laccase in the Ligninolytic Fungus *Pleurotus eryngii*. *Current Microbiology* 34 (1): 1–5. doi: 10.1007/s002849900134.

Nakamura, M., T. Nakajima, Y. Ohba, S. Yamauchi, B. R. Lee and E. Ichishima. 2000. Identification of copper ligands in *Aspergillus oryzae* tyrosinase by site-directed mutagenesis. *Biochemical Journal* 350 (2): 537.

Ng, T. B. and H. X. Wang. 2004. A homodimeric laccase with unique characteristics from the yellow mushroom *Cantharellus cibarius*. *Biochemical and Biophysical Research Communications* 313 (1): 37–41. doi: https://doi.org/10.1016/j.bbrc.2003.11.087.

Nicolucci, C., S. Rossi, C. Menale, T. Godjevargova, Y. Ivanov, M. Bianco, L. Mita, U. Bencivenga, D. G. Mita and N. Diano. 2011. Biodegradation of bisphenols with immobilized laccase or tyrosinase on polyacrylonitrile beads. *Biodegradation* 22 (3): 673–683. doi: 10.1007/s10532-010-9440-2.

Niku-Paavola, M. L., and L. Viikari. 2000. Enzymatic oxidation of alkenes. *Journal of Molecular Catalysis B: Enzymatic* 10 (4): 435–444. doi: https://doi.org/10.1016/S1381-1177(99)00117-4.

Nitheranont, T., A. Watanabe and Y. Asada. 2011. Extracellular laccase produced by an edible basidiomycetous mushroom, *Grifola frondosa*: Purification and characterization. *Bioscienc., Biotechnolog., and Biochemistry* 75 (3): 538–543. doi: 10.1271/bbb.100790.

Nyanhongo, G. S., J. Gomes, G. Gübitz, R. Zvauya, J. S. Read and W. Steiner. 2002. Production of laccase by a newly isolated strain of *Trametes modesta*. *Bioresource Technology* 84 (3): 259–263. doi: https://doi.org/10.1016/S0960-8524(02)00044-5.

Osipov, E. M., K. M. Polyakov, T. V. Tikhonova, R. Kittl, P. V. Dorovatovskii, S. V. Shleev, V. O. Popov and R. Ludwig. 2015. Incorporation of copper ions into crystals of T2 copper-depleted laccase from *Botrytis aclada*. *Acta Crystallographica. Section, Structural Biology Communications* 71 (Pt 12): 1465–1469. doi: 10.1107/s2053230x1502052x.

Otsuka Saito, K., S. Kurose, Y. Tsujino, T. Osakai, K. Kataoka, T. Sakurai and E. Tamiya. 2013. Electrochemical characterization of a unique, "neutral" laccase from Flammulina velutipes. *Journal of Bioscience and Bioengineering* 115(2): 159–167.

Palmeiri, G., P. Giardina, L. Marzullo, B. Desiderio, G. Nittii, R. Cannio and G. Sannia. 1993. Stability and activity of a phenol oxidase from the ligninolytic fungus *Pleurotus ostreatus*. *Applied Microbiology and Biotechnology* 39 (4): 632–636. doi: 10.1007/bf00205066.

Palmieri, G., P. Giardina, C. Bianco, A. Scaloni, A. Capasso and G. Sannia. 1997. A novel white laccase from *Pleurotus ostreatus*. *Journal of Biological Chemistry* 272 (50): 31301–31307. doi: 10.1074/jbc.272.50.31301.

Palmieri, G., P. Giardina, C. Bianco, B. Fontanella and G. Sannia. 2000. Copper induction of laccase isoenzymes in the ligninolytic fungus *Pleurotus ostreatus*. *Applied and Environmental Microbiology* 66 (3): 920–924.

Palmieri, G., G. Cennamo, V. Faraco, A. Amoresano, G. Sannia and P. Giardina. 2003. Atypical laccase isoenzymes from copper supplemented *Pleurotus ostreatus* cultures. *Enzyme and Microbial Technology* 33 (2): 220–230. doi: https://doi.org/10.1016/S0141-0229(03)00117-0.

Palonen, H., M. Saloheimo, L. Viikari and K. Kruus. 2003. Purification, characterization and sequence analysis of a laccase from the ascomycete *Mauginiella* sp. *Enzyme and Microbial Technology* 33 (6): 854–862. doi: https://doi.org/10.1016/S0141-0229(03)00247-3.

Park, K. M. and P. Sang-Shin. 2008. Purification and characterization of laccase from basidiomycete fomitella fraxinea. *Journal of Microbiology and Biotechnology* 18 (4): 670–675.

Périé, F. H., G. V. B. Reddy, N. J. Blackburn, and M. H. Gold. 1998. Purification and Characterization of Laccases from the White-Rot Basidiomycete *Dichomitus squalens*. *Archives of Biochemistry and Biophysics* 353 (2): 349–355. doi: https://doi.org/10.1006/abbi.1998.0625.

Perry, C. R., M. Smith, C. H. Britnell, D. A. Wood and C. F Thurston. 1993. Identification of two laccase genes in the cultivated mushroom *Agaricus bisporus*. *Journal of General Microbiology* 139 (Pt 6): 1209–1218. doi: 10.1099/00221287-139-6-1209.

Pigatto, G., A. Lodi, B. Aliakbarian, A. Converti, R. M. G. da Silva and M. S. A. Palma. 2013. Phenol oxidation by mushroom waste extracts: A kinetic and thermodynamic study. *Bioresource Technology* 143: 678–681. doi: https://doi.org/10.1016/j.biortech.2013.06.069.

Piontek, K., M. Antorini and T. Choinowski. 2002. Crystal structure of a laccase from the fungus trametes versicolor at 1.90-Å resolution containing a full complement of coppers. *Journal of Biological Chemistry* 277 (40): 37663–9. doi: 10.1074/jbc.M204571200.

Qasemian, L., C. Billette, D. Guiral, E. Alazard, M. Moinard and A.-M. Farnet. 2012. Halotolerant laccases from *Chaetomium* sp., *Xylogone sphaerospora*, and *Coprinopsis* sp. isolated from a Mediterranean coastal area. *Fungal Biology* 116 (10): 1090–1098. doi: https://doi.org/10.1016/j.funbio.2012.08.002.

Ramsden, C. A. and P. A. Riley. 2014. Tyrosinase: The four oxidation states of the active site and their relevance to enzymatic activation, oxidation and inactivation. *Bioorganic & Medicinal Chemistry* 22 (8): 2388–2395. doi: https://doi.org/10.1016/j.bmc.2014.02.048.

Rezende, M. I., A. Barbosa, M. A.-F. Vasconcelos, D. R. Haddad and R. F. H. Dekker. 2005. Growth and production of laccases by the ligninolytic fungi, *Pleurotus ostreatus* and *Botryosphaeria rhodina*, cultured on basal medium containing the herbicide, Scepter® (imazaquin). *Journal of Basic Microbiology* 45 (6): 460–469. doi: 10.1002/jobm.200410552.

Rich, J. O., T. D. Leathers, A. M. Anderson, K. M. Bischoff and P. Manitchotpisit. 2013. Laccases from *Aureobasidium pullulans*. *Enzyme and Microbial Technology* 53 (1): 33–37. doi: https://doi.org/10.1016/j.enzmictec.2013.03.015.

Rigling, D. and N. K. Van Alfen. 1991. Regulation of laccase biosynthesis in the plant-pathogenic fungus *Cryphonectria parasitica* by double-stranded RNA. *Journal of Bacteriology* 173 (24): 8000–8003.

Riva, S. 2006. Laccases: blue enzymes for green chemistry. *Trends in Biotechnology* 24 (5): 219–226. doi: 10.1016/j.tibtech.2006.03.006.

Robb, D. A. 1984. Tyrosinase. In: *Copper Proteins and Copper Enzymes: Volume II*, edited by Rene Lontie. Boca Raton, FL: CRC Press.

Rodríguez Couto, S., E. Rosales and M. A. Sanromán. 2006. Decolourization of synthetic dyes by *Trametes hirsuta* in expanded-bed reactors. *Chemosphere* 62 (9): 1558–1563. doi: https://doi.org/10.1016/j.chemosphere.2005.06.042.

Rodríguez-López, J., P. Serna-Rodríguez, J. Tudela, R. Varón and F. Garcia-Cánovas. 1991. A continuous spectrophotometric method for the determination of diphenolase activity of tyrosinase using 3,4-dihydroxymandelic acid. *Analytical Biochemistry* 195 (2): 369–374. doi: https://doi.org/10.1016/0003-2697(91)90343-R.

Rodríguez-López, J. N., M. Bañón-Arnao, F. Martinez-Ortiz, J. Tudela, M. Acosta, R. Varón and F. García-Cánovas. 1992. Catalytic oxidation of 2,4,5-trihydroxyphenylalanine by tyrosinase: identification and evolution of intermediates. *Biochimica et Biophysica Acta (BBA)—Protein Structure and Molecular Enzymology* 1160 (2): 221–228. doi: https://doi.org/10.1016/0167-4838(92)90011-2.

Ros, J. R., J. N. Rodríguez-López and F. García-Cánovas. 1993. Effect of L-ascorbic acid on the monophenolase activity of tyrosinase. *Biochemical Journal* 295 (Pt 1): 309–312.

Ryu, S.-H., A. Y. Lee and M. Kim. 2008. Molecular characteristics of two laccase from the basidiomycete fungus *Polyporus brumalis*. *The Journal of Microbiology* 46 (1): 62–69. doi: 10.1007/s12275-007-0110-y.

Sahay, R., R. S. S. Yadav, S. Yadava and K. D. S. Yadav. 2012. A Laccase of *Fomes durissimus* MTCC-1173 and Its Role in the Conversion of Methylbenzene to Benzaldehyde. *Applied Biochemistry and Biotechnology* 166 (3): 563–575. doi: 10.1007/s12010-011-9448-z.

Salony, M., S. and V. S. Bisaria. 2006. Production and characterization of laccase from *Cyathus bulleri* and its use in decolourization of recalcitrant textile dyes. *Applied Microbiology and Biotechnology* 71 (5): 646–653. doi: 10.1007/s00253-005-0206-4.

Sánchez-Ferrer, Á., J. Neptuno Rodríguez-López, F. García-Cánovas and F. García-Carmona. 1995. Tyrosinase: a comprehensive review of its mechanism. *Biochimica et Biophysica Acta (BBA) - Protein Structure and Molecular Enzymology* 1247 (1): 1–11. doi: https://doi.org/10.1016/0167-4838(94)00204-T.

Sani, S., M. N. Mohd Muhid and H. Hamdan. 2011. Design, synthesis and activity study of tyrosinase encapsulated silica aerogel (TESA) biosensor for phenol removal in aqueous solution. *Journal of Sol-Gel Science and Technology* 59 (1): 7–18. doi: 10.1007/s10971-011-2454-3.

Sannia, G., P. Giardina, M. Luna, M. Rossi and V. Buonocore. 1986. Laccase fromPleurotus ostreatus. *Biotechnology Letters* 8 (11): 797–800. doi: 10.1007/bf01020827.

Saparrat, M. C. N., F. Guillén, A. M. Arambarri, A. T. Martínez and M. J. Martínez. 2002. Induction, Isolation, and Characterization of Two Laccases from the White Rot Basidiomycete *Coriolopsis rigida*. *Applied and Environmental Microbiology* 68 (4): 1534–1540. doi: 10.1128/aem.68.4.1534-1540.2002.

Sapmak, A., J. Kaewmalakul, J. D. Nosanchuk, N. Vanittanakom, A. Andrianopoulos, K. Pruksaphon and S. Youngchim. 2016. *Talaromyces marneffei* laccase modifies THP-1 macrophage responses. *Virulence* 7 (6): 702–717. doi: 10.1080/21505594.2016.1193275.

Şaşmaz, S., S. Gedikli, P. Aytar, G. Güngörmedi, A. Çabuk, E. Hür, A. Ünal and N. Kolankaya. 2011. Decolorization potential of some reactive dyes with crude laccase and laccase-mediated system. *Applied Biochemistry and Biotechnology* 163 (3): 346–361. doi: 10.1007/s12010-010-9043-8.

Schückel, J., A. Matura and K.-H. van Pée. 2011. One-copper laccase-related enzyme from *Marasmius* sp.: Purification, characterization and bleaching of textile dyes. *Enzyme and Microbial Technology* 48 (3): 278–284. doi: https://doi.org/10.1016/j.enzmictec.2010.12.002.

Schurink, M., W. J. H. van Berkel, H. J. Wichers and C. G. Boeriu. 2007. Novel peptides with tyrosinase inhibitory activity. *Peptides* 28 (3): 485–495. doi: https://doi.org/10.1016/j.peptides.2006.11.023.

Selinheimo, E., C. Gasparetti, M.-L. Mattinen, C. L. Steffensen, J. Buchert and K. Kruus. 2009. Comparison of substrate specificity of tyrosinases from *Trichoderma reesei* and *Agaricus bisporus*. *Enzyme and Microbial Technology* 44 (1): 1–10. doi: https://doi.org/10.1016/j.enzmictec.2008.09.013.

Seo, S.-Y., V. K. Sharma and N. Sharma. 2003. Mushroom tyrosinase: Recent prospects. *Journal of Agricultural and Food Chemistry* 51 (10): 2837–2853. doi: 10.1021/jf020826f.

Sethuraman, A., D. E. Akin and K. E. L. Eriksson. 1999. Production of ligninolytic enzymes and synthetic lignin mineralization by the bird's nest fungus *Cyathus stercoreus*. *Applied Microbiology and Biotechnology* 52 (5): 689–697. doi: 10.1007/s002530051580.

Shraddha, R. Shekher, S. Sehgal, M. Kamthania and A. Kumar. 2011. Laccase: Microbial sources, production, purification, and potential biotechnological applications. *Enzyme Research* 2011: 11. doi: 10.4061/2011/217861.

Soden, D. M. and A. D. W. Dobson. 2001. Differential regulation of laccase gene expression in *Pleurotus sajor-caju*. *Microbiology* 147 (7): 1755–1763. doi: doi:10.1099/00221287-147-7-1755.

Solomon, E. I., M. J. Baldwin and M. D. Lowery. 1992. Electronic structures of active sites in copper proteins: contributions to reactivity. *Chemical Reviews* 92 (4): 521–542. doi: 10.1021/cr00012a003.

Solomon, E. I., U. M. Sundaram and T. E. Machonkin. 1996. Multicopper oxidases and oxygenases. *Chemical Reviews* 96 (7): 2563–2606. doi: 10.1021/cr950046o.

Solomon, E. I., D. E. Heppner, E. M. Johnston, J. W. Ginsbach, J. Cirera, M. Qayyum, M. T. Kieber-Emmons, C. H. Kjaergaard, R. G. Hadt and L. Tian. 2014. Copper active sites in biology. *Chemical Reviews* 114 (7): 3659–3853. doi: 10.1021/cr400327t.

Sterjiades, R., J. F. D. Dean and K.-E. L. Eriksson. 1992. Laccase from Sycamore Maple (*Acer pseudoplatanus*) Polymerizes Monolignols. *Plant Physiology* 99 (3):1162.

Stoj, C. and D. J. Kosman. 2003. Cuprous oxidase activity of yeast Fet3p and human ceruloplasmin: implication for function. *FEBS Letters* 554 (3): 422–426. doi: 10.1016/s0014-5793(03)01218-3.

Strothkamp, K. G., R. L. Jolley and H. S. Mason. 1976. Quaternary structure of mushroom tyrosinase. *Biochemical and Biophysical Research Communications* 70 (2): 519–524. doi: https://doi.org/10.1016/0006-291X(76)91077-9.

Sugumaran, M. 2010. Chemistry of cuticular sclerotization. In: *Advances in Insect Physiology*, edited by Stephen J. Simpson, 151–209. Academic Press.

Sun, J., Q. J. Chen, Q. Q. Cao, Y. Y. Wu, L. J. Xu, M. J. Zhu, T. B. Ng, H. X. Wang and G. Q. Zhang. 2012. A Laccase with Antiproliferative and HIV-I Reverse Transcriptase Inhibitory Activities from the Mycorrhizal Fungus *Agaricus placomyces*. *Journal of Biomedicine and Biotechnology*. doi:10.1155/2012/736472.

Tamaru, H. and H. Inoue. 1989. Isolation and characterization of a laccase-derepressed mutant of *Neurospora crassa*. *Journal of Bacteriology* 171 (11): 6288–6293. doi: 10.1128/jb.171.11.6288-6293.1989.

Tetsch, L., J. Bend and U. Hölker. 2006. Molecular and enzymatic characterisation of extra- and intracellular laccases from the acidophilic ascomycete *Hortaea acidophila*. *Antonie van Leeuwenhoek* 90 (2): 183–194. doi: 10.1007/s10482-006-9064-z.

Thurston, C. F. 1994. The structure and function of fungal laccases. *Microbiology* 140 (1): 19–26. doi: doi:10.1099/13500872-140-1-19.

Tomšovský, M. and L. Homolka. 2004. Tyrosinase activity discovered in *Trametes* spp. *World Journal of Microbiology and Biotechnology* 20 (5): 529–530. doi: 10.1023/B:WIBI.0000040404.83282.5c.

Tsai, H.-F., M. H. Wheeler, Y. C. Chang and K. J. Kwon-Chung. 1999. A developmentally regulated gene cluster involved in conidial pigment biosynthesis in *Aspergillus fumigatus*. *Journal of Bacteriology* 181 (20): 6469–6477.

Ulčnik, A., I. Kralj Cigić and F. Pohleven. 2013. Degradation of lindane and endosulfan by fungi, fungal and bacterial laccases. *World Journal of Microbiology and Biotechnology* 29 (12): 2239–2247. doi: 10.1007/s11274-013-1389-y.

Uzan, E., P. Nousiainen, V. Balland, J. Sipila, F. Piumi, D. Navarro, M. Asther, E. Record and A. Lomascolo. 2010. High redox potential laccases from the ligninolytic fungi *Pycnoporus coccineus* and *Pycnoporus sanguineus* suitable for white biotechnology: from gene cloning to enzyme characterization and applications. *Journal of Applied Microbiology* 108 (6): 2199–2213. doi: 10.1111/j.1365-2672.2009.04623.x.

Valle, J. S., L. P. Vandenberghe, T. T. Santana, P. H. Almeida, A. M. Pereira, G. A. Linde, N. B. Colauto and C. R. Soccol. 2014. Optimum conditions for inducing laccase production in *Lentinus crinitus*. *Genetics and Molecular Research* 13 (4): 8544–8551. doi: http://dx.doi.org/10.4238/2014.

Valle, J. S., L. P. S. Vandenberghe, A. C. C. Oliveira, M. F. Tavares, G. A. Linde, N. B. Colauto and C. R. Soccol. 2015. Effect of different compounds on the induction of laccase production by *Agaricus blazei*. *Genetics and Molecular Research* 14 (4): 15882–15891. doi: 10.4238/2015.December.1.40.

Vieira, A. C., C. Marschalk, D. C. Biavatti, C. A. Lorscheider, R. M. Peralta and F. A. V. Seixas. 2015. Modeling Based Structural Insights into Biodegradation of the Herbicide Diuron by Laccase-1 from *Ceriporiopsis subvermispora*. *Bioinformation* 11 (5): 224–228. doi: 10.6026/97320630011224.

Viswanath, B., B. Rajesh, A. Janardhan, A. P. Kumar and G. Narasimha. 2014. Fungal Laccases and Their Applications in Bioremediation. *Enzyme Research* 2014:21. doi: 10.1155/2014/163242.

Vivekanand, V., P. Dwivedi, N. Pareek and R. P. Singh. 2011. Banana Peel: A Potential Substrate for Laccase Production by *Aspergillus fumigatus* VkJ2.4.5 in Solid-State Fermentation. *Applied Biochemistry and Biotechnology* 165 (1): 204. doi: 10.1007/s12010-011-9244-9.

Wahleithner, J. A., F. Xu, K. M. Brown, S. H. Brown, E. J. Golightly, T. Halkier, S. Kauppinen, A. Pederson and P. Schneider. 1996. The identification and characterization of four laccases from the plant pathogenic fungus *Rhizoctonia solani*. *Current Genetics* 29 (4): 395–403. doi: 10.1007/bf02208621.

Wang, H. X. and T. B. Ng. 2006. A laccase from the medicinal mushroom *Ganoderma lucidum*. *Applied Microbiology and Biotechnology* 72 (3): 508–513. doi: 10.1007/s00253-006-0314-9.

Wichers, H. J., Y. A. M. Gerritsen and C. G. J. Chapelon. 1996. Tyrosinase isoforms from the fruitbodies of *Agaricus bisporus*. *Phytochemistry* 43 (2): 333–337. doi: https://doi.org/10.1016/0031-9422(96)00309-3.

Wichers, H. J., K. Recourt, M. Hendriks, C. E. M. Ebbelaar, G. Biancone, F. A. Hoeberichts, H. Mooibroek and C. Soler-Rivas. 2003. Cloning, expression and characterisation of two tyrosinase cDNAs from *Agaricus bisporus*. *Applied Microbiology and Biotechnology* 61 (4): 336–341. doi: 10.1007/s00253-002-1194-2.

Wong, K.-S., Q. Huang, C.-H. Au, J. Wang and H.-S. Kwan. 2012. Biodegradation of dyes and polyaromatic hydrocarbons by two allelic forms of *Lentinula edodes* laccase expressed from *Pichia pastoris*. *Bioresource Technology* 104: 157–164. doi: https://doi.org/10.1016/j.biortech.2011.10.097.

Wösten, H. A. B. and J. G. H. Wessels. 2006. The Emergence of Fruiting Bodies in Basidiomycetes. In: *Growth, Differentiation and Sexuality*, edited by Ursula Kües and Reinhard Fischer, 393–414. Berlin, Heidelberg: Springer Berlin Heidelberg.

Wu, Y.-R. and D.-L. Nian. 2014. Production optimization and molecular structure characterization of a newly isolated novel laccase from *Fusarium solani* MAS2, an anthracene-degrading fungus. *International Biodeterioration & Biodegradation* 86: 382–389. doi: https://doi.org/10.1016/j.ibiod.2013.10.015.

Xu, D.-Y. and Z. Yang. 2013. Cross-linked tyrosinase aggregates for elimination of phenolic compounds from wastewater. *Chemosphere* 92 (4): 391–398. doi: https://doi.org/10.1016/j.chemosphere.2012.12.076.

Xu, L., M. Zhu, X. Chen, H. Wang and G. Zhang. 2015. A novel laccase from fresh fruiting bodies of the wild medicinal mushroom *Tricholoma matsutake*. *Acta ABP Biochimica Polonica* 62 (1): 35–40. doi: http://dx.doi.org/10.18388/abp.2014_713.

Yamada, K., T. Inoue, Y. Akiba, A. Kashiwada, K. Matsuda and M. Hirata. 2006. Removal of *p*-Alkylphenols from aqueous solutions by combined use of mushroom tyrosinase and chitosan beads. *Bioscienc, Biotechnolog, and Biochemistry* 70 (10): 2467–2475. doi: 10.1271/bbb.60205.

Yamazaki, S.-i. and S. Itoh. 2003. Kinetic Evaluation Of Phenolase Activity Of Tyrosinase Using Simplified Catalytic Reaction System. *Journal of the American Chemical Society* 125 (43): 13034–13035. doi: 10.1021/ja036425d.

Yaver, D. S. and E. J. Golightly. 1996. Cloning and characterization of three laccase genes from the white-rot basidiomycete *Trametes villosa*: genomic organization of the laccase gene family. *Gene* 181 (1): 95–102. doi: https://doi.org/10.1016/S0378-1119(96)00480-5.

Yoshida, H. 1883. LXIII.-Chemistry of lacquer (Urushi). Part I. Communication from the Chemical Society of Tokio. *Journal of the Chemical Societ,, Transactions* 43 (0): 472–486. doi: 10.1039/ct8834300472.

Yoshitake, A., Y. Katayama, M. Nakamura, Y. Iimura, S. Kawai and N. Morohoshi. 1993. N-linked carbohydrate chains protect laccase-III from proteolysis in *Coriolus versicolor*. *Microbiology* 139 (1): 179–185. doi: doi:10.1099/00221287-139-1-179.

Yuan, X., G. Tian, Y. Zhao, L. Zhao, H. Wang and T. B. Ng. 2016. Biochemical Characteristics of Three Laccase Isoforms from the Basidiomycete *Pleurotus nebrodensis*. *Molecules* 21 (2). doi: 10.3390/molecules21020203.

Zapata-Castillo, P., L. Villalonga-Santana, I. Islas-Flores, G. Rivera-Muñoz, W. Ancona-Escalante and S. Solís-Pereira. 2015. Synergistic action of laccases from *Trametes hirsuta* Bm2 improves decolourization of indigo carmine. *Letters in Applied Microbiology* 61 (3): 252–258. doi: 10.1111/lam.12451.

Zawistowski, J., C. G. Biliaderis and N. A. M. Eskin. 1991. Polyphenol oxidase. In: *Oxidative Enzymes in Foods*, edited by D. S. Robinson and N. A. Michael Eskin, 217–273. London: Elsevier Science Publishers.

Zhang, J., H. Chen, M. Chen, A. Ren, J. Huang, H. Wang, M. Zhao and Z. Feng. 2015. Cloning and functional analysis of a laccase gene during fruiting body formation in *Hypsizygus marmoreus*. *Microbiological Research* 179: 54–63. doi: https://doi.org/10.1016/j.micres.2015.06.005.

Zhao, J. and H. S. Kwan. 1999. Characterization, molecular cloning, and differential expression analysis of laccase genes from the edible mushroom *Lentinula edodes*. *Applied and Environmental Microbiology* 65 (11): 4908–4913.

Zhu, M., G. Zhang, L. Meng, H. Wang, K. Gao and T. Ng. 2016. Purification and Characterization of a white laccase with pronounced dye decolorizing ability and HIV-1 reverse transcriptase inhibitory activity from *Lepista nuda*. *Molecules* 21 (4). doi: 10.3390/molecules21040415.

Zille, A., B. Górnacka, A. Rehorek and A. Cavaco-Paulo. 2005. Degradation of azo dyes by *Trametes villosa* laccase over long periods of oxidative conditions. *Applied and Environmental Microbiology* 71 (11): 6711–6718. doi: 10.1128/aem.71.11.6711-6718.2005.

Zou, Y.-J., H.-X. Wang, T.-B. Ng, C.-Y. Huang and J.-X. Zhang. 2012. Purification and characterization of a novel laccase from the edible mushroom *Hericium coralloides*. *The Journal of Microbiology* 50 (1): 72–78. doi: 10.1007/s12275-012-1372-6.

PART II

APPLICATIONS

CHAPTER-4

Chlorophenols Removal by Fungi

H. Hugo León-Santiesteban[1],* and *Araceli Tomasini*[2]

1. Introduction

Throughout our history, the chemical sciences have played a preponderant role in the development of our civilization—upgrading our quality of life, notably. Nowadays, it is impossible to imagine our life without drugs for curing or decreasing diseases, vaccines for eradicating negative effects of viruses, antibiotics for preventing the propagation of pathogenic bacteria, fuels for generating energy, plastic polymers for manufacturing innumerable items of everyday use, preservatives for increasing the storage period of putrescible foods, pesticides for preventing, destroying, repelling, or mitigating a great variety of pests, etc.

The enormous efficiency of many of these chemical products, in their respective field of application, has generated a certain dependency, which has led to their excessive use and, in some cases, their accumulation in the environment. The organic compounds that historically have tended to accumulate easier in the environment are those of anthropogenic sources, such as plastic polymers, some drugs, and pesticides. The latter are, due to their molecular complexity and their particular physicochemical properties, slowly oxidized in environmental conditions, causing their persistence in ground and water sources.

Pesticides have been, since their production in the last century, one of the most effective compounds for preventing the proliferation of harmful organisms in field crops. Thanks to the use of pesticides, it has been possible to avoid economic losses caused by pests that invade the different crops. Epidemics caused by mosquitoes that transmit viral or bacterial pathogens were controlled with the use of pesticides. These

---

[1] Departamento de Energía, Universidad Autónoma Metropolitana-Azcapotzalco, C.P. 02200, Mexico City, Mexico.
[2] Departamento de Biotecnología, Universidad Autónoma Metropolitana-Iztapalapa, C.P. 09340, Mexico City, Mexico.
* Corresponding author: hels@azc.uam.mx

epidemics caused the death of thousands of people before pesticides production. During the last century, halogenated compounds such as chlorophenols were excessively used as pesticides.

The aromatic compounds belonging to the group of chlorophenols have three distinctive molecular characteristics: (1) they own a benzene ring as base molecule, (2) at least one of the functional groups linked to an aromatic ring is a hydroxyl, (3) they can contain from 1 to 5 chlorine atoms linked to an aromatic ring. There are tens of chlorinated aromatic compounds classified as chlorophenols, many of which are chloro-derivatives of methyl- and ethyl-phenols. However, all of them stem from only 19 main congeners that are clustered in 5 classes, depending on the number of chlorine atoms linked to the phenol structure. There are three monochlorophenol isomers, six dichlorophenol isomers, six trichlorophenol isomers, four tetrachlorophenol isomers, and one pentachlorophenol (Czaplicka 2004). Table 4.1 shows the classification of chlorophenols, as well as their IUPAC name.

Table 4.1 Classification of the 19 main chlorophenol congeners.

Type of chlorophenol	Isomers
Monochlorophenols	2-chlorophenol (2-CP o *ortho*-CP o *o*-CP) 3-chlorophenol (3-CP o *para*-CP o *p*-CP) 4-chlorophenol (4-CP o *meta*-CP o *m*-CP)
Dichlorophenols	2,3-dichlorophenol (2,3-DCP) 2,4-dichlorophenol (2,4-DCP) 2,5-dichlorophenol (2,5-DCP) 2,6-dichlorophenol (2,6-DCP) 3,4-dichlorophenol (3,4-DCP) 3,5-dichlorophenol (3,5-DCP)
Trichlorophenols	2,3,4-trichlorophenol (2,3,4-TCP) 2,3,5-trichlorophenol (2,3,5-TCP) 2,3,6-trichlorophenol (2,3,6-TCP) 2,4,5-trichlorophenol (2,4,5-TCP) 2,4,6-trichlorophenol (2,4,6-TCP) 3,4,5-trichlorophenol (3,4,5-TCP)
Tetrachlorophenols	2,3,4,5-tetrachlorophenol (2,3,4,5-TeCP) 2,3,4,6-tetrachlorophenol (2,3,4,6-TeCP) 2,3,5,6-tetrachlorophenol (2,3,5,6-TeCP)
Pentachlorophenol	2,3,4,5,6-pentachlorophenol (2,3,4,5,6-PCP or PCP)

Some fungal strains, from diverse taxonomic origins, have demonstrated to be capable of removing, degrading, even mineralizing chlorophenols in liquid and solid matrices; mainly by means of cometabolism. Fungi utilize one or multiple mechanisms to remove chlorophenols, among which stand out sorption, oxidative and reductive dechlorination, conjugation, ring cleavage, mineralization, and polymerization.

In order to better understand the biological role of fungi in the field of degradation of xenobiotic compounds, this chapter deals with the state of the art of the fungal degradation of chlorophenols and it reports the main fungal strains used to remove chlorophenols. Moreover, the review describes the degradation pathways of the more widely used chlorophenols, as well as the enzymes responsible for transforming the chlorophenols and their metabolites.

2. Physicochemical Properties of Chlorophenols

The key knowledge to understand how chlorophenols are distributed, accumulated, and degraded in the numerous substrates that constitute our environment is related to the physical and chemical properties of these compounds. Both molecular structure and content of chlorine atoms are the most important variables that can modify the physicochemical properties of a chlorophenol; temperature and pH are also factors with great impact on these properties. An abundant compendium of physicochemical properties, and some thermodynamic ones, of the main chlorophenols can be consulted in reviews published by Shiu et al. (1994) and Czaplicka (2004).

Normally, chlorophenols are in the solid state at room temperature because their melting point lies between 33° and 190°C, except 2-CP that is found in the liquid state. They are soluble in organic solvents and slightly soluble in aqueous solution. Indeed, the water solubility of chlorophenols decreases considerably when increasing the chlorine atoms on the aromatic ring. For instance, at 20°C, the average water solubility of monochlorophenols is 27 g L^{-1}, in contrast to the water solubility of PCP, which is three orders of magnitude lower. The higher chlorinated phenols (Cl $\geq$ 4) tend to be fatter soluble than the lower ones, thus these tend to accumulate more easily in hydrophobic substrates, such as organic matter. Hydrophobicity of a chlorophenol can be estimated through the octanol-water partition coefficient (Log K_{ow}); the higher the value of this coefficient, the chlorophenol is more hydrophobic. At 20°C, PCP has an average Log K_{ow} of 5.4, whereas the lower chlorinated phenols have a Log K_{ow} between 2.12 and 2.50 (Czaplicka 2004).

In solution, chlorophenols behave like weak acids; therefore, they tend to dissociate and form salts. The degree of dissociation of a chlorophenol depends on its constant *Ka*, or rather the negative logarithm of *Ka* (p*Ka*), and the pH value of the aqueous phase. If the pH value is equal to the p*Ka* value of a chlorophenol, only 50% of molecules of this chlorophenol will be dissociated, whereas the other 50% will remain without dissociation, as neutral molecules. On the other hand, if the pH value is greater than the p*Ka* value (pH > p*Ka*), dissociation and, thereby, the chlorophenol solubility will be increased considerably. Indeed, the p*Ka* value of chlorophenols tends to decrease when the degree of chlorination increases; for instance, the higher chlorinated phenols have p*Ka* values between 4.7 and 7.0, whereas the lower ones have values between 8.3 and 9.4 (Czaplicka 2004).

3. Degradation of Monochlorophenols

The presence of monochlorophenols in the environment is not due to their application *in situ* as biocides. Indeed, they have been applied as antiseptics, but this has not been their primary use. Monochlorophenols have been mostly used as precursors to synthesize other chlorophenols and chlorophenoxyacetic acids. It has been observed that a relevant source of monochlorophenol contamination is the partial biodegradation of the herbicides 2,4-dichlorophenoxacetic acid and 2,4,5-trichlorophenoxacetic acid. Although, the incomplete dechlorination of some di- and trichlorophenols have also contributed to accumulation of monochlorophenols in soil and water. During the process of water purification with chlorine, chlorophenols, such as 2-CP and 4-CP, can be formed if there are phenols as water pollutants.

Due to their high water solubility, monochlorophenols are more easily in contact with humans than the other chlorophenols. It is well-known that monochlorophenols are the least toxic chlorinated phenols, nevertheless an acute interaction with them would provoke polar narcosis (Penttinen 1995). In liquid phase, where monochlorophenols are more bioavailable, some fungi are able to oxidize, dechlorinate, and/or polymerize them. Few fungi have been able to mineralize monochlorophenols, i.e., degrade them up to CO_2. *Aspergillus awamori* NRRL3112 and *Trichosporon cutaneum* R57 immobilized in modified polyamide beads were capable of using either 2-CP, 3-CP, or 4-CP as sole carbon source (Yordanova et al. 2013). On the contrary, the yeasts *Candida tropicalis* HP 15, *Trichosporon cutaneum* KUY-6A, and the deuteromycete *Penicillium camemberti* could not metabolize any monochlorophenols without an alternative carbon source (Hasegawa et al. 1990, Krug et al. 1985, Taseli and Gokcay 2005).

Usually, fungal degradation of monochlorophenols is by co-metabolism, i.e., monochlorophenols are considered non-growth substrates, whereas either a sugar or phenol is used as a (primary) substrate for growth. Walker (1973) utilized the yeast *Rhodotorula glutinis* for the co-metabolic degradation of 3-CP and 4-CP in a phenol-containing mineral medium. *R. glutinis* oxidized both monochlorophenols to 4-chlrocatechol, nevertheless it was not able to release chlorines during oxidation. A similar biodegradation behavior of 3-CP and 4-CP was observed by Marr et al. (1996) in *Penicillium simplicissimum* SK9117 cultures with mineral salt medium supplemented with phenol; 4-chlorocatechol was a common metabolite during degradation of both monochlorophenols. Nonetheless, chlorohydroquinone, 4-chloro-1,2,3-trihydroxybenzene, and 5-chloro-1,2,3-trihydroxybenzene (5-chloropyrogallol) were also identified as metabolites from the 3-CP degradation by *P. simplicissimum*. Polnisch et al. (1991) found that the yeast *Candida maltosa* SBUG 700, in mineral medium with both glucose and phenol, was capable of oxidizing 3-CP and 4-CP, and was also able to dechlorinate their derivatives. 4-chlorocatechol, 5-chloropyrogallol (5-chloro-1,2,3-trihydroxybenzene), and 4-carboxymethylenebut-2-en-4-olide were identified as products from the degradation of 3-CP and 4-CP. Moreover, *C. maltose* formed 3-chlorocatechol and 2-chloro-cis, cis-muconic acid during the 2-CP degradation, which occurred via oxidation and *ortho* cleavage, but not by dechlorination. According to Polnisch et al. (1991), the extracellular phenol hydroxylase is the enzyme responsible for catalyzing the *ortho*-hydroxylation of monochlorophenols. A similar conclusion was drawn by Krug and Straube (1986) during degradation of monochlorophenols by the yeast *Candida tropicalis* HP 15. Figure 4.1 schematizes the monochlorophenol congeners identified during the degradation of 2-CP, 3-CP and 4-CP by several fungi.

Grey et al. (1998) correlated the *para*-oxidation of 2-CP to the extracellular laccase activity produced, in nitrogen-starved liquid cultures, by the white-rot fungus *Trametes versicolor* DSM 11269, which transformed 2-CP to 2-chloro-1,4-benzoquinone (Fig. 4.1). Laccase not only oxidizes 2-CP, but is also able to polymerize

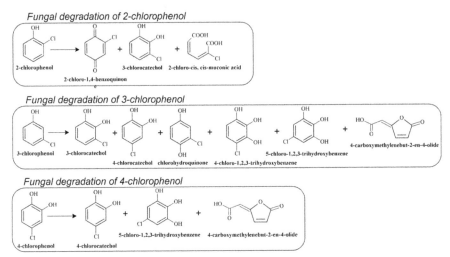

Figure 4.1 Metabolites identified during the fungal degradation of 2-CP, 3-CP, and 4-CP.

monochlorophenols, as Sjoblad and Bollag (1977) and Bollag et al. (1979) stated. Purified laccase from *Rhizoctonia praticola* was able to synthesize some dimeric, trimeric, and tetrameric compounds from 2-CP and 4-CP by means of oxidative coupling reactions (Sjoblad and Bollag 1977). Oxidative polymerization was also present during co-metabolic degradation of 4-CP by *Penicillium frequentans* Bi 7/2, in Czapek-Dox medium supplemented with either glucose or phenol as carbon and energy source. Here, this deuteromycete first oxidized 4-CP to 4-chlorocatechol, and then it polymerized 4-chlorocatechol to a dark brown polymeric product; up to 35% of chlorine covalently bound to 4-CP was released at the same time as oxidative polymerization happened. Indeed, *P. frequentans* cultured in Czapek-Dox medium was also able to degrade 2-CP and 3-CP, although no metabolites were identified (Hofrichter et al. 1992). The phenol-induced biomass of *P. frequentans* biomass (resting cells) resuspended in 0.05 M potassium-phosphate buffer (pH 7.8) remained with a catalytic activity for metabolizing monochlorophenols; without any carbon source. In potassium-phosphate buffer, 4-CP was oxidized to 4-chlorocatechol, but 35% of the latter was degraded and dechlorinated up to 4-carboxymethylene-but-2-ene-4-olide. The remaining 4-chlorocatechol (65%) was oxidatively polymerized to a dark brown product. 2-CP was mainly oxidized to 3-chlorocatechol, although traces of 2-chloromuconic acid were identified. On the other hand, 3-CP was broken down into 4-chlorocatechol, 3-chlorocatechol, and 4-carboxymethylene-but-2-ene-4-olide; in fact, 3-CP metabolism was a combination of two degradation pathways, that of 4-CP and 2-CP (Fig. 4.1) (Hofrichter et al. 1994). As reported by Hofrichter et al. (1994), at least two enzymes were related to the degradation of monochlorophenols by *P. frequentans*. In the first degradation step, phenol hydroxylase catalyzed the oxidation of monochlorophenols to chlorocatechols. In the second step, catechol-1,2-dioxygenase catalyzed the aromatic ring cleavage to form chloromuconic acid, which was probably cycled by a cic,cic-muconic acid cyclase and, subsequently, dechlorinated to yield a dienlactone, 4-carboxymethylene-but-2-ene-4-olide.

Armenante et al. (1992) found high dechlorination rates of 2-CP in bioreactors with immobilized biomass of *Phanerochaete chrysosporium* ATCC 24725. In a 10-L batch reactor with silica-based particles (JGP 73-4), as immobilization support, *P. chrysosporium* released from 73 to 96% of chlorine bound to approximately 325 mg L^{-1} 2-CP. Meanwhile, in a 4.5-L packed-bed reactor with Balsa wood chips, this basidiomycete released about 80% of chlorine present in the feed concentration of 2-CP (255 mg L^{-1}). Later, Rubio Pérez et al. (1997) showed that the monochlorophenol degradation capacity of *P. chrysosporium* depended on both the position of the chlorine atom in the aromatic ring and the biomass acclimation to monochlorophenols. The *ortho* isomer (2-CP) was more efficiently degraded than *meta* and *para* isomers (3- and 4-CP) by non-acclimated *P. chrysosporium* biomass. However, the acclimated mycelium was capable of tolerating and degrading higher concentrations of monochlorophenols than the non-acclimated one. With acclimated *P. chrysosporium* mycelium, the three chlorophenol isomers were degraded at similar rates. Table 4.2 summarizes some successful applications of fungi for degrading monochlorophenols.

4. Degradation of Dichlorophenols

The accumulation of dichlorophenols in the environment historically is due to three main causes: (1) their use as reagents for the synthesis of herbicides, for instance, the production of 2,4-dichlorophenoxyacetic acid from 2,4-dichlorophenol (2,4-DCP); (2) their appearance, as by-products, during biological and physicochemical degradation processes of chlorinated aromatic pesticides. The partial degradation of 2,4-dichlorophenoxyacetic acid has fostered the concentration of 2,4-DCP in polluted soil and water; and (3) their accidental synthesis during the chlorination processes (water disinfection and wood pulp bleaching), as well as the incineration of municipal and/or agroindustrial wastes (Wei et al. 2014).

Mineralization of dichlorophenols has been largely documented, especially with ascomycetes and basidiomycetes. Stoilova et al. (2006) observed that the ascomycete *Aspergillus awamori* NRRL 3112 is metabolically adapted to mineralize 2,4-dichlorophenol (2,4-DCP) in Czapek-Dox medium without any additional carbon source. Under non-cometabolic conditions, *A. awamori* could completely mineralize 1 and 2 g L^{-1} of 2,4-DCP in only 5 days. Nonetheless, the mineralization rate decreased up to 85% in 6 days, when the initial concentration of 2,4-DCP was increased to 3 g L^{-1}.

According to Stoilova et al. (2007), not only is *A. awamori* able to mineralize 2,4-DCP, but it is also able to metabolize other phenolic compounds, such as phenol, catechol, and 2,6-dimethoxyphenol. In Czapek-Dox medium, under cometabolic conditions (i.e., 0.5 g L^{-1} 2,4-DCP plus 0.5 g L^{-1} of another phenolic compound), *A. awamori* always metabolizes 2,4-DCP first and then the other phenolic compound; 2,4-DCP was completely exhausted on the 4th day of incubation. Immediately afterward, the phenolic compounds began to metabolize. On the 6th day of incubation, in liquid cultures with a 2,4-DCP-phenol mixture, *A. awamori* had

Table 4.2 Successful studies about fungal degradation of monochlorophenols.

Fungus	Degradation system	Chlorophenol	Concentration	Enzyme	Removal	References
P. chrysosporium*	Packed-bed reactor	2-chlorophenol	460 mg L^{-1}	UE	64–100%	Lewandowski et al. (1990)
P. chrysosporium*	Stirring tank	2-chlorophenol	520 mg L^{-1}	UE	59–97%	Lewandowski et al. (1990)
P. chrysosporium	Batch	2-chlorophenol	50–150 mg L^{-1}	UE	100%	Rubio Pérez et al. (1997)
T. versicolor	Batch	2-chlorophenol	64.3 mg L^{-1}	Laccase	100%	Grey et al. (1998)
P. camemberti	Batch	2-chlorophenol	128.6 mg L^{-1}	UE	26–53%	Taseli and Gokcay (2005)
Trametes pubescens	Batch	2-chlorophenol	15 mg L^{-1}	Possibly laccase	73–95%	González et al. (2010)
A. awamori* and *T. cutaneum**	Batch	2-chlorophenol	100–300 mg L^{-1}	UE	26.3–50%	Yordanova et al. (2013)
P. simplicissimum	Batch	3-chlorophenol	64.3 mg L^{-1}	UE	83%	Marr et al. (1996)
P. chrysosporium	Batch	3-chlorophenol	50–100 mg L^{-1}	UE	100%	Rubio Pérez et al. (1997)
A. awamori* and *T. cutaneum**	Batch	3-chlorophenol	100–300 mg L^{-1}	UE	29.3–54%	Yordanova et al. (2013)
P. simplicissimum	Batch	4-chlorophenol	64.3 mg L^{-1}	UE	100%	Marr et al. (1996)
P. chrysosporium	Batch	4-chlorophenol	50–100 mg L^{-1}	UE	100%	Rubio Pérez et al. (1997)
Candida tropicalis mutant	Batch	4-chlorophenol	400 mg L^{-1}	UE	100%	Yan et al. (2008)
P. chrysosporium*	Batch	4-chlorophenol	100 mg L^{-1}	Possibly MnP and intracellular enzymes	90–100%	Zouari et al. (2002)
P. chrysosporium*	Rotaring biological contactor	4-chlorophenol	100–200 mg L^{-1}	Possibly MnP and intracellular enzymes	80–100%	Zouari et al. (2002)
Paraconiothyrium variable	Batch	4-chlorophenol	40 mg L^{-1}	Laccase	86%	Forootanfar et al. (2012)
A. awamori* and *T. cutaneum**	Batch	4-chlorophenol	100–300 mg L^{-1}	UE	18.6–44%	Yordanova et al. (2013)

* Fungus immobilized
UE = Unidentified enzyme

degraded 24% of the initial concentration of phenol. In contrast, the ascomycete had removed 45% of the initial catechol, in cultures with a 2,4-DCP-catechol mixture. In the case of cultures with a 2,4-DCP-2,6-dimethoxyphenol mixture, only 5–6% of the initial concentration of 2,6-dimethoxyphenol had been mineralized after 6 days of incubation.

In regard to the 2,4-DCP mineralization by basidiomycetes, Valli and Gold (1991) quantified an extensive release of $^{14}CO_2$ when *P. chrysosporium* OGC101 was cultured in nitrogen-limiting medium (1.2 mM ammonium tartrate) with 0.1 Ci/µmol 2,4-[^{14}C]dichlorophenol; about 50% of the 2,4-DCP supplemented to cultures was mineralized after 30 days of incubation. Under nitrogen starvation, the ligninolytic enzymes (LiP and MnP) were partially responsible for degrading 2,4-DCP. Although, approximately 8% of the initial 2,4-DCP was mineralized in nitrogen-sufficient medium (12 mM ammonium tartrate) after 30 days of incubation.

Fahr et al. (1999) demonstrated that brown rot fungi belonging to the genus *Gloeophyllum* were capable of mineralizing 2,4-DCP in both wheat-straw solid state fermentation and liquid cultures with a mineral medium bare of carbon, nitrogen, and phosphate sources. However, the depletion of 2,4-DCP was different in both growing conditions, since, in wheat-straw cultures, 2,4-DCP was mineralized in the presence of a second substrate (cellulose), whereas in liquid cultures, 2,4-DCP was used as sole carbon and energy source. Indeed, the greatest mineralization of 2,4-DCP was quantified as $^{14}CO_2$ release, in the solid state cultures at 24°C after 6 weeks, where 53.7 ± 2.6, 33.4 ± 3.6, and 45.6 ± 1.6% of 0.1 mM 2,4-DCP (labeled with 13.5 nCi) was mineralized by *Gloeophyllum striatum* DSM 9592, *G. striatum* DSM 10335, and *Gloeophyllum trabeum* WP 11269, respectively. In contrast, in nutrient-deficient liquid cultures, the mineralization was considerably less than in solid-state cultures; only *G. striatum* DSM 9592 and *G. striatum* DSM 10335 could mineralize 2,4-DCP. After 19 days, 17.8 ± 0.2 and 21.0 ± 0.1% of 20 µM 2,4-DCP (labeled with 8.1 nCi) was degraded to $^{14}CO_2$ by *G. striatum* DSM 9592 and *G. striatum* DSM 10335, respectively. As specified by Fahr et al. (1999), none of the ligninolytic enzymes were involved in any step of the 2,4-DCP degradation by any strains of *Gloeophyllum*. Nonetheless, 4-chlorocatechol and 3,5-dichlorocatechol (3,5-DCC) were identified as breakdown products from the degradation of 2,4-DCP.

The *ortho*-oxidation of 2,4-DCP (i.e., the formation of 3,5-DCC) has also been documented as a metabolic strategy followed by *P. frequentans* Bi 7/2 to mitigate the toxic effect of 2,4-DCP. Hofrichter et al. (1994) discovered that the formation of 3,5-DCC was the first oxidation step in the degradation pathway of 2,4-DCP by the phenol-grown *P. frequentans* cultured in Czapek-Dox medium. The second step in the pathway was the methylation of 3,5-DCC, which yielded two dichloroguaiacol isomers: 4,6-dichloroguaiacol (4,6-DCG) and 3,5-dichloroguaiacol (3,5-DCG). Both biotransformation steps in the degradation pathway were catalyzed by intracellular enzymes. Figure 4.2 illustrates the degradation 2,4-DCP by *P. frequentans* Bi 7/2, *Mortierella* sp. (FERM p-17687), and fungal laccases.

Like *P. frequentans*, the soil fungus *Mortierella* sp. (FERM p-17687) has been capable of degrading 2,4-DCP via *ortho*-oxidation and methylation, forming the same metabolites stated by Hofrichter et al. (1994). However, this zygomycete has shown an alternative pathway for degrading 2,4-DCP, which involves both

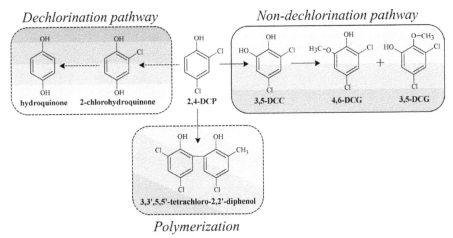

Polymerization

Figure 4.2 2,4-DCP congeners identified during the degradation of 2,4-DCP by *P. frequentans* Bi 7/2, *Mortierella* sp. (FERM p-17687), and fungal laccases.

oxidative and reductive dechlorination. Nakagawa et al. (2006) showed that the dechlorination pathway of 2,4-DCP by *Mortierella* sp. begins with the formation of 2-chlorohydroquinone via *para*-oxidation. Afterwards, 2-chlorohydroquinone is reductively dechlorinated to yield hydroquinone, which is the dead-end metabolite (Fig. 4.2, dechlorination pathway).

Valli and Gold (1991) found that the extracellular enzymes LiP and MnP were responsible for catalyzing some reactions during the degradation of 2,4-DCP by *P. chrysosporium* OGC101 in nitrogen-limited agitated cultures; particularly those reactions related to oxidations. On the other hand, the transformation of 2,4-DCP congeners, by reduction and methylation, was related to intracellular enzymes. As stated by Valli and Gold (1991), the first reaction in the degradation pathway of 2,4-DCP is the oxidative dechlorination of 2,4-DCP to 2-chloro-1,4-benzoquinone, which is catalyzed by either LiP or MnP; however, MnP oxidizes 2,4-DCP more efficiently than LiP. Then, 2-chloro-1,4-benzoquinone is transported to the intracellular environment, where it is reduced to 2-chloro-1,4-hydroquinone, which is subsequently methylated to yield 2-chloro-1,4-dimethoxybenzene. The latter metabolite is transported to the extracellular environment, where it is oxidatively dechlorinated to 2,5-dimethoxy-1,4-benzoquionone by LiP. Because 2,5-dimethoxy-1,4-benzoquionone is not a substrate for any extracellular peroxidase, this non-chlorinated metabolite is taken up by the fungal biomass and reduced to 2,5-dihydroxy-1,4-dimethoxybenzene. The metabolite 2,5-dihydroxy-1,4-dimethoxybenzene is then excreted to the extracellular medium, where it is oxidized to 2,5-dihydroxy-1,4-benzoquinone by either LiP or MnP. Since 2,5-dihydroxy-1,4-benzoquinone is not a substrate anymore for extracellular enzymes, it is sorbed into the biomass and thereupon reduced to 1,2,4,5-tetrahydroxybenzene. Finally, 1,2,4,5-tetrahydroxybenzene is ring-cleaved to produce malonic acid, which is oxidized to CO_2. Figure 4.3 exemplifies the fungal degradation pathway of 2,4-DCP, as stated by Valli and Gold (1991).

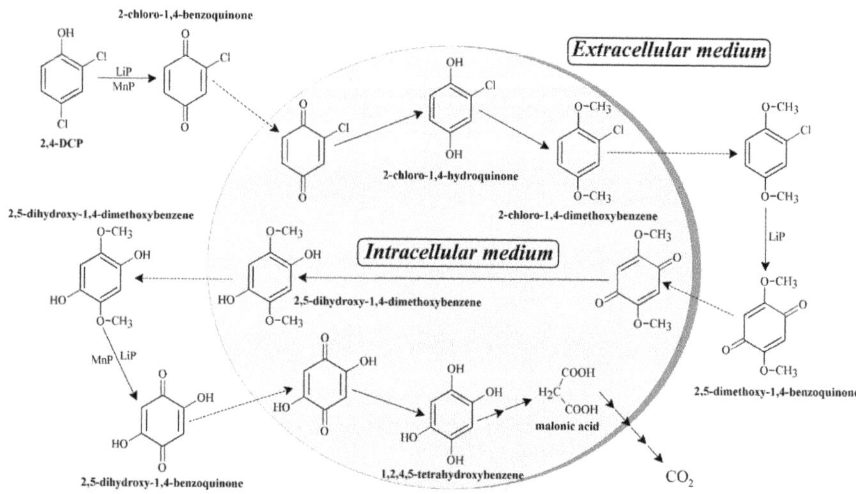

Figure 4.3 Degradation pathway of 2,4-DCP by *P. chrysosporium* OGC101.

Forootanfar et al. (2012) correlated the 2,4-DCP degradation to the production of laccase in cultures of *Paraconiothyrium variable* in Sabouraud Dextrose Broth (SDB). In fact, purified laccase from *P. variable* cultures was able to remove 72% of 100 mg L^{-1} 2,4-DCP after 5 hours of incubation at pH 5.0 and 30°C. Gaitan et al. (2011) removed 99% of 15 mg L^{-1} 2,4-DCP after 8 hours with a purified laccase from *Trametes pubescens* at pH 6.0 and 40°C. A chlorinated dimer (3,3',5,5'-tetrachloro-2,2'-diphenol) was synthetized, via oxidative coupling polymerization, when 2,4-DCP was incubated with a purified laccase from *Rhizoctonia praticola* at pH 6.9 and 23°C (Fig. 4.2, polymerization) (Sjoblad and Bollag 1977, Bollag et al. 1979). Semi-purified laccase from *Coriolus versicolor* IFO 30340 catalyzed the transformation of 50 mg L^{-1} 2,3-dichlorophenol (2,3-DCP), 2,4-DCP, and 2,6-dichlorophenol (2,6-DCP), possibly by oxidative coupling reactions, in a universal buffer at pH 5.0 and 30°C; only 16, 32, and 46% of the initial concentration of 2,3-DCP, 2,4-DCP, and 2,6-DCP were degraded, respectively, after 45 minutes of incubation (Itoh et al. 2000). Purified laccase from *Trametes* sp. strain AH 28-2 completely oxidized 100 µmol 2,6-DCP in only 6 hours (Xiao et al. 2003).

Makkar et al. (2001) managed to remove 2,5-dichlorophenol (2,5-DCP) through an enzymatic reaction with a protein complex that contained MnP, laccase, and β-glucosidase activities, in a pH 4.0 McIlvaine buffer. This 660-kDa enzyme complex was isolated from *Lentinula edodes* TMI 800 cultures with rice bran and beech sawdust, and it removed 23% of the initial concentration of 2,5-DCP (8.2 µg ml^{-1}) after 16 hours at 30°C.

Regarding the degradation of 3,4-dichlorophenol (3,4-DCP), the phenol-grown *P. frequentans* Bi 7/2 was capable of oxidizing 3,4-DCP to 4,5-dichlorocatechol (4,5-DCC) when the fungal biomass was incubated in 0.05 M potassium-phosphate buffer (pH 7.8), as well as polymerizing 4,5-DCC via oxidative coupling (Hofrichter et al. 1994). Figure 4.4 schematizes the fungal degradation of 3,4-DCP.

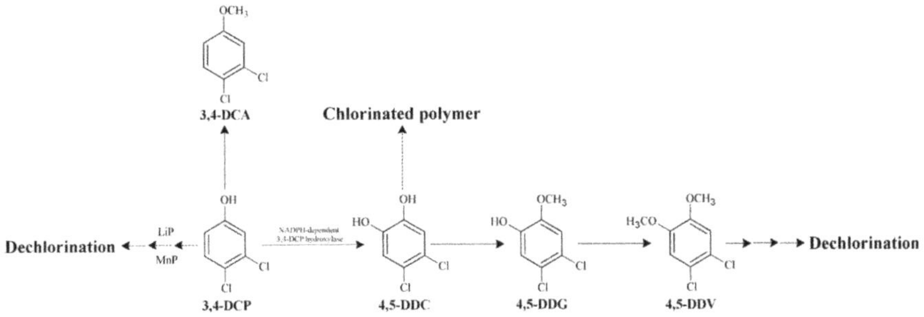

Figure 4.4 Fungal degradation pathway of 3,4-DCP.

Deschler et al. (1998) found that the NADPH-dependent 3,4-DCP hydroxylase was the intracellular enzyme responsible for catalyzing the *ortho*-oxidation of 3,4-DCP to 4,5-DCC during the growth of *P. chrysosporium* ATCC 24725 in a carbon-limiting medium with glycerol as the carbon source. In cultures under carbon starvation and with 3,4-DCP (added at the time of inoculation), the synthesis of extracellular peroxidases from *P. chrysosporium* was suppressed; therefore, only a methylated congeners of 3,4-DCP and 4,5-DCC were identified by Deschler et al. (1998). After the oxidation of 3,4-DCP, the 4,5-DCC formed was methylated to 4,5-dichloroguaiacol (4,5-DCG). In the end, 4,5-dichloroveratrol (4,5-DCV) was produced as a result of the methylation of 4,5-DCG. 3,4-dichloroanisole (3,4-DCA) was also detected as a degradation by-product, demonstrating that the methylation of 3,4-DCP is another metabolic mechanism utilized by *P. chrysosporium* to decrease the bioavailability of this dichlorophenol (Fig. 4.4). Indeed, 3,4-DCA was the main metabolite observed during the fungal growth phase.

Despite the fact that non-monochlorinated compounds were not identified during the degradation process, Deschler et al. (1998) quantified large amounts of chlorine ions when the 3,4-DCP-acclimated biomass of *P. chrysosporium* was incubated with 250 µM 3,4-DCP, or its metabolites, in 10 mM acetate buffer (pH 4.5) for 7 days. The amount of chlorine ions recovered was 64, 78, 14, and 6% for 3,4-DCP, 4,5-DCC, 4,5-DCG, and 4,5-DCV, respectively.

Furthermore, Deschler et al. (1998) stated that 3,4-DCP can also be oxidized by peroxidases secreted from *P. chrysosporium*; producing a degradation pathway similar to that of 2,4-DCP degradation reported by (Valli and Gold 1991). Duran et al. (2002) studied the repressor effect of 3,4-DCP on the enzymatic activities of peroxidases in carbon-limiting cultures of *P. chrysosporium*. The culture medium was not an inhibiting factor by itself. Indeed, the addition time of 3,4-DCP to the cultures was the key factor that either favored or inhibited the activity of those enzymes. 3,4-DCP inhibits, at biosynthesis level, the presence of peroxidases in *P. chrysosporium* cultures when it is added at the time of fungal inoculation (time cero). Nevertheless, when 3,4-DCP is added at the time of the maximum activity of peroxidases, on day 7, the activities of both LiP and MnP are stabilized and remain constant for at least 8 more days. In the latter case, 3,4-DCP did not inhibit the biosynthesis of peroxidases but decreased the activity of proteases, which favored the ligninolytic activities. Table 4.3 summarizes other outstanding works about fungal degradation of dichlorophenols.

Table 4.3 Successful studies about fungal degradation of dichlorophenols

Fungus	Degradation system	Dichlorophenol	Concentration	Enzyme	Removal	Reference
P. chrysosporium*	Batch	2,4-dichlorophenol	32.6 mg L^{-1}	LiP	40.8%	Choi et al. (2002)
T. versicolor*	Stirred tank reactor	2,4-dichlorophenol	2000 mg L^{-1}	Laccase and MnP	85%	Sedarati et al. (2003)
T. versicolor	Stirred tank reactor	2,4-dichlorophenol	2000 mg L^{-1}	Laccase and MnP	20%	Sedarati et al. (2003)
Aspergillus penicilloides	Batch	2,4-dichlorophenoxyacetic acid	100 mg L^{-1}	UE	62%	Vroumsia et al. (2005)
Mortierella isabellina	Batch	2,4-dichlorophenoxyacetic acid	100 mg L^{-1}	UE	54%	Vroumsia et al. (2005)
Chrysosporium pannorum	Batch	2,4-dichlorophenol	100 mg L^{-1}	UE	70%	Vroumsia et al. (2005)
Mucor genevensis	Batch	2,4-dichlorophenol	100 mg L^{-1}	UE	81%	Vroumsia et al. (2005)
C. versicolor	Batch	2,4-dichlorophenol	2.5–12.6 mg L^{-1}	Laccase	85–98%	Zhang et al. (2008)
T. pubescens	Batch	2,4-dichlorophenol	15 mg L^{-1}	Possibly laccase	69.6–92.4%	González et al. (2010)
A. awamori* and *T. cutaneum**	Batch	2,4-dichlorophenol	100–300 mg L^{-1}	UE	100%	Yordanova et al. (2013)
A. awamori* and *T. cutaneum**	Batch	2,4-dichlorophenol	800–1500 mg L^{-1}	UE	69.6–84.7%	Yordanova et al. (2013)

* Fungus immobilized

UE = Unidentified enzyme

5. Degradation of Trichlorophenols

There are six chlorophenols classified as trichlorophenols, but only two isomers, 2,4,5-trichloropheno (2,4,5-TCP) and 2,4,6-trichloropheno (2,4,6-TCP), are considered priority pollutants. The United States Environment Protection Agency (US EPA) has classified 2,4,5-TCP and 2,4,6-TCP within Group D (not classifiable as human carcinogen) and Group B2 (probable human carcinogen), respectively (EPA 1992a, b).

Traditionally, 2,4,5-TCP has been utilized as fungicide and herbicide, although nowadays it is mainly used as an intermediate in both the synthesis of other pesticides, such as 2,4,5-trichlorophenoxyacetic acid and Silvex® and the manufacture of the insecticide Ronnel®. On the other hand, the 2,4,6-isomer was excessively used, during the last century, as fungicide, pesticide of wood and leather, antiseptic, and glue preservative. Currently, 2,4,6-TCP is forbidden in many developed countries due to its probable carcinogen effects (EPA 1992a, b).

Fungi identified as trichlorophenol degraders have been mainly classified as either basidiomycetes or ascomycetes. Despite the taxonomic, physiological, and metabolic differences between both classes of fungi, they have in common some degradation mechanisms to detoxify trichlorophenols, among which are oxidative and reductive dechlorination, conjugation, polymerization, as well as mineralization; although, the latter mechanism has been found only in basidiomycetes.

Ligninolytic fungi have proven to be excellent degraders of 2,4,5-trichlorophenoxyacetic acid (2,4,5-T), they have been able even to mineralize it. The fungal degradation of 2,4,5-T depends on the one hand on the concentration of the nitrogen source, and on the other hand on the concentration of easily assimilable carbon sources. Ryan and Bumpus (1989) revealed that the ligninolytic-degrading system of *P. chrysosporium* BKM-F-1767 was partially responsible for cleaving the chlorinated aromatic ring of 2,4,5-T, and, thereby, causing the mineralization of the carbon atoms in the aromatic structure. However, a marginal mineralization of 2,4,5-T was measured when *P. chrysosporium* BKM-F-1767 was cultured under nitrogen-sufficient conditions; a fact that showed that non-ligninolytic enzymes were also related to the degradation of 2,4,5-T. Later, Yadav and Reddy (1992) evidenced that the non ligninolytic-degrading system of *P. chrysosporium* ATCC 34541, expressed in high nitrogen cultures, was the main responsible for mineralizing the side chain of 2,4,5-T, and not the extracellular peroxidases.

Reddy et al. (1997), during the degradation of $[^{14}C]$2,4,5-T (3×10^4 cpm/flask) in 31-day liquid cultures of *P. chrysosporium* OGC101, estimated that roughly 22% and 19% of the initial 1-^{14}C-side-chain-labelled 2,4,5-T concentration was mineralized to $^{14}CO_2$ in high carbon-high nitrogen (HCHN) and high carbon-low nitrogen (HCLN) conditions, respectively. About 4% and 13% of the initial uniformly-^{14}C-ring-labelled 2,4,5-T concentration was converted to $^{14}CO_2$ in HCHN and HCLN conditions, respectively.

Like *P. chrysosporium*, the fungus *Dichomitus squalens* is able to degrade 2,4,6-T through two degradative systems; the system that depends on ligninolytic enzymes, which is expressed during the secondary metabolism and the system that does not depend on ligninolytic enzymes, which is expressed during primary metabolism.

D. squalens produces MnP and laccase, but not LiP, in a HCHN-medium with Mn^{2+} cations. However, if Mn^{2+} cations are not added to the culture media, ligninolytic enzymes will not be generated.

Reddy et al. (1997) observed that *D. squalens* is a 2,4,5-T more efficient degrader than *P. chrysosporium*. After 31 days of incubation, in HCHN conditions, *D. squalens* mineralized 65% and 54% of the initial 1-^{14}C-side-chain-labelled 2,4,5-T concentration in a Mn^{2+}-deficient medium and a Mn^{2+}-sufficient medium, respectively. Instead, 8% and 32% of the initial uniformly-^{14}C-ring-labelled 2,4,5-T concentration were mineralized to $^{14}CO_2$ in a Mn^{2+}-deficient and a Mn^{2+}-sufficient medium, respectively.

As per Reddy et al. (1997), the non-oxidative steps in the 2,4,5-T degradation pathway in *D. squalens* are catalyzed by non-ligninolytic enzymes, whereas the oxidative steps are carried out by ligninolytic enzymes. In the first reaction of the pathway, 2,4,5-T is intracellularly cleaved to 2,4,5-TCP, which is later xylosylated to yield a xylose-conjugated 2,4,5-TCP. The latter conjugate is transported to the extracellular environment where it is hydrolyzed to 2,4,5-TCP. Finally, 2,4,5-TCP is oxidatively dechlorinated to 2,5-dichloro-1,4-benzoquinone, which is presumably ring-cleaved with subsequent oxidation to CO_2. Figure 4.5 shows the pathway for the degradation of 2,4,5-T by *D. squalens*; this pathway is equally valid for the degradation of 4-chlorophenoxyacetic acid and 2,4-dichlorophenoxyacetic acid.

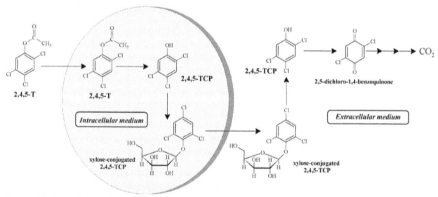

Figure 4.5 Fungal degradation pathway of 2,4,5-T proposed by Reddy et al. (1997).

The white rot fungus *P. chrysosporium* has also been proved to have a great metabolic capacity for degrading 2,4,5-TCP, mostly during its secondary metabolism. In 36-day-cultures with HCLN medium, *P. chrysosporium* OGC101 mineralized approximately 61% of the initial 2,4,5-[14C]TCP amount (10^5 cpm/flask, 0.01 µCi/µmol); only 8% of mineralization was measured when the fungus was cultured in HCHN conditions (Joshi and Gold 1993).

Joshi and Gold (1993) pointed out that, in nitrogen-limiting cultures of *P. chrysosporium* OGC101, the *para*-oxidative dechlorination of 2,4,5-TCP is the first reaction in the degradation pathway. This reaction is catalyzed by either LiP or MnP, forming 2,5-dichloro-1,4-benzoquinone, which is reduced to 2,5-dichloro-1,4-hydroquinone by either biomass-associated enzymes or non-enzymatic reactions; although the latter to a lesser extent. 2,5-dichloro-1,4-hydroquinone is then oxidized to 5-chloro-4-hydroxy-1,2-benzoquinone by MnP. In the next step, 5-chloro-4-hydroxy-1,2-benzoquinone is reduced to 5-chloro-1,2,4-trihydroxybenzene, which is oxidatively dechlorinated to form 4,5-dihydroxy-1,2-benzoquinone by MnP. Finally, 4,5-dihydroxy-1,2-benzoquinone is reduced to 1,2,4,5-tetrahydroxybenzene, which is subsequently ring-cleaved and intracellularly oxidized to CO_2. Figure 4.6 schematizes the degradation pathway of 2,4,5-TCP deduced by Joshi and Gold (1993) during the secondary metabolism of *P. chrysosporium* OGC101.

Regardless of the MnP-catalyzed oxidation of 2,5-dichloro-1,4-hydroquinone, this intermediate may be methylated to 2,5-dicloro-4-methoxyphenol and 2,5-dichloro-1,4-dimethoxyphenol by a mycelium-associated methyltransferase. The 2,5-dicloro-4-methoxyphenol is also oxidized by MnP to produce 2,5-dichloro-1,4-benzoquinone. Unfortunately, a polychlorinated biphenyl is also able to become synthetized during the oxidation of 2,4,5-TCP by either LiP or MnP (Fig. 4.6).

Regarding the detoxification of 2,4,6-TCP, Reddy et al. (1998) showed that the white-rotting fungus *P. chrysosporium* OGC101 was metabolically capable of mineralizing 2,4,6-TCP in liquid cultures under both nitrogen abundance and nitrogen starvation conditions. However, the release of $^{14}CO_2$ was more favorable in nitrogen-limiting medium than in nitrogen-sufficient one; about 58% and 8% of the initial 2,4,6-[^{14}C]trichlorophenol (10^5 cpm/flask, 0.01 µCi/µmol) was mineralized after 30 days of incubation at 37°C, respectively.

Given the metabolic limitations of most 2,4,6-TCP-degrading fungi to use this trichlorophenol as carbon and energy source, the addition of an easily assimilable carbon source to the growth media is an indispensable requirement to foster degradation. Aranciaga et al. (2012) fostered dechlorination of 2,4,6-TCP by *Penicillium chrysogenum* ERK 1 by adding, as the chief carbon source, 2 g L^{-1} of sodium acetate to the mineral salt medium. Cultured under cometabolic conditions, *P. chrysogenum* ERK 1 attained an 85% degradation of 10 mg L^{-1} of 2,4,6-TCP in a 26-day culture, forming hydroquinone and benzoquinone as the main breakdown products; non dechlorination step was related to either laccase or peroxidase activities. *P. chrysogenum* ERK 1 was unable to dechlorinate 2,4,6-TCP when glucose was supplied as carbon source.

On the other hand, Reddy et al. (1998) demonstrated that the degradation metabolism of 2,4,6-TCP by *P. chrysosporium* OGC10 is stimulated in a liquid mineral medium with a high concentration of glucose (2%, HC) and either a low concentration of ammonium tartrate (1.2 mM, LN) or a high concentration of ammonium tartrate (12 mM, HN). Nevertheless, the 2,4,6-TCP degradation efficiency, as well as the amount of by-products degradation, was higher in nitrogen-limiting cultures than in a nitrogen-rich one due to the presence of extracellular peroxidases. Armenante et al. (1994) assert that the concentration of extracellular enzymes produced by *P. chrysosporium* is the limiting factor in the degradation process of 2,4,6-TCP rather than the biomass. However, both biocatalysts must act in synergy to efficiently degrade trichlorophenol.

According to Reddy et al. (1998), the cometabolic degradation of 2,4,6-TCP, observable during the secondary metabolism of *P. chrysosporium* OGC10 under nitrogen scarcity, begins when either of the extracellular peroxidases (LiP or MnP) oxidatively dechlorinate 2,4,6-TCP to produce 2,6-dichloro-1,4-benzoquinone,

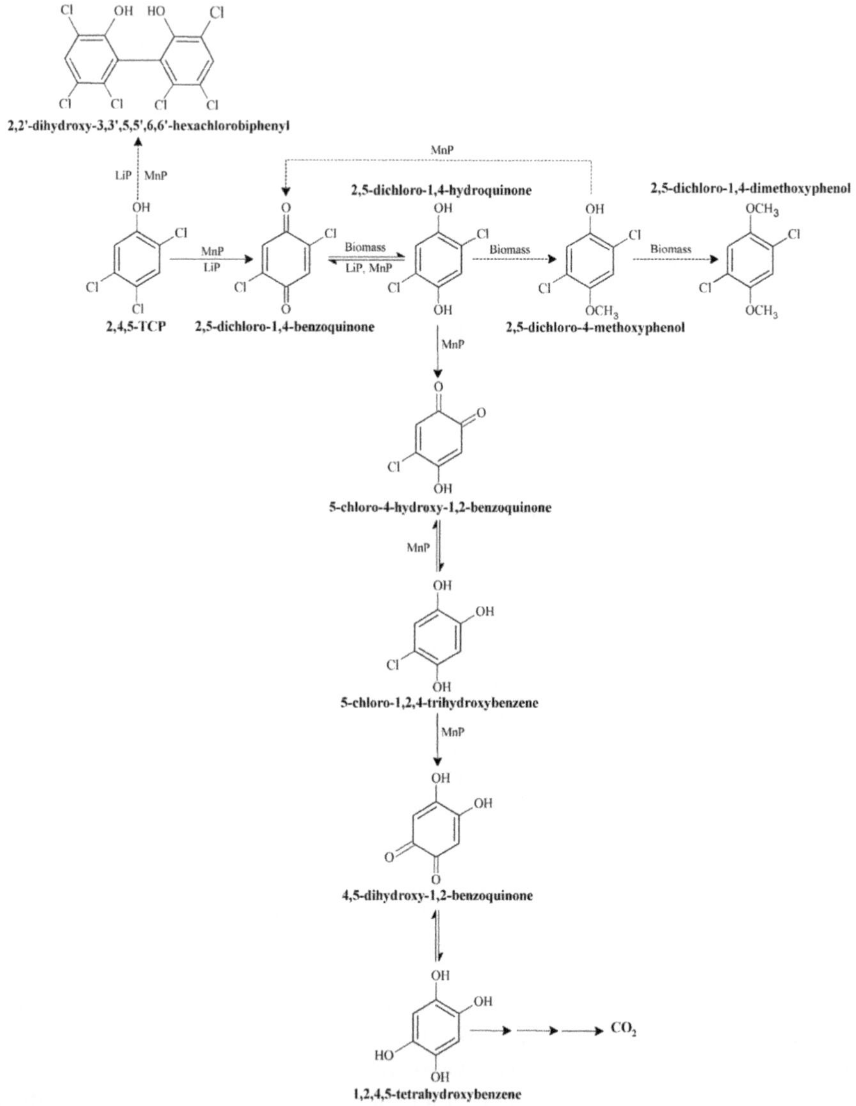

Figure 4.6 Degradation pathway of 2,4,5-TCP by *P. chrysosporium* OGC101.

a metabolite that is afterward reduced to 2,6-dichloro-1,4-dihydroxybenzene, possibly due to intracellular enzymes. Then, 2,6-dichloro-1,4-dihydroxybenzene is reductively dechlorinated to 2-chloro-1,4-dihydroxybenzene. In the next step in the degradation pathway, 2-chloro-1,4-dihydroxybenzene is either hydroxylated to 5-chloro-1,2,4-trihydroxybenzene, which undergoes reductive dechlorination to yield 1,2,4-trihydroxybenzene, or is reductively dechlorinated to 1,4-dihydroxybenzene, which undergoes hydroxylation to yield 1,2,4-trihydroxybenzene. Finally, the aromatic metabolite 1,2,4-trihydroxybenzene is ring-cleaved and oxidized to CO_2

by a set of intracellular enzymes. Figure 4.7 is a schematic representation of the degradation pathway of 2,4,6-TCP, as stated by Reddy et al. (1998).

Leontievsky et al. (2000) observed that the main by-products formed during the cometabolic degradation of 2,4,6-TCP, in mineral medium with low nitrogen concentration, by either *Panus tigrinus* 8/18 or *Coriolus* (*Trametes*) *versicolor* VKM F-116, were 2,6-dichloro-1,4-benzoquinone and 2,6-dichloro-1,4-dihydroxybenzene, as well as a chlorinated oligomeric compound (360 kDa). MnP was the enzyme responsible for oxidizing 2,4,6-TCP to dichloroquinones in maltose-supplemented cultures of *P. tigrinus* 8/18, while laccase brought about the *para*-oxidation of 2,4,6-TCP in glucose-supplemented cultures of *C. versicolor* VKM F-116. Figure 4.8 shows the degradation pathway of 2,4,6-TCP by *P. tigrinus* 8/18 and *C. versicolor* VKM F-116.

Further investigations performed by Leontievsky and coworkers (Leontievsky et al. 2001, Leontievsky et al. 2002) showed that both *C. versicolor* laccase and *P. tigrinus* MnP are also able to oxidatively dechlorinate 2,4,6-TCP in *ortho* position to yield 3,5-dichlorocatechol, which may be the key metabolite that favors the spontaneous polymerization of 2,4,6-TCP metabolites (Fig. 4.8). A chlorinated polymer, with a molecular weight of about 1000 kDa, was synthetized when both free and immobilized *C. versicolor* laccases were incubated with 400 ppm of 2,4,6-TCP at pH 5.0 and 20°C during 3 hours; a 100% removal of 2,4,6-TCP was achieved (Leontievsky et al. 2001).

Parallel to the fungal oxidation of 2,4,6-TCP, the fungal methylation of some 2,4,6-TCP congeners, including 2,4,6-TCP itself, has been widely documented. For instance, Reddy et al. (1998) quantified trace amounts of 2,6-dichloro-4-methoxyphenol, 2,6-dichloro-1,4-dimethoxybenzene, and 2,4,6-trichloromethoxybenzene (2,4,6-trichloroanisol, 2,4,6-TCA) during the degradation of 2,4,6-TCP by *P. chrysosporium* OGC10. Although, 2,4,6-TCA was only detectable under nitrogen abundance condition (Fig. 4.7). Leontievsky et al. (2001) only determined traces of 2,6-dichloro-4-methoxyphenol and 2,6-dichloro-4-methoxy-1,3-dihydroxybenzene during the degradation process of 2,4,6-TCP with *P. trigris* 8/18 and *C. versicolor* VKM F-116 (Fig. 4.8). Both Reddy et al. (1998) and Leontievsky et al. (2001) concluded that the transfer of methyl groups to 2,4,6-TCP and its congeners must be carried out by cell-associated enzymes. Later, Álvarez-Rodríguez et al. (2002) found that the mycelium-associated *S*-adenosyl-*L*-methionine (SAM)-dependent methyltransferase was the enzyme responsible for catalyzing the *ortho*-methylation of 2,4,6-TCP in cultures of *Trichoderma longibrachiatum* CECT 20431; indeed, this chlorophenol-inducible enzyme is capable of methylating other chlorophenols (Coque et al. 2003). Álvarez-Rodríguez et al. (2002) and Prak et al. (2007) stated that the formation of 2,4,6-TCA, from 2,4,6-TCP, is a distinctive feature of some fungal strains isolated from cork samples. Table 4.4 depicts some important works about fungal degradation of trichlorophenols.

6. Degradation of Tetrachlorophenols

Tetrachlorophenol (TeCP) has been produced in large scale to be used as pesticide. However, the most produced is pentachlorophenol. TeCP is mixed with

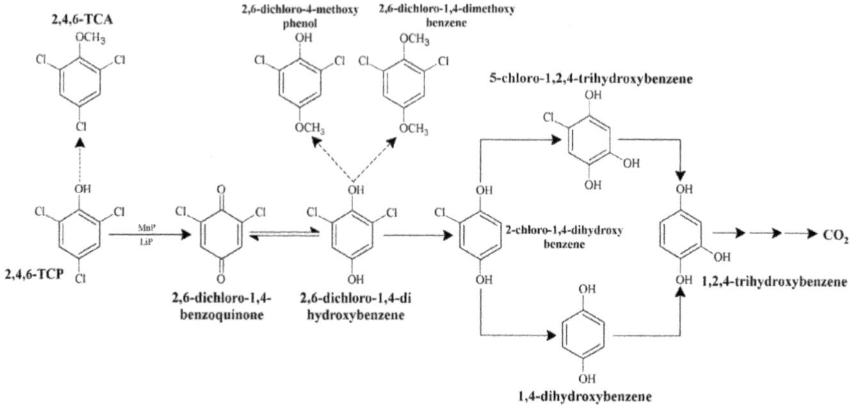

Figure 4.7 Degradation pathway of 2,4,6-TCP by *P. chrysosporium* OGC10.

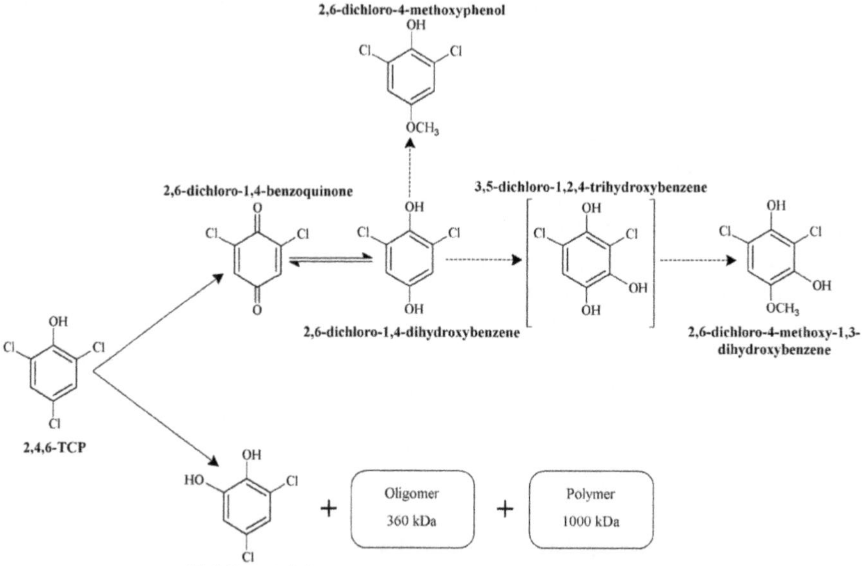

Figure 4.8 2,5,6-TCP congeners identified during the degradation of 2,4,6-TCP by *P. tigrinus* 8/18 and *C. versicolor* VKM F-116.

pentachlorophenol and the mixture is used at sawmills as wood preservative. Soil contamination by chlorophenols around sawmills has been reported in Finland. A solution containing PCP, TeCP, and TCP, called Ky-5, was widely used as wood preservative, and its use was banned in many countries since 1984 (Kitunen et al. 1987, Laine et al. 2003). Another way that TeCP became immersed in the environment, mainly soil and water, is from the degradation of pentachlorophenol. One of the most abundant metabolites produced from pentachlorophenol degradation, regardless of the process (chemical or biological) is TeCP (Czaplicka 2004).

Table 4.4 Successful studies about fungal degradation of trichlorophenols.

Fungus	Degradation system	Trichlorophenol	Concentration	Enzyme	Removal	References
*P. chrysosporium***	Batch	2,4,5-TCP	39.5 mg/L	LiP and MnP	79%	Choi et al. (2002)
*P. chrysosporium*** *T. versicolor*** *L. edones***	Pinewood chip bead-packed trickling reactor	2,4,6-TCP	43–85 mg/L	LiP, MnP and Laccase	90–98%	Ehlers and Rose (2005)
*P. chrysosporium*** *T. versicolor*** *L. edones***	Foam glass bead-packed trickling reactor	2,4,6-TCP	43–85 mg/L	LiP, MnP and Laccase	100%	Ehlers and Rose (2005)
T. pubescens	Batch	2,4,6-TCP	15 mg/L	Laccase	82%	Gaitan et al. (2011)
T. pubescens	Batch	2,4,6-TCP	15 mg/L	Laccase	37.9–38.2%	González et al. (2010)
P. tigrinus	Batch	2,4,6-TCP	25–50 mg/L	MnP	80–100%	Leontievsky et al. (2000)
C. versicolor	Batch	2,4,6-TCP	15–25 mg/L	Laccase	80–100%	Leontievsky et al. (2000)
C. versicolor	Batch	2,4,6-TCP	400 mg/L	Laccase	100%	Leontievsky et al. (2001)
C. versicolor	Flow-through column	2,4,6-TCP	400 mg/L	Immobilized Laccase	100%	Leontievsky et al. (2001)
P. tigrinus	Feed batch reactor	2,4,6-TCP	100–2000 mg/L	MnP	90–100%	Leontievsky et al. (2002)
P. chrysogenum	Batch	2,4,6-TCP	10 mg/L	UE	85%	Aranciaga et al. (2012)

* Fungus immobilized

UE = Unidentified enzyme

There are three TeCP isomers, 2,3,4,5-, 2,3,5,6-, and 2,3,4,6-tetrachlorophenol, the last one is the most common. TeCP is a recalcitrant compound. It is barely soluble in water, 0.1 mg mL^{-1}, and very soluble in ethanol, and has an octanol/water partition coefficient (log W_{ow}) of 4.45. This toxicant has been found in soil and in downstream water. Due to its low solubility in water, TeCP, as well as pentachlorophenol, can be bioaccumulated in aquatic organisms (Folke and Birklund 1986).

Human contamination by this compound occurs mainly in farmers and in manufacturing workers, as also happens with pentachlorophenol.

Studies of TeCP toxicity were conducted in rats, showing that TeCP has a hepatoxicity effect. The toxicant was administered in an oral dose, and results showed that TeCP administration caused an increment in liver weight, cell necrosis, hepatocyte hypertrophy, and vacuolation (Dodd et al. 2012). Recent studies reported the toxicity of tetrachlorophenol to willow trees, authors showed that 2,3,5,6-tetrachlorophenol was toxic to willows, *Salix viminalis*; at neutral and acid conditions, transpiration was inhibited (Clausen et al. 2018). As TeCP has been less used than pentachlorophenol, its degradation has also been scarcely studied. The microorganisms that degrade pentachlorophenol, in some cases produce TeCP as an intermediate and, in some cases, it is degraded to other compounds, as quinones and trichlorophenols, when pentachlorophenol is not completely mineralized.

Very few studies have been published about the degradation of TeCP by fungi; for example, Gee and Peel (1974) found that isolates of *Penicillium corylophilum*, *Aspergillus versicor*, and *Paecilomyces varioti* methylated 2,3,4,6-TCP and produced 2,3,4,6-tetrachloroanisole. These authors reported methylation of approximately 80%. They proposed that TeCP could be degraded by two different pathways, one of them is methylation. TeCP and pentachlorophenol degradation by fungi was studied simulating natural conditions. Experiments were carried out using non-sterile soil and wood wastes taken from a contaminated sawmill area. Other authors found nine fungi, among them *Stropharia rugosoannulata*, *Merillius tremellosus*, and *Agrocybe praecox*, that could degrade 50% of the initial TeCP and PCP in wood. They reported that methylation was negligible (Valentín et al. 2013).

Degradation and removal TeCP have not been studied much, because it has hardly been used, and it is not a common intermediate from PCP degradation, as will be mentioned in the following section.

7. Degradation of Pentachlorophenol

Pentachlorophenol (PCP) is a xenobiotic compound that was synthetized, for the first time, in 1841. Its production and commercialization began in 1936 by Dow Chemical Company and Monsanto Chemical Company, USA (Fisher 1991). PCP could be synthetized by two processes, the first by direct chlorination of the phenol molecule, described by Boehringer, and the second one by alkaline hydrolysis of hexachlorobenzene, described by Dow (Williams 1982). During the phenol chlorination process, some intermediate molecules are formed, such as tri- and tetrachlorophenols, dioxins, and furans (Johnson et al. 1973).

PCP is a weak acid and could be an anion when it occurs as sodium pentachlorophenate. The acid PCP is soluble in organic solvents, and the salt of sodium pentachlorophenate is soluble in water. It is very stable even at high temperatures, like 300°C, because of the binding of chlorine atoms to phenol; the hydroxyl groups are not reactive. This characteristic makes PCP a recalcitrant compound, which means it is not easy to degrade.

PCP has been widely used as herbicide, bactericide, fungicide, algicide, defoliant, germicide, and molluscicide. Their main use is as a wood preserver. PCP can be released to the environment in various ways. It could be present in industrial wastewater and be deposited in sediments or consumed by aquatic organisms. Once inside the organisms, it is deposited in their adipose cells and this is how humans can ingest it. PCP can contaminate soils as a result of the use of pesticides, washing of treated woods, industrial spills, and hazardous waste containment sites. Plants can adsorb PCP and store it in their roots, hence, they can be consumed by livestock and reach humans through the food chain; this is another form of ingesting this compound by humans (Fisher 1991).

The skin of humans can adsorb PCP, or it can be inhaled when it is found in the air. These are other forms of human intoxication with this compound (Horstman et al. 1989).

Prolonged exposure to PCP can cause conjunctivitis, chronic sinusitis, chronic respiratory tract diseases, chronic skin diseases, and increased temperature. People occasionally exposed to PCP, like those living in houses where PCP was used as a wood preservative, present numerous symptoms, ranging from "mild" to "severe"; among the most frequent are dizziness, headache, irritation of the respiratory tract, abdominal pains, neurological disorders, and skin diseases (Seiler 1991).

At the cellular level, the toxic effect of PCP is related to the decrease in the synthesis of ATP through the decoupling of oxidative phosphorylation (Weinbach 1954, 1957, Escher et al. 1996). It has also been proven that it can form adducts with nucleosides and nucleotides in DNA (Vaidyanathan et al. 2007). It is for this latter cytotoxic activity, that PCP is classified as a probable human carcinogen (Group B2) by the US EPA (EPA 1992c, IARC 1991).

Due to its high toxicity, the use of PCP as a biocide agent has been prohibited since 1988 in several countries (Table 4.5). However, its use is allowed in some cases, as a preservative of wood (EPA 1990). Because of the recalcitrant characteristic of PCP and that it had been used indiscriminately, many places in the world, including water and soil, are still contaminated by PCP, although its use is currently restricted.

Pentachlorophenol degradation has been studied for several years. PCP degradation could occur by a physicochemical process, such as photodegradation or volatilization, but the degradation rate is too slow and low. Biodegradation has been extensively reported; it could be either anaerobic or aerobic, by bacteria, algae, or fungi. Fungi do not use PCP as carbon or energy source, as bacteria do; they are capable of degrading PCP in co-metabolism conditions; usually, peroxidases, phenoloxidases, and cytochrome are involved in the degradation process.

Fungal PCP biodegradation has been studied widely, and published works have reported different aspects. For example, the PCP degradation pathway and the intermediates produced by degradation (Lamar et al. 1990, Lin et al. 1990, Bosso and Cristinzio 2014) have been reported, as well as the isolation and identification of strains able to tolerate and degrade PCP have been accomplished (Seigle-Murandi et al. 1992, Tomasini et al. 2001, Szewczyk et al. 2003, Rubilar et al. 2007, Carvalho et al. 2011). Studies on the genetics and proteomics of fungi capable to degrade PCP have also been carried out. These works include studies on the peroxidase- and phenoloxidase-encoding gene involved in PCP degradation, as well as on the enzymes expressed during PCP degradation by fungi (Jensen et al. 2002, Wymelenberg et al. 2006, Johansson et al. 2002, Alves et al. 2004, Carvalho et al. 2013). Some works reported the possible application of PCP degradation by fungi, including submerged and solid cultures; other works reported the use of immobilized enzymes to degrade PCP and the use of different types of bioreactors (Marcial et al. 2006, Jiang et al. 2006, Field and Sierra-Alvarez 2007). The first studies about PCP degradation by fungi were published in the 1980s, using the basidiomycete *P. chrysosporium*, showing PCP mineralization by this fungus and the role of ligninolytic enzymes (Mileski et al. 1988). Lin et al. (1990) proposed that PCP mineralization by *P. chrysosporium* could be due to extracellular and ligninolytic enzymes, as well as by intracellular enzymes.

Reddy and Gold (2000) proposed a PCP degradation pathway by *P. chrysosporium* under nitrogen starvation, and showed that the first intermediates are tetrachloro-1,4-dihydroxybenzene (tetrachlorohydroquinone), tetrachloro-*p*-benzoquinone, and traces of pentachloroanisole. Under non-nitrogen starvation conditions, the principal metabolite is pentachloroanisole; tetrachlorohydroquinone is dechlorinated by reduction reactions to produce trichlorohydroquinone (2,3,5-trichloro-1,4-

Table 4.5 Legal use of PCP in some countries.

Legal status	Country	Year	Allowed Used	References
Prohibited	Austria	1991	Research only	FAO (1996)
	India	1991	None	
	Indonesia	1980	None	
	New Zealand	1991	None	
	Sweden	1978	None	
	Switzerland	1988	None	
	Germany	1989	Research only	Heudorf et al. (2000)
Rigorously restricted	EU/EEA	1991–2	Wood preservation for industrial use and in the treatment of cultural buildings and of historical interest	FAO (1996) Muir and Eduljee (1999)
	Belize	1985	Wood preservation	FAO (1996)
	China	1997	Wood germicide	Ge et al. (2007)
	United States	1984	Wood preservation for industrial use	U.S. EPA (2010)
	Canada	1981	Wood preservation for industrial use	Demers et al. (2006)
Restricted	Mexico	____	Wood preservation, material for the synthesis of pesticides and use as a pesticide	CICOPLAFEST (2004)

dihydroxybenzene) (Fig. 4.9). The dehalogenation reaction is carried out in two steps and is catalyzed by two intracellular enzymes, glutathione transferase and glutathione reductase conjugation, which are enzymes bound to membranes and belonging to the soluble protein fraction of the biomass of *P. chrysosporium* (Reddy and Gold 1999). *T. versicolor*, another white rot fungus, produces laccase and can degrade PCP to form tetrachloro-*p* and *o*-bezoquinone (Ricotta et al. 1996, Walter et al. 2004). *Coriolus versicolor* FPRL-28A produces a chlorinated polymer, of 8 kDa, from PCP degradation by an enzymatic reaction with laccase (Ullah et al. 2000). Other fungi belonging to the genus *Trametes* have also shown the ability to degrade PCP. Boyle (2006) indicated that *Trametes hirsuta* (Wulf.:Fr.) was able to mineralize PCP under nitrogen limiting and non-limiting conditions in liquid culture. Gonzáles et al. (2010) showed that in a liquid culture limited in nitrogen, *Trametes pubescens* (CBS 696.94) degraded PCP and other chlorophenols. Authors reported that PCP degradation by *T. pubescens* was improved by supplementing the culture medium with glucose. *Pleurotus pulmonarius*, *Anthracophyllum discolor*, *Gloeophyllum striatum*, *Gloeophyllum trabeum*, *Pycnosporum*, among other basidiomycetes, have been reported as fungi capable of degrading PCP; in some cases, intermediates were found. For example, it has been reported that pentachloroanisole, tetrachlorohydroquinone, tetrachlorodimethoxybenzene, 2,5-dicloro-1,4-dimethoxybenzene, and 2-chloro-1,4-dimethoxybenzene were found from PCP degradation by *A. discolor* (Rubilar et

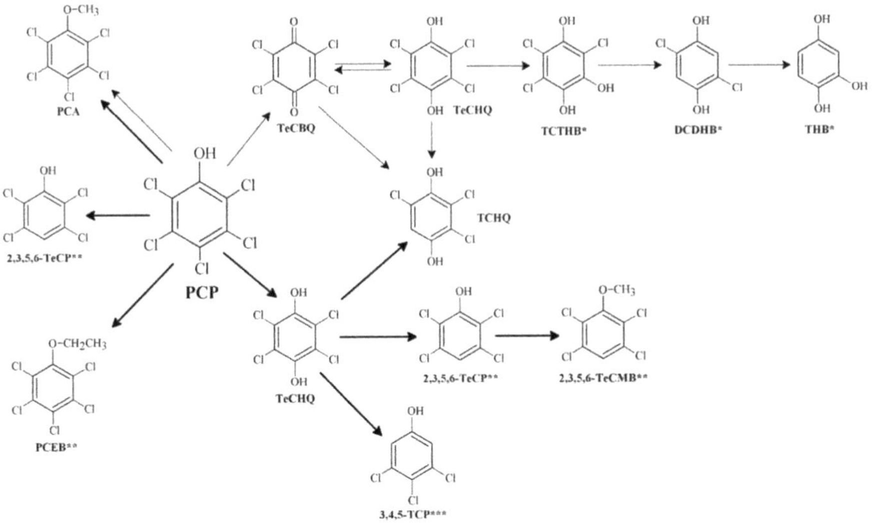

Figure 4.9 Main metabolites produced from PCP degradation by filamentous fungi, like basidiomycetes (thin line) and zygomycetes (thick line). PCP = pentachlorophenol, PCA = pentachloroanisole, TeCBQ = tetrachlorobenzoquinone, TeCHQ = tetrachlorohydroquinone, TCTHB = trichlorotrihydroxybenzene, DCDHB = dichlorodihydroxybenzene, THB = trihydroxybenzene, PCEB = pentachloroethybenzene, TeCMB = tetrachloromethylbenzene, TeCP = tetrachlorophenol, TCP = trichlorophenol. *Metabolites produced by *P. chrysosporium*, reported by Reddy and Gold (2000). **Metabolites produced by *Mucor ramosissimus* reported by Szewczyk and Długoński (2009) and ***Metabolites produced by *Rhizopus oryzae*, reported by León-Santiesteban et al. (2014). Metabolites without asterisk are the common metabolites produced by fungi from PCP degradation.

al. 2007). These authors mention that the degradation does not depend completely on the ligninolytic activities, whereas other authors (Rubilar et al. 2011, Cea et al. 2010) found a correlation between PCP degradation and the activity of Mn peroxidase. Fahr et al. (1999) reported PCP mineralization by the brown-rot fungi *G. striatum* and *G. trabeum*.

The metabolic capacity to degrade and/or mineralize chlorophenols is not an exclusive faculty of basidiomycetes. Some fungi belonging to ascomycetes, zygomycetes, and even deuteromycetes have tolerated, degraded, and mineralized organochlorine compounds. Sage et al. (1993) isolated 61 strains of micromycetes from sediments of a freshwater reservoir and studied the ability of these microorganisms to tolerate and grow in the presence of PCP. They found that the strains most tolerant to PCP (1000 mg L^{-1}) were four ascomycetes—*Eurotium repens*, *Pseudeurotium zonatum*, *Talaromyces trachyspermus*, and *Talaromyces wortmanii*, and one deuteromycete, *Scytalidium lignicola*.

On the other hand, Steiman et al. (1994) found that some ascomycetes and yeasts could degrade PCP (100 mg L^{-1}) in liquid culture; however, this ability was not directly related to the ability of fungi to produce extracellular phenoloxidases, since some fungi were good producers of phenoloxidases but poor PCP removers, and others were good removers but not producers of phenoloxidases. Ascomycetes, such as *Neurospora sitophila*, *Pyronema omphalodes*, *Sporormiella minimoides*, *Emericella variecolor* var. *variecolor*, *Sclerotinia sclerotiorum*, *Otthia spireae*, *Trichophaea abundans*, and *Emericella nidulans* degrade between 75–96% of PCP; yeasts are not capable of degrading the toxicant, except *Rhodototula*, which degraded approximately 40%.

Rigot and Matsumura (2002) studied PCP degradation by *Trichoderma harzianum* T.h.2023, an ascomycete, in liquid culture and grown in a rhizosphere medium. Authors showed that approximately 80% of initial PCP was transformed to pentachloroanisole. When the fungus was grown in the rhizosphere of corn seedlings, transformation of the initial PCP was of 90%, but pentachloroanisole was detected at a low concentration, and only 13%. *T. harziamun* CBMAI 1677 was able to degrade PCP. The intermediates were pentachloroanisole and 2,3,4,6-tetrachloroanisole. These intermediates were also degraded by the fungus (Vacondio et al. 2015).

Zygomycetes as *Mucor ramosissimus*, *Mucor plumbeus*, and *Rhizopus oryzae* have been reported as fungi capable to degrade PCP. In cultures of *M. ramisissimus* in Sabouraud medium and with PCP, 2,3,5-6-tetrachlorophenol, 2,3,4-6-tetrachlorophenol, and pentachloroanisole were found at 7 days of culture. During PCP degradation there is an inclusion of hydroxyl groups in the aromatic ring with the simultaneous removal of chlorine ions (an oxidative dechlorination reaction). When *M. ramisissimus* is cultured in the same medium in the presence on pentachlorobenzene, PCP degradation is carried out by another pathway; in this case, the hydroxyl groups of the chlorophenols are methylated producing anisole, such as penta- and tetrachloroanisole (Szewczyk et al. 2003, Szewczyk and Długoński 2009). Szewczyk and Długoski (2009) also showed that when *M. ramisissimus* is grown in a mineral medium with glucose as carbon source and 10 mg L^{-1} of PCP, 0.76 mg L^{-1} of tetrachlorohydroquinoe was formed. When the fungus is grown in the same medium, but with spent engine oil as carbon source, 0.89 mg L^{-1} of pentachlorohydroquinone was produced (Fig. 4.9). Carvalho et al. (2009) reported that *M. plumbeus* tolerates, grows, and degrades PCP in metabolic and co-metabolic conditions. In the first condition, PCP removal is slow, between 50 and 60 days of culture are needed to remove 85% of 5 mg of initial PCP per liter. However, in co-metabolic conditions with 10 g L^{-1} of glucose, the fungus removes 100% of 10 mg of initial PCP in 60 days. The identified intermediates produced from PCP degradation by *M. plumbeus* were conjugated molecules, such as glucose-PCP, glucose-tetrachlorohydroquinone, ribose-tetrachlorohydroquinone, and sulfate-tetrachlorohydroquinone. Some di-conjugated molecules such as sulfate-glucose-PCP and sulfate-glucose-tetrachlorohydroquinoe were also identified (Carvalho et al. 2011). Authors reported that oxidative dechlorination to form tetrachlorohydroquinone and then reductive dechlorination of the latter to produce 2,3,6-trichlorophenol are the pathways used by *M. plumbeaus*. Carvalho et al. (2013) did not find a correlation between PCP degradation and ligninolytic activity, or with cytochrome P-450. Leon-Santiesteban et al. (2008) reported a *R. oryzae* ENHE CDBB-H-1877, isolated from soil polluted with PCP, capable to tolerate, grow, and degrade PCP. Authors found that *R. oryzae* produces peroxidases and phenoloxidases involved in PCP degradation. The *R. oryzae* ENHE CDBB-H-1877 strain, immobilized in nylon fiber, removed 88 and 92% of 12.5 and 25 mg L^{-1} of the initial PCP, in 48 and 72 hours of culture (León-Santiestebán et al. 2011). The study of different carbon and nitrogen sources in co-metabolic conditions to degrade PCP by *R. oryzae* ENHE showed that the fungus produce different metabolites. These results indicate that fungi use two or more PCP degradation pathways. 3,4,5-trichlorophenol, tetrahydroquinone, and one chlorinated compound were found from PCP degradation in cultures with glutamic acid and sodium nitrate. However, in cultures of *R. oryzae* with glucose and ammonium sulfate, pentachloroanisole, 3,4,5-trichlorophenol, tetrachlorohydroquinone and four unidentified chlorinated compounds were found (Fig. 4.9). This means that when fungi grow in co-metabolism with glucose-PCP, there is more metabolic activity and more intermediates are produced than when fungi are grown in metabolism with glutamic acid-PCP (León-Santiesteban et al. 2014). Authors suggest that *R. oryzae* could present methylation reactions and oxidative and reductive dechlorinated reactions, as well as polymerization reactions, as reported by Carvalho et al. (2011) and Szewczyk and Długńoski (2009).

It can be observed that PCP degradation pathways and the metabolites produced mainly depend on the fungus used, the carbon source, and the culture conditions.

Conclusions

The use of filamentous fungi to degrade chlorophenol compounds has been widely studied. There is knowledge on how fungi degrade toxic compounds, as reported by biochemical, molecular, and proteomics studies. However, until now, the relationship between the degradation capacity and the enzymatic activities produced by fungi has not been understood very well. The main obstacle to the application of fungi in the remediation of contaminated soil and water is the technological aspect. The application *in situ* is still very complicated due to the competition among microorganisms in natural environments. A better understanding of the degradation mechanisms and the role of the enzymes involved is still needed. Thus, bioaugmentation processes and addition of enzymes could be proposed to degrade these toxic compounds.

References

Álvarez-Rodríguez, M. L., L. López-Ocaña, J. M. López-Coronado, E. Rodríguez, M. J. Martínez, G. Larriba and J.-J. R. Coque. 2002. Cork taint of wines: Role of the filamentous fungi isolated from cork in the formation of 2,4,6-Trichloroanisole by O methylation of 2,4,6-Trichlorophenol. *Applied and Environmental Microbiology* 68 (12): 5860–5869. doi: 10.1128/aem.68.12.5860-5869.2002.
Alves, A. M. C. R., E. Record, A. Lomascolo, K. Scholtmeijer, M. Asther, J. G. H. Wessels and H. A. B. Wösten. 2004. Highly efficient production of laccase by the basidiomycete *Pycnoporus cinnabarinus*. *Applied and Environmental Microbiology* 70 (11): 6379.
Aranciaga, N., I. Durruty, J. Gonzalez and E. Wolski. 2012. Aerobic biotransformation of 2,4,6–trichlorophenol by *Penicillium chrysogenum* in aqueous batch culture: Degradation and residual phytotoxicity. *African Journals Online* 38 (5): 683–687. doi: 10.4314/wsa.v38i5.5.
Armenante, P. M., G. Lewandowski and I. U. Haq. 1992. Mineralization of 2-Chlorophenol by *P. Chrysosporium* using different reactor designs. *Hazardous waste and hazardous materials* 9 (3): 213–229. doi: 10.1089/hwm.1992.9.213.
Armenante, P. M., N. Pal and G. Lewandowski. 1994. Role of mycelium and extracellular protein in the biodegradation of 2,4,6-trichlorophenol by *Phanerochaete chrysosporium*. *Applied and Environmental Microbiology* 60 (6): 1711–1718.
Bollag, J. M., S. Y. Liu and R. D. Minard. 1979. Asymmetric diphenol formation by a fungal laccase. *Applied and Environmental Microbiology* 38 (1): 90–92.
Bosso, L. and G. Cristinzio. 2014. A comprehensive overview of bacteria and fungi used for pentachlorophenol biodegradation. *Reviews in Environmental Science and Bio/Technology* 13 (4): 387–427. doi: 10.1007/s11157-014-9342-6.
Boyle, D. 2006. Effects of pH and cyclodextrins on pentachlorophenol degradation (mineralization) by white-rot fungi. *Journal of Environmental Management* 80 (4): 380–386. doi: https://doi.org/10.1016/j.jenvman.2005.09.017.
Carvalho, M. B., I. Martins, M. C. Leitão, H. Garcia, C. Rodrigues, V. San Romão, I. McLellan, A. Hursthouse and C. Silva Pereira. 2009. Screening pentachlorophenol degradation ability by environmental fungal strains belonging to the phyla Ascomycota and Zygomycota. *Journal of Industrial Microbiology & Biotechnology* 36 (10): 1249–1256. doi: 10.1007/s10295-009-0603-2.
Carvalho, M. B., S. Tavares, J. Medeiros, O. Núñez, H. Gallart-Ayala, M. C. Leitão, M. T. Galceran, A. Hursthouse and C. Silva Pereira. 2011. Degradation pathway of pentachlorophenol by *Mucor plumbeus* involves phase II conjugation and oxidation–reduction reactions. *Journal of Hazardous Materials* 198: 133–142. doi: https://doi.org/10.1016/j.jhazmat.2011.10.021.
Carvalho, M. B., I. Martins, J. Medeiros, S. Tavares, S. Planchon, J. Renaut, O. Núñez, H. Gallart-Ayala, M. T. Galceran, A. Hursthouse and C. Silva Pereira. 2013. The response of *Mucor plumbeus* to pentachlorophenol: A toxicoproteomics study. *Journal of Proteomics* 78: 159–171. doi: https://doi.org/10.1016/j.jprot.2012.11.006.
Cea, M., M. Jorquera, O. Rubilar, H. Langer, G. Tortella and M. C. Diez. 2010. Bioremediation of soil contaminated with pentachlorophenol by *Anthracophyllum discolor* and its effect on soil microbial community. *Journal of Hazardous Materials* 181 (1): 315–323. doi: https://doi.org/10.1016/j.jhazmat.2010.05.013.
CICOPLAFEST. 2004. *Catalogo de plaguicidas, SSA, SAGARPA, SEMARNAT, SE*. México.
Clausen, L. P. W., C. K. Jensen and S. Trapp. 2018. Toxicity of 2,3,5,6-tetrachlorophenol to willow trees (*Salix viminalis*). *Human and Ecological Risk Assessment: An International Journal* 24 (4): 941–948. doi: 10.1080/10807039.2017.1403280.
Coque, J. J. R., M. L. Álvarez-Rodríguez and G. Larriba. 2003. Characterization of an inducible chlorophenol *O*-methyltransferase from *Trichoderma longibrachiatum* involved in the formation of chloroanisoles and determination of its role in cork taint of wines. *Applied and Environmental Microbiology* 69 (9): 5089.

Czaplicka, M. 2004. Sources and transformations of chlorophenols in the natural environment. *Science of the Total Environment* 322 (1): 21–39. doi: http://dx.doi.org/10.1016/j.scitotenv.2003.09.015.

Choi, S. H., S. H. Moon and M. B. Gu. 2002. Biodegradation of chlorophenols using the cell-free culture broth of *Phanerochaete chrysosporium* immobilized in polyurethane foam. *Journal of Chemical Technology & Biotechnology* 77 (9): 999–1004. doi: 10.1002/jctb.667.

Demers, P. A., H. W. Davies, M. C. Friesen, C. Hertzman, A. Ostry, R. Hershler and K. Teschke. 2006. Cancer and occupational exposure to pentachlorophenol and tetrachlorophenol (Canada). *Cancer Causes & Control* 17 (6): 749. doi: 10.1007/s10552-006-0007-9.

Deschler, C., R. Duran, M. Junqua, C. Landou, J.-C. Salvado and P. Goulas. 1998. Involvement of 3,4-dichlorophenol hydroxylase in degradation of 3,4-dichlorophenol by the white rot fungus *Phanerochaete chrysosporium*. *Journal of Molecular Catalysis B: Enzymatic* 5 (1): 423–428. doi: https://doi.org/10.1016/S1381-1177(98)00056-3.

Dodd, D. E., L. J. Pluta, M. A. Sochaski, D. A. Banas and R. S. Thomas. 2012. Subchronic hepatotoxicity evaluation of 2,3,4,6-tetrachlorophenol in sprague dawley rats. *Journal of Toxicology* 2012: 10. doi: 10.1155/2012/376246.

Duran, R., C. Deschler, S. Precigou and P. Goulas. 2002. Degradation of chlorophenols by *Phanerochaete chrysosporium*: effect of 3,4-dichlorophenol on extracellular peroxidase activities. *Applied Microbiology and Biotechnology* 59 (2): 284–288. doi: 10.1007/s00253-002-0988-6.

Ehlers, G. A. and P. D. Rose. 2005. Immobilized white-rot fungal biodegradation of phenol and chlorinated phenol in trickling packed-bed reactors by employing sequencing batch operation. *Bioresource Technology* 96 (11): 1264–1275. doi: https://doi.org/10.1016/j.biortech.2004.10.015.

EPA, U. S. 1990. *Suspended, Cancelled, and Restricted Pesticides*. Washington, D.C.: Office of Pesticides and Toxic Substances.

EPA, U. S. 1992a. 2,4,5-Trichlorophenol (fact sheet). U.S. Environmental Protection Agency.

EPA, U. S. 1992b. 2,4,6-Trichlorophenol (fact sheet). U.S. Environmental Protection Agency.

EPA, U. S. 1992c. *Pentachlorophenol, hazard summary-created in April 1992; Revised in January 2000,* http://www.epa.gov/ttnatw01/hlthef/pentachl.html.

EPA, U. S. 2010. *IRIS Toxicological review of pentachlorophenol, U.S. Environmental Protection Agency.* Washington, DC, EPA/635/R-09/004F.

Escher, B. I., M. Snozzi and R. P. Schwarzenbach. 1996. Uptake, speciation, and uncoupling activity of substituted phenols in energy transducing membranes. *Environmental Science & Technology* 30 (10): 3071–3079. doi: 10.1021/es960153f.

Fahr, K., H.-G. Wetzstein, R. Grey and D. Schlosser. 1999. Degradation of 2,4-dichlorophenol and pentachlorophenol by two brown rot fungi. *FEMS Microbiology Letters* 175 (1): 127–132. doi: 10.1111/j.1574-6968.1999.tb13611.x.

FAO. 1996. Decision guidance documents: Pentachlorophenol and its salts and esters, Joint FAO/UNEP programme for the operation of prior informed consent. Rome-Geneva: Food and Agriculture Organization of the United Nations, United Nations Environment Program.

Field, J. A. and R. Sierra-Alvarez. 2007. Microbial degradation of chlorinated phenols. *Reviews in Environmental Science and Bio/Technology* 7 (3): 211–241. doi: 10.1007/s11157-007-9124-5.

Fisher, S. W. 1991. Changes in the toxicity of three pesticides as a function of environmental pH and temperature. *Bulletin of Environmental Contamination and Toxicology* 46 (2): 197–202. doi: 10.1007/bf01691937.

Folke, J. and J. Birklund. 1986. Danish coastal water levels of 2,3,4,6-tetrachlorophenol, pentachlorophenol, and total organohalogens in blue mussels (*Mytilus edulis*). *Chemosphere* 15 (7): 895–900. doi: https://doi.org/10.1016/0045-6535(86)90054-8.

Forootanfar, H., M. M. Movahednia, S. Yaghmaei, M. Tabatabaei-Sameni, H. Rastegar, A. Sadighi and M. A. Faramarzi. 2012. Removal of chlorophenolic derivatives by soil isolated ascomycete of *Paraconiothyrium variabile* and studying the role of its extracellular laccase. *Journal of Hazardous Materials* 209: 199–203. doi: http://dx.doi.org/10.1016/j.jhazmat.2012.01.012.

Gaitan, I. J., S. C. Medina, J. C. González, A. Rodríguez, Á. J. Espejo, J. F. Osma, V. Sarria, C. J. Alméciga-Díaz and O. F. Sánchez. 2011. Evaluation of toxicity and degradation of a chlorophenol mixture by the laccase produced by *Trametes pubescens*. *Bioresource Technology* 102 (3):3632-3635. doi: https://doi.org/10.1016/j.biortech.2010.11.040.

Ge, J., J. Pan, Z. Fei, G. Wu and J. P. Giesy. 2007. Concentrations of pentachlorophenol (PCP) in fish and shrimp in Jiangsu Province, China. *Chemosphere* 69 (1): 164–169. doi: https://doi.org/10.1016/j.chemosphere.2007.04.025.

Gee, J. M. and J. L. Peel. 1974. Metabolism of 2,3,4,6-tetrachlorophenol by micro-organisms from broiler house litter. *Microbiology* 85 (2): 237–243. doi: doi:10.1099/00221287-85-2-237.

González, L. F., V. Sarria and O. F. Sánchez. 2010. Degradation of chlorophenols by sequential biological-advanced oxidative process using *Trametes pubescens* and TiO_2/UV. *Bioresource Technology* 101 (10): 3493–3499. doi: https://doi.org/10.1016/j.biortech.2009.12.130.

Grey, R., C. Höfer and D. Schlosser. 1998. Degradation of 2-chlorophenol and formation of 2-chloro-1,4-benzoquinone by mycelia and cell-free crude culture liquids of *Trametes versicolor* in relation to extracellular laccase activity. *Journal of Basic Microbiology* 38 (5–6): 371–382. doi: 10.1002/(sici)1521-4028(199811)38:5/6<371::aid-jobm371>3.0.co;2-v.

Hasegawa, Y., T. Okamoto, H. Obata and T. Tokuyama. 1990. Utilization of aromatic compounds by *Trichosporon cutaneum* KUY-6A. *Journal of Fermentation and Bioengineering* 69 (2): 122–124. doi: http://dx.doi.org/10.1016/0922-338X(90)90199-7.

Heudorf, U., S. Letzel, M. Peters and J. Angerer. 2000. PCP in the blood plasma: Current exposure of the population in Germany, based on data obtained in 1998. *International Journal of Hygiene and Environmental Health* 203 (2): 135–139. doi: https://doi.org/10.1078/S1438-4639(04)70018-8.

Hofrichter, M., T. Günther and W. Fritsche. 1992. Metabolism of phenol, chloro- and nitrophenols by the *Penicillium strain* Bi 7/2 isolated from a contaminated soil. *Biodegradation* 3 (4): 415–421. doi: 10.1007/bf00240363.

Hofrichter, M., F. Bublitz and W. Fritsche. 1994. Unspecific degradation of halogenated phenols by the soil fungus *Penicillium frequentans* Bi 7/2. *Journal of Basic Microbiology* 34 (3): 163–172. doi: 10.1002/jobm.3620340306.

Horstman, S. W., A. Rossner, D. A. Kalman and M. S. Morgan. 1989. Penetration of pentachlorophenol and tetrachlorophenol through human skin. *Journal of Environmental Science and Health . Part A: Environmental Science and Engineering* 24 (3): 229–242. doi: 10.1080/10934528909375477.

IARC. 1991. IARC monographs on the evaluation of carcinogenic risks to humans, Occupational exposures in insecticide application, and some pesticides. Lyon. 53.

Itoh, K., M. Fujita, K. Kumano, K. Suyama and H. Yamamoto. 2000. Phenolic acids affect transformations of chlorophenols by a *Coriolus versicolor* laccase. *Soil Biology and Biochemistry* 32 (1): 85–91. doi: https://doi.org/10.1016/S0038-0717(99)00133-9.

Jensen, J. K. A., Z. C. Ryan, A. Vanden Wymelenberg, D. Cullen and K. E. Hammel. 2002. An NADH: quinone oxido reductase active during biodegradation by the brown-rot basidiomycete *Gloeophyllum trabeum*. *Applied and Environmental Microbiology* 68 (6): 2699–2703. doi: 10.1128/aem.68.6.2699-2703.2002.

Jiang, X. Y., G. M. Zeng, D. L. Huang, Y. Chen, F. Liu, G. H. Huang, J. B. Li, B. D. Xi and H. L. Liu. 2006. Remediation of pentachlorophenol-contaminated soil by composting with immobilized *Phanerochaete chrysosporium*. *World Journal of Microbiology and Biotechnology* 22 (9): 909–913. doi: 10.1007/s11274-006-9134-4.

Johansson, T., P. O. Nyman and D. Cullen. 2002. Differential regulation of *mnp2*, a new manganese peroxidase-encoding gene from the ligninolytic fungus *Trametes versicolor* PRL 572. *Applied and Environmental Microbiology* 68 (4): 2077.

Johnson, R. L., P. J. Gehring, R. J. Kociba and B. A. Schwertz. 1973. Chlorinated dibenzodioxins and pentachlorophenol. *Environmental Health Perspectives* 5: 171–175.

Joshi, D. K., and M. H. Gold. 1993. Degradation of 2,4,5-trichlorophenol by the lignin-degrading basidiomycete *Phanerochaete chrysosporium*. *Applied and Environmental Microbiology* 59 (6): 1779–1785.

Kitunen, V. H., R. J. Valo and M. S. Salkinoja-Salonen. 1987. Contamination of soil around wood-preserving facilities by polychlorinated aromatic compounds. *Environmental Science & Technology* 21 (1): 96–101. doi: 10.1021/es00155a012.

Krug, M., and G. Straube. 1986. Degradation of phenolic compounds by the yeast *Candida tropicalis* HP 15. II. Some properties of the first two enzymes of the degradation pathway. *Journal of Basic Microbiology* 26 (5):271-281. doi: 10.1002/jobm.3620260505.

Krug, M., H. Ziegler and G. Straube. 1985. Degradation of phenolic compounds by the yeast *Candida tropicalis* HP 15 I. Physiology of growth and substrate utilization. *Journal of Basic Microbiology* 25 (2): 103–110. doi: 10.1002/jobm.3620250206.

Laine, M. M., X00E, Nnist, X00F, M. K., M. S. Salkinoja-Salonen and J. A. Puhakka. 2003. Chlorinated organic contaminants from mechanical wood processing and their bioremediation.

In: *Dehalogenation: Microbial Processes and Environmental Applications*, edited by Max M. Häggblom and Ingeborg D. Bossert, 421–442. Boston, MA: Springer US.

Lamar, R. T., M. J. Larsen and T. K. Kirk. 1990. Sensitivity to and degradation of pentachlorophenol by *Phanerochaete* spp. *Applied and Environmental Microbiology* 56 (11): 3519–3526.

León-Santiesteban, H., R. Bernal, F. J. Fernández and A. Tomasini. 2008. Tyrosinase and peroxidase production by *Rhizopus oryzae* strain ENHE obtained from pentachlorophenol-contaminated soil. *Journal of Chemical Technology & Biotechnology* 83 (10): 1394–1400. doi: 10.1002/jctb.1955.

León-Santiesteban, H., M. Meraz, K. Wrobel and A. Tomasini. 2011. Pentachlorophenol sorption in nylon fiber and removal by immobilized *Rhizopus oryzae* ENHE. *Journal of Hazardous Materials* 190 (1): 707–712. doi: https://doi.org/10.1016/j.jhazmat.2011.03.101.

León-Santiesteban, H. H., K. Wrobel, S. Revah and A. Tomasini. 2014. Pentachlorophenol removal by *Rhizopus oryzae* CDBB-H-1877 using sorption and degradation mechanisms. *Journal of Chemical Technology & Biotechnology* 91 (1): 65–71. doi: 10.1002/jctb.4566.

Leontievsky, A. A., N. M. Myasoedova, B. P. Baskunov, C. S. Evans and L. A. Golovleva. 2000. Transformation of 2,4,6-trichlorophenol by the white rot fungi *Panus tigrinus* and *Coriolus versicolor*. *Biodegradation* 11 (5): 331–340. doi: 10.1023/a:1011154209569.

Leontievsky, A., N. Myasoedova, B. Baskunov, L. Golovleva, C. Bucke and C. Evans. 2001. Transformation of 2,4,6-trichlorophenol by free and immobilized fungal laccase. *Applied Microbiology and Biotechnology* 57 (1): 85–91. doi: 10.1007/s002530100756.

Leontievsky, A., N. Myasoedova, L. Golovleva, M. Sedarati and C. Evans. 2002. Adaptation of the white-rot basidiomycete *Panus tigrinus* for transformation of high concentrations of chlorophenols. *Applied Microbiology and Biotechnology* 59 (4): 599–604. doi: 10.1007/s00253-002-1037-1.

Lewandowski, G. A., P. M. Armenante and D. Pak. 1990. Reactor design for hazardous waste treatment using a white rot fungus. *Water Research* 24 (1): 75–82. doi: https://doi.org/10.1016/0043-1354(90)90067-G.

Lin, J. E., H. Y. Wang and R. F. Hickey. 1990. Degradation kinetics of pentachlorophenol by *Phanerochaete chrysosporium*. *Biotechnology and Bioengineering* 35 (11): 1125–1134. doi: 10.1002/bit.260351108.

Makkar, R. S., A. Tsuneda, K. Tokuyasu and Y. Mori. 2001. *Lentinula edodes* produces a multicomponent protein complex containing manganese (II)-dependent peroxidase, laccase and β-glucosidase. *FEMS Microbiology Letters* 200 (2): 175–179. doi: 10.1111/j.1574-6968.2001.tb10711.x.

Marcial, J. J. Barrios-González and A. Tomasini. 2006. Effect of medium composition on pentachlorophenol removal by *Amylomyces rouxii* in solid-state culture. *Process Biochemistry* 41 (2): 496–500. doi: https://doi.org/10.1016/j.procbio.2005.07.010.

Marr, J., S. Kremer, O. Sterner and H. Anke. 1996. Transformation and mineralization of halophenols by *Penicillium simplicissimum* SK9117. *Biodegradation* 7 (2): 165–171. doi: 10.1007/bf00114628.

Mileski, G. J. J. A. Bumpus, M. A. Jurek and S. D. Aust. 1988. Biodegradation of pentachlorophenol by the white rot fungus *Phanerochaete chrysosporium*. *Applied and Environmental Microbiology* 54 (12): 2885.

Muir, J. and G. Eduljee. 1999. PCP in the freshwater and marine environment of the European Union. *Science of the Total Environment* 236 (1): 41–56. doi: https://doi.org/10.1016/S0048-9697(99)00281-8.

Nakagawa, A., S. Osawa, T. Hirata, Y. Yamagishi, J. Hosoda and T. Horikoshi. 2006. 2,4-Dichlorophenol Degradation by the soil fungus *Mortierella* sp. *Bioscience, Biotechnology, and Biochemistry* 70 (2): 525–527. doi: 10.1271/bbb.70.525.

Penttinen, O.-P. 1995. Chlorophenols in aquatic environments: Structure-activity correlations. *Annales Zoologici Fennici* 32 (3): 287–294.

Polnisch, E., H. Kneifel, H. Franzke and K. H. Hofmann. 1991. Degradation and dehalogenation of monochlorophenols by the phenol-assimilating yeast *Candida maltosa*. *Biodegradation* 2 (3): 193–199. doi: 10.1007/bf00124493.

Prak, S., Z. Gunata, J.-P. Guiraud and S. Schorr-Galindo. 2007. Fungal strains isolated from cork stoppers and the formation of 2,4,6-trichloroanisole involved in the cork taint of wine. *Food Microbiology* 24 (3): 271–280. doi: https://doi.org/10.1016/j.fm.2006.05.002.

Reddy, G. V. B. and M. H. Gold. 1999. A two-component tetrachlorohydroquinone reductive dehalogenase system from the lignin-degrading basidiomycete *Phanerochaete chrysosporium*. *Biochemical and Biophysical Research Communications* 257 (3): 901–905. doi: https://doi.org/10.1006/bbrc.1999.0561.

Reddy, G. V. B., D. K. Joshi and M. H. Gold. 1997. Degradation of chlorophenoxyacetic acids by the lignin-degrading fungus *Dichomitus squalens*. *Microbiology* 143 (7): 2353–2360. doi: doi:10.1099/00221287-143-7-2353.

Reddy, G. V. B., M. D. Sollewijn Gelpke and M. H. Gold. 1998. Degradation of 2,4,6-trichlorophenol by *Phanerochaete chrysosporium*: Involvement of reductive dechlorination. *Journal of Bacteriology* 180 (19): 5159–5164.

Reddy, G. V. B. and M. H. Gold. 2000. Degradation of pentachlorophenol by *Phanerochaete chrysosporium*: intermediates and reactions involved. *Microbiology* 146 (2): 405–413. doi: doi:10.1099/00221287-146-2-405.

Ricotta, A., R. F. Unz and J. M. Bollag. 1996. Role of a laccase in the degradation of pentachlorophenol. *Bulletin of Environmental Contamination and Toxicology* 57 (4): 560–567. doi: 10.1007/s001289900227.

Rigot, J. and F. Matsumura. 2002. Assessment of the rhizosphere competency and pentachlorophenol-metabolizing activity of a pesticide-degrading strain of *Trichoderma harzianum* introduced into the root zone of corn seedlings. *Journal of Environmental Science and Health, Part B* 37 (3): 201–210. doi: 10.1081/pfc-120003098.

Rubilar, O., G. Feijoo, C. Diez, T. A. Lu-Chau, M. T. Moreira and J. M. Lema. 2007. Biodegradation of pentachlorophenol in soil slurry cultures by *Bjerkandera adusta* and *Anthracophyllum discolor*. *Industrial & Engineering Chemistry Research* 46 (21): 6744–6751. doi: 10.1021/ie061678b.

Rubilar, O., G. Tortella, M. Cea, F. Acevedo, M. Bustamante, L. Gianfreda and M. C. Diez. 2011. Bioremediation of a chilean andisol contaminated with pentachlorophenol (PCP) by solid substrate cultures of white-rot fungi. *Biodegradation* 22 (1): 31–41. doi: 10.1007/s10532-010-9373-9.

Rubio Pérez, R., G. González Benito and M. Peña Miranda. 1997. Chlorophenol degradation by *Phanerochaete chrysosporium*. *Bioresource Technology* 60 (3): 207–213. doi: http://dx.doi.org/10.1016/S0960-8524(97)00022-9.

Ryan, T. P. and J. A. Bumpus. 1989. Biodegradation of 2,4,5-trichlorophenoxyacetic acid in liquid culture and in soil by the white rot fungus *Phanerochaete chrysosporium*. *Applied Microbiology and Biotechnology* 31 (3): 302–307. doi: 10.1007/bf00258414.

Sage, L., R. Steiman, F. Seigle-Murandi, J. L. Benoit-Guyod and G. Merlin. 1993. Toxicity of pentachlorophenol for micromycetes isolated from freshwater sediments. *Journal of Aquatic Ecosystem Health* 2 (4): 335–340. doi: 10.1007/bf00044035.

Sedarati, M. R., T. Keshavarz, A. A. Leontievsky and C. S. Evans. 2003. Transformation of high concentrations of chlorophenols by the white-rot basidiomycete *Trametes versicolor* immobilized on nylon mesh. *Electronic Journal of Biotechnology* 6 (2): 104–114.

Seigle-Murandi, F., R. Steiman, J. L. Benoit-Guyod and P. Guiraud. 1992. Biodegradation of pentachlorophenol by micromycetes. I. Zygomycetes. *Environmental Toxicology and Water Quality* 7 (2): 125–139. doi: 10.1002/tox.2530070204.

Seiler, J. P. 1991. Pentachlorophenol. *Mutation Research/Reviews in Genetic Toxicology* 257 (1): 27–47. doi: https://doi.org/10.1016/0165-1110(91)90018-Q.

Shiu, W. Y., K. C. Ma, D. Varhaníčková and D. Mackay. 1994. Chlorophenols and alkylphenols: A review and correlation of environmentally relevant properties and fate in an evaluative environment. *Chemosphere* 29 (6): 1155–1224. doi: https://doi.org/10.1016/0045-6535(94)90252-6.

Sjoblad, R. D. and J. M. Bollag. 1977. Oxidative coupling of aromatic pesticide intermediates by a fungal phenol oxidase. *Applied and Environmental Microbiology* 33 (4): 906–910.

Steiman, R., J. L. Benoit-Guyod, F. Seigle-Murandi, L. Sage and A. Toe. 1994. Biodegradation of pentachlorophenol by micromycetes II. Ascomycetes, basidiomycetes, and yeasts. *Environmental Toxicology and Water Quality* 9 (1): 1–6. doi: 10.1002/tox.2530090102.

Stoilova, I., A. Krastanov, V. Stanchev, D. Daniel, M. Gerginova and Z. Alexieva. 2006. Biodegradation of high amounts of phenol, catechol, 2,4-dichlorophenol and 2,6-dimethoxyphenol by Aspergillus awamori cells. *Enzyme and Microbial Technology* 39 (5): 1036–1041. doi: https://doi.org/10.1016/j.enzmictec.2006.02.006.

Stoilova, I., A. Krastanov, I. Yanakieva, M. Kratchanova and H. Yemendjiev. 2007. Biodegradation of mixed phenolic compounds by *Aspergillus awamori* NRRL 3112. *International Biodeterioration & Biodegradation* 60 (4): 342–346. doi: https://doi.org/10.1016/j.ibiod.2007.05.011.

Szewczyk, R., P. Bernat, K. Milczarek and J. Długoński. 2003. Application of microscopic fungi isolated from polluted industrial areas for polycyclic aromatic hydrocarbons and pentachlorophenol reduction. *Biodegradation* 14 (1): 1–8. doi: 10.1023/a:1023522828660.

Szewczyk, R. and J. Długoński. 2009. Pentachlorophenol and spent engine oil degradation by *Mucor ramosissimus*. *International Biodeterioration & Biodegradation* 63 (2): 123–129. doi: https://doi.org/10.1016/j.ibiod.2008.08.001.

Taseli, B. K. and C. F. Gokcay. 2005. Degradation of chlorinated compounds by *Penicillium camemberti* in batch and up-flow column reactors. *Process Biochemistry* 40 (2): 917–923. doi: http://dx.doi.org/10.1016/j.procbio.2004.02.006.

Tomasini, A., V. Flores, D. Cortés and J. Barrios-González. 2001. An isolate of *Rhizopus nigricans* capable of tolerating and removing pentachlorophenol. *World Journal of Microbiology and Biotechnology* 17 (2): 201–205. doi: 10.1023/a:1016694720608.

Ullah, M. A., C. T. Bedford and C. S. Evans. 2000. Reactions of pentachlorophenol with laccase from *Coriolus versicolor*. *Applied Microbiology and Biotechnology* 53 (2): 230–234. doi: 10.1007/s002530050013.

Vacondio, B., W. G. Birolli, I. M. Ferreira, M. H. R. Seleghim, S. Gonçalves, S. P. Vasconcellos and A. L. M. Porto. 2015. Biodegradation of pentachlorophenol by marine-derived fungus *Trichoderma harzianum* CBMAI 1677 isolated from ascidian *Didemnun ligulum*. *Biocatalysis and Agricultural Biotechnology* 4 (2): 266–275. doi: https://doi.org/10.1016/j.bcab.2015.03.005.

Vaidyanathan, V. G., P. W. Villalta and S. J. Sturla. 2007. Nucleobase-dependent reactivity of a quinone metabolite of pentachlorophenol. *Chemical Research in Toxicology* 20 (6): 913–919. doi: 10.1021/tx600359d.

Valentín, L., H. Oesch-Kuisma, K. T. Steffen, M. A. Kähkönen, A. Hatakka and M. Tuomela. 2013. Mycoremediation of wood and soil from an old saw mill area contaminated for decades. *Journal of Hazardous Materials* 260: 668–675. doi: https://doi.org/10.1016/j.jhazmat.2013.06.014.

Valli, K. and M. H. Gold. 1991. Degradation of 2,4-dichlorophenol by the lignin-degrading fungus *Phanerochaete chrysosporium*. *Journal of Bacteriology* 173 (1): 345–352.

Vroumsia, T., R. Steiman, F. Seigle-Murandi, J. L. Benoit-Guyod and I. E. Groupe pour l'Étude du Devenir des Xénobiotiques dans. 2005. Fungal bioconversion of 2,4-dichlorophenoxyacetic acid (2,4-D) and 2,4-dichlorophenol (2,4-DCP). *Chemosphere* 60 (10): 1471–1480. doi: https://doi.org/10.1016/j.chemosphere.2004.11.102.

Walker, N. 1973. Metabolism of chlorophenols by *Rhodotorula glutinis*. *Soil Biology and Biochemistry* 5 (5): 525–530. doi: http://dx.doi.org/10.1016/0038-0717(73)90042-4.

Walter, M., L. Boul, R. Chong and C. Ford. 2004. Growth substrate selection and biodegradation of PCP by New Zealand white-rot fungi. *Journal of Environmental Management* 71 (4): 361–369. doi: https://doi.org/10.1016/j.jenvman.2004.04.002.

Wei, Y., J. Zhu and A. Nguyen. 2014. Urinary concentrations of dichlorophenol pesticides and obesity among adult participants in the U.S. national health and nutrition examination survey (NHANES) 2005–2008. *International Journal of Hygiene and Environmental Health* 217 (2): 294–299. doi: https://doi.org/10.1016/j.ijheh.2013.07.003.

Weinbach, E. C. 1954. The effect of pentachlorophenol on oxidative phosphorylation. *Journal of Biological Chemistry* 210 (2): 545–550.

Weinbach, E. C. 1957. Biochemical basis for the toxicity of pentachlorophenol. *Proceedings of the National Academy of Sciences of the United States of America* 43 (5): 393–397.

Williams, P. L. 1982. Pentachlorophenol, an assessment of the occupational hazard. *American Industrial Hygiene Association Journal* 43 (11): 799–810. doi: 10.1080/15298668291410602.

Wymelenberg, A. V., G. Sabat, M. Mozuch, P. J. Kersten, D. Cullen and R. A. Blanchette. 2006. Structure, organization, and transcriptional regulation of a family of copper radical oxidase genes in the lignin-degrading basidiomycete *Phanerochaete chrysosporium*. *Applied and Environmental Microbiology* 72 (7): 4871.

Xiao, Y., X. Tu, J. Wang, M. Zhang, Q. Cheng, W. Zeng and Y. Shi. 2003. Purification, molecular characterization and reactivity with aromatic compounds of a laccase from basidiomycete *Trametes* sp. strain AH28-2. *Applied Microbiology and Biotechnology* 60 (6): 700–707. doi: 10.1007/s00253-002-1169-3.

Yadav, J. S. and C. A. Reddy. 1992. Non-involvement of lignin peroxidases and manganese peroxidases in 2,4,5-trichlorophenoxyacetic acid degradation by *Phanerochaete chrysosporium*. *Biotechnology Letters* 14 (11): 1089–1092. doi: 10.1007/bf01021065.

Yan, J., R. Nanqi, C. Xun, W. Di, Q. Liyan and L. Sen. 2008. Biodegradation of phenol and 4-chlorophenol by the mutant strain CTM 2. *Chinese Journal of Chemical Engineering* 16 (5): 796–800. doi: https://doi.org/10.1016/S1004-9541(08)60158-5.

Yordanova, G., T. Godjevargova, R. Nenkova and D. Ivanova. 2013. Biodegradation of phenol and phenolic derivatives by a mixture of immobilized cells of *Aspergillus awamori* and *Trichosporon cutaneum*. *Biotechnology & Biotechnological Equipment* 27 (2): 3681–3688. doi: 10.5504/bbeq.2013.0003.
Zhang, J., X. Liu, Z. Xu, H. Chen and Y. Yang. 2008. Degradation of chlorophenols catalyzed by laccase. *International Biodeterioration & Biodegradation* 61 (4): 351–356. doi: https://doi.org/10.1016/j.ibiod.2007.06.015.
Zouari, H., M. Labat and S. Sayadi. 2002. Degradation of 4-chlorophenol by the white rot fungus *Phanerochaete chrysosporium* in free and immobilized cultures. *Bioresource Technology* 84 (2): 145–150. doi: https://doi.org/10.1016/S0960-8524(02)00032-9.

CHAPTER-5

Azo Dye Decoloration by Fungi

M. R. Vernekar,[1] *J. S. Gokhale*[2] *and S. S. Lele*[2,*]

1. Introduction

Synthetic dyes are widely used for textile dyeing and many other industrial applications, such as cosmetics, leather, pharmaceutical, and food. Approximately 800,000 tons of dyes are produced annually (Palmieri et al. 2005). Due to a high demand for bright colored fabrics, the textile industry accounts for about two-thirds of the total dyestuff market (Aksu and Dönmez 2003). Dyes used in textile industry are resistant to microbial attacks, light, water, sweat, and many chemicals. Due to inefficiency in the dyeing process, at least 10% of the used dye gets disposed of as waste in the environment (Zollinger 1987), adversely affecting the dissolved oxygen, biological oxygen demand (BOD), chemical oxygen demand (COD), color, etc. (Saratale et al. 2011). Effluents of various industries, such dyeing, textile, paper, and pulp are markedly colored and the release of these effluents into water bodies causes environmental damage. Dye at a concentration of 10–50 mg/L affects the aesthetic value and gas solubility of water bodies (Banat et al. 1996). In addition, it also disturbs the natural growth of aquatic life due to reduction in light penetration.

A study by Zaharia and Suteu (2012) showed that 98% of dyes have a lethal concentration value (LC50) for fishes > 1 mg/L. In addition to their toxicity, these dyes are mutagenic and carcinogenic (Wesenberg et al. 2003), and pose potential threat to public health. Because of their recalcitrant nature, they are difficult to remove, and hence decoloration of industrial effluents is a major environmental concern. Biological methods using several microorganisms, such as algae, bacteria, fungi have been widely studied for decoloration and degradation of azo dyes. An efficient biodecoloration system should be functional upon repeated exposure to different dyes, and fungi seem to be promising in this case. This chapter deals with the

---
[1] School of Biotechnology and Bioinformatics, D. Y. Patil University, Navi Mumbai 400614 INDIA.
[2] Department of Food Engineering and Technology, Institute of Chemical Technology, Mumbai 400019 INDIA.
* Corresponding author: ss.lele@ictmumbai.edu.in

use of fungi for decoloration and degradation of azo dyes, with special emphasis on the fungal decoloration systems, such as use of yeast, filamentous fungi, ligninolytic fungi, immobilized fungi and their enzymes, genetically modified strains, etc., as well as process parameters affecting the dye removal process, mechanism involved, and different bioreactors used for dye removal processes.

1.1 Azo dyes: structure and classification

Dyes are colored substances that on application on fibers give permanent color resistance to fading upon exposure to light, sweat, water, chemicals, oxidizing agents, and microbial attacks (Rai et al. 2005). There are different ways of dyes classification. Based on the source materials, they are classified as natural dyes and synthetic dyes. Depending upon the method of application to the substrate, they are classified as direct dyes, reactive dyes, vat dyes, disperse dyes, azoic dyes, etc. Based on the nature of chromophore, they are classified as azo dyes, anthraquinone dyes, acridine dyes, arylmethane dyes, nitroso dyes, phthalocyanine dyes, indophenol dyes, oxazone dyes, etc. Depending on the origin and complex molecular structure, dyes can be classified as nonionic: disperse dyes; anionic: acid, direct and reactive dyes; and cationic: basic dyes (McMullan et al. 2001).

Among the different dyes, azo dyes are the main class, comprising almost 50% of all commercial dyes. These are characterized by the presence of one azo group (-N=N-), but can have two (disazo), three (trisazo), more rarely, four (tetrakisazo), or more (polyazo) azo groups. These azo bonds (-N=N-) are attached to two other groups, out of which, at least one, or more usually both, are aromatic groups. In comparison to natural colors, synthesis of these dyes is easy, available in a wide array of colors, with good fixative and permanency properties at an affordable cost; hence, they are widely used in textile industry. In addition, azo dyes are also used in fur dyeing, photographic dyes, paper printing, hair dye, indicators, oil industry, etc. Toxicological studies show that most of these dyes are carcinogenic, genotoxic, and mutagenic, causing severe health hazards. Because of their recalcitrant nature, they pose further concern regarding accumulation in the environment. Therefore, degradation of these dyes from the industrial effluent requires significant attention.

1.2 Biological methods for azo dye decoloration

According to Jin et al. (2007), worldwide, approximately 2.8×10^5 tons of textile dyes are discharged annually in industrial textile effluents. The release of these dyes to the ecosystem has detrimental effects because most of the azo dyes and their degraded products are lethal and/or mutagenic to living organisms. A wide range of physical, chemical, and biological methods have been employed for removal of synthetic dyes from waters and wastewaters. Figure 5.1 lists various methods used in decoloration and degradation of azo dyes. However, because of the complexity and varied chemical structure, no single method is effective for wastewater treatment. Currently, physicochemical methods are used for removal of these dyes. These methods are not only costly, but also inefficient in reducing chemical oxygen demand (COD), color, toxicity, and generates a large amount of sludge that further raises disposal problems

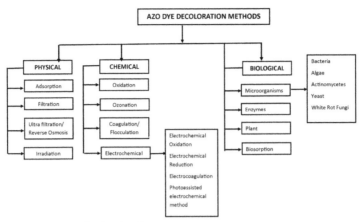

Figure 5.1 Methods for azo dye decoloration.

(Slokar and Majcen Le Marechal 1998). Hence, bioremediation seems to be an attractive alternative in effluent treatment. Biological methods (microorganisms and enzymes) for complete decoloration and degradation of azo dyes are more efficient, consume less water, and produce less amount of sludge (Banat et al. 1996, Saratale et al. 2011). These processes are economical, more eco-friendly than the end products produced by physicochemical methods. The end products formed using biological methods are nontoxic and cause complete mineralization of dyes to simpler inorganic compounds that are not lethal to living organisms. Moreover, microorganisms have an indigenous ability to adapt themselves to the toxic wastes, resulting in development of resistant strains that convert various toxic compounds into less harmful products. Bacteria, actinomycetes, fungi, algae, and various other microorganisms have been widely studied and used in decoloration and degradation of azo dyes. However, fungi seem to be the most promising organisms capable of decolorizing azo dyes and degrading their toxic intermediaries.

2. Decoloration of Azo Dyes

2.1 Decoloration of azo dyes by fungi

Filamentous fungi are used for bioremediation of textile effluents and dye removal because of their capability to quickly adapt their metabolism in the presence of different carbon and nitrogen sources. In comparison to single cell organisms, fungal mycelium produces extracellular enzymes that dissolve the insoluble substrates and help in tolerating high concentrations of toxicants. Thus, bioremediation using fungi seems to be the most appropriate in the treatment of colored effluents (Ezeronye and Okerentugba 1999). Many researchers have described studies on fungal decoloration and degradation of dyes in wastewater (Hao et al. 2000, Fu and Viraraghavan 2001). Table 5.1 summarizes various fungi used for dye decoloration. *Aspergillus* sp., seems to be the most promising strain capable of decolorizing a large variety of azo dyes. Decoloration of dyes by these fungi is mainly achieved by adsorption and, to a lesser extent, by enzymatic degradation. Adsorption is due to electrostatic

Table 5.1 Dye decoloration studies using various fungi.

Organism	Mechanism	Medium	Parameters (Static or Agitated/rpm/pH/Temperature/Others, if any)	Dye (Concentration)	% Decoloration (Time)	References
Achaetomium strumarium	Biosorption and biodegradation by enzyme NADH-DCIP reductase and laccase	Mineral salt medium	−/−/4/40°C	Acid red 88 (10 mg/l)	99 (96 h)	Bankole et al. (2018)
Aspergillus foetidus	Biosorption	1% Glucose	Agitated/−/8/30°C	Remazol Red (50 mg/l)	98.4 (2 d)	Sumathi and Phatak (1999)
				Remazol Dark Blue (50 mg/l)	99.2 (2 d)	
				Remazol Brilliant Orange (50 mg/l)	68 (2 d)	
				Remazol Brown GK (50 mg/l)	98.7 (2 d)	
		Mineral Salt medium with 1% glucose	Agitated/200 rpm/8/30°C	Drimarene Red BFF3B1 (50 mg/l)	> 95 (48 h)	Sumathi and Manju (2000)
				Drimarene Navy Blue BF F2G1	> 95 (48 h)	
Aspergillus fumigatus	Decoloration	MSM medium with glucose & yeast extract	Agitated/150 rpm/−/28°C	Direct violet (100 μg/ml)	85.8 (2 d)	Abd El-Rahim et al. (2017)
				Methyl Red (10 μg/ml)	91.3 (2 d)	
Aspergillus fumigatus XC6	–	NH₄Cl, Sucrose	Agitated/150 rpm/3–8/30°C	Dye industry effluent	100 (72 h)	Jin et al. (2007)

Fungi	Mechanism	Medium	Conditions	Dye	Decoloration (%)	Reference
Aspergillus niger	Biosorption	Sabouraud's dextrose broth medium	Agitated/250 rpm/9/–	Direct red azo dye (1000 mg/l)	–	Mahmoud et al. (2017)
	Decoloration due to laccase	Glucose, Peptone, Yeast extract, Ammonium tartarate, KH_2PO_4, $MgSO_4$, Trace element	Agitated/ 150 rpm/ 5.5/ 30°C	Synozol RedHF-6BN (20 mg/l)	88 (24 d)	Ilyas and Rehman (2013)
Aspergillus niger 180	Biosorption	–	–/–/30°C	Procion Red MX-5B (200 µg/ml)	30 (3 h)	Almeida and Corso (2014)
	Biodegradation and adsorption	Molasses wastewater, Sucrose, $MgSO_4$, KH_2PO_4, NH_4NO_3	Agitated/200 rpm/5/30°C	Molasses wastewater	69 (4 d)	Miranda et al. (1996)
Aspergillus oryzae	Decoloration	MSM medium with glucose and yeast extract	Agitated/–/–/28°C	Direct violet (100 µg/ ml)	83.7 (2 d)	Abd El-Rahim et al. (2017)
				Methyl Red (10 µg/ml)	87.1 (2 d)	
A. sojae B-10	Biodegradation and Adsorption	Nitrogen limited medium (Glucose, Sodium nitrate, KH_2PO_4, $MgSO_4$)	Agitated/ 220 rpm/–/–	Amaranth (10 mg/l)	97.8 (5 d)	Ryu and Weon (1992)
			–/–/5.0/37°C	Sudan III (10 mg/l)	97.4 (5 d)	
				Congo Red (10 mg/l)	93 (5 d)	

Table 5.1 contd. ...

...Table 5.1 contd.

Organism	Mechanism	Medium	Parameters (Static or Agitated/rpm/pH/Temperature/Others, if any)	Dye (Concentration)	% Decoloration (Time)	References
Aspergillus terreus	Biodegradation	–	–/–/4/30°C	Procion Red MX-5B (200 µg/ml)	100 (336 h)	Almeida and Corso (2014)
Cunninghamella elegans ATCC 36112	Biodegradation	Modified Sabouraud broth medium	Agitated/50 rpm/5.8/28°C	Reactive Red 198 (0.025 mM)	48 (72 h)	Ambrósio and Campos-Takaki (2004)
Myrothecium verrucaria	Adsorption	Potato dextrose broth	–	Orange II (0.025 mM)	83 (96 h)	Brahimi-Horn et al. (1992)
				Orange II (0.25g/l)	70 (–)	
				RS (H/C Red) (0.26 g/l)	86 (–)	
				10B(H/C blue) (0.22 g/l)	95 (–)	
Nigrospora sp.	Decoloration due to laccase	Glucose, Peptone, Yeast extract, Ammonium tartarate, KH_2PO_4, $MgSO_4$, Trace element	Agitated/150 rpm/5.5/30°C	Synozol RedHF-6BN (20 mg/l)	96 (24 d)	Ilyas and Rehman (2013)
Penicillium ATCC74414	Adsorption and Biodegradation	Potato dextrose broth	Agitated/150 rpm/–/25°C	Poly R 478 (0.01% w/v)	100 (4 d)	Zheng et al. (1999)
		Half Murashige Skoog Medium	Agitated/150 rpm/–/25°C	Poly R 478 (0.01% w/v)	100 (4 d)	
		Schenk and Hildebrandt-K_2SO_4 medium	Agitated/150 rpm/–/25°C	Poly S119 (0.01% w/v)	100 (4 d)	

Organism	Mechanism	Media	Conditions	Dye	Decoloration	Reference
Penicillium sp., QQ	Bioadsorption	Potato dextrose broth	Agitated/150 rpm/–/ 25°C	Poly S119 (0.01% w/v)	100 (4 d)	Gou et al. (2009)
		Half Murashige Skoog Medium	Agitated/150 rpm/–/ 25°C	Poly S119 (0.01% w/v)	100 (4 d)	
		Modified Martin Broth	Agitated/150 rpm/–/ 30°C	Reactive Brilliant Red X-3B (100 mg/l)	97.7 (90 h)	

attraction between charged dye molecules and the charged cell surfaces (Kaushik and Malik 2009, Erden et al. 2011, Sen et al. 2016). The binding rate and efficiency depends on the structure of the dye, its concentration, pH, temperature, and the species of the microorganism used.

2.2 Decoloration of azo dyes by wood decaying fungi

Wood decaying fungi are extensively studied microorganisms, and by far the most efficient class of ligninolytic organisms able to degrade different classes of dyes, including azo, reactive, heterocyclic, and polymeric dyes (Novotný et al. 2004). They offer several advantages, making them the organism of choice in wastewater treatment. First, these fungi are ubiquitously found in nature. The substrates used for their growth are cheap and can be easily incorporated during effluent treatment. In addition to their natural substrate, they can degrade a diverse range of recalcitrant organic pollutants, such as polyaromatic hydrocarbons, chlorophenol, polychlorinated biphenyls, and dyes. This distinguishes them from biodegradative bacteria that are more substrate-specific (Reddy 1995). Secondly, the enzymes involved in the oxidation of pollutants are expressed under nutrient limiting conditions and, by being extracellular, they can efficiently oxidize poorly soluble substrates (Reddy 1995). White rot fungi, such as *Phanerochaete chyrsosporium* and *Trametes versicolor*, are the most widely studied organisms for dye decoloration studies. Table 5.2 summarizes dye decoloration studies with various wood decaying fungi. These fungi produce extracellular polyphenol oxidases, such as laccases, lignin peroxidases, and manganese peroxidases that are capable of degrading lignin, azo dyes, and other aromatic compounds.

2.3 Decoloration of azo dyes by yeast

Yeasts are ubiquitous organisms that possess mechanisms to overcome various physical and chemical stresses. They can resist adverse conditions, such as high salt concentration, low pH, and high-strength organic wastewater, etc. Such characteristics make yeasts useful and important organisms for decoloration and degradation of azo dyes. In contrast to filamentous fungi, decoloration by yeast occurs during the exponential growth phase and is associated with primary metabolism (Ramalho et al. 2002, Jafari et al. 2014), and hence requires a short incubation period. Recently, extensive studies have been carried out investigating the potential of yeast in dye decoloration and degradation. Table 5.3 summarizes decoloration of azo dyes by various yeast species. Dye decoloration by yeast is mainly due to biodegradation or biosorption. Biodegradation is mainly by enzymes, such as NADH-dependent reductases and azo reductase (Saratale et al. 2009, Phugare et al. 2010). Phugare et al. (2010) reported 78% decoloration of textile industry effluents in 48 hours with enhanced activities of DCIP reductase and azoreductase; thereby suggesting the prominent role of these enzymes in dye decoloration. Lucas et al. (2006) reported azoreductase-like activity in azo bond cleavage of Reactive black 5 by a wild isolate, *Candida oleophilia*. Similar observations were made by Ramalho et al. (2002), who studied azo dye decoloration with *Candida zeylanoides*. In addition to the above

Table 5.2 Decoloration of various dyes by wood decaying fungi.

Organism	Mechanism	Medium	Parameters (Static or Agitated/rpm/pH/ Temperature/Others, if any)	Dye (Concentration)	% Decoloration (time)	References
Bjerkandera adjusta	Decoloration due to ligninolytic enzymes LiP, MnP	Nitrogen limited medium	Agitated/–/4.5/30°C	Reactive Black 5 (200/l)	100 (7 d)	Mohorčič et al. (2004)
	Decoloration due to ligninolytic enzymes LiP, MnP	Glucose, Yeast extract, Veratryl alcohol, Tartaric acid, Tween 80, $(NH_4)_2HPO_4$, KH_2PO_4, $CaCl_2$, $MgSO_4$, $ZnSO_4$, $MnSO_4$, $CoCl_2$, $CuCl_2$	Agitated/90/4.5/30°C	Artificial dye bath	94 (7 d)	Mohorčič et al. (2006)
Bjerkandera fumosa	Adsorption on fungal biomass or biodegradation	Glucose, Yeast extract, KH_2PO_4, $MgSO_4$, $CaCl_2$, Ammonium tartarate, 2,2 dimethyl succinate, $MnCl_2$	Agitated/180/5/25°C	Acid Red 183 (0.1 g/l)	68 (7 d)	Jarosz-Wilkołazka et al. (2002)
Bjerkandera sp., Strain BOL 13	Decoloration Biodegradation	Glucose, Yeast extract, Asparagine, NH_4NO_3, KH_2PO_4, K_2HPO_4, $MnCl_2$	Static/–/6/–	Reactive Red 2 (100 mg/l) Reactive Blue 4 (100 mg/l)	99 (10 d) 99 (10 d)	Axelsson et al. (2006)
Coriolopsis sp., Strain arf5	Biodegradation	Nitrogen limited medium	Static/–/5/–	Ponceu 2R (0.2 g/l) Orange G Direct Blue 71	100 (10 d) 100 (7 d) 100 (12 d)	Cheng et al. (2016)

Table 5.2 contd. ....

...Table 5.2 contd.

Organism	Mechanism	Medium	Parameters (Static or Agitated/rpm/pH/ Temperature/Others, if any)	Dye (Concentration)	% Decoloration (time)	References
Coriolus versicolor	Decoloration and degradation	MgSO$_4$, CaCl$_2$, KH$_2$PO$_4$, Glucose, NH$_4$H$_2$PO$_4$	Static/–/5/30°C	Biebrich Scarlet	100 (12 d)	Mazmanci et al. (2002)
				Methylene blue (5 ppm)	98 (8 d)	
	Biodegradation and Adsorption	Nitrogen limited medium	Static/–/5/27°C	Acid Green 27 (0.5 g/l)	100 (7 d)	Knapp et al. (1995)
				Cu-phthalocyanine tetrasulfonic acid (0.5 g/l)	99 (7 d)	
				Indigocarmine (0.5 g/l)	100 (7 d)	
				Neutral Red (0.5 g/l)	91 (7 d)	
				Acid red106 (0.5 g/l)	100 (7 d)	
				Mordant Yellow (0.5 g/l)	84 (7 d)	
				Brilliant Yellow (0.5 g/l)	99 (7 d)	
				Reactive Red 4 (0.5 g/l)	90 (7 d)	
				Orange II (0.5 g/l)	100 (7 d)	
				Brilliant green (0.5 g/l)	100 (7 d)	
Datronia sp. KAP10039	Biodegradation with ligninolytic enzymes MnP, Lac	Glucose, K$_2$HPO$_4$, MgSO$_4$, KCl, FeSO$_4$, NH$_4$NO$_3$	Agitated/150/5.5/30°C	Remazol brilliant blue R (200 mg/l)	99.6 (72 h)	Vaithanomsat et al. (2010)
Dichomitus squalens	Ligninolytic enzymes laccase, MnP	N-limited Kirk medium containing 0.1% glucose	Static/–/–/27°C	Reactive black 5	> 90 (40 h)	Eichlerová et al. (2006)
				Orange G (0.5 g/l)	80 (14 d)	

Fungus	Method	Medium	Conditions	Dye (concentration)	% (time)	Reference
Echinodontium taxodii	Biodegradation with ligninolytic enzymes Laccase	Malt Extract medium containing 0.1% glucose	Static/–/–/27°C	Remazol Brilliant Blue R (0.5 g/l)	98 (14 d)	Han et al. (2014)
		Glucose, Ammonium tartarate, KH_2PO_4, gSO_4, $CaCl_2$, Trace element, Lignin	Static/–/–/28°C	Remazol Brilliant Violet 5R (600 mg/l)	97.75 (6 d)	
				Direct Red 5B (600 mg/l)	76.89 (6 d)	
				Direct Black 38 (600 mg/l)	43.44 (6 d)	
				Direct Black 22 (600 mg/l)	44.75 (6 d)	
Funalia trogii	Adsorption and decoloration through microbial metabolism	Sabouraud Dextrose broth	Agitated/150/6-11/30°C	Astrazon red FBL	72 (2 h)	Yesilada et al. (2002)
	Ligninolytic enzyme laccase	$MgSO_4$ $CaCl_2$ Glucose $(NH_4)_2PO_4$	Static/–/5/30°C	Remazol brilliant blue royal (100 mg/l)	90 (48 h)	Erkurt et al. (2007)
				Drimarene blue CL-BR (100 mg/l)	90 (48 h)	
Fungus F29	Decoloration	Nitrogen limited medium	Agitated/–/5.5-6.3/27°C	Orange II (1000 mg/l)	98 (2 d)	Knapp et al. (1997)
Ganoderma cupreum AG-1	Biodegradation with ligninolytic enzymes MnP, Lac	Mannose, Yeast Extract, $MgSO_4$, KH_2PO_4, KCl	Agitated/–/4.5/30°C	Reactive violet-1 (5 g/l)	93 (4 d)	Gahlout et al. (2013)
Ganoderma sp. WR-1	Biodegradation with ligninolytic enzyme laccase	Starch, Yeast extract, KH_2PO_4, Na_2HPO_4, $MgSO_4$, $CaCl_2$, $FeSO_4$, $MnSO_4$, $ZnSO_4$, $CuSO_4$	Agitated/150 rpm/5.5/28°C	Amaranth (100 ppm)	97 (8 h)	Revankar and Lele (2007)

Table 5.2 contd. ...

...Table 5.2 contd.

Organism	Mechanism	Medium	Parameters (Static or Agitated/rpm/pH/ Temperature/Others, if any)	Dye (Concentration)	% Decoloration (time)	References
Irpex lacteus			Agitated/150 rpm/5.5/ 28°C	Reactive Orange 16 (100 ppm)	65 (8 h)	Novotný et al. (2001)
			Agitated/150 rpm/5.5/ 28°C	Cibacron Brilliant Red (100 ppm)	50 (8 h)	
	Biodegradation using enzymes MnP, LIP, laccase	Low nitrogen mineral medium	Static/–/4.5/22°C	Methyl Red (150 µg g^{-1})	56 (2 w)	
				Congo Red (150 µg g^{-1})	58 (2 w)	
				Remazol Brilliant blue R (150 µg g^{-1})	92 (2 w)	
			Static/–/4.5/28°C	Reactive Orange 16 (150 mg/l)	86 (7 d)	Novotný et al. (2004)
	Dye decolourization due to enzymes other than LMEs	Mineral medium with 10 g/l Glucose Diammonium tartarate	Agitated/120/5/28°C	Reactive Black 5 (100 mg/l)	90 (10 d)	Máximo and Costa-Ferreira (2004)
	Decoloration by oxidative enzyme (laccase)	Nitrogen free medium	Agitated/100/–/28°C	Reactive orange 16 (150 mg/l)	88.5 (24 h)	Svobodová et al. (2008)
				Remazol brilliant blue R (150 mg/l)	98.6 (24 h)	

Organism	Mechanism	Medium	Conditions	Dye (concentration)	% (time)	Reference
Kuehneromyces mutabilis	Adsorption on fungal biomass or biodegradation	Glucose, Yeast extract, KH_2PO_4, $MgSO_4$, $CaCl_2$, Ammonium tartarate, 2,2 dimethyl succinate, $MnCl_2$	Agitated/180/5/25°C	Acid Red 183 (0.1 g/l)	100 (7 d)	Jarosz-Wilkolazka et al. (2002)
Phanerochaete chrysosporium	Biosorption and Biodegradation	Nitrogen Limited medium	Static/−/−/39°C	Direct Green 6 (120 mg/l)	100 (6 d)	Pazarlioglu et al. (2005)
				Direct Brown 2 (120 mg/l)	100 (6 d)	
				Direct Blue 15 (120 mg/l)	100 (6 d)	
				Direct Red 23 (120 mg/l)	100 (8 d)	
				Direct Orange 26 (120mg/l)	100 (8 d)	
				Congo red (120 mg/l)	57 (10 d)	
				Direct Black 38 (120 mg/l)	77 (10 d)	
	Biodegradation with lignin peroxidases and adsorption	N-limited medium	−/−/4.5/39°C	Orange II (57 μM)	100 (5 d)	Cripps et al. (1990)
				Azure B (16 μM)	100 (1 d)	
				Congo Red (76 μM)	97 (5 d)	
				Tropaeolin O (63 μM)	96 (2 d)	
	Adsorption, biodegradation with lignin degrading enzymes	N-limited medium	Agitated/−/−/−	Congo Red (500 mg/l)	70 (2 d)	Tatarko and Bumpus (1998)

Table 5.2 contd. ...

...Table 5.2 contd.

Organism	Mechanism	Medium	Parameters (Static or Agitated/rpm/pH/Temperature/Others, if any)	Dye (Concentration)	% Decoloration (time)	References
	Adsorption and biodegradation with MnP	Glucose, Ammonium tartarate, BIII Mineral medium, Thiamine	Agitated/–/–/37°C	Acid Black 1 (100 mg/l)	>90 (–)	Urra et al. (2006)
				Reactive black 5 (100 mg/l)	>90 (–)	
				Reactive Orange 16 (100 mg/l)	>90 (–)	
	Biodegradation with ligninolytic enzymes MnP, Lac	Glucose, Ammonium tartarte, KH_2PO_4, $MgSO_4$, $CaCl_2$, Veratryl alcohol, Vitamin B1	Agitated/160/–/37°C	Reactive Brilliant Red K-2BP (30 mg/l)	89 (1 d)	Gao et al. (2008)
	Biodegrdation with ligninolytic enzymes MnP, LiP, Lac	Wheat straw, $MnSO_4$, Sodium lactate, Minimal medium	Agitated/135/4.5/30°C	Reactive Brilliant blue R (500 mg/l)	99 (14 h)	González Ramírez et al. (2014)
	Decoloration and Degradation	Minimal medium	Static/–/–/25–30°C	Indigo blue dye (0.02% v/v)	75 (4 d)	Balan and Monteiro (2001)
P. chrysosporium KCCM 60256	Ligninolytic enzymes laccase, LiP, MnP	Kirk Basal Salt medium	Agitated/100/4.5/28°C	Acid Blue 350	60 (48 h)	Chulhwan et al. (2004)
				Acid yellow 99 (100 mg/l)	20 (48 h)	
				Acid Red 114 (100 mg/l)	55 (48 h)	
Phellinus gilvus (CCB 254)	Decoloration and Degradation	Minimal medium	Static/–/–/25–30°C	Indigo blue dye (0.02% v/v)	100 (4 d)	Balan and Monteiro (2001)

Organism	Mechanism/Enzyme	Medium	Conditions	Dye (conc.)	Decoloration (%)	Reference
Pleurotus ostreatus	Ligninolytic enzyme laccase	$MgSO_4$, $CaCl_2$, Glucose, $(NH_4)_2PO_4$	Static/–/5/30°C	Remazol brilliant blue royal (100 mg/l)	44.9 (48 h)	Erkurt et al. (2007)
				Drimarene blue CL-BR (100 mg/l)	57.79 (48 h)	
Pleurotus sajor-caju (CCB020)	Decoloration and Degradation	Minimal medium	Static/–/–/25–30°C	Indigo blue dye (0.02% v/v)	94 (4 d)	Balan and Monteiro (2001)
Pycnoporus sanguineus (Morr) (CCB 458)	Decoloration and Degradation	Minimal medium	Static/–/–/25–30°C	Indigo blue dye (0.02% v/v)	91 (4 d)	Balan and Monteiro (2001)
Schizophyllum commune	Biodegradation with ligninolytic enzymes MnP, Lac	Kirk's basal salt medium	Agitated/120/4.5/35°C	Solar golden yellow R (0.1 mg/ml)	73 (6 d)	Asgher et al. (2008)
Stropharia rugoso-annulata	Adsorption on fungal biomass or biodegradation	Glucose, Yeast extract, KH_2PO_4, $MgSO_4$, $CaCl_2$, Ammonium tartarate, 2,2 dimethyl succinate, $MnCl_2$	Agitated/180/5/25°C	Acid Red 183 (0.1 g/l)	100 (7 d)	Jarosz-Wilkołazka et al. (2002)
T. versicolor ATCC 200801	Ligninolytic enzymes laccase, LiP, MnP	Kirk Basal Salt medium	Agitated/100/4.5/28°C	Acid Blue 350	95 (48 h)	Chulhwan et al. (2004)
				Acid yellow 99 (100 mg/l)	89 (48 h)	
				Acid Red 114 (100 mg/l)	95 (48 h)	

Table 5.2 contd....

...Table 5.2 contd.

Organism	Mechanism	Medium	Parameters (Static or Agitated/rpm/pH/ Temperature/Others, if any)	Dye (Concentration)	% Decoloration (time)	References
T. versicolor KCTC 16781	Ligninolytic enzymes laccase, LiP, MnP	Kirk Basal Salt medium	Agitated/100/4.5/28°C	Acid Blue 350	98.4 (48 h)	
				Acid yellow 99 (100 mg/l)	99 (48 h)	
				Acid Red 114 (100 mg/l)	100 (48 h)	
Trametes versicolor	–	Modified Kirk Medium	Agitated/–/–/21°C	Carpet dye effluent	95 (10 h)	Ramsay et al. (2005)
Thelephora sp.	Ligninolytic enzymes laccase, LiP, MnP	Carbon limited medium	Agitated/–/–/30°C	Orange G (50 μM)	33.3 (9 d)	Selvam et al. (2003)
				Congo Red (50 μM)	97.1 (8 h)	
				Amido black 10B (25 μM)	98.8 (24 h)	
Trametes pubescens Cui 7571	Adsorption and degradation by laccase, LiP, MnP, tyrosinase	Yeast extract, Glucose, KH_2PO_4, $MgSO4. 7H_2O$, $ZnSO4.7H_2O$, Vitamin B1, $CuSO_4$, Tween 80, Ferulic acid	Agitated/150/5/28°C	Congo Red (50 mg/l)	98.8 (7 d)	Si et al. (2013)

Table 5.3 Decoloration of various azo dyes by different yeast cultures.

Organism	Mechanism	Medium	Parameters (Static or Agitated/rpm/ pH/Temperature)	Dye (Concentration)	% Decoloration (Time)	References
Candida albicans	Biosorption and biodegradation	Mineral salt medium	Agitation/150/7.0/ 37°C	Direct violet 51 (100 μg/ml)	87 (96 h)	Vitor and Corso (2008)
Candida krusei	Biodegradation	Glucose, KH_2PO_4, $(NH_4)_2SO_4$, $MgSO_4$, Yeast extract	Agitation/200/5–6/ 28°C	Reactive Brilliant Red K-2BP (200 mg/l)	99 (24 h)	Yu and Wen (2005)
Candida oleophila	Biodegradation with azo reductase enzyme	Glucose, $(NH_4)_2SO_4$, KH_2PO_4, $MgSO_4$, Yeast extract, $CaCl_2$	Agitation/150/–/ 26°C	Reactive Black 5 (200 mg/l)	> 99 (24 h)	Lucas et al. (2006)
Candida rugosa INCQS 71011	Biodegradation	Potato dextrose broth	Agitation/140/–/–	Reactive Red 198 (100 mg/l)	100 (48 h)	Nascimento et al. (2013)
				Reactive Red 141 (100 mg/l)	100 (144 h)	
				Reactive Blue 214 (100 mg/l)	76 (168 h)	
Candida zeylanoides	Biodegradation	Yeast extract, Bactopeptone, Glucose	Agitation/150/5/25°C	Azo dye	90 (7 d)	Martins et al. (1999)
Candida zeylanoides UM2		Glucose, $(NH_4)_2SO_4$, Yeast extract, KH_2PO_4, $MgSO_4$, $CaCl_2$	Agitation/120/–/25°C	Orange 52 (0.2 mM)	100 (40 h)	Ramalho et al. (2002)
				Orange 7 (0.2 mM)	100 (40 h)	

Table 5.3 contd. ...

...Table 5.3 contd.

Organism	Mechanism	Medium	Parameters (Static or Agitated/rpm/ pH/Temperature)	Dye (Concentration)	% Decoloration (Time)	References
Debaryomyces polymorphus	Biodegrdation with MnP	Glucose, KH_2PO_4, $(NH_4)_2SO_4$, $MgSO_4$, Yeast extract	Agitation/140/5-6/ 28°C	Reactive Black 5 (200mg/l)	98 (24 h)	Yang et al. (2005)
Issatchenkia orientalis JKS6	Biodegradation	Glucose, Yeast extract, NH_4Cl, KH_2PO_4,K_2HPO_4, $MgCl_2$	Agitation/150/7.0 ± 0.2/32°C	Reactive Black 5 (200 mg/l)	90 (18 h)	Jafari et al. (2014)
Kluyveromyces marxianus IMB3	Adsorption/Biosorption	Yeast growth medium with 5% glucose	Agitation/150/5/45°C	Remazol Black B	37 (–)	Bustard et al. (1998)
				Remazol Red	68 (–)	
				Remazol golden Yellow	33 (–)	
		Glucose mineral medium	Agitation/150/5/37°C	Remazol Black B (25-200 mg/l)	> 98 (24 h)	Meehan et al. (2000)
Magnusiomyces ingens LH-F1	Biodegradation	K_2HPO_4, $MgSO_4$, Ammonium sulphate, Glucose	Agitation/–/7/30°C	Acid Red B 20 mg/l	97 (8 h)	Tan et al. (2014)
				Acid Red 3R 20 mg/l	97 (18 h)	
				Acid Brilliant Scarlet GR (20 mg/l)	95 (10 h)	
				Acid orange II (20 mg/l)	91 (30 h)	

Organism	Mechanism	Medium	Condition	Dye (concentration)	% Decoloration (time)	Reference
Pseudozyma rugolosa Y-48	Biodegradation	Glucose, KH_2PO_4, $(NH_4)_2SO_4$, $MgSO_4$, Yeast extract	Agitation/200/5-6/28°C	Reactive Yellow 3RS(20 mg/l)	97 (14 h)	Yu and Wen (2005)
				Reactive Green KE-4BD	96 (14 h)	
				Reactive Brilliant Red K-2BP (200 mg/l)	99 (24 h)	
Saccharomyces cerevisae	Bioaccumulation	Beet Molasses, $(NH_4)_2SO_4$, KH_2PO_4	Agitation/100/-/25°C	Remazol Blue	84.6 mg/g accumulated	Aksu (2003)
				Remazol Black B	88.5 mg/g accumulated	
Saccharomyces cerevisae MTCC 463	Biodegradation with azoreductase	Plain water	Stationary/-/-/30°C	Remazol Red RB	48.8 mg/g accumulated	Jadhav et al. (2007)
				Methyl Red (100 mg/l)	100 (16 min)	
Trichosporon beigelii NCIM 3326	Biodegradation due to NADH-DCIP reductase and azo reductase	$(NH_4)_2SO_4$, NH_4Cl, KH_2PO_4, $MgSO_4$, $CaCl_2$, $FeCl_3$, Yeast extract, NaCl, $NaHCO_3$, Trace element	Static/-/7/37°C	Navy Blue HER (50 mg/l)	100 (24 h)	Saratale et al. (2009)
Trichosporon cutaneum	Biodegradation	Mineral medium	Agitation/150/6.5/28°C	Olive mill water waste	80 (8 d)	Chtourou et al. (2004)
Yarrowia lipolytica	Decoloration	Nitrogen limited medium	Agitation/180/4/28°C	Olive mill water waste	36 (100 h)	Papanikolaou et al. (2008)

enzymes, few reports also suggest the production of the ligninolytic enzyme MnP (Yang et al. 2005, Martorell et al. 2017), LiP (Jadhav et al. 2007) by yeast and its involvement in azo dye decoloration.

Another mechanism responsible for dye decoloration in yeast is biosorption, which involves nonspecific binding of the dye to the periphery of the cell, followed by specific accumulation in the wall and inside the cell (Vitor and Corso 2008). Aksu and Dönmez (2003) studied the biosorption capacities of nine different kinds of dried yeast for reactive azo dye, Remazol Blue. They suggested that the difference in dye binding capacity is due to variation in the structure, functional groups, surface area, and morphological differences among various genera and species used. They also proposed that polysaccharides, proteins, lipids, which are the building blocks in the cells wall of yeasts, contain large numbers of functional groups, such as amino, carboxylic, sulfhydryl, phosphate, thiol, etc., capable of binding dye molecules. Based on their affinity and specificity, variation in dye binding efficiency is observed (Fu and Viraraghavan 2001, Aksu and Dönmez 2003). Various other studies on dye biosorption using yeast have been carried out and proved yeast as a good adsorbent with high adsorption capacities, suggesting biosorption as an alternative for dye removal (Bustard et al. 1998, Meehan et al. 2000, Aksu and Dönmez 2003).

3. Parameters Affecting Azo Dye Decoloration by Fungi

3.1 Carbon source

Physiological differences among the cultures and the enzymes may account for differences in their decolorizing ability. Ambrósio and Campos-Takaki (2004) observed that medium composition was important for dye decoloration, and suggested that selective decoloration can be achieved by combining different parameters. The ligninolytic enzymes responsible for decoloration are expressed when the carbon source becomes limiting. Swamy and Ramsay (1999a) studied the effect of glucose on sequential dye decoloration by *T. versicolor*. They observed that decoloration was not stimulated by carbon limitation, but a minimum amount of glucose is required for dye decoloration. With glucose depletion, the rate of decoloration was found to decrease, and was restored with glucose replenishment. Kapdan et al. (2000) reported glucose as the most efficient carbon source for decoloration of Everzol Turquoise Blue G, as compared to fructose and starch as sole carbon sources. Zhang et al. (1999a) observed the effect of varying carbon sources on the decoloration of cotton bleaching effluents by an unidentified white-rot fungus. Glucose, starch, maltose, and cellobiose were found to be good carbon sources as compared to sucrose, lactose, xylan, xylose, methanol, and glyoxal. Even though glucose is the simplest and the most readily available and used carbon source for fungi, it is expensive and generally not preferred in wastewater treatment. Revankar and Lele (2007) studied the effect of varying carbon sources on Amaranth decoloration. They found lowest dye decoloration with fructose, while starch was found to be the best with 79% decoloration in 8 hours of incubation. This was because of the enhanced laccase activity in the presence of starch. In accordance with this, Sumathi and Phatak (1999) reported 98% decoloration of Drimarene Red and Drimarene blue by *Aspergillus foetidus* after 72 hours of incubation with starch as carbon source.

3.2 Nitrogen source

The nature and amount of the nitrogen source exert a prominent effect on ligninolytic enzyme production by wood decaying fungi, altering their effect on the dye decoloration. In several species, the ligninolytic enzymes responsible for dye decoloration are secreted by the fungal cells under nitrogen limiting conditions (Glenn and Gold 1983, Spadaro et al. 1992, Zhen and Yu 1998). For example, *Aspergillus sojae B-1* decolorized the azo dyes Amaranth, Congo red, and Sudan III in a nitrogen poor medium after 3–5 days of incubation (Rodríguez et al. 1999). Tatarko and Bumpus (1998) observed that supplementation of nitrogen inhibited decoloration of Congo red. In accordance with this, our research group also observed that the lowest concentration of yeast extract was optimum for Amaranth decoloration by *Ganoderma* sp. (Revankar and Lele 2007). This is mainly because the ligninolytic enzyme activity responsible for dye decoloration by wood decaying fungi is triggered at low nitrogen concentration. Similar observations were made with many other species, such as *Phanerochaete chrysosporium*, *Phlebia brevispora* (Leatham and Kirk 1983), *T. versicolor*, *Phlebia radiata* (Zhao et al. 1996), *Pleurotus ostreatus* (Fu et al. 1997). It can also be hypothesized that at high concentrations, nitrogen is easily available for metabolism and, hence, there is no need to find an alternative nitrogen source from the dye molecule causing slow decoloration rate (Srikanlayanukul et al. 2008). However, for some fungi, such as *Bjerkandera adjusta* (Kaal et al. 1993), *P. ostreatus* (Kaal et al. 1995), *Ceriporiopsis subvermispora* (Rüttimann-Johnson et al. 1993), high nitrogen concentration stimulated ligninolytic enzyme production.

The nature of the nitrogen source also affects dye decoloration efficiency. Organic nitrogen sources influence production of ligninolytic enzymes, affecting dye decoloration rate. For example, tryptone and peptone improved laccase production by *Trametes gallica* (Dong et al. 2005). Similar results were reported by our group with maximum Amaranth decoloration using yeast extract as organic nitrogen source, and minimum in the presence of ammonium sulphate and urea (Revankar and Lele 2007). However, some reports support the use of inorganic nitrogen sources in dye decoloration, such as ammonium tartrate (Cripps et al. 1990, Swamy and Ramsay 1999b), ammonium chloride (Hatvani and Mécs 2002), and ammonium oxalate (Srikanlayanukul et al. 2008).

Extensive research is being carried out to check if fungi can utilize azo dye as carbon or nitrogen source. This is possible if these organisms are adapted by gradually increasing the dye concentration and decreasing the carbon/nitrogen source until they survive only in the presence of the dye, for example, decoloration of azo dye Methyl red by *Saccharomyces cerevisiae* in plain distilled water (Jadhav et al. 2007). However more studies need to be performed for the development of such strains.

3.3 Aeration and agitation

Aeration and agitation play a significant role in dye decoloration. Azo dye degradation involves hydroxylation and ring opening to form simple aromatic compounds. Hence for complete degradation, aerobic conditions are rather preferred than anaerobic (Pandey et al. 2007). In fungi, agitation results in change in morphology from mat growth to pellets, improving the nutrient and oxygen mass transfer between the cells and the medium. Knapp et al. (1997) reported 45% decoloration of Orange II dye after 23 hours of incubation in static cultures, whereas 97.5% was achieved in shaking conditions. Decoloration of Amaranth with two strains *Coriolus versicolor* and indigenous isolate *Ganodermasp WR-1* was studied in our lab. Agitated cultures of *C. versicolor* showed 96% decoloration, while only 26% was observed after 6 hours of incubation under stationary conditions. With the indigenous isolate *Ganoderma* sp., *WR-1* after 8 hours of incubation, 75% decoloration was observed in agitated cultures, whereas only 27% decoloration was observed in static cultures (Revankar and Lele 2007).

Similar observations were made with *Schizophyllum commune* (Belsare and Prasad 1988), *Phanerochaete chrysosporium* (Glenn and Gold 1983), *Trametes villosa* (Soares and Durán 1998, Machado et al. 2006), *Pleurotus sanguineus* (Machado et al. 2006), and *T. versicolor* (Pilatin and Kunduhoğlu 2011). Static cultures result in thick mycelial growth on the surface, which limits oxygen transfer to the cell beneath the surface and in the medium, further decreasing the enzyme activity and reducing decoloration (Swamy and Ramsay 1999b, Jarosz-Wilkołazka et al. 2002). In addition, reports also suggest that, in static cultures, aerial mycelia which are rich in hydrophobins, are produced, also affecting adsorption (Prigione et al. 2008).

3.4 Temperature

Temperature is a key factor for every process associated with growth of microorganisms. Dye decoloration rate is maximum at optimal growth temperature. Decoloration rate of azo dyes increases up to optimal temperature, after which there is a substantial decrease in the decoloration ability. For most fungi the optimal growth temperature is about 25°C to 30°C. Increasing temperature has a negative effect on fungal growth and enzyme activity, thereby affecting dye decoloration. For example, decoloration of Reactive brilliant blue R by *T. versicolor* was studied at increasing temperature. It was observed that 79% decoloration was achieved at 25°C, 97% at 30°C, 51% at 35°C, and 37% at 40°C. This is because for microbial or enzyme-based systems, high incubation temperature leads to inhibition of growth and denaturation of enzymes, decreasing dye decoloration efficiency. Temperature also affects biosorption of the dye. Erden et al. (2011) observed that *T., versicolor* absorbs 44 mg/g of Sirius Blue K-FCN at 7°C, 58 mg/g at 26°C, and 48 mg/g at 45°C, thus suggesting that biosorption increases with temperature because of higher surface activity and kinetic energy of the dye molecules at high temperature.

3.5 Buffer and pH

The choice of buffer has a pronounced effect on pH stability after dye incorporation and, consequently, dye decoloration. Swamy and Ramsay (1999c) studied the effect of the buffer and pH on dye decoloration. They observed that in absence of dye, changes in pH due to growth of fungi were insignificant. After dye addition, with time, the pH was found to increase, resulting in fragmentation of mycelial pellets and no dye decoloration. However, in the absence of dye, buffers maintained a stable pH. Therefore, it was concluded that the presence of dyes reduces the buffering capacity of buffers; pH is also a key factor for fungal growth and dye decoloration. Ligninolytic enzymes show maximum activity at low pH, hence dye decoloration is better at low pH. Knapp et al. (1997) observed that the initial pH of the medium affected decoloration of Orange II. Media with initial pH values in the range of 5.5–6.3 gave similarly good results, whereas at pH greater than 7, no decoloration was observed. Kapdan et al. (2000) studied decoloration of Everzol Turquoise Blue G by *C. versicolor* at different pH. Maximum decoloration (99%) was observed at pH 4.5, which was reduced to 80% at pH 5, and to about 50% at pH between 6 and 7.

Similar observations were made by Asgher et al. (2008) in decoloration of Solar golden yellow R by *Schizophyllum commune,* and Parshetti et al. (2007) in decoloration of Reactive Blue-25 by *Aspergillis ochraceus* NCIM-1146. Various studies have been carried out with different fungi to determine the optimum pH for dye decoloration. The optimum pH for decoloration with *A. niger* was 5 (Miranda et al. 1996), with *S. commune* was 4–5 (Belsare and Prasad 1988), and with marine fungi was 4.5 (Raghukumar et al. 1996). Thus, it can be concluded that ligninolytic enzymes show maximum activity at lower pH, which improves dye decoloration.

3.6 Dye concentration and structure

Slower dye decoloration rates are observed with increased dye concentrations (Selvam et al. 2003). Decreased decoloration rates with high dye concentration was reported by Mou et al. (1991) and Young and Yu (1997). A similar observation was made by Zhang et al. (1999a) with increasing concentrations of cotton bleaching effluents. Aksu et al. (2007) reported 92% decoloration of Remazol Black B by *T. versicolor* within 8 days of incubation, whereas 77% decoloration was achieved after 14 days, when the initial dye concentration was increased. Thus, it is concluded that decoloration of dyes decreases with increasing dye concentrations due to the reduction in fungal growth at toxic concentrations. Many other studies have been carried out with varying concentrations of dye, and lower concentrations were reported to be best decolorized (Levin et al. 2004, Wells et al. 2006, Niebisch et al. 2010).

Along with dye concentration, dye decoloration capacity of each fungus is dependent on the molecular structure of the dye. As compared to azo dyes, fungi can decolorize anthraquinone dyes easily and at a much higher rate (Jarosz-Wilkołazka et al. 2002). Since microbial and enzyme decoloration mainly involve binding of dyes to enzymes, small structural differences of dyes are expected to markedly affect the decoloration rate. The number, position, and type of functional groups influence the decoloration process (Spadaro et al. 1992, Martins et al. 2002, Urra et al. 2006). Spadaro et al. (1992) reported that *P. chrososporium* could mineralize dyes with hydroxyl, amino, acetamide, or nitro functional groups on the aromatic ring rather than dyes with unsubstituted rings. On the other hand, the relationship between dye substitution pattern and biodegradation of sulfonated azo dyes was studied by Paszczynski et al. (1992). They reported that *P. chryrsosporium* could mineralize all sulfonated azo dyes and the susceptibility to degradation was not significantly influenced by the substitution pattern of the dyes. Knapp et al. (1995) observed that Brilliant yellow was easily decolorized in comparison to structurally similar Chrysophenine, and concluded that even relatively small structural differences, such as electron distribution, charge density, and stearic factors have a pronounced effect on dye decoloration. In contrast to this, Podgornik et al. (1999) reported that decoloration does not depend on the chromophore, sign, and charge distribution, but on the auxochromic group. However, more studies are needed to establish a relationship between dye structure and degradation by fungi.

3.7 *Metal salt*

Metal salts increase the binding of the dye with the fiber. Hence, along with dyes, large amount of these salts are released in the effluent that affect the dye decolorizing capacity of the fungi. Si et al. (2013) studied the effect of copper sulfate, ferric chloride, calcium chloride, manganese sulfate, potassium chloride, and sodium chloride on decoloration of Congo red by *T. pubescens*. In the presence of copper and manganese, dye decoloration was enhanced, whereas in the presence of iron and calcium, it decreased. The presence of sodium and potassium did not show any effect on dye decoloration. Moreira et al. (2000) also reported enhanced decoloration of Poly R-478, Orange II, and Reactive Blue 38 by *Bjerkandera* sp., *BOS55*, *P. chrysosporium, and Phanerochaete sordida* in the presence of manganese. This might be due to the inductive effect of copper and manganese on laccase and manganese peroxidase production, respectively, resulting in a synergistic effect on dye decoloration (Revankar and Lele 2006, Kaushik and Malik 2009). However, at higher concentration, these salts are toxic to fungi, affecting growth and dye decoloration capacity. The presence of heavy metals has a deleterious effect on dye decoloration. Aksu et al. (2007) reported that 15 mg/L of Cr(VI) inhibited the growth and decreased the dye decoloration rate of Remazol Black B by *T. versicolor*. Similarly, Sumathi and Manju (2000) observed an inhibitory effect of Cr(VI) on decoloration of Drimarene Red by *A. foetidus*.

3.8 *Surfactant*

Surfactants are occasionally used during the dyeing process, and, hence, may be present in effluents. Si et al. (2013) studied the effect of surfactants, such as SDS, Tween 40, Tween 60, and Tween 80 on decoloration of Congo red by *T. pubscens*. The presence of these surfactants enhanced dye decoloration by increasing the amount of dissolved oxygen available and the permeability of the cell membrane. Several authors have reported an improvement in enzyme excretion ability by increasing the permeability of plasma membranes (Asther et al. 1988, Ürek and Pazarlioğlu 2005), thereby improving the dye decoloration rate. However, increase in surfactant concentration results in low adsorption of dye, thus decreasing the dye decoloration (Ürek and Pazarlioğlu 2005, Brahimi-Horn et al. 1992).

3.9 *Mediators*

The role of ligninolytic enzymes, such as laccase, in bioremediation is limited to phenolic compounds because of their low oxidation potential. Application of this enzyme in presence of mediators resulted in increased oxidation of phenolic and nonphenolic compounds. Mediators are small molecules that, after oxidation by the enzyme, diffuse away from the active site and, in turn, oxidize any substrate that could not enter the active site. Thus, they act as electron shuttles and increase the range of substrates (Fabbrini et al. 2002). In addition, mediators can diffuse far away from the mycelium to sites that are difficult to reach by the enzyme itself (Camarero et al. 2005).

The first synthetic laccase mediator used was 2,2-azinobis-(3-ethylbenzothiazoline-6-sulfonic acid) (ABTS). Other synthetic mediators include, 1-hydroxybenzotriazole (HBT), violuric acid, N-hydroxyacetanilide (NHA), 3-hydroxyanthranilic acid (HAA), etc. Natural mediators include syringaldehyde, acetosyringone, vanillyl alcohol, vanillin, ferulic acid, p-coumaric acid, etc. The effect of redox mediators on the decoloration rates of recalcitrant azo dye Reactive black 5 was investigated by Camarero et al. (2005). They concluded that the natural mediators, syringaldehyde and acetosyringone, can be used as eco-friendly, potentially cheap laccase mediators for various environmental applications. However, because of the wide difference in properties and redox potentials of azo dyes in effluents, the effect of mediators in the decoloration of an effluent is still not clear.

4. Mechanism for Decoloration of Azo Dyes by Fungi

The basic step in dye decoloration is destruction of the chromophore. Generally, the chromophore in azo dyes has conjugated double bonds that can be cleaved into smaller molecules (Peralta-Zamora et al. 1999). The breaking of azo bonds depends on the identity, number, and position of the functional group in the aromatic region and its resultant interactions with the azo group (Mechsner and Wuhrmann 1982). The next step in degradation of azo dyes is the breakdown of aromatic rings, which depends on the type of the substituent. The presence of easily biodegradable functional groups, such as phenolic, amino, acetamide, 2-methoxyphenol, etc., enhances the rate of degradation (Paszczynski et al. 1992, Swamy and Ramsay 1999c).

Studies suggest that, in fungi, the major steps for dye decoloration include binding of dyes to fungal mycelium, physical adsorption, and enzymatic degradation by extracellular and intracellular enzymes (Young and Yu 1997, Saratale et al. 2007). Physical adsorption on the mycelium surface results in close contact among the degrading enzymes, which are bound to the surface of the hyphae and the chromophore group (Evans et al. 1991, Yang et al. 2003). Thus, an added effect is observed between adsorption in the biomass and further degradation by the extracellular ligninolytic enzymes (Urra et al. 2006). The extent of decoloration is based on the adsorption ability and affinity to fungal hyphae, whereas the rate of decoloration depends on dye structure, nitrogen level, and enzymatic activity in the medium. Many researchers (Chivukula and Renganathan 1995, McMullan et al. 2001, Campos et al. 2001, Si et al. 2013, Almeida and Corso 2014) have attempted to explain the degradation pathway for different dyes based on their structure and the metabolites formed. However, due to the inherent complexity of both the dye structure and the enzymatic transformation, elucidation of the exact degradation pathways for all dyes becomes a difficult task.

Two enzyme families, azoreductases and laccases, show a great potential in enzyme decoloration and degradation of azo dyes. Generally, breakdown of azo compounds by azoreductases takes place under anaerobic conditions (Chen et al. 2005, Sen et al. 2016) except for *Pseudomonas* strains reported by Kulla (1981), which can aerobically degrade certain azo dyes. Laccases are copper-containing polyphenol oxidases that catalyze the oxidation of both phenolic and nonphenolic compounds (Bourbonnais et al. 1995). In addition to these, there are one or more extracellular lignin-modifying enzymes (LME) produced by wood decaying fungi, such as manganese peroxidases (MnP), lignin peroxidases (LiP), and phenol oxidase that also play an important role in decoloration and degradation of azo dyes.

4.1 *Azoreductases*

Azoreductases (EC 1.7.1.6) are a significant group of enzymes expressed in bacteria and fungi capable of azo dye degradation. These enzymes undergo reductive breakdown of the azo bond to yield corresponding aromatic amines (colorless product), resulting in complete decoloration of azo dyes (Pandey et al. 2007). They catalyze these reactions in presence of cofactors, such as NADH, NADPH, and $FADH_2$ that act as electron donors.

4.2 *Laccases*

Laccases (EC 1.10.3.2) are copper-containing ligninolytic enzymes that oxidize various aromatic and non-aromatic compounds by removing hydrogen ions and creating radicals. They possess low substrate specificity, which can be extended by addition of redox mediators, such as ABTS (2,2'-azino-bis(3-ethylbenzthiazoline) sulphonic acid), 1-hydroxybenzotriazole (Reyes et al. 1999, Abadulla et al. 2000, Soares et al. 2001, Singh et al. 2015) and, thus can degrade a variety of synthetic dyes (Swamy and Ramsay 1999c). Many researchers have studied laccases in large scale for their ability to mineralize azo dyes (Chivukula and Renganathan 1995, Blánquez et al. 2004, Novotný et al. 2004). Laccases are considered an attractive and potential option for the development of new methodologies to treat dye effluents.

4.3 *Lignin peroxidases*

Lignin peroxidases (LiPs) (EC 1.11.1.7) are extracellular glycosylated heme proteins. They catalyze H_2O_2-dependent one-electron oxidation of a variety of lignin related aromatic structures, resulting in aryl cation radicals that undergo various non-enzymatic reactions, yielding multiple end products. These enzymes are non-specific and, hence, can oxidize xenobiotics having structural similarity to lignin substructures (Reddy 1995). In addition, they possess very high oxidation-reduction potential and can oxidize compounds that are not attacked by other peroxidases.

4.4 *Manganese peroxidases*

Manganese peroxidases (MnPs) (EC 1.11.1.13) are glycosylated glycoproteins with a prosthetic heme group. They were first discovered in *Phanerochaete chyrsosporium* and later in various other white-rot fungi. MnPs can oxidize and polymerize lignin as well as other xenobiotics. The reaction mainly involves conversion of Mn^{+2} to Mn^{+3}, which is stabilized by chelating agents such as oxalic acids. The chelated Mn^{+3} acts as low molecular weight redox mediator that can diffuse and further oxidize secondary substrates (Glenn et al. 1986, Mester and Field 1998).

Another group of peroxidases, versatile peroxidase (VP), has been recently isolated and purified from *Bjerkandera* sp., strain B33/3. This enzyme is considered a hybrid between MnP and LiP, as it can oxidize manganese as well as phenolic and non-phenolic aromatic compounds, including dyes. Heinfling et al. (1998) and Mester and Field (1998) have also described VP in *Pleurotus* and *Bjerkandera* sp.

The cooperative interactions of these enzymes play a key role in dye degradation. The relative contributions of these enzymes differ for each fungus. Some wood decaying fungi produce all three LME, whereas others produce one or two of them (Hatakka 1994). In *P. chyrsosporium*, LiP was found to be responsible for the decoloration of the dyes (Glenn and Gold 1983, Paszczynski et al. 1992). Similar observations were made by Ollikka et al. (1993), who reported that MnP is not essential for degradation of azo dyes, whereas LiP played a major role in it. However, Zhang et al. (1999a) reported the prominent role of MnP in the decoloration of a cotton bleaching effluent.

5. Dye Decoloration using Immobilized Fungi

Dye decoloration using free cell cultures has certain disadvantages, such as shear stress and cell stability under agitated conditions. Hence, along with free cell cultures, immobilized fungi have also been used for dye decoloration in both static and agitated conditions. Immobilized fungal biomass can easily sustain the toxicity of dye effluents, hence, it is preferred. In addition, immobilized cells can be reused and show higher decoloration efficiency, cell stability, and decreased clogging in continuous flow systems. Entrapment and attachment are the two commonly used techniques of immobilization for dye decoloration. In entrapment, microorganisms are entrapped in the interstices of fibrous or porous materials, such as agar, chitosan, alginate, cellulose derivatives, gelatin, collagen, polyvinyl alcohol, etc., whereas the attachment method involves adherence of microorganisms on surfaces, such as polyurethane foams, nylon sponge, stainless steel sponge due to chemical bonding or self-adhesion. Natural substrates, such as wheat straw, jute, wood chips, pine wood, corncob, etc., have also been used to immobilize fungi. These materials mimic the natural environment, and thus stimulate the production of ligninolytic enzymes. In addition, utilization of these agricultural wastes makes the process more economical and ecological (Pandey 1992). Table 5.4 summarizes the various immobilization support materials used for dye decoloration.

Another approach in decoloration of dyes is by using dehydrated biomass. Recently, Cano et al. (2011) reported the use of lyophilized mycelium of *T. versicolor* in decoloration of indigo carmine. The problems associated with wet biomass, such as microbial contamination, preservation, stability, could be easily overcome using lyophilized mycelia. Enzymatic activity in these lyophilized cultures was retained and was responsible for biodegradation. However, the feasibility of this needs to be assessed.

6. Decoloration of Dyes in Bioreactors

Extensive studies on dye decoloration have been carried out with bioreactor systems, using both free and immobilized biomass. These bioreactors provide a favorable environment for growth and enhance the ability of decoloration (Diorio et al. 2008). Packed bed reactors are the most commonly used reactors for effluent treatment and bioremediation, where the effluent is passed through a column packed with immobilized cells (Srikanlayanukul et al. 2008). It offers various advantages, such as high reaction rate, simple design, easy to construct, and operate. Zhang et al. (1999b) reported 95% decoloration of Orange II by *fungus F29* in a continuous packed bed reactor. Similarly, 90% decoloration of Orange II was achieved by *Coriolus versicolor* in a continuous packed bed reactor (Srikanlayanukul et al. 2008). Many other researchers have used packed bed reactors for effluent and azo dye decoloration using free and immobilized mycelia (Schliephake et al. 1993, Mielgo et al. 2001, Urra et al. 2006, Sedighi et al. 2009) in batch and continuous mode. However, in continuous mode, intensive mycelial growth results in formation of a thick mat, thereby impeding the movement of fluids in the reactor, and causing a decrease in the oxygen mass transfer rate and in the dye decoloration efficiency. Thus, the performance of the reactor in dye decoloration is hindered (Zhang et al. 1999b, Rodríguez-Couto et al. 2006). Studies have also been conducted for dye decoloration using stirred tank bioreactors. Borchert and Libra (2001) reported about 99% decoloration of azo dyes Reactive Black 5 and Reactive Red 198 using *T. versicolor* in a stirred tank bioreactor. To overcome contamination and stress on the mycelia using stirred tank reactors, Leidig et al. (1999) used polyvinylalcohol-hydrogel (PVAL) encapsulated *T. versicolor* for decoloration of dye Poly R-478, and was able to achieve 89% decoloration, partially, due to biotransformation and adsorption. Similarly, Mohorčič et al. (2004) immobilized *Bjerkandera adjusta* on a plastic net in the form of cylinder inside the vessel, and achieved 95% decoloration of diazo dye Reactive Black 5. Rotating biological contactor reactors (RBCR) have also been efficiently used in dye decoloration studies. Axelsson et al. (2006) reported the decoloration of Reactive Red 2 and Reactive Blue 4 using *Bjerkandera* sp. in a continuous RBCR. Various other submerged bioreactors have been used for dye decoloration studies, such as air lift reactor (Buckley and Dobson 1998, Shahvali et al. 2000, Assadi and Jahangiri 2001), bubble reactor (Binupriya et al. 2008), biofilm reactor (Huang et al. 2001), fluidized reactor (Blánquez et al. 2004, Casas et al. 2007), trickle bed reactor (Tavčar et al. 2006), membrane bioreactor (López et al. 2002), etc. Table 5.5 summarizes different bioreactor studies used for dye decoloration and degradation by fungi.

Along with submerged reactors, solid-state bioreactors have also been explored to assess their efficiency in decoloration and degradation of dyes. Solid-state fermentation simulates the natural environment of fungi, thus encouraging ligninolytic enzyme production (Ürek and Pazarlioğlu 2005), which helps in degradation of pollutants. Enhanced activities of MnP and LiP were reported by Dominguez

Table 5.4 Different immobilization support materials used for fungal dye decoloration.

Support Material	Organism	Mechanism	Parameters (Static or Agitated/rpm/pH/temperature/others/if any)	Dye (Concentration)	% Decoloration (time)	References
Alginate beads	*Trametes versicolor*	Decoloration	Agitated/–/–/–/	Amaranth 50 mg/l	75 (4.5 h)	Ramsay et al. (2005)
				Reactive blue 19 50 mg/l	75 (4.5 h)	
	Phanerochaete chrysosporium	Decoloration due to MnP	Agitated/150 rpm/–/30°C	Direct Violet 51 (65 mg/l)	85 (120 h)	Enayatzamir et al. (2009)
				Reactive Black 5 (120 mg/l)	85 (120 h)	
				Ponceau Xylidine (100 mg/l)	100 (96 h)	
				Bismark Brown R (100 mg/l)	86.7 (144 h)	
		Biodegradation with ligninolytic enzymes MnP/LiP/Lac	Agitated/–/–/–	Brilliant blue R (500 mg/l)	85 (130 h)	González–Ramírez et al. (2014)
Carboxymethyl cellulose	*Aspergillus fumigatus*	Biosorption	Agitated/150 rpm/–/30°C	Reactive brilliant Red K–2BP (–)	–	Wang et al. (2008)
Corn cob	*Pleurotus pulmonarius*	Decoloration due to Lac	Static/–/–/–	Congo red (200 ppm)	93 (6 d)	Tychanowicz et al. (2004)
				Trypan Blue (200 ppm)	95 (6 d)	

Table 5.4 contd. ...

...Table 5.4 contd.

Support Material	Organism	Mechanism	Parameters (Static or Agitated/ rpm/pH/temperature/ others/if any)	Dye (Concentration)	% Decoloration (time)	References
				Amido black (200 ppm)	89 (6 d)	
Kissiris	*Phanerochaete chrysosporium*	Decoloration due to MnP	Agitated/–/–/–	Methylene Blue (60 ppm)	94 (8 d)	Karimi et al. (2006)
Luffa Sponge	*Phanerochaete chrysosporium*	Biosorption	Agitated/100 rpm/–/–	Remazol Brilliant Blue R (10–500 mg/l)	101 mg. g^{-1} biomass	Iqbal and Saeed (2007)
	Trametes versicolor	Decoloration due to fungal enzymes	Agitated/120 rpm/–/–	Reactive Red 2 (0.2 g/l)	70 (15 d)	Nilsson et al. (2006)
Lyophilized mycelia	*Trametes versicolor*	Biodegradation and Decoloration due to Lac	Agitated/120 rpm/–/ Room temperature	Indigocarmine (100 mg/l)	99.9 (1 h)	Cano et al. (2011)
Polyurethane foam	*Coriolus versicolor* RC3	Decoloration due to Lac	Agitated/120 rpm/–/ 37°C	Orange II (20 ppm)	90 (120 h)	Srikanlayanukul et al. (2008)
	Phanerochaete chrysosporium	Biodegradation with ligninolytic enzymes MnP/Lac	Agitated/160 rpm/–/ 37°C	Reactive Brilliant Red K–2BP (30 mg/l)	93.5 (3 d)	Gao et al. (2008)
Sawdust	*Coriolopsis gallica*	Biodegradation and adsorption	Static/–/–/–	Reactive Black 5 (50 mg/l)	67 (24 h)	Daâssi et al. (2016)
				Acid Orange 51 (50 mg/l)	75 (24 h)	
Sodium alginate	*Funalia trogii ATCC 200800*	Decoloration with ligninolytic enzymes lac/LiP/MnP	Agitated/130 rpm/–/–	Acid Black 52 (100 mg/l)	93.8 (–)	Park et al. (2006)

ZrOCl$_2$–activated Pumice	*Phanerochaete chysosporium*	Biosorption and biodegradation	Static/–/39°C	Direct Green 6 (120 mg/l)	70 (8 d)	Pazarlioglu et al. (2005)
				Direct Brown 2 (120 mg/l)	64 (10 d)	
				Direct Blue 15 (120 mg/l)	80.5 (10 d)	
				Direct Red 23 (120 mg/l)	35 (8 d)	
				Direct Orange 26 (120 mg/l)	80.3 (10 d)	
				Congo red (120 mg/l)	57.8 (10 d)	
				Direct Black 38 (120 mg/l)	65 (10 d)	

Table 5.5 Different reactors used for dye decoloration by fungi.

Reactors	Organism	Mechanism	Dye (Concentration)	Support material	% Decoloration (Time)	References
Agitated Sequencing batch	*Trametes versicolor*	Biodegradation	Reactive Black 5 (100 mg/l)	–	99 (200 d)	Borchert and Libra (2001)
Air lift reactor	*Aspergillus niger*	Adsorption	Textile effluent (–)	–	97 (20 h)	Assadi et al. (2001)
	Chrysosporium lignorum CL1	Biodegradation with ligninolytic enzymes LIP and MnP	Poly R-478 (50 mg/l)	Circular plastic packing material	80 (12 d)	Buckley and Dobson (1998)
	Phanerochaete chrysosporium	Decoloration	Poly S-119 (50 mg/l)	–Circular plastic packing material	95 (10 d)	Shahvali et al. (2000)
			Textile effluent (–)	–	97 (28 h)	
		Biodegradation with ligninolytic enzymes MnP, LiP, Lac	Reactive Brilliant blue R (500 mg/l)	Alginate beads	98 (120 h)	González-Ramírez et al. (2014)
Cylindrical Bioreactor	*Irpex lacteus*	Biodegradation with ligninolytic enzymes Lac, MnP	Reactive Orange 16 (–)	Polyurethane foam	80 (24 h)	Svobodová et al. (2008)
	Trametes versicolor	Decoloration due to fungal enzymes	Reactive Red 2 (0.5 g/l)	Natural Luffa Sponge	80 (15 d)	Nilsson et al. (2006)
Expanded Bed Tubular bioreactor (Batch mode)	*Trametes hirsuta*	Decoloration and degradation	Indigo carmine (56 mg/l)	Crushed Orange peeling	100 (3 h)	Rodríguez-Couto et al. (2006)
			Bromophenol blue (670 mg/l)	Crushed Orange peeling	85 (7 h)	
Expanded Bed Tubular bioreactor (Continuous mode)	*Trametes hirsuta*	Decoloration and degradation	Methyl Orange (60 mg/l)	Crushed Orange peeling	81 (3 d)	
			Poly R-478 (180 mg/l)	Crushed Orange peeling	47 (3 d)	
Fixed Bed Tubular Bioreactor	*Trametes pubescens*	Biodegradation with ligninolytic enzyme Lac	Reactive Black 5 (30 mg/l)	Stainless steel sponge	74 (24 h)	Enayatzamir et al. (2009)

Reactor	Fungus	Mechanism	Dye (concentration)	Support	% (time)	Reference
Fluidized bioreactor	Phanerochaete chrysosporium	Decoloriztion due to ligninolytic enzymes MnP	Reactive Black 5 (80 mg/l)	Nylon Sponge	47 (96 h)	Enayatizamir et al. (2011)
	Trametes versicolor	Biodegradation	Reactive Black 5 (100 mg/l)	Sunflower seed shells	90.3 (72 h)	Blánquez et al. (2004)
		Adsorption in degradation due to laccase	Grey Lanaset G (–)	–	90 (5 d)	Casas et al. (2007)
			Orange G (150 mg/l)	–	97 (20 h)	
Fluidized Bed reactor (Fed batch)	Fungus F29	–	Orange II (2000 mg/l)	Free mycelium pellets	95 (30 d)	Zhang et al. (1999b)
Fluidized bed reactor (Continuous)	Fungus F29	–	Orange II (2000 mg/l)	Alginate beads	97 (24 h)	Zhang et al. (1999a)
Large Trickle Bed Reactor	Irpex lacteus	Decoloration due to ligninolytic enzymes MnP and Lac	Azo Dye RO 16 (150 mg/l)	Polyurethane foam	90 (4 d)	Tavčar et al. (2006)
Packed Bed Reactor	Coriolus versicolor RC3	Decoloration due to ligninolytic enzyme Lac	Orange II (20 ppm)	Polyurethane foam	90 (120 h)	Srikanlayanukul et al. (2008)
	Irpex lacteus	Biodegradation with ligninolytic enzymes MnP, Lac	Remazol Brilliant Blue R (150 mg/l)	Polyurethane foam	97 (10 d)	Novotný et al. (2004)
				Pinewood cubes	100 (6 d)	
	Phanerochaete chrysosporium	Decoloration due to ligninolytic enzymes MnP	Astrazon Red FBL (–)	Kissiris	87 (40 h)	Sedighi et al. (2009)
		Adsorption and Biodegrdation with MnP	Acid Black 1 (400 mg/l)	Polyurethane foam	88 (11 d)	Urra et al. (2006)
			Reactive Black 5 (400 mg/l)	Polyurethane foam	70 (9 d)	

Table 5.5 contd. ...

...Table 5.5 contd.

Reactors	Organism	Mechanism	Dye (Concentration)	Support material	% Decoloration (Time)	References
	Pycnoporus cinnabarinus	Decoloration and degradation	Reactive Orange 16 (200 mg/l)	Polyurethane foam	92 (10 d)	
			Pigment plant effluent (−)	Nylon web cubes	−(72 h)	Schliephake et al. (1993)
	Schizophyllum sp., F17	Decoloration due to ligninolytic enzymes MnP and biosorption by rice hull	Congo Red (142.63 mg/l)	Rice hull	89.7 (41 h)	Li and Jia (2008)
Packed bed reactor (Continuous)	Fungus F29	—	Orange II (1000 mg/l)	Homogenized mycelium	95 (90 d)	Zhang et al. (1999b)
	Phanerochaete chysosporium	Biosorption and biodegradation	Direct Blue 15 (−)	$ZrOCl_2$-activated Pumice	100 (3 d)	Pazarlioglu et al. (2005)
		Decoloration and degradation by MnP	Orange II (0.1 g/l)	Polyurethane foam	99.5 (24 h)	Mielgo et al. (2001)
			Reactive Blue 4 (50 mg/l)	Birch wood	94 (3 d)	
Pulsed packed bed reactor	Phanerochaete chrysosporium	Biodegradation with ligninolytic enzymes MnP	Poly R-478 (100 mg/l)	Polyurethane foam	73 (20 d)	Palma et al. (1999)
				Polyurethane foam	80 (24 h)	Mielgo et al. (2002)
Rotating Biological Contactor	Trametes versicolor	Adsorption	Carpet dye effluent (−)	Jute twine	100 (12 h)	Ramsay and Goode (2004)
Rotating Biological Contactor (Continuous)	Bjerkandera sp. strain BOL 13	Decoloration and degradation	Reactive Red 2 (50 mg/l)	Birch wood	96 (3 d)	Axelsson et al. (2006)

Reactor	Fungus	Mechanism	Dye (concentration)	Support	% (time)	Reference
Rotating Disc Contactors	*Irpex lacteus*	Decoloration due to ligninolytic enzymes MnP and Lac	Azo Dye RO 16 (150 mg/l)	Polyurethane foam	90 (6 d)	Tavčar et al. (2006)
Rotating drum bioreactor	*Phanerochaete chyrsosporium*	Biodegradation with ligninolytic enzymes laccase, LIP and MnP	Poly R-478 (–)	Fibrous nylon sponge	19 (15 min)	Domínguez et al. (2001)
Semicontinous process bioreactor	*Coriolus versicolor f. antarcticus*	Biodegradation with ligninolytic enzymes MnP, Lac	Malachite green (100 μM)	Wheat bran	50 (48 min)	Diorio et al. (2008)
Separating funnel Reactor	*Trametes MUT2295 pubescens*	Decoloration with ligninolytic enzymes	R 243 (2000 ppm)	Polyurethane foam	65.4 (–)	Casieri et al. (2008)
	Pleurotus ostreatus MUT 2976	Decoloration with ligninolytic enzymes	R 243 (2000 ppm)	Polyurethane foam	44.8 (–)	
Small Trickle Bed Reactor	*Irpex lacteus*	Decoloration due to ligninolytic enzymes MnP and Lac	Azo Dye RO 16 (150 mg/l)	Polyurethane foam	90 (3 d)	Tavčar et al. (2006)
Stirred Bioreactor	*Bjerkandera adjusta*	Decoloration due to lignin peroxidases and manganese peroxidase	Reactive Black 5 (0.2 g/l)	Plastic net	–(20 d)	Mohorčič et al. (2004)
	Trametes veresiolor	Biotransformation and Adsorption	Poly R-478 (200 mg/l)	Encapsulation in PVAL hydrogel beads	89 (41 d)	Leidig et al. (1999)
		Biodegradation	Reactive Black 5 (52% w/w)	Encapsulation in PVAL hydrogel beads	99.5 (200 d)	Borchert and Libra (2001)
			Reactive Red 198 (65% w/w)	Encapsulation in PVAL hydrogel beads	97 (200 d)	
			Reactive blue 19 (50% w/w)	Encapsulation in PVAL hydrogel beads	97 (200 d)	

et al. (2001) using rotating drum bioreactors for cultivation of *Phanerochaete chrysosporium*. Rodríguez-Couto et al. (2003) compared three different reactors, immersion, expanded bed, and tray reactor, for laccase production by *T. versicolor* using nylon sponge and barley bran as support. The cellulose content of barley bran induced laccase production, thus, giving higher laccase yield, while among the reactors tested, the tray reactor was found to be the best for laccase production. Decoloration studies have also been carried out using other solid-state reactors. Rodríguez et al. (2006) reported 81% decoloration of methyl orange with *Trametes hirsuta* grown on ground orange peel as support in an expanded bed bioreactor. Many other bioreactor designs have been developed and studied for efficient dye removal at lab and industrial scale, both for submerged and solid-state fermentation, but their commercial application is still limited.

7. Decoloration of Industrial Effluents

Decoloration of the textile and dyestuff manufacturing industry effluents is a major problem and a matter of environmental concern. About 0.6–0.8 g/L dye is present in dye industry effluents (Jadhav et al. 2007). Most of these dyes are recalcitrant and are considered as one of the main pollutants in wastewaters. Along with an array of dyes, the actual textile dye effluent contains large amounts of salts, chelating agents, surfactant, byproducts, etc., and has high biological oxygen demand (BOD) and chemical oxygen demand (COD). In addition, the composition of these industrial effluents varies considerably. Hence, there is a need for newer and effective techniques for the decoloration of dyes in effluents. In the present scenario, microbial decoloration is preferred as it is environmentally friendly, produces less sludge, and is cost effective (Singh et al. 2015). Biological decoloration of industrial effluents requires an appropriate choice of strain and proper environmental conditions. Intensive research is being carried out to find out the best possible strain for industrial effluent treatment.

Revankar and Lele (2007) studied decoloration of dyestuff industry effluents by an indigenously isolated strain *Ganoderma WR-1*. After 5 days of incubation, 50% decoloration was achieved and complete decoloration was observed within 12 days. Selvam et al. (2003) reported 61% decoloration of a dye industry effluent in 3 days in batch mode and 50% decoloration within 7 days in continuous mode using white-rot fungi *Thelephora* sp., Nilsson et al. (2006) studied decoloration of actual textile industry wastewater by *Pleurotus flabellatus* grown on luffa sponge packed in a continuous reactor. Extensive research on dye decoloration has been carried out in laboratory conditions using simulated textile effluents. Martins et al. (2003) reported 90% decoloration of a mixture of eight reactive azo dyes by *T. versicolor*. Immobilized *Irpex lacteus* decolorized textile effluents containing different dye mixtures, such as Drimarene Blue (100%), Drimarene Red (80%), Remazol Green (45%), and acid black (35%) within 3–5 days (Novotný et al. 2004). Nilsson et al. (2006) reported 70% decoloration of synthetic wastewater using *T, versicolor*. Along with a complex mixture of dyes, textile industry effluents often contain many other components, such as variable concentrations of salts, heavy metals, etc. The influence of these components on the uptake and metabolism of azo dyes was investigated by Sumathi and Manju (2000). They observed that, in presence of 5 mg/L of chromium and 1% sodium chloride, respectively, the time required for decoloration of Drimarene Red and Drimarene Blue by *A. foetidus* was delayed 60–70 hours, as compared to 48 hours in the absence of salt and heavy metals. The extent of effluent dilution also plays an important role in dye decoloration. Assadi et al. (2001) observed that decoloration decreased from 98% to 35% when the effluent was changed from undiluted to a 10-times diluted effluent. They concluded that more than 50% dilution significantly lowers the enzyme production responsible for dye decoloration.

Not many studies have reported using yeast biomass. Phugare et al. (2010) reported the use of yeast biomass, a waste from the distillery industry, for decoloration and degradation of textile effluents. Almost 69% decoloration was achieved within 48 hours. On laboratory scale, a continuous effluent treatment plant was developed by immobilizing the yeast biomass on Plaster of Paris (POP). POP is easily available and inexpensive; hence, it can be a cost-effective alternative for remediation of textile effluents. The success of efficient effluent decoloration by fungi clearly indicates their potential in wastewater treatment.

8. Decoloration of Azo Dyes by Genetically Modified Fungi

Diverse techniques, suchas cloning, heterologous expression, random mutagenesis, site directed mutagenesis, and gene recombination, have been used to develop superior strains that can be used for bioremediation purposes. For example, random mutagenesis was used to improve the biodegradation of Congo red using a mutant *Bacillus* sp. (Gopinath et al. 2009). Miura et al. (1997) isolated a mutant of the hyper-lignolytic fungus IZU-154 with enhanced MnP productivity and ability to decolorize synthetic melanin. Due to low levels of expression in their native systems and mutants, studies have been conducted to clone and express gene in different hosts. The major genes encoding the lignin-degrading enzymes systems, such as LiP, MnP, and laccases, have been cloned and sequenced in a variety of expression systems. For example, lignin peroxidase gene from *Streptomyces viridosporus* in *Streptomyces lividans* (Wang et al. 1990); laccase gene from *T. villosa* in *A. oryzae* (Yaver et al. 1996); laccase gene from *Phlebia radiata* in *Trichoderma resei* (Saloheimo et al. 1991); ligninolytic phenoloxidase gene from *Coriolus hirsutus* in *Saccharomyces cerevisiae* (Kojima et al. 1990); laccase gene from *Myceliophthora thermophilia* in *S. cerevisiae* (Bulter et al. 2003); laccase gene from basidiomycete PM1 HRPL in *S. cerevisiae* (Maté et al. 2010); laccase gene from *Lentinula edodes* in *Pichia pastoris* (Jönsson et al. 1997, Wong et al. 2012); laccase gene from *Pleurotus sajor-caju* in *P. pastoris* (Soden et al. 2002); manganese peroxidase from *Phanerchaete chrysosporium* in yeast *P. pastoris* (Gu et al. 2003); laccase gene from *T. versicolor* in *Pichia methanolica* (Guo et al. 2006); laccase gene from *T. versicolor* in the yeast *Yarrowia lipolytica* (Theerachat et al. 2012), etc. Studies show that the productivity of recombinant enzymes produced by transformants is higher and have been found to possess enhanced ability to decolorize different dyes (Fan et al. 2011, Zhuo et al. 2011). These recombinant enzymes would facilitate their application in decolorizing and detoxifying azo dyes and effluents. Yet, more research needs to be carried out to develop recombinant strains with higher degradation capacities.

Conclusions

In the current scenario, decoloration and degradation of azo dyes from wastewater effluents is a huge challenge and a major environmental concern. Due to stringent government legislations, development of commercially viable techniques is becoming a necessity. The composition of actual industrial effluents is variable and complex. Hence, intensive efforts are still needed to achieve success at large scale. From this chapter it is evident that fungal decoloration systems by themselves can be effective in effluent treatment or can work synergistically with the existing chemical/biological methods used in wastewater treatment. Immobilization of fungi or fungal enzymes on a biosorbents system can be developed for effluent treatment. Use of genetically modified thermotolerant fungal strains able to tolerate a wide range of pH could be an added advantage that will significantly lower the cost. More research needs to be carried out in developing bioengineered fungi that can utilize dyes as sole substrates and efficiently degrade them. Systematic studies on the characterization of the intermediates and metabolites must be done and the treated wastewater should be evaluated with proper toxicity studies, so that it can be further used for irrigation purposes. With the advent of nanotechnology, in the near future, membrane-bound enzymes might be used for the treatment of industrial effluents, but their feasibility and viability need to be evaluated.

References

Abadulla, E., T. Tzanov, S. Costa, K.-H. Robra, A. Cavaco-Paulo and G. M. Gübitz. 2000. Decolorization and detoxification of textile dyes with a laccase from *Trametes hirsuta*. *Applied and Environmental Microbiology* 66 (8): 3357–3362.

Abd El-Rahim, W. M., H. Moawad, A. Z. Abdel Azeiz and M. J. Sadowsky. 2017. Optimization of conditions for decolorization of azo-based textile dyes by multiple fungal species. *Journal of Biotechnology* 260: 11–17. doi: https://doi.org/10.1016/j.jbiotec.2017.08.022.

Aksu, Z. 2003. Reactive dye bioaccumulation by *Saccharomyces cerevisiae*. *Process Biochemistry* 38 (10): 1437–1444. doi: https://doi.org/10.1016/S0032-9592(03)00034-7.

Aksu, Z. and G. Dönmez. 2003. A comparative study on the biosorption characteristics of some yeasts for remazol blue reactive dye. *Chemosphere* 50 (8): 1075–1083. doi: https://doi.org/10.1016/S0045-6535(02)00623-9.

Aksu, Z., N. K. Kılıç, S. Ertuğrul and G. Dönmez. 2007. Inhibitory effects of chromium(VI) and remazol black B on chromium(VI) and dyestuff removals by *Trametes versicolor*. *Enzyme and Microbial Technology* 40 (5): 1167–1174. doi: https://doi.org/10.1016/j.enzmictec.2006.08.024.

Almeida, E. J. R. and C. R. Corso. 2014. Comparative study of toxicity of azo dye Procion Red MX-5B following biosorption and biodegradation treatments with the fungi *Aspergillus niger* and *Aspergillus terreus*. *Chemosphere* 112: 317–322. doi: https://doi.org/10.1016/j.chemosphere.2014.04.060.

Ambrósio, S. T. and G. M. Campos-Takaki. 2004. Decolorization of reactive azo dyes by *Cunninghamella elegans* UCP 542 under co-metabolic conditions. *Bioresource Technology* 91 (1): 69–75. doi: https://doi.org/10.1016/S0960-8524(03)00153-6.

Asgher, M., S. Kausar, H. N. Bhatti, S. A. Hassan Shah and M. Ali. 2008. Optimization of medium for decolorization of solar golden yellow R direct textile dye by *Schizophyllum commune* IBL-06. *International Biodeterioration & Biodegradation* 61 (2): 189–193. doi: https://doi.org/10.1016/j.ibiod.2007.07.009.

Assadi, M. M. and M. R. Jahangiri. 2001. Textile wastewater treatment by *Aspergillus niger*. *Desalination* 141 (1): 1–6. doi: https://doi.org/10.1016/S0011-9164(01)00383-6.

Assadi, M. M., K. Rostami, M. Shahvali and M. Azin. 2001. Decolorization of textile wastewater by *Phanerochaete chrysosporium*. *Desalination* 141 (3): 331–336. doi: https://doi.org/10.1016/S0011-9164(01)85010-4.

Asther, M., L. Lesage, R. Drapron, G. Corrieu and E. Odier. 1988. Phospholipid and fatty acid enrichment of *Phanerochaete chrysosporium* INA-12 in relation to ligninase production. *Applied Microbiology and Biotechnology* 27 (4): 393–398. doi: 10.1007/bf00251775.

Axelsson, J., U. Nilsson, E. Terrazas, T. Alvarez-Aliaga and U. Welander. 2006. Decolorization of the textile dyes reactive red 2 and reactive blue 4 using *Bjerkandera* sp. strain BOL 13 in a continuous rotating biological contactor reactor. *Enzyme and Microbial Technology* 39 (1): 32–37. doi: https://doi.org/10.1016/j.enzmictec.2005.09.006.

Balan, D. S. L. and R. T. R. Monteiro. 2001. Decolorization of textile indigo dye by ligninolytic fungi. *Journal of Biotechnology* 89 (2): 141–145. doi: https://doi.org/10.1016/S0168-1656(01)00304-2.

Banat, I. M., P. Nigam, D. Singh and R. Marchant. 1996. Microbial decolorization of textile-dyecontaining effluents: A review. *Bioresource Technology* 58 (3): 217–227. doi: https://doi.org/10.1016/S0960-8524(96)00113-7.

Bankole, P. O., A. A. Adekunle and S. P. Govindwar. 2018. Enhanced decolorization and biodegradation of acid red 88 dye by newly isolated fungus, *Achaetomium strumarium*. *Journal of Environmental Chemical Engineering* 6 (2): 1589–1600. doi: https://doi.org/10.1016/j.jece.2018.01.069.

Belsare, D. K. and D. Y. Prasad. 1988. Decolorization of effluent from the bagasse-based pulp mills by white-rot fungus, *Schizophyllum commune*. *Applied Microbiology and Biotechnology* 28 (3): 301–304. doi: 10.1007/bf00250460.

Binupriya, A. R., M. Sathishkumar, K. Swaminathan, C. S. Kuz and S. E. Yun. 2008. Comparative studies on removal of congo red by native and modified mycelial pellets of *Trametes versicolor* in various reactor modes. *Bioresource Technology* 99 (5): 1080–1088. doi: https://doi.org/10.1016/j.biortech.2007.02.022.

Blánquez, P., N. Casas, X. Font, X. Gabarrell, M. Sarrà, G. Caminal and T. Vicent. 2004. Mechanism of textile metal dye biotransformation by *Trametes versicolor*. *Water Research* 38 (8): 2166–2172. doi: https://doi.org/10.1016/j.watres.2004.01.019.

Borchert, M. and J. A. Libra. 2001. Decolorization of reactive dyes by the white rot fungus *Trametes versicolor* in sequencing batch reactors. *Biotechnology and Bioengineering* 75 (3): 313–321. doi: doi:10.1002/bit.10026.

Bourbonnais, R., M. G. Paice, I. D. Reid, P. Lanthier and M. Yaguchi. 1995. Lignin oxidation by laccase isozymes from *Trametes versicolor* and role of the mediator 2,2'-azinobis(3-ethylbenzthiazoline-6-sulfonate) in kraft lignin depolymerization. *Applied and Environmental Microbiology* 61 (5): 1876–80.

Brahimi-Horn, M. C., K. K. Lim, S. L. Liang and D. G. Mou. 1992. Binding of textile azo dyes by *Myrothecium verrucaria*. *Journal of Industrial Microbiology* 10 (1): 31–36. doi: 10.1007/bf01583631.

Buckley, K. F. and A. D. W. Dobson. 1998. Extracellular ligninolytic enzyme production and polymeric dye decolourization in immobilized cultures of *Chrysosporium lignorum* CL1. *Biotechnology Letters* 20 (3): 301–306. doi: 10.1023/a:1005398423674.

Bulter, T., M. Alcalde, V. Sieber, P. Meinhold, C. Schlachtbauer and F. H. Arnold. 2003. Functional expression of a fungal laccase in *Saccharomyces cerevisiae* by directed evolution. *Applied and Environmental Microbiology* 69 (2): 987–995. doi: 10.1128/AEM.69.2.987-995.2003.

Bustard, M., G. McMullan and A. P. McHale. 1998. Biosorption of textile dyes by biomass derived from *Kluyveromyces marxianus* IMB3. *Bioprocess Engineering* 19 (6): 427–430. doi: 10.1007/pl00009028.

Camarero, S., D. Ibarra, M. J. Martínez and Á. T. Martínez. 2005. Lignin-derived compounds as efficient laccase mediators for decolorization of different types of recalcitrant dyes. *Applied and Environmental Microbiology* 71 (4): 1775–1784. doi: 10.1128/AEM.71.4.1775-1784.2005.

Campos, R., A. Kandelbauer, K. H. Robra, A. Cavaco-Paulo and G. M. Gübitz. 2001. Indigo degradation with purified laccases from *Trametes hirsuta* and *Sclerotium rolfsii*. *Journal of Biotechnology* 89 (2): 131–139. doi: https://doi.org/10.1016/S0168-1656(01)00303-0.

Cano, M., M. Solis, J. Diaz, A. Solis, O. Loera and M. Teutli. 2011. Biotransformation of indigo carmine to isatin sulfonic acid by lyophilized mycelia from *Trametes versicolor*. *African Journal of Biotechnology* 10 (57): 12224–12231.

Casas, N., P. Blánquez, X. Gabarrell, T. Vicent, G. Caminal and M. Sarrà. 2007. Degradation of orange G by laccase: Fungal versus enzymatic process. *Environmental Technology* 28 (10): 1103–1110. doi: 10.1080/09593332808618874.

Casieri, L., G. C. Varese, A. Anastasi, V. Prigione, K. Svobodová, V. Filippelo Marchisio and Č. Novotný. 2008. Decolorization and detoxication of reactive industrial dyes by immobilized fungi *Trametes pubescens* and *Pleurotus ostreatus*. *Folia Microbiologica* 53 (1): 44. doi: 10.1007/s12223-008-0006-1.

Chen, H., S. L. Hopper and C. E. Cerniglia. 2005. Biochemical and molecular characterization of an azoreductase from *Staphylococcus aureus*, a tetrameric NADPH-dependent flavoprotein. *Microbiology (Reading, England)* 151 (Pt 5): 1433–1441. doi: 10.1099/mic.0.27805-0.

Cheng, W. N., H. K. Sim, S. A. Ahmad, M. Syed, M. Y. Shukor and M. T. Yusof. 2016. Characterization of an azo-dye-degrading white rot fungus isolated from Malaysia. *Mycosphere* 7 (5): 560–569. doi: 10.5943/mycosphere/7/5/3.

Chivukula, M. and V. Renganathan. 1995. Phenolic azo dye oxidation by laccase from *Pyricularia oryzae*. *Applied and Environmental Microbiology* 61 (12): 4374–4377.

Chtourou, M., E. Ammar, M. Nasri and K. Medhioub. 2004. Isolation of a yeast, *Trichosporon cutaneum*, able to use low molecular weight phenolic compounds: application to olive mill waste water treatment. *Journal of Chemical Technology & Biotechnology* 79 (8): 869–878. doi: doi:10.1002/jctb.1062.

Chulhwan, P., Y. Lee, T. H. Kim, B. Lee, J. Lee and S. Kim. 2004. Decolorization of three acid dyes by enzymes from fungal strains. *Jorunal of Microbiology and Biotechnology* 14 (6): 1190–1195.

Cripps, C., J. A. Bumpus and S. D. Aust. 1990. Biodegradation of azo and heterocyclic dyes by *Phanerochaete chrysosporium*. *Applied and Environmental Microbiology* 56 (4): 1114–1118.

Daâssi, D., H. Zouari-Mechichi, F. Frikha, S. Rodríguez-Couto, M. Nasri and T. Mechichi. 2016. Sawdust waste as a low-cost support-substrate for laccases production and adsorbent for azo dyes decolorization. *Journal of Environmental Health Science and Engineering* 14: 1. doi: 10.1186/s40201-016-0244-0.

Diorio, L. A., A. A. Mercuri, D. E. Nahabedian and F. Forchiassin. 2008. Development of a bioreactor system for the decolorization of dyes by *Coriolus versicolor* f. antarcticus. *Chemosphere* 72 (2): 150–156. doi: https://doi.org/10.1016/j.chemosphere.2008.02.011.

Domínguez, A., I. Rivela, S. R. Couto and M. A. Sanromán. 2001. Design of a new rotating drum bioreactor for ligninolytic enzyme production by *Phanerochaete chrysosporium* grown on an inert support. *Process Biochemistry* 37 (5): 549–554. doi: https://doi.org/10.1016/S0032-9592(01)00233-3.

Dong, J. L., Y. W. Zhang, R. H. Zhang, W. Z. Huang and Y. Z. Zhang. 2005. Influence of culture conditions on laccase production and isozyme patterns in the white-rot fungus *Trametes gallica*. *Journal of Basic Microbiology* 45 (3): 190–198. doi: doi:10.1002/jobm.200410511.

Eichlerová, I., L. Homolka and F. Nerud. 2006. Synthetic dye decolorization capacity of white rot fungus *Dichomitus squalens*. *Bioresource Technology* 97 (16): 2153–2159. doi: https://doi.org/10.1016/j.biortech.2005.09.014.

Enayatzamir, K., H. A. Alikhani and S. Rodríguez-Couto. 2009. Simultaneous production of laccase and decolouration of the diazo dye reactive black 5 in a fixed-bed bioreactor. *Journal of Hazardous Materials* 164 (1): 296–300. doi: https://doi.org/10.1016/j.jhazmat.2008.08.032.

Enayatzamir, N., F. Tabandeh, S. Rodríguez-Couto, B. Yakhchali, H. A. Alikhani, and L. Mohammadi. 2011. Biodegradation pathway and detoxification of the diazo dye reactive black 5 by *Phanerochaete chrysosporium*. *Bioresource Technology* 102 (22): 10359–10362. doi: https://doi.org/10.1016/j.biortech.2011.08.130.

Erden, E., Y. Kaymaz and N. K. Pazarlioglu. 2011. Biosorption kinetics of a direct azo dye sirius blue K-CFN on *Trametes versicolor*. *Electronic Journal of Biotechnology* 14 (2): 8. doi: 10.2225/vol14-issue2-fulltext-8.

Erkurt, E. A., A. Ünyayar and H. Kumbur. 2007. Decolorization of synthetic dyes by white rot fungi, involving laccase enzyme in the process. *Process Biochemistry* 42 (10): 1429–1435. doi: https://doi.org/10.1016/j.procbio.2007.07.011.

Evans, C. S., I. M. Gallagher, P. T. Atkey and D. A. Wood. 1991. Localisation of degradative enzymes in white-rot decay of lignocellulose. *Biodegradation* 2 (2): 93–106. doi: 10.1007/bf00114599.

Ezeronye, O. U. and P. O. Okerentugba. 1999. Performance and efficiency of a yeast biofilter for the treatment of a nigerian fertilizer plant effluent. *World Journal of Microbiology and Biotechnology* 15 (4): 515–516. doi: 10.1023/a:1008931824416.

Fabbrini, M., C. Galli and P. Gentili. 2002. Comparing the catalytic efficiency of some mediators of laccase. *Journal of Molecular Catalysis B: Enzymatic* 16 (5): 231–240. doi: https://doi.org/10.1016/S1381-1177(01)00067-4.

Fan, F., R. Zhuo, S. Sun, X. Wan, M. Jiang, X. Zhang and Y. Yang. 2011. Cloning and functional analysis of a new laccase gene from *Trametes* sp. 48424 which had the high yield of laccase and strong ability for decolorizing different dyes. *Bioresource Technology* 102 (3): 3126–3137. doi: https://doi.org/10.1016/j.biortech.2010.10.079.

Fu, S. Y., H. S. Yu and J. A. Buswell. 1997. Effect of nutrient nitrogen and manganese on manganese peroxidase and laccase production by *Pleurotus sajor*-caju. *FEMS Microbiology Letters* 147 (1): 133–137. doi: 10.1111/j.1574-6968.1997.tb10232.x.

Fu, Y. and T. Viraraghavan. 2001. Fungal decolorization of dye wastewaters: a review. *Bioresource Technology* 79 (3): 251–262. doi: https://doi.org/10.1016/S0960-8524(01)00028-1.

Gahlout, M., S. Gupte and A. Gupte. 2013. Optimization of culture condition for enhanced decolorization and degradation of azo dye reactive violet 1 with concomitant production of ligninolytic enzymes by *Ganoderma cupreum* AG-1. *3 Biotech* 3 (2): 143–152. doi: 10.1007/s13205-012-0079-z.

Gao, D., Y. Zeng, X. Wen and Y. Qian. 2008. Competition strategies for the incubation of white rot fungi under non-sterile conditions. *Process Biochemistry* 43 (9): 937–944. doi: https://doi.org/10.1016/j.procbio.2008.04.026.

Glenn, J. K. and M. H. Gold. 1983. Decolorization of several polymeric dyes by the lignin-degrading basidiomycete *Phanerochaete chrysosporium*. *Applied and Environmental Microbiology* 45 (6): 1741–1747.

Glenn, J. K., L. Akileswaran and M. H. Gold. 1986. Mn(II) oxidation is the principal function of the extracellular Mn-peroxidase from *Phanerochaete chrysosporium*. *Archives of Biochemistry and Biophysics* 251 (2): 688–696. doi: https://doi.org/10.1016/0003-9861(86)90378-4.

González-Ramírez, D. F., C. R. Muro-Urista, A. Arana-Cuenca, A. Téllez-Jurado and A. E. González-Becerra. 2014. Enzyme production by immobilized *Phanerochaete chrysosporium* using airlift reactor. *Biologia* 69 (11): 1464–1471. doi: 10.2478/s11756-014-0453-x.

Gopinath, K. P., S. Murugesan, J. Abraham and K. Muthukumar. 2009. *Bacillus* sp. mutant for improved biodegradation of congo red: Random mutagenesis approach. *Bioresource Technology* 100 (24): 6295–6300. doi: https://doi.org/10.1016/j.biortech.2009.07.043.

Gou, M., Y. Qu, J. Zhou, F. Ma and L. Tan. 2009. Azo dye decolorization by a new fungal isolate, *Penicillium* sp. QQ and fungal-bacterial cocultures. *Journal of Hazardous Materials* 170 (1): 314–319. doi: https://doi.org/10.1016/j.jhazmat.2009.04.094.

Gu, L., C. Lajoie and C. Kelly. 2003. Expression of a *Phanerochaete chrysosporium* manganese peroxidase gene in the yeast *Pichia pastoris*. *Biotechnology Progress* 19 (5): 1403–1409. doi: doi:10.1021/bp025781h.

Guo, M., F. Lu, L. Du, J. Pu and D. Bai. 2006. Optimization of the expression of a laccase gene from *Trametes versicolor* in *Pichia methanolica*. *Applied Microbiology and Biotechnology* 71 (6): 848–852. doi: 10.1007/s00253-005-0210-8.

Han, Y., L. Shi, J. Meng, H. Yu and X. Zhang. 2014. Azo dye biodecolorization enhanced by *Echinodontium taxodii* cultured with lignin. *PLoS ONE* 9 (10): e109786. doi: 10.1371/journal.pone.0109786.

Hao, O. J., H. Kim and P.-C. Chiang. 2000. Decolorization of wastewater. *Critical Reviews in Environmental Science and Technology* 30 (4): 449–505. doi: 10.1080/10643380091184237.

Hatakka, A. 1994. Lignin-modifying enzymes from selected white-rot fungi: production and role from in lignin degradation. *FEMS Microbiology Reviews* 13 (2–3): 125–135. doi: doi:10.1111/j.1574-6976.1994.tb00039.x.

Hatvani, N. and I. Mécs. 2002. Effect of the nutrient composition on dye decolorisation and extracellular enzyme production by *Lentinus edodes* on solid medium. *Enzyme and Microbial Technology* 30 (3): 381–386. doi: https://doi.org/10.1016/S0141-0229(01)00512-9.

Heinfling, A., M. J. Martínez, A. T. Martínez, M. Bergbauer and U. Szewzyk. 1998. Transformation of industrial dyes by manganese peroxidases from *Bjerkandera adusta* and *Pleurotus eryngii* in a manganese-independent reaction. *Applied and Environmental Microbiology* 64 (8): 2788–2793.

Huang, M. S., R. Huang, Y. Q. Cheng and G. Y. Zhang. 2001. Experimental study on decolorization and degradation of reactive brilliant red X-3B in a white rot fungal biofilm reactor. *Journal of Shanghai University (English Edition)* 5 (3): 260–264. doi: 10.1007/s11741-996-0036-3.

Ilyas, S. and A. Rehman. 2013. Decolorization and detoxification of synozol red HF-6BN azo dye, by *Aspergillus niger* and *Nigrospora* sp. *Iranian Journal of Environmental Health Science & Engineering* 10 (1): 12–12. doi: 10.1186/1735-2746-10-12.

Iqbal, M. and A. Saeed. 2007. Biosorption of reactive dye by loof a sponge-immobilized fungal biomass of *Phanerochaete chrysosporium*. *Process Biochemistry* 42 (7): 1160–1164. doi: https://doi.org/10.1016/j.procbio.2007.05.014.

Jadhav, J. P., G. K. Parshetti, S. D. Kalme and S. P. Govindwar. 2007. Decolourization of azo dye methyl red by *Saccharomyces cerevisiae* MTCC 463. *Chemosphere* 68 (2): 394–400. doi: https://doi.org/10.1016/j.chemosphere.2006.12.087.

Jafari, N., M. R. Soudi and R. Kasra-Kermanshahi. 2014. Biodecolorization of textile azo dyes by isolated yeast from activated sludge: *Issatchenkia orientalis* JKS6. *Annals of Microbiology* 64 (2): 475–482. doi: 10.1007/s13213-013-0677-y.

Jarosz-Wilkołazka, A., J. Kochmańska-Rdest, E. Malarcžyk, W. Wardas and A. Leonowicz. 2002. Fungi and their ability to decolourize azo and anthraquinonic dyes. *Enzyme and Microbial Technology* 30 (4): 566–572. doi: https://doi.org/10.1016/S0141-0229(02)00022-4.

Jin, X.-C., G.-Q. Liu, Z.-H. Xu and W.-Y. Tao. 2007. Decolorization of a dye industry effluent by *Aspergillus fumigatus* XC6. *Applied Microbiology and Biotechnology* 74 (1): 239–243. doi: 10.1007/s00253-006-0658-1.

Jönsson, L. J., M. Saloheimo and M. Penttilä. 1997. Laccase from the white-rot fungus *Trametes versicolor*: cDNA cloning of *lcc*1 and expression in Pichia pastoris. *Current Genetics* 32 (6): 425–430. doi: 10.1007/s002940050298.

Kaal, E. E. J., E. de Jong and J. A. Field. 1993. Stimulation of ligninolytic peroxidase activity by nitrogen nutrients in the white rot fungus *Bjerkandera* sp. strain BOS55. *Applied and Environmental Microbiology* 59 (12): 4031–4036.

Kaal, E. E. J. J. A. Field and T. W. Joyce. 1995. Increasing ligninolytic enzyme activities in several white-rot basidiomycetes by nitrogen-sufficient media. *Bioresource Technology* 53 (2): 133–139. doi: https://doi.org/10.1016/0960-8524(95)00066-N.

Kapdan, I. K., F. Kargia, G. McMullan and R. Marchant. 2000. Effect of environmental conditions on biological decolorization of textile dyestuff by *C, versicolor*. *Enzyme and Microbial Technology* 26 (5): 381–387. doi: https://doi.org/10.1016/S0141-0229(99)00168-4.

Karimi, A., F. Vahabzadeh and B. Bonakdarpour. 2006. Use of *Phanerochaete chrysosporium* immobilized on kissiris for synthetic dye decolourization: Involvement of manganese peroxidase. *World Journal of Microbiology and Biotechnology* 22 (12): 1251–1257. doi: 10.1007/s11274-006-9169-6.

Kaushik, P. and A. Malik. 2009. Fungal dye decolourization: Recent advances and future potential. *Environment International* 35 (1): 127–141. doi: https://doi.org/10.1016/j.envint.2008.05.010.

Knapp, J. S., P. S. Newby and L. P. Reece. 1995. Decolorization of dyes by wood-rotting basidiomycete fungi. *Enzyme and Microbial Technology* 17 (7): 664–668. doi: https://doi.org/10.1016/0141-0229(94)00112-5.

Knapp, J. S., F. M. Zhang and K. N. Tapley. 1997. Decolourisation of orange II by a wood-rotting fungus. *Journal of Chemical Technology & Biotechnology* 69 (3): 289–296. doi: doi:10.1002/(SICI)1097-4660(199707)69:3<289::AID-JCTB702>3.0.CO;2-H.

Kojima, Y., Y. Tsukuda, Y. Kawai, A. Tsukamoto, J. Sugiura, M. Sakaino and Y. Kita. 1990. Cloning, sequence analysis, and expression of ligninolytic phenoloxidase genes of the white-rot basidiomycete *Coriolus hirsutus*. *Journal of Biological Chemistry* 265 (25): 15224–30.

Kulla, H.G. 1981. Aerobic bacterial degradation of azo dyes. In: *Microbial degradation of xenobiotics and recalcitrant compounds*, edited by Leisinger, T., Cook, A.M., Hutter, R and Nuesch, J. pp. 387–399. Academic Press, London.

Leatham, G. F. and T. K. Kirk. 1983. Regulation of ligninolytic activity by nutrient nitrogen in white-rot basidiomycetes. *FEMS Microbiology Letters* 16 (1): 65–67.

Leidig, E., U. Prüsse, K. D. Vorlop and J. Winter. 1999. Biotransformation of Poly R-478 by continuous cultures of PVAL-encapsulated *Trametes versicolor* under non-sterile conditions. *Bioprocess Engineering* 21 (1): 5–12. doi: 10.1007/pl00009064.

Levin, L., L. Papinutti and F. Forchiassin. 2004. Evaluation of argentinean white rot fungi for their ability to produce lignin-modifying enzymes and decolorize industrial dyes. *Bioresource Technology* 94 (2): 169–176. doi: https://doi.org/10.1016/j.biortech.2003.12.002.

Li, X. and R. Jia. 2008. Decolorization and biosorption for congo red by system rice hull- *Schizophyllum* sp. F17 under solid-state condition in a continuous flow packed-bed bioreactor. *Bioresource Technology* 99 (15): 6885–6892. doi: https://doi.org/10.1016/j.biortech.2008.01.049.

López, C., I. Mielgo, M. T. Moreira, G. Feijoo and J. M. Lema. 2002. Enzymatic membrane reactors for biodegradation of recalcitrant compounds. Application to dye decolourisation. *Journal of Biotechnology* 99 (3): 249–257. doi: https://doi.org/10.1016/S0168-1656(02)00217-1.

Lucas, M. S., C. Amaral, A. Sampaio, J. A. Peres and A. A. Dias. 2006. Biodegradation of the diazo dye reactive black 5 by a wild isolate of *Candida oleophila. Enzyme and Microbial Technology* 39 (1): 51–55. doi: https://doi.org/10.1016/j.enzmictec.2005.09.004.

Machado, K. M. G., L. C. A. Compart, R. O. Morais, L. H. Rosa and M. H. Santos. 2006. Biodegradation of reactive textile dyes by basidiomycetous fungi from brazilian ecosystems. *Brazilian Journal of Microbiology* 37: 481–487.

Mahmoud, M. S., M. K. Mostafa, S. A. Mohamed, N. A. Sobhy and M. Nasr. 2017. Bioremediation of red azo dye from aqueous solutions by *Aspergillus niger* strain isolated from textile wastewater. *Journal of Environmental Chemical Engineering* 5 (1): 547–554. doi: https://doi.org/10.1016/j.jece.2016.12.030.

Martins, M. A. M. M. H. Cardoso, M. J. Queiroz, M. T. Ramalho and A. M. O. Campus. 1999. Biodegradation of azo dyes by the yeast *Candida zeylanoides* in batch aerated cultures. *Chemosphere* 38 (11): 2455–2460. doi: https://doi.org/10.1016/S0045-6535(98)00448-2.

Martins, M. A. M. M. J. Queiroz, A. J. D. Silvestre and N. Lima. 2002. Relationship of chemical structures of textile dyes on the pre-adaptation medium and the potentialities of their biodegradation by *Phanerochaete chrysosporium. Research in Microbiology* 153 (6): 361–368. doi: https://doi.org/10.1016/S0923-2508(02)01332-3.

Martins, M. A. M., N. Lima, A. J. D. Silvestre and M. J. Queiroz. 2003. Comparative studies of fungal degradation of single or mixed bioaccessible reactive azo dyes. *Chemosphere* 52 (6): 967–973. doi: https://doi.org/10.1016/S0045-6535(03)00286-8.

Martorell, M. M., H. F. Pajot, P. M. Ahmed and L. I. C. de Figueroa. 2017. Biodecoloration of reactive black 5 by the methylotrophic yeast *Candida boidinii* MM 4035. *Journal of Environmental Sciences* 53: 78–87. doi: https://doi.org/10.1016/j.jes.2016.01.033.

Maté, D., C. García-Burgos, E. García-Ruiz, A. O. Ballesteros, S. Camarero and M. Alcalde. 2010. Laboratory evolution of high-redox potential laccases. *Chemistry & Biology* 17 (9): 1030–1041. doi: https://doi.org/10.1016/j.chembiol.2010.07.010.

Máximo, C. and M. Costa-Ferreira. 2004. Decolourisation of reactive textile dyes by *Irpex lacteus* and lignin modifying enzymes. *Process Biochemistry* 39 (11): 1475–1479. doi: https://doi.org/10.1016/S0032-9592(03)00293-0.

Mazmanci, M., A. Ünyayar and h. i. Ekiz. 2002. Decolorization of methylene blue by white rot fungus *Coriolus versicolor. Fresenius Environmental Bulletin* 11 (5).

McMullan, G., C. Meehan, A. Conneely, N. Kirby, T. Robinson, P. Nigam, I. Banat, R. Marchant and W. Smyth. 2001. Microbial decolourisation and degradation of textile dyes. *Applied Microbiology and Biotechnology* 56 (1): 81–87. doi: 10.1007/s002530000587.

Mechsner, K. and K. Wuhrmann. 1982. Cell permeability as a rate limiting factor in the microbial reduction of sulfonated azo dyes. *European journal of applied microbiology and biotechnology* 15 (2): 123–126. doi: 10.1007/bf00499518.

Meehan, C., I. M. Banat, G. McMullan, P. Nigam, F. Smyth and R. Marchant. 2000. Decolorization of remazol black-B using a thermotolerant yeast, *Kluyveromyces marxianus* IMB3. *Environment International* 26 (1): 75–79. doi: https://doi.org/10.1016/S0160-4120(00)00084-2.

Mester, T. and J. A. Field. 1998. Characterization of a novel manganese peroxidase-lignin peroxidase hybrid isozyme produced by *Bjerkandera* species strain BOS55 in the absence of manganese. *Journal of Biological Chemistry* 273 (25): 15412–15417. doi: 10.1074/jbc.273.25.15412.

Mielgo, I., M. T. Moreira, G. Feijoo and J. M. Lema. 2001. A packed-bed fungal bioreactor for the continuous decolourisation of azo-dyes (orange II). *Journal of Biotechnology* 89 (2): 99–106. doi: https://doi.org/10.1016/S0168-1656(01)00319-4.

Mielgo, I., M. T. Moreira, G. Feijoo and J. M. Lema. 2002. Biodegradation of a polymeric dye in a pulsed bed bioreactor by immobilised *Phanerochaete chrysosporium. Water Research* 36 (7): 1896–1901. doi: https://doi.org/10.1016/S0043-1354(01)00384-0.

Miranda, M. P., G. G. Benito, N. S. Cristobal and C. H. Nieto. 1996. Color elimination from molasses wastewater by *Aspergillus niger. Bioresource Technology* 57 (3): 229–235. doi: https://doi.org/10.1016/S0960-8524(96)00048-X.

Miura, M., T. Deguchi, M. Matsubara and M. Kakezawa. 1997. Isolation of manganese peroxidase-producing mutants of the hyper-lignolytic fungus IZU-154 under nitrogen nonlimiting conditions.

Journal of Fermentation and Bioengineering 83 (2): 191–193. doi: https://doi.org/10.1016/S0922-338X(97)83581-7.

Mohorčič, M., J. Friedrich and A. Pavko. 2004. Decoloration of the diazo dye reactive black 5 by immobilised *Bjerkandera adusta* in a stirred tank bioreactor. *Acta Chimica Slovenica* 51 (4): 619–628.

Mohorčič, M., S. Teodorovič, V. Golob and J. Friedrich. 2006. Fungal and enzymatic decolourisation of artificial textile dye baths. *Chemosphere* 63 (10): 1709–1717. doi: https://doi.org/10.1016/j.chemosphere.2005.09.063.

Moreira, M. T., I. Mielgo, G. Feijoo and J. M. Lema. 2000. Evaluation of different fungal strains in the decolourisation of synthetic dyes. *Biotechnology Letters* 22 (18): 1499–1503. doi: 10.1023/a:1005606330152.

Mou, D. G., K. K. Lim and H. P. Shen. 1991. Microbial agents for decolorization of dye wastewater. *Biotechnology Advances* 9 (4): 613–622. doi: https://doi.org/10.1016/0734-9750(91)90734-D.

Nascimento, C. R. S., M. M. Nishikawa, A. B. M. Vaz, C. A. Rosa and M. DaSilva. 2013. Textile azo dye degradation by *Candida rugosa* INCQS 71011 isolated from a non-impacted area in semi-arid region of brazilian northeast. *African Journal of Biotechnology* 12 (47): 6636–6642. doi: 10.5897/AJB2013.12859.

Niebisch, C. H., A. K. Malinowski, R. Schadeck, D. A. Mitchell, V. Kava-Cordeiro and J. Paba. 2010. Decolorization and biodegradation of reactive blue 220 textile dye by *Lentinus crinitus* extracellular extract. *Journal of Hazardous Materials* 180 (1): 316–322. doi: https://doi.org/10.1016/j.jhazmat.2010.04.033.

Nilsson, I., A. Möller, B. Mattiasson, M. S. T. Rubindamayugi and U. Welander. 2006. Decolorization of synthetic and real textile wastewater by the use of white-rot fungi. *Enzyme and Microbial Technology* 38 (1): 94–100. doi: https://doi.org/10.1016/j.enzmictec.2005.04.020.

Novotný, Č., B. Rawal, M. Bhatt, M. Patel, V. Šašek and H. P. Molitoris. 2001. Capacity of irpex lacteus and *Pleurotus ostreatus* for decolorization of chemically different dyes. *Journal of Biotechnology* 89 (2): 113–122. doi: https://doi.org/10.1016/S0168-1656(01)00321-2.

Novotný, Č., K. Svobodová, A. Kasinath and P. Erbanová. 2004. Biodegradation of synthetic dyes by *Irpex lacteus* under various growth conditions. *International Biodeterioration & Biodegradation* 54 (2): 215–223. doi: https://doi.org/10.1016/j.ibiod.2004.06.003.

Ollikka, P., K. Alhonmäki, V. M. Leppänen, T. Glumoff, T. Raijola, and I. Suominen. 1993. Decolorization of azo, triphenyl methane, heterocyclic, and polymeric dyes by lignin peroxidase isoenzymes from *Phanerochaete chrysosporium*. *Applied and Environmental Microbiology* 59 (12): 4010–4016.

Palma, C., M. T. Moreira, I. Mielgo, G. Feijoo and J. M. Lema. 1999. Use of a fungal bioreactor as a pretreatment or post-treatment step for continuous decolorisation of dyes. *Water Science and Technology* 40 (8): 131–136. doi: 10.2166/wst.1999.0404.

Palmieri, G. G. Cennamo and G. Sannia. 2005. Remazol brilliant blue R decolourisation by the fungus *Pleurotus ostreatus* and its oxidative enzymatic system. *Enzyme and Microbial Technology* 36 (1): 17–24. doi: https://doi.org/10.1016/j.enzmictec.2004.03.026.

Pandey, A. 1992. Recent process developments in solid-state fermentation. *Process Biochemistry* 27 (2): 109–117. doi: https://doi.org/10.1016/0032-9592(92)80017-W.

Pandey, A., P. Singh and L. Iyengar. 2007. Bacterial decolorization and degradation of azo dyes. *International Biodeterioration & Biodegradation* 59 (2): 73–84. doi: https://doi.org/10.1016/j.ibiod.2006.08.006.

Papanikolaou, S., M. Galiotou-Panayotou, S. Fakas, M. Komaitis and G. Aggelis. 2008. Citric acid production by *Yarrowia lipolytica* cultivated on olive-mill wastewater-based media. *Bioresource Technology* 99 (7): 2419–2428. doi: https://doi.org/10.1016/j.biortech.2007.05.005.

Park, C., B. Lee, E. J. Han, J. Lee and S. Kim. 2006. Decolorization of acid black 52 by fungal immobilization. *Enzyme and Microbial Technology* 39 (3): 371–374. doi: https://doi.org/10.1016/j.enzmictec.2005.11.045.

Parshetti, G. K., S. D. Kalme, S. S. Gomare and S. P. Govindwar. 2007. Biodegradation of reactive blue-25 by *Aspergillus ochraceus* NCIM-1146. *Bioresource Technology* 98 (18): 3638–3642. doi: https://doi.org/10.1016/j.biortech.2006.11.017.

Paszczynski, A., M. B. Pasti-Grigsby, S. Goszczynski, R. L. Crawford and D. L. Crawford. 1992. Mineralization of sulfonated azo dyes and sulfanilic acid by *Phanerochaete chrysosporium* and *Streptomyces chromofuscus*. *Applied and Environmental Microbiology* 58 (11): 3598–3604.

Pazarlioglu, N. K., R. O. Urek and F. Ergun. 2005. Biodecolourization of direct blue 15 by immobilized *Phanerochaete chrysosporium*. *Process Biochemistry* 40 (5): 1923–1929. doi: https://doi.org/10.1016/j.procbio.2004.07.005.

Peralta-Zamora, P., A. Kunz, S. G. de Moraes, R. Pelegrini, P. de Campos Moleiro, J. Reyes and N. Duran. 1999. Degradation of reactive dyes I. A comparative study of ozonation, enzymatic and photochemical processes. *Chemosphere* 38 (4): 835–852. doi: https://doi.org/10.1016/S0045-6535(98)00227-6.

Phugare, S., P. Patil, S. Govindwar and J. Jadhav. 2010. Exploitation of yeast biomass generated as a waste product of distillery industry for remediation of textile industry effluent. *International Biodeterioration & Biodegradation* 64 (8): 716–726. doi: https://doi.org/10.1016/j.ibiod.2010.08.005.

Pilatin, S. and B. Kunduhoğlu. 2011. Decolorization of textile dyes by newly isolated *trametes versicolor* strain. *Anadolu University Journal of Science and Technology–C, Life Sciences and Biotechnology* 1 (2): 125–135.

Podgornik, H., I. Grgić and A. Perdih. 1999. Decolorization rate of dyes using lignin peroxidases of *Phanerochaete chrysosporium*. *Chemosphere* 38 (6): 1353–1359. doi: https://doi.org/10.1016/S0045-6535(98)00537-2.

Prigione, V., G. C. Varese, L. Casieri and V. F. Marchisio. 2008. Biosorption of simulated dyed effluents by inactivated fungal biomasses. *Bioresource Technology* 99 (9): 3559–3567. doi: https://doi.org/10.1016/j.biortech.2007.07.053.

Raghukumar, C., D. Chandramohan, J. F. C. Michel and C. A. Redd. 1996. Degradation of lignin and decolorization of paper mill bleach plant effluent (BPE) by marine fungi. *Biotechnology Letters* 18 (1): 105–106. doi: 10.1007/bf00137820.

Rai, H. S., M. S. Bhattacharyya, J. Singh, T. K. Bansal, P. Vats and U. C. Banerjee. 2005. Removal of dyes from the effluent of textile and dyestuff manufacturing industry: A review of emerging techniques with reference to biological treatment. *Critical Reviews in Environmental Science and Technology* 35 (3): 219–238. doi: 10.1080/10643380590917932.

Ramalho, P. A., H. Scholze, M. H. Cardoso, M. T. Ramalho and A. M. Oliveira-Campos. 2002. Improved conditions for the aerobic reductive decolourisation of azo dyes by *Candida zeylanoides*. *Enzyme and Microbial Technology* 31 (6): 848–854. doi: https://doi.org/10.1016/S0141-0229(02)00189-8.

Ramsay, J. A. and C. Goode. 2004. Decoloration of a carpet dye effluent using *trametes versicolor*. *Biotechnology Letters* 26 (3): 197–201. doi: 10.1023/B:BILE.0000013711.32890.5d.

Ramsay, J. A., W. H. W. Mok, Y. S. Luu and M. Savage. 2005. Decoloration of textile dyes by alginate-immobilized *Trametes versicolor*. *Chemosphere* 61 (7): 956–964. doi: https://doi.org/10.1016/j.chemosphere.2005.03.070.

Reddy, C. A. 1995. The potential for white-rot fungi in the treatment of pollutants. *Current Opinion in Biotechnology* 6 (3): 320–328. doi: https://doi.org/10.1016/0958-1669(95)80054-9.

Revankar, M. S. and S. S. Lele. 2006. Enhanced production of laccase using a new isolate of white rot fungus WR-1. *Process Biochemistry* 41 (3): 581–588. doi: https://doi.org/10.1016/j.procbio.2005.07.019.

Revankar, M. S. and S. S. Lele. 2007. Synthetic dye decolorization by white rot fungus, *Ganoderma* sp. WR-1. *Bioresource Technology* 98 (4): 775–780. doi: https://doi.org/10.1016/j.biortech.2006.03.020.

Reyes, P., M. A. Pickard and R. Vazquez-Duhalt. 1999. Hydroxybenzotriazole increases the range of textile dyes decolorized by immobilized laccase. *Biotechnology Letters* 21 (10): 875–880. doi: 10.1023/a:1005502906890.

Rodríguez, E., M. A. Pickard and R. Vazquez-Duhalt. 1999. Industrial dye decolorization by laccases from ligninolytic fungi. *Current Microbiology* 38 (1): 27–32. doi: 10.1007/pl00006767.

Rodríguez-Couto, S., D. Moldes, A. Liébanas and A. Sanromán. 2003. Investigation of several bioreactor configurations for laccase production by *Trametes versicolor* operating in solid-state conditions. *Biochemical Engineering Journal* 15 (1): 21–26. doi: https://doi.org/10.1016/S1369-703X(02)00180-8.

Rodríguez-Couto, S., E. Rosales and M. A. Sanromán. 2006. Decolourization of synthetic dyes by *Trametes hirsuta* in expanded-bed reactors. *Chemosphere* 62 (9): 1558–1563. doi: https://doi.org/10.1016/j.chemosphere.2005.06.042.

Rüttimann-Johnson, C., L. Salas, R. Vicuña and T. K. Kirk. 1993. Extracellular enzyme production and synthetic lignin mineralization by *Ceriporiopsis subvermispora*. *Applied and Environmental Microbiology* 59 (6): 1792–1797.

Ryu, B. H. and Y. D. Weon. 1992. Decolorization of azo dyes by *Aspergillus sojae* B-10. *Journal of Microbiology and Biotechnology* 2: 1792–1797.

Saloheimo, M. M. L. Niku-Paavola and J. K. C. Knowles. 1991. Isolation and structural analysis of the laccase gene from the ligninegrading fungus *Phlebia radiata*. *Microbiology* 137 (7): 1537–1544. doi: doi:10.1099/00221287-137-7-1537.

Saratale, G., S. Kalme, S. Bhosale and S. Govindwar. 2007. Biodegradation of kerosene by *Aspergillus ochraceus* NCIM-1146. *Journal of Basic Microbiology* 47 (5): 400–405. doi: doi:10.1002/jobm.200710337.

Saratale, R. G. G. D. Saratale, J. S. Chang and S. P. Govindwar. 2009. Decolorization and biodegradation of textile dye navy blue HER by *Trichosporon beigelii* NCIM-3326. *Journal of Hazardous Materials* 166 (2): 1421–1428. doi: https://doi.org/10.1016/j.jhazmat.2008.12.068.

Saratale, R. G. G. D. Saratale, J. S. Chang and S. P. Govindwar. 2011. Bacterial decolorization and degradation of azo dyes: A review. *Journal of the Taiwan Institute of Chemical Engineers* 42 (1): 138–157. doi: https://doi.org/10.1016/j.jtice.2010.06.006.

Schliephake, K., G. T. Lonergan, C. L. Jones and D. E. Mainwaring. 1993. Decolourisation of a pigment plant effluent by *Pycnoporus cinnabarinus* in a packed-bed bioreactor. *Biotechnology Letters* 15 (11): 1185–1188. doi: 10.1007/bf00131213.

Sedighi, M., A. Karimi and F. Vahabzadeh. 2009. Involvement of ligninolytic enzymes of *Phanerochaete chrysosporium* in treating the textile effluent containing astrazon red FBL in a packed-bed bioreactor. *Journal of Hazardous Materials* 169 (1): 88–93. doi: https://doi.org/10.1016/j.jhazmat.2009.03.070.

Selvam, K., K. Swaminathan and K. S. Chae. 2003. Decolourization of azo dyes and a dye industry effluent by a white rot fungus *Thelephora* sp. *Bioresource Technology* 88 (2): 115–119. doi: https://doi.org/10.1016/S0960-8524(02)00280-8.

Sen, S. K., S. Raut, P. Bandyopadhyay and S. Raut. 2016. Fungal decolouration and degradation of azo dyes: A review. *Fungal Biology Reviews* 30 (3): 112–133. doi: https://doi.org/10.1016/j.fbr.2016.06.003.

Shahvali, M., M. M. Assadi and K. Rostami. 2000. Effect of environmental parameters on decolorization of textile wastewater using *Phanerochaete chrysosporium*. *Bioprocess Engineering* 23 (6): 721–726. doi: 10.1007/s004490000196.

Si, J., B. K. Cui and Y. C. Dai. 2013. Decolorization of chemically different dyes by white-rot fungi in submerged cultures. *Annals of Microbiology* 63 (3): 1099–1108. doi: 10.1007/s13213-012-0567-8.

Singh, R. L., P. K. Singh and R. P. Singh. 2015. Enzymatic decolorization and degradation of azo dyes—A review. *International Biodeterioration & Biodegradation* 104: 21–31. doi: https://doi.org/10.1016/j.ibiod.2015.04.027.

Slokar, Y. M. and A. Majcen Le Marechal. 1998. Methods of decoloration of textile wastewaters. *Dyes and Pigments* 37 (4): 335–356. doi: https://doi.org/10.1016/S0143-7208(97)00075-2.

Soares, C. H. L. and N. Durán. 1998. Degradation of low and high molecular mass fractions of kraft E1 effluent by *Trametes Villosa*. *Environmental Technology* 19 (9): 883–891. doi: 10.1080/09593331908616746.

Soares, G. M. B., M. T. P. de Amorim and M. Costa-Ferreira. 2001. Use of laccase together with redox mediators to decolourize remazol brilliant blue R. *Journal of Biotechnology* 89 (2): 123–129. doi: https://doi.org/10.1016/S0168-1656(01)00302-9.

Soden, D. M., J. O'Callaghan and A. D. W. Dobson. 2002. Molecular cloning of a laccase isozyme gene from *Pleurotus sajor-caju* and expression in the heterologous *Pichia pastoris* host. *Microbiology* 148 (12): 4003–4014. doi: doi:10.1099/00221287-148-12-4003.

Spadaro, J. T., M. H. Gold and V. Renganathan. 1992. Degradation of azo dyes by the lignin-degrading fungus *Phanerochaete chrysosporium*. *Applied and Environmental Microbiology* 58 (8): 2397–2401.

Srikanlayanukul, M., W. Kitwechkun, T. Watanabe and C. Khanongnuch. 2008. Decolorization of orange II by immobilized thermotolerant white rot fungus *Coriolus versicolor* RC3 in packed-bed bioreactor. *Biotechnology* 7 (2): 280–286.

Sumathi, S. and V. Phatak. 1999. Fungal treatment of bagasse based pulp and paper mill wastes. *Environmental Technology* 20 (1): 93–98. doi: 10.1080/09593332008616797.

Sumathi, S. and B. S. Manju. 2000. Uptake of reactive textile dyes by *Aspergillus foetidus*. *Enzyme and Microbial Technology* 27 (6): 347–355. doi: https://doi.org/10.1016/S0141-0229(00)00234-9.

Svobodová, K., A. Majcherczyk, Č. Novotný and U. Kües. 2008. Implication of mycelium-associated laccase from *Irpex lacteus* in the decolorization of synthetic dyes. *Bioresource Technology* 99 (3): 463–471. doi: https://doi.org/10.1016/j.biortech.2007.01.019.

Swamy, J. and J. A. Ramsay. 1999a. Effects of glucose and NH_4^+ concentrations on sequential dye decoloration by *Trametes versicolor*. *Enzyme and Microbial Technology* 25 (3): 278–284. doi: https://doi.org/10.1016/S0141-0229(99)00058-7.

Swamy, J. and J. A. Ramsay. 1999b. Effects of Mn^{2+} and NH_4^+ concentrations on laccase and manganese peroxidase production and Amaranth decoloration by *Trametes versicolor*. *Applied Microbiology and Biotechnology* 51 (3): 391–396. doi: 10.1007/s002530051408.

Swamy, J. and J. A. Ramsay. 1999c. The evaluation of white rot fungi in the decoloration of textile dyes. *Enzyme and Microbial Technology* 24 (3): 130–137. doi: https://doi.org/10.1016/S0141-0229(98)00105-7.

Tan, L., H. Li, S. Ning and B. Xu. 2014. Aerobic decolorization and degradation of azo dyes by suspended growing cells and immobilized cells of a newly isolated yeast *Magnusiomyces ingens* LH-F1. *Bioresource Technology* 158: 321–328. doi: https://doi.org/10.1016/j.biortech.2014.02.063.

Tatarko, M. and J. A. Bumpus. 1998. Biodegradation of congo red by *Phanerochaete chrysosporium*. *Water Research* 32 (5): 1713–1717. doi: https://doi.org/10.1016/S0043-1354(97)00378-3.

Tavčar, M. B., J. Babič, Č. Novotný, K. Svobodová and A. Pavko. 2006. Biodegradation of azo dye RO16 in different reactors by immobilized *Irpex lacteus*. *Acta chimica slovenica* 3 (53): 338–343.

Theerachat, M., S. Emond, E. Cambon, F. Bordes, A. Marty, J.-M. Nicaud, W. Chulalaksananukul, D. Guieysse, M. Remaud-Siméon and S. Morel. 2012. Engineering and production of laccase from *Trametes versicolor* in the yeast *Yarrowia lipolytica*. *Bioresource Technology* 125: 267–274. doi: https://doi.org/10.1016/j.biortech.2012.07.117.

Tychanowicz, G. K., A. Zilly, C. G. M. de Souza and R. M. Peralta. 2004. Decolourisation of industrial dyes by solid-state cultures of *Pleurotus pulmonarius*. *Process Biochemistry* 39 (7): 855–859. doi: https://doi.org/10.1016/S0032-9592(03)00194-8.

Ürek, R. Ö. and N. K. Pazarlioğlu. 2005. Production and stimulation of manganese peroxidase by immobilized *Phanerochaete chrysosporium*. *Process Biochemistry* 40 (1): 83–87. doi: https://doi.org/10.1016/j.procbio.2003.11.040.

Urra, J., L. Sepúlveda, E. Contreras and C. Palma. 2006. Screening of static culture and comparison of batch and continuous culture for the textile dye biological decolorization by *Phanerochaete chrysosporium*. *Brazilian Journal of Chemical Engineering* 23: 281–290.

Vaithanomsat, P., W. Apiwatanapiwat, O. Petchoy and J. Chedchant. 2010. Production of ligninolytic enzymes by white-rot fungus *Datronia* sp. KAPI0039 and their application for reactive dye removal. *International Journal of Chemical Engineering* 2010: 6. doi: 10.1155/2010/162504.

Vitor, V. and C. R. Corso. 2008. Decolorization of textile dye by *Candida albicans* isolated from industrial effluents. *Journal of Industrial Microbiology & Biotechnology* 35 (11): 1353–1357. doi: 10.1007/s10295-008-0435-5.

Wang, B. E., Y. Y. Hu, L. Xie and K. Peng. 2008. Biosorption behavior of azo dye by inactive CMC immobilized *Aspergillus fumigatus* beads. *Bioresource Technology* 99 (4): 794–800. doi: https://doi.org/10.1016/j.biortech.2007.01.043.

Wang, Z., B. H. Bleakley, D. L. Crawford, G. Hertel and F. Rafii. 1990. Cloning and expression of a lignin peroxidase gene from *Streptomyces viridosporus* in *Streptomyces lividans*. *Journal of Biotechnology* 13 (2): 131–144. doi: https://doi.org/10.1016/0168-1656(90)90099-W.

Wells, A., M. Teria and T. Eve. 2006. Green oxidations with laccase–mediator systems. *Biochemical Society Transactions* 34 (2): 304.

Wesenberg, D., I. Kyriakides and S. N. Agathos. 2003. White-rot fungi and their enzymes for the treatment of industrial dye effluents. *Biotechnology Advances* 22 (1): 161–187. doi: https://doi.org/10.1016/j.biotechadv.2003.08.011.

Wong, K. S., Q. Huang, C. H. Au, J. Wang and H. S. Kwan. 2012. Biodegradation of dyes and polyaromatic hydrocarbons by two allelic forms of *Lentinula edodes* laccase expressed from *Pichia pastoris*. *Bioresource Technology* 104: 157–164. doi: https://doi.org/10.1016/j.biortech.2011.10.097.

Yang, Q., M. Yang, K. Pritsch, A. Yediler, A. Hagn, M. Schloter and A. Kettrup. 2003. Decolorization of synthetic dyes and production of manganese-dependent peroxidase by new fungal isolates. *Biotechnology Letters* 25 (9): 709–713. doi: 10.1023/a:1023454513952.

Yang, Q., A. Yediler, M. Yang and A. Kettrup. 2005. Decolorization of an azo dye, reactive black 5 and MnP production by yeast isolate: *Debaryomyces polymorphus*. *Biochemical Engineering Journal* 24 (3): 249–253. doi: https://doi.org/10.1016/j.bej.2004.12.004.

Yaver, D. S., F. Xu, E. J. Golightly, K. M. Brown, S. H. Brown, M. W. Rey, P. Schneider, T. Halkier, K. Mondorf and H. Dalboge. 1996. Purification, characterization, molecular cloning, and expression of two laccase genes from the white rot basidiomycete *Trametes villosa. Applied and Environmental Microbiology* 62 (3): 834–841.

Yesilada, O., S. Cing and D. Asma. 2002. Decolourisation of the textile dye astrazon red FBL by *Funalia trogii* pellets. *Bioresource Technology* 81 (2): 155–157. doi: https://doi.org/10.1016/S0960-8524(01)00117-1.

Young, L. and J. Yu. 1997. Ligninase-catalysed decolorization of synthetic dyes. *Water Research* 31 (5): 1187–1193. doi: https://doi.org/10.1016/S0043-1354(96)00380-6.

Yu, Z. and X. Wen. 2005. Screening and identification of yeasts for decolorizing synthetic dyes in industrial wastewater. *International Biodeterioration & Biodegradation* 56 (2): 109–114. doi: https://doi.org/10.1016/j.ibiod.2005.05.006.

Zaharia, C. and D. Suteu. 2012. Textile organic dyes—Characteristics, polluting effects and separation/elimination procedures from industrial effluents—A critical overview. In: *Organic Pollutants Ten Years After the Stockholm Convention—Environmental and Analytical Update*, edited by Tomasz Puzyn. InTech.

Zhang, F., J. S. Knapp and K. N. Tapley. 1999a. Decolourisation of cotton bleaching effluent with wood rotting fungus. *Water Research* 33 (4): 919–928. doi: https://doi.org/10.1016/S0043-1354(98)00288-7.

Zhang, F., J. S. Knapp and K. N. Tapley. 1999b. Development of bioreactor systems for decolorization of orange II using white rot fungus. *Enzyme and Microbial Technology* 24 (1): 48–53. doi: https://doi.org/10.1016/S0141-0229(98)00090-8.

Zhao, J., T. H. de Koker and B. J. H. Janse. 1996. Comparative studies of lignin peroxidases and manganese-dependent peroxidases produced by selected white rot fungi in solid media. *FEMS Microbiology Letters* 145 (3): 393–399. doi: https://doi.org/10.1016/S0378-1097(96)00438-7.

Zhen, Z. and J. Yu. 1998. Stresses on immobilized *Phanerochaete chrysosporium* hyphae in submerged cultures for ligninase production. *The Canadian Journal of Chemical Engineering* 76 (4): 784–789. doi: doi:10.1002/cjce.5450760414.

Zheng, Z., R. E. Levin, J. L. Pinkham and K. Shetty. 1999. Decolorization of polymeric dyes by a novel *Penicillium* isolate. *Process Biochemistry* 34 (1): 31–37. doi: https://doi.org/10.1016/S0032-9592(98)00061-2.

Zhuo, R., L. Ma, F. Fan, Y. Gong, X. Wan, M. Jiang, X. Zhang and Y. Yang. 2011. Decolorization of different dyes by a newly isolated white-rot fungi strain *Ganoderma* sp. En3 and cloning and functional analysis of its laccase gene. *Journal of Hazardous Materials* 192 (2): 855–873. doi: https://doi.org/10.1016/j.jhazmat.2011.05.106.

Zollinger, H. 1987. Azo dyes and pigments. In *Color Chemistry: Synthesis, Properties and Applications of Organic Dyes and Pigments*, 92–100. VCH, New York, USA: Wiley-Blackwell.

CHAPTER-6

Fungal Processes of Interaction with Chromium

P. Romo-Rodríguez and *J. F. Gutiérrez-Corona**

1. Introduction

1.1 Chromium in the environment

Among metals, chromium has a double aspect of interest; on the one hand, it is of economic importance due to its extensive use in different industrial activities and processes, on the other hand, it is of interest to regulatory agencies related to the surveyance and monitoring of the quality of the environment, due to the toxicity that chromium exerts on the ecosystems.

Although chromium can exist in several oxidation states, the most common and stable forms in the environment are the trivalent Cr[Cr(III)] and the hexavalent Cr[Cr(VI)], which have different chemical properties. Cr(VI), considered the most toxic form of chromium, is usually associated with oxygen in the form of chromate (CrO_4^{2-}) and dichromate ($Cr_2O_7^{2-}$), which due to their high solubility are highly mobile in the soil and in aquatic environments. On the other hand, Cr(III) is found in the form of poorly soluble oxides, hydroxides, or sulfates, which is why it is much less mobile, and is bound to organic matter in the soil and aquatic environments (Palmer and Wittbrodt 1991).

Chromium compounds are environmental contaminants present in groundwater, soil, and industrial effluents due to their extensive use in various industries (such as leather tanning, pigments and dyes production, metal finishing, electroplating, metallurgy, battery manufacturing, and wood preservation) and that some of these activities do not have appropriate waste management procedures (Guertin et al. 2004, Madhavi et al. 2013). Cr(VI) compounds are highly soluble and therefore mobile in the environment, and can be transported through cell membranes; the common Cr(VI) species present in aqueous solution are hydrochromate ($HCrO_4^-$), chromate

Departamento de Biología, DCNE, Universidad de Guanajuato, Gto., CP 36050, Mexico.
* Corresponding author: xilefgu@gmail.com

(CrO_4^{2-}), and dichromate ($Cr_2O_7^{2-}$). Instead, Cr(III) derivatives (hydroxides [$Cr(OH)_3$] or hydrated oxides [$Cr_2O \cdot H_2O$]) are insoluble and therefore much less mobile, exist in the environment mostly forming stable complexes with both organic and inorganic ligands, and they cannot be transported inside the cells (Ramírez-Díaz et al. 2008, Barrera-Díaz et al. 2012).

1.2 Chromium toxicity

The biological effects of Cr depend on its oxidation state. Cr(VI) is considered the most toxic form of the metal, because it easily crosses biological membranes and can be transported actively to the interior of the cells by means of the sulfate transporter. Cr(VI) is highly toxic to all life forms, being mutagenic and carcinogenic in animals, and mutagenic in microorganisms and plants. It has been proposed that the toxicity of Cr(VI) is due to the fact that within the cells, Cr(VI) reduction by metabolites or flavoenzymes produces the reduced chromium intermediates Cr(IV)/Cr(V) and the reactive oxygen species (ROS), superoxide, hydrogen peroxide, and hydroxyl radicals, with Cr(III) as the final product. ROS production causes oxidative damage, producing lipid peroxidation, oxidation of proteins, and nucleic acids (Viti et al. 2014, Ramírez-Díaz et al. 2008). It has been considered that the genotoxic effects of Cr cannot be solely explained by the action of ROS derived of Cr(VI) reduction, since intracellular Cr(III) complexes also interact with phosphate groups of DNA, which could affect replication, transcription, and cause mutagenesis (Cervantes et al. 2001). The compounds of Cr(III) are considered relatively innocuous, because they are insoluble and cannot cross the biological membranes. It has been proposed that Cr(III) functions as an essential trace element for animals and humans, although there is controversy in this respect, and caution has been expressed regarding the use of Cr(III) as a micronutrient or antidiabetic agent (Viti et al. 2014).

The toxicity of chromium has been studied in yeasts and filamentous fungi. In the yeast *Saccharomyces cerevisiae*, the toxicity of chromium is higher in cells grown in non-fermentable carbon sources than in those grown in fermentable sources; additional effects described include inhibition of oxygen uptake, induction of mutations, and gene conversion (Cervantes et al. 2001). A recent study reported that, in *S. cerevisiae*, both Cr(III) and Cr(VI) increase mutational rates and cause DNA degradation, finding that Cr(III) is more genotoxic than Cr(VI); additional observations indicated that these two chromium species interact with DNA differently (Fang et al. 2014). Another study observed that, in *S. cerevisiae*, protein oxidation is the main toxicity mechanism of Cr(VI), with glycolytic and heat-shock proteins being mainly oxidized, and oxidation occurring in specific isoforms. Till date, it is unknown whether this mechanism explains the toxicity of Cr(VI) in other biological systems (Sumner et al. 2005). In *S. cerevisiae*, mutation of the gene encoding the transcriptional activator MSN1 caused chromate-tolerance and reduced Cr accumulation in the cells, and overexpression of MSN1 led to increased accumulation of Cr; in another study, it was shown that Cr(VI) exposure induces intracellular signal transduction processes (Poljsak et al. 2010). In *S. cerevisiae*, metal-responsive genomic profiles were obtained that provided information on the transcriptional changes (transcriptome) after the exposure of chromate or other metals and on the relationship between the expression of 4700 non-essential genes and sensitivity to metal exposure (deletome). These studies showed that a total of 1341 genes were differentially expressed among different conditions. In the case of chromate, different genes were identified that participate in the resistance to the toxicity caused by the oxyanion, whose products participate in different cellular processes, from the organization and biogenesis of vacuoles, sulfur amino acid metabolism, transcriptional regulators, up to cytoskeletal organization and cytokinesis (Jin et al. 2008). In other yeasts, the effect of Cr(VI) toxicity has also been investigated (Poljsak et al. 2010): it was observed that the incubation of the fission yeast *Schizosaccharomyces pombe* with Cr(VI) affects the specific activities of antioxidant enzymes and the production of various ROS species; in *Candida intermedia*, Cr(VI) exposure increases the level of the Hsp 104 chaperone protein and elevates the intracellular level of the reduced form of glutathione. The effects of Cr(VI) on the plasma membrane of *Sch. pombe* have been studied by electron paramagnetic resonance (EPR) spectroscopy, observing an increase in membrane fluidity; membrane perturbations due to Cr(VI) are mostly located at the lipid-water interfaces. On the other hand, in *Candida albicans*, treatment with Cr(III) caused an increased fluidity of the membrane. Although this effect did not result in any visible changes in the plasma membrane structure, it was proposed that Cr(III) stress caused lipid peroxidation. The latter might contribute to both the fluidification and the decreased membrane barrier function in Cr(III)-treated cells, which was verified by the loss of intracellular metabolites after treatment with Cr(III) (Poljsak et al. 2010).

Traditional methods for the treatment of chromium-contaminated wastewaters include solvent extraction, adsorption using polymeric resins, clays, biopolymers, biomass, activated carbon, and graphene oxide. Nevertheless, these technologies have significant disadvantages, including incomplete or low Cr(VI) removal, high operating costs and capital, high consumption of reagent and energy, and generation of toxic secondary pollutants which are difficult to dispose of (Kalidhasan et al. 2016). Due to these disadvantages of conventional techniques, the use of bioremediation processes for this purpose is currently being considered. Different microbial mechanisms can influence Cr speciation, toxicity, and mobility, serving as the basis for the development of new technologies to remove Cr from contaminated settings (Cheung and Gu 2007, Ramírez-Díaz et al. 2008, Poljsak et al. 2010, Viti et al. 2014, Thatoi et al. 2014, Joutey et al. 2015, Gutiérrez-Corona et al. 2016).

Based on the relevance of fungi as organisms that predominate in the soil, their fundamental properties as bioconverters, and their role in the geochemical cycles of metals (Gadd 2013, Burford et al. 2016), this chapter focuses on the underlying mechanisms of fungal interaction with chromium and on the biotechnological applications of some of these mechanisms.

2. Fungal Processes of Interaction with Chromium

The interaction of fungi with chromium involves active or passive processes, which are schematically shown in Fig. 6.1, and include transport, accumulation, biosorption, and the enzymatic and non-enzymatic reduction of the oxyanion to Cr(III).

176   *Fungal Bioremediation: Fundamentals and Applications*

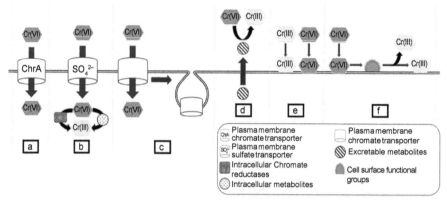

Figure 6.1 Schematic summaries of microbial interactions with chromium. (a) Plasma membrane Cr(VI) transport, through the CHR-1 protein (b) plasma membrane Cr(VI) transport through the sulfate transporter followed by enzymatic and/or non-enzymatic reduction, (c) plasma membrane Cr(VI) transport inactivated by a ubiquitin-dependent endocytic process, (d) extracellular non-enzymatic Cr(VI) reduction, (e) Cr(III) and Cr(VI) biosorption, and (f) Cr(VI) biosorption coupled to reduction.

2.1 Transport and bioaccumulation

The incorporation of metals in microorganisms occurs through two processes, one called "active uptake" and another called "passive uptake" or "biosorption". The former process occurs only in living cells and requires metabolism and energy for the transport of metals through the cell membrane, whereas the second process is the accumulation of metals that do not require metabolism and energy and occurs in living or dead cells or in biological materials. The combination of active and passive modes of metal uptake has been called "bioaccumulation" (Mohan and Pittman 2006, Wang and Chen 2009).

In fungi, chromate is transported to the interior of the cells through the permease system, a non-specific anion carrier that transports different anions, such as sulfate and phosphate (Cervantes et al. 2001); this has also been suggested for the transport of other oxyanions of similar structure, such as molybdate and selenate (Tamás et al. 2006). In the filamentous fungus, *Neurospora crassa* (Marzluf 1970), and in *S. cerevisiae* (Cherest et al. 1997), two genes encoding sulfate permeases have been identified using, as a genetic screen, the resistance to toxic analogs of sulfate (chromate in *N. crassa* and selenate and chromate in *S. cerevisiae*). In both organisms, single gene mutants grow slowly in medium with low sulfate concentration, although the double mutant is unable to grow in that condition and lacks the capacity to transport this anion. Two other alternative chromate transport systems have been described in recent years. The first involves the participation of the ChrA transport protein, which belongs to the chromate ion transporter (CHR) superfamily with hundreds of homologs from all three domains of life, and subdivided into two families of sequences: the short-chain monodomain family and the long-chain bidomain family (Díaz-Pérez et al. 2007, Viti et al. 2014). Although CHR function in bacteria is described as a chromate transporter, transferring chromate from the cytoplasm to the extracellular medium, in fungi the opposite function has been described for the CHR-1 protein in the filamentous fungus *Neurospora crassa*, which transports chromate from the extracellular medium to the cytoplasm, causing chromate toxicity (Flores-Alvarez et al. 2012). It is important to mention that although CHR-1 homologs are present in fungal species belonging to Ascomycota, Basidiomycota, Chytridiomycota, and Zygomycota, CHR-1-protein encoding genes were absent in the genome of the vast majority of yeasts, including *S. cerevisiae* (Díaz-Pérez et al. 2007, Flores-Alvarez et al. 2012). It was hypothesized that, in addition to chromate transport, other physiological function(s) could be performed by fungal CHR-1 proteins, as has been suggested for bacterial ChrA proteins (Díaz-Pérez et al. 2007). The other mechanism by which Cr(VI) is internalized in the cells consists of plasma membrane Cr(VI) transport inactivated by a ubiquitin-dependent endocytic process, which is required for the accumulation of chromium and seems to be a mechanism independent of the sulfate and chromate transporters studied till date; this process was described in *S. cerevisiae* (Holland and Avery 2009).

As mentioned, the process of bioaccumulation has been described as a two-step process. In living biomass of *Termitomyces clypeatus*, it was demonstrated that the first step consists of the rapid interaction of chromium with components of the cell surface, followed by a relatively slow accumulation of metal inside the cells. The uptake of chromium was reduced in presence of the sulfate ion, a chromate analog, and different metabolic inhibitors, which indicated that intracellular accumulation of chromium occurs through sulfate active transport. Transmission electron microscopy equipped with energy dispersive X-ray analysis (TEM-EDX) revealed that chromium was localized in the cell wall as well as in the cytoplasm. Fourier transform infrared spectroscopy (FTIR) analysis indicated that chromate ions accumulated into the cytoplasm and then were reduced to less toxic trivalent chromium compounds (Das and Guha 2009). Similar studies have also been conducted on *Phanerochaete chrysosporium*, where the use of FTIR allowed the elucidation of the functional groups involved in the first step of the process; SEM and TEM were used to determine the cellular location of chromium (Murugavelh and Mohanty 2014).

Another process for chromium incorporation in fungal cells is biosorption (passive uptake), which can occur either in growing cells or in inactive non-living biomass; this process is carried out by the functional groups present in the cell wall (Wang and Chen 2009). The anionic species of Cr(VI) can be linked to the fungal surface through electrostatic interactions with functional groups, such as amines, present in chitin and chitosan, which are positively charged (Saha and Orvig 2010). Other functional groups involved in the absorption of chromium include amide, hydroxyl, carboxyl, and carbonyl; the different analytical techniques employed for the identification of these groups have been previously summarized (Mukherjee et al. 2013, Gutiérrez-Corona et al. 2016).

2.2 Biotransformation

The transformation of Cr(VI) to Cr(III) using the metabolic properties of fungi is a process of fundamental and technological relevance, since the genetic, biochemical, and physiological bases of this capacity must be understood, to use this knowledge in the design and implementation of biotechnological processes for the removal of chromium from contaminated settings. In bacteria, once inside the cell, Cr(VI) can be reduced to Cr(III) by oxidoreductases, nitroreductases, or YieF-like reductases (Cervantes and Campos-García 2007); this would suggest that similar enzymes are present in fungi. In the yeasts *Cyberlindnera jadinii* M9 and *Wickerhamomyces anomalus* M10, isolated from textile-dye liquid effluents using enzymatic and proteomic approaches, it was observed that the major responses against acute chromium stress were related to stress response proteins, methionine synthase, energy and degradation proteins, and oxide-reduction proteins (Irazusta et al. 2018). Chromate reductase activity has been studied in cell-free extracts of the filamentous fungi *Penicillium* sp. (Arévalo-Rangel et al. 2013), *Aspergillus niger* (Sallau et al. 2014, Gu et al. 2015), and *Trichoderma asperellum* (Chang et al. 2016). In *Aspergillus flavus*, the increase in the amount of extracellular protein is directly related to the activity of chromate reductase and this is accompanied by a biosorption process, indicating that there is a complex remediation mechanism of Cr(VI), which includes different interactions with the metal (Singh et al. 2016). The adsorption-coupled reduction of Cr(VI) illustrates the interaction between biosorption and reduction mechanisms in the transformation of Cr(VI) to Cr(III) using dead fungal biomass (Park et al. 2005, Park et al. 2007); in this process Cr(VI) is reduced to Cr(III) in the aqueous phase or in the biomass by contact with electron-donor groups of the biomass under an acidic pH (Park et al. 2006).

Indirectly, chromium can be reduced through the activity of enzymes, such as glucose oxidase that uses glucose as a substrate to produce gluconolactone and hydrogen peroxide, which can reduce Cr(VI) to Cr(III) (Romo-Rodríguez et al. 2015).

In addition to Cr(VI) reduction mediated by enzymatic reactions, this process can occur by non-enzymatic reactions with chemical compounds derived from fungal metabolism; these compounds include amino acids, nucleotides, sugars, vitamins, organic acids, or glutathione (Cervantes et al. 2001, Dhal et al. 2013). The extracellular fungal metabolites citrate (Barrera-Díaz et al. 2012) and oxalate (Barrera-Díaz et al. 2012, Wrobel et al. 2015), have been reported to participate in the reduction of Cr(VI) through the photocatalytic effect of Fe(III) or by Mn without light. The biological importance of oxalic acid production in Cr(VI) reduction in the presence of Mn^{2+} was indicated by the observation that oxalic acid-non-producing mutants of *Aspergillus tubingensis* Ed8 showed a decreased capacity to perform the biotransformation of Cr(VI) (Gutiérrez-Corona et al. 2016). In some yeasts, Cr(VI) detoxification occurs by the production and excretion of metabolites, such as riboflavin or sulfate to the extracellular medium (Joutey et al. 2015). In *Schwanniomyces occidentalis*, production of the extracellular polymeric substance (EPS) has been described as an important factor for the reduction of Cr(VI), as well as for the formation of Cr_2O_3 nanoparticles (Mohite et al. 2016).

3. Biosorption and Bioreduction as Fungal Processes of Interest for the Implementation of Chromium Removal Technologies

Of the different processes of fungal interaction with chromium, biosorption and bioreduction have been considered for the development of biotechnological strategies to eliminate chromium from the environment. The use of fungal species to remove inorganic pollutants from wastewater has shown to be a good alternative to traditional wastewater treatment technologies, and the advantages of their use have been proposed (Reviewed by Espinosa-Ortiz et al. (2016)).

The use of fungal strains in biotechnological processes for chromium removal, based on biosorption or bioreduction processes, can be carried out through a variety of strategies, as shown in Table 6.1. Two types of bioreactors, a stirred tank bioreactor and a concentric tube airlift bioreactor, were compared for the removal efficiency of Cr(VI) through bioreduction by the filamentous fungus *Trichoderma viride*, and results revealed that the airlift bioreactor might be a promising alternative, especially when shear-sensitive microorganisms are used (Morales-Barrera and Cristiani-Urbina 2006). In the same way, using a strain of *A. niger*, two types of bioreactors, an airlift reactor (Sepehr et al. 2012a) and a stirred tank reactor (Sepehr et al. 2012b) were tested, comparing the removal efficiency of Cr(III) by biosorption (88 and 72%, respectively); as observed the use of the airlift bioreactor performed better.

The ability of the yeast *Candida tropicalis* to remove Cr(VI) was tested in artificially contaminated soils in a microcosm system, and simulating natural environmental conditions, a reduction efficiency of 72.2% was obtained (Bahafid et al. 2013). Using a batch bioreactor and the yeast *Wickerhamomyces anomalus*, a 100% of Cr(VI) reduction efficiency was achieved (Fernández et al. 2017). The use of *Pleurotus mutilis* biomass as an adsorptive material of Cr(VI) in a torus reactor showed promising results (60% efficiency). This is the first report on the use of this type of reactors in Cr(VI) removal processes (Alouache et al. 2017).

Most of the studies found in the literature describing chromium removal by fungal strains use synthetic chromium solutions, although some reports describe the use of more complex mixtures, such as industrial effluents. Fungal isolates, such as *Aspergillus* sp. (Srivastava and Thakur 2006) and *Saccharomyces cerevisiae* (Benazir et al. 2010), or fungal consortiums (Sharma and Malaviya 2016) were used in bioreactor experiments of Cr(VI) removal from industrial effluents and the results showed good Cr(VI) removal efficiencies. These observations demonstrated the ability of fungal cells to carry out the reduction of chromium in complex conditions.

4. *In situ* Bioremediation of Chromium-contaminated Settings

Two main strategies for the *in situ* reduction of chromium have been described. The first one, biostimulation, involves an adjustment in the chemical and physical conditions of the environment to allow optimal growth conditions for resistant microorganisms adapted to grow in the toxicity conditions imposed by the presence of chromium (Chambers 1991). The second, bioaugmentation, is based on the acceleration and increase in pollutant decontamination efficiency by the insertion of populations able to carry out the Cr(VI) reduction process with greater efficiency (Das and Mishra 2010). In a soil contaminated with chromium and polycyclic aromatic hydrocarbons, bioaugmentation with *Penicillium frequentans* increased the lability and mobility of chromium, with potential consequences for bioremediation by endogenous microorganisms with bioaccumulation or bioreduction capacities (Amezcua-Allieri et al. 2005); similar results were obtained using a bioaugmentation process with *A. niger* and *Penicillium simplicissimum* (Braud et al. 2006). The

Table 6.1 Biotechnological approaches for the removal of Cr(VI).

Organism	Process	Removal efficiency (%)	Removal mechanism	References
Trichoderma viride	Stirred tank bioreactor	60	Reduction	(Morales-Barrera and Cristiani-Urbina 2006)
Trichoderma viride	Concentric tube airlift bioreactor	94.3	Reduction	(Morales-Barrera and Cristiani-Urbina 2006)
Carndida sp.	Concentric Draft-Tube Airlift Bioreactor	100	Reduction	(Guillén-Jiménez et al. 2009)
Aspergillus niger	Airlift bioreactor	88	Adsorption	(Sepehr et al. 2012a)
Aspergillus niger	Stirred tank bioreactor	72	Adsorption	(Sepehr et al. 2012b)
Candida tropicalis	Microcosm	72.2	Reduction	(Bahafid et al. 2013)
Bacterial-Fungal consortium *(Raoultella* sp., *Citrobacter* sp., *Klebsiella* sp., *Salmonella* sp., *Achromobacter* sp. and *Kerstersia* sp. and *Pichia jadinii)*	Suspended Sludge bioreactor	100	Reduction	(Tekerlekopoulou et al. 2013)
Aspergillus tubingensis	Bubble column bioreactor	100	Reduction	(Coreño-Alonso et al. 2014)
Hypocrea tawa	Concentric draft-tube airlift bioreactor	100	Reduction	(Morales-Barrera and Cristiani-Urbina 2015)
Consortium (*Cladosporium perangustum, Penicillium commune, Paecilomyces lilacinus, Fusarium equiseti)*	Stirred tank bioreactor	99.9	Adsorption	(Sharma and Malaviya 2016)
Wickerhamomyces anomalus	Batch bioreactor	100	Reduction	(Fernández et al. 2017)
Pleurotus mutilis	Torus reactor	60	Adsorption	(Alouache et al. 2017)

bioaugmentation of chromium-contaminated soil has also been studied using a microcosm system in which chromium removal was achieved with cultures of *Candida tropicalis* (Bahafid et al. 2013) and *Cyberlindnera fabianii* (Bahafid et al. 2017).

Concluding remarks

Cr(VI) causes toxicity and mutagenesis in fungi. In response to the toxicity of chromium, fungi readjust the processes of gene expression regulation that manifest in changes in several cellular processes; so far, most studies at the molecular level related to chromium responses have been carried out with the yeast *S. cerevisiae*. In yeasts and filamentous fungi, the transport of Cr(VI) is mediated by the sulfate permease system; however, these organisms differ with respect to the presence of the CHR-1 chromate transporter proteins and it remains to be established whether the actin-mediated endocytosis chromate accumulation is operative in filamentous fungi. Biosorption and biotransformation are important fungal processes of interaction with chromium; different technological strategies have been implemented for the removal of chromium, in which the biosorption and/or bioreduction capacity of fungi are exploited, with most reports related to *ex situ* treatments and a few studies to *in situ* treatments. Current research lines are focused on establishing more efficient and profitable processes for chromium removal. It is expected that more work will be done with molecular approaches and with an increased use of powerful analytical methodologies to study the interaction of fungi with chromium. The latter can lead to the discovery of new molecules (proteins, metabolites) interacting with chromium that can be instrumental to improve the biotechnological processes based on biosorption and/or bioreduction.

Acknowledgments

This work was funded by CONACyT, Mexico, Project CB-2014-01 registry number 239575 and by DAIP Universidad de Guanajuato, Mexico, Project FO-DAI-05.

References

Alouache, A., A. Selatnia and F. Halet. 2017. Biosorption of Cr(VI) from aqueous solutions by dead biomass of pleurotus mutilus in torus reactor. Cham.

Amezcua-Allieri, M. A., J. R. Lead and R. Rodríguez-Vázquez. 2005. Changes of chromium behavior in soil during phenanthrene removal by *Penicillium frequentans*. *Biometals* 18 (1): 23–29. doi: 10.1007/s10534-004-5771-y.

Arévalo-Rangel, D. L., J. F. Cárdenas-González, V. M. Martínez-Juárez and I. Acosta-Rodríguez. 2013. Hexavalent chromate reductase activity in cell free extracts of *Penicillium* sp. *Bioinorganic Chemistry and Applications* 2013: 6. doi: 10.1155/2013/909412.

Bahafid, W., J. N. Tahri, H. Sayel, I. Boularab and E. G. N. 2013. Bioaugmentation of chromium-polluted soil microcosms with *Candida tropicalis* diminishes phytoavailable chromium. *Journal of Applied Microbiology* 115 (3): 727–734. doi: doi:10.1111/jam.12282.

Bahafid, W., N. Tahri Joutey, H. Sayel, M. Asri, F. Laachari and N. el ghachtouli. 2017. Soil bioaugmentation with Cyberlindnera fabianii diminish phytotoxic effects of chromium(VI) on *Phaseolus vulgaris* L. *Journal of materials and Environmental Sciences* 8 (2): 438–443.

Barrera-Díaz, C. E., V. Lugo-Lugo and B. Bilyeu. 2012. A review of chemical, electrochemical and biological methods for aqueous Cr(VI) reduction. *Journal of Hazardous Materials* 223–224: 1–12. doi: https://doi.org/10.1016/j.jhazmat.2012.04.054.

Benazir, J. F., S. Ramasamy, D. Rajvel, M. Padmini Pooja and B. Mathithumilan. 2010. Bioremediation of chromium in tannery effluent by microbial consortia. *African Journal of Biotechnology* 9 (21): 3140–3143.

Braud, A., K. Jézéquel, E. Vieille, A. Tritter and T. Lebeau. 2006. Changes in extractability of Cr and Pb in a polycontaminated soil after bioaugmentation with microbial producers of biosurfactants, organic acids and siderophores. *Water, Air, & Soil Pollution: Focus* 6 (3): 261–279. doi: 10.1007/s11267-005-9022-1.

Burford, E. P., M. Fomina and G. M. Gadd. 2016. Fungal involvement in bioweathering and biotransformation of rocks and minerals. *Mineralogical Magazine* 67 (6): 1127–1155. doi: 10.1180/0026461036760154.

Cervantes, C., J. Campos-García, S. Devars, F. Gutiérrez-Corona, H. Loza-Tavera, J. C. Torres-Guzmán and R. Moreno-Sánchez. 2001. Interactions of chromium with microorganisms and plants. *FEMS Microbiology Reviews* 25 (3): 335–347. doi: doi:10.1111/j.1574-6976.2001.tb00581.x.

Cervantes, C. and J. Campos-García. 2007. Reduction and efflux of chromate by bacteria. In: *Molecular Microbiology of Heavy Metals*, edited by Dietrich H. Nies and Simon Silver, 407–419. Berlin, Heidelberg: Springer Berlin Heidelberg.

Coreño-Alonso, A., A. Solé, E. Diestra, I. Esteve, J. F. Gutiérrez-Corona, G. E. Reyna López, F. J. Fernández and A. Tomasini. 2014. Mechanisms of interaction of chromium with *Aspergillus niger* var tubingensis strain Ed8. *Bioresource Technology* 158: 188–192. doi: https://doi.org/10.1016/j.biortech.2014.02.036.

Chambers, C. D. 1991. *In situ* treatment of hazardous waste contaminated soils. *Soil Science* 155: 71.

Chang, F., C. Tian, S. Liu and J. Ni. 2016. Discrepant hexavalent chromium tolerance and detoxification by two strains of *Trichoderma asperellum* with high homology. *Chemical Engineering Journal* 298: 75–81. doi: https://doi.org/10.1016/j.cej.2016.04.023.

Cherest, H., J.-C. Davidian, D. Thomas, V. Benes, W. Ansorge and Y. Surdin-Kerjan. 1997. Molecular Characterization of two high affinity sulfate transporters in *Saccharomyces cerevisiae*. *Genetics* 145 (3): 627.

Cheung, K. H. and J.-D. Gu. 2007. Mechanism of hexavalent chromium detoxification by microorganisms and bioremediation application potential: A review. *International Biodeterioration & Biodegradation* 59 (1): 8–15. doi: https://doi.org/10.1016/j.ibiod.2006.05.002.

Das, A. P. and S. Mishra. 2010. Biodegradation of the metallic carcinogen hexavalent chromium Cr(VI) by an indigenously isolated bacterial strain. *Journal of Carcinogenesis* 9: 6. doi: 10.4103/1477-3163.63584.

Das, S. K. and A. K. Guha. 2009. Biosorption of hexavalent chromium by *Termitomyces clypeatus* biomass: Kinetics and transmission electron microscopic study. *Journal of Hazardous Materials* 167 (1): 685–691. doi: https://doi.org/10.1016/j.jhazmat.2009.01.037.

Dhal, B., H. N. Thatoi, N. N. Das and B. D. Pandey. 2013. Chemical and microbial remediation of hexavalent chromium from contaminated soil and mining/metallurgical solid waste: A review. *Journal of Hazardous Materials* 250–251: 272–291. doi: https://doi.org/10.1016/j.jhazmat.2013.01.048.

Díaz-Pérez, C., C. Cervantes, J. Campos-García, A. Julián-Sánchez and H. Riveros-Rosas. 2007. Phylogenetic analysis of the chromate ion transporter (CHR) superfamily. *The FEBS Journal* 274 (23): 6215–6227. doi: doi:10.1111/j.1742-4658.2007.06141.x.

Espinosa-Ortiz, E. J., E. R. Rene, K. Pakshirajan, E. D. van Hullebusch and P. N. L. Lens. 2016. Fungal pelleted reactors in wastewater treatment: Applications and perspectives. *Chemical Engineering Journal* 283: 553–571. doi: https://doi.org/10.1016/j.cej.2015.07.068.

Fang, Z., M. Zhao, H. Zhen, L. Chen, P. Shi and Z. Huang. 2014. Genotoxicity of tri- and hexavalent chromium compounds *in vivo* and their modes of action on DNA damage *in vitro*. *PLoS ONE* 9 (8): e103194. doi: 10.1371/journal.pone.0103194.

Fernández, P. M., E. L. Cruz, S. C. Viñarta and L. I. Castellanos de Figueroa. 2017. Optimization of culture conditions for growth associated with Cr(VI) removal by *Wickerhamomyces anomalus* M10. *Bulletin of Environmental Contamination and Toxicology* 98 (3): 400–406. doi: 10.1007/s00128-016-1958-5.

Flores-Alvarez, L. J., A. R. Corrales-Escobosa, C. Cortés-Penagos, M. Martínez-Pacheco, K. Wrobel-Zasada, K. Wrobel-Kaczmarczyk, C. Cervantes and F. Gutiérrez-Corona. 2012. The *Neurospora crassa* chr-1 gene is up-regulated by chromate and its encoded CHR-1 protein causes chromate sensitivity and chromium accumulation. *Current Genetics* 58 (5): 281–290. doi: 10.1007/s00294-012-0383-5.

Gadd, G. M. 2013. Fungi and their role in the biosphere. In: *Reference Module in Earth Systems and Environmental Sciences*, 1–9. Elsevier.

Gu, Y., W. Xu, Y. Liu, G. Zeng, J. Huang, X. Tan, H. Jian, X. Hu, F. Li and D. Wang. 2015. Mechanism of Cr(VI) reduction by *Aspergillus niger*: enzymatic characteristic, oxidative stress response, and reduction product. *Environmental Science and Pollution Research* 22 (8): 6271–6279. doi: 10.1007/s11356-014-3856-x.

Guertin, J. J. A. Jacobs, and C. P. Avakian. 2004. *Chromium(VI) Handbook*. Boca Raton: CRC Press Inc.

Guillén-Jiménez, F. d. M., A. R. Netzahuatl-Muñoz, L. Morales-Barrera and E. Cristiani-Urbina. 2009. Hexavalent chromium removal by *Candida* sp. in a concentric draft-tube airlift bioreactor. *Water, Air, and Soil Pollution* 204 (1): 43. doi: 10.1007/s11270-009-0024-x.

Gutiérrez-Corona, J. F., P. Romo-Rodríguez, F. Santos-Escobar, A. E. Espino-Saldaña and H. Hernández-Escoto. 2016. Microbial interactions with chromium: basic biological processes and applications in environmental biotechnology. *World Journal of Microbiology and Biotechnology* 32 (12): 191. doi: 10.1007/s11274-016-2150-0.

Holland, S. L. and S. V. Avery. 2009. Actin-mediated endocytosis limits intracellular Cr accumulation and Cr toxicity during chromate stress. *Toxicological Sciences* 111 (2): 437–446. doi: 10.1093/toxsci/kfp170.

Irazusta, V., A. R. Bernal, M. C. Estévez and L. I. C. de Figueroa. 2018. Proteomic and enzymatic response under Cr(VI) overload in yeast isolated from textile-dye industry effluent. *Ecotoxicology and Environmental Safety* 148: 490–500. doi: https://doi.org/10.1016/j.ecoenv.2017.10.076.

Jin, Y. H., P. E. Dunlap, S. J. McBride, H. Al-Refai, P. R. Bushel and J. H. Freedman. 2008. Global transcriptome and deletome profiles of yeast exposed to transition metals. *PLoS Genetics* 4 (4): e1000053. doi: 10.1371/journal.pgen.1000053.

Joutey, N. T., H. Sayel, W. Bahafid and N. El Ghachtouli. 2015. Mechanisms of hexavalent chromium resistance and removal by microorganisms. In: *Reviews of Environmental Contamination and Toxicology Volume 233*, edited by David M. Whitacre, 45–69. Cham: Springer International Publishing.

Kalidhasan, S., A. Santhana Krishna Kumar, V. Rajesh and N. Rajesh. 2016. The journey traversed in the remediation of hexavalent chromium and the road ahead toward greener alternatives—A perspective. *Coordination Chemistry Reviews* 317: 157–166. doi: https://doi.org/10.1016/j.ccr.2016.03.004.

Madhavi, V., A. V. B. Reddy, K. G. Reddy, G. Madhavi and T. N. V. K. V. Prasad. 2013. An overview on research trends in remediation of chromium. *Research Journal of Recent Sciences* 2 (1): 71–83.

Marzluf, G. A. 1970. Genetic and metabolic controls for sulfate metabolism in *Neurospora crassa*: Isolation and study of chromate-resistant and sulfate transport-negative mutants. *Journal of Bacteriology* 102 (3): 716–721.

Mohan, D. and C. U. Pittman. 2006. Activated carbons and low cost adsorbents for remediation of tri- and hexavalent chromium from water. *Journal of Hazardous Materials* 137 (2): 762–811. doi: https://doi.org/10.1016/j.jhazmat.2006.06.060.

Mohite, P. T., A. R. Kumar and S. S. Zinjarde. 2016. Biotransformation of hexavalent chromium into extracellular chromium(III) oxide nanoparticles using *Schwanniomyces occidentalis*. *Biotechnology Letters* 38 (3): 441–446. doi: 10.1007/s10529-015-2009-8.

Morales-Barrera, L. and E. Cristiani-Urbina. 2006. Removal of hexavalent chromium by *Trichoderma viride* in an airlift bioreactor. *Enzyme and Microbial Technology* 40 (1): 107–113. doi: https://doi.org/10.1016/j.enzmictec.2005.10.044.

Morales-Barrera, L. and E. Cristiani-Urbina. 2015. Bioreduction of hexavalent chromium by *Hypocrea tawa* in a concentric draft-tube airlift bioreactor. *Journal of Environment and Biotechnology Research* 1 (1): 37–44.

Mukherjee, K., R. Saha, A. Ghosh and B. Saha. 2013. Chromium removal technologies. *Research on Chemical Intermediates* 39 (6): 2267–2286. doi: 10.1007/s11164-012-0779-3.

Murugavelh, S. and K. Mohanty. 2014. Mechanism of Cr(VI) bioaccumulation by *Phanerochaete chrysosporium*. *Environmental Engineering and Management Journal* 13 (2): 281–287.

Palmer, C. D. and P. R. Wittbrodt. 1991. Processes affecting the remediation of chromium-contaminated sites. *Environmental Health Perspectives* 92: 25–40.

Park, D., Y.-S. Yun and J. M. Park. 2005. Use of dead fungal biomass for the detoxification of hexavalent chromium: screening and kinetics. *Process Biochemistry* 40 (7): 2559–2565. doi: https://doi.org/10.1016/j.procbio.2004.12.002.

Park, D., Y.-S. Yun and J. M. Park. 2006. Mechanisms of the removal of hexavalent chromium by biomaterials or biomaterial-based activated carbons. *Journal of Hazardous Materials* 137 (2): 1254–1257. doi: https://doi.org/10.1016/j.jhazmat.2006.04.007.

Park, D., S.-R. Lim, Y.-S. Yun and J. M. Park. 2007. Reliable evidences that the removal mechanism of hexavalent chromium by natural biomaterials is adsorption-coupled reduction. *Chemosphere* 70 (2): 298–305. doi: https://doi.org/10.1016/j.chemosphere.2007.06.007.

Poljsak, B., I. Pócsi, P. Raspor and M. Pesti. 2010. Interference of chromium with biological systems in yeasts and fungi: a review. *Journal of Basic Microbiology* 50 (1): 21–36. doi: doi:10.1002/jobm.200900170.

Ramírez-Díaz, M. I., C. Díaz-Pérez, E. Vargas, H. Riveros-Rosas, J. Campos-García and C. Cervantes. 2008. Mechanisms of bacterial resistance to chromium compounds. *BioMetals* 21 (3): 321–332. doi: 10.1007/s10534-007-9121-8.

Romo-Rodríguez, P., F. J. Acevedo-Aguilar, A. Lopez-Torres, K. Wrobel, K. Wrobel and J. F. Gutiérrez-Corona. 2015. Cr(VI) reduction by gluconolactone and hydrogen peroxide, the reaction products of fungal glucose oxidase: Cooperative interaction with organic acids in the biotransformation of Cr(VI). *Chemosphere* 134: 563–570. doi: https://doi.org/10.1016/j.chemosphere.2014.12.009.

Saha, B. and C. Orvig. 2010. Biosorbents for hexavalent chromium elimination from industrial and municipal effluents. *Coordination Chemistry Reviews* 254 (23): 2959–2972. doi: https://doi.org/10.1016/j.ccr.2010.06.005.

Sallau, A. B., H. M. Inuwa, S. Ibrahim and A. J. Nok. 2014. Isolation and properties of chromate reductase from *Aspergillus niger*. *International Journal of Modern Cellular and Molecular Biology* 3 (1): 10–21.

Sepehr, M. N., S. Nasseri, M. Zarrabi, M. R. Samarghandi and A. Amrane. 2012a. Removal of Cr(III) from tanning effluent by *Aspergillus niger* in airlift bioreactor. *Separation and Purification Technology* 96: 256–262. doi: https://doi.org/10.1016/j.seppur.2012.06.013.

Sepehr, M. N., M. Zarrabi and A. Amrane. 2012b. Removal of Cr(III) from model solutions by isolated *Aspergillus niger* and *Aspergillus oryzae* living microorganisms: Equilibrium and kinetic studies. *Journal of the Taiwan Institute of Chemical Engineers* 43 (3): 420–427. doi: https://doi.org/10.1016/j.jtice.2011.12.001.

Sharma, S. and P. Malaviya. 2016. Bioremediation of tannery wastewater by *chromium* resistant novel fungal consortium. *Ecological Engineering* 91: 419–425. doi: https://doi.org/10.1016/j.ecoleng.2016.03.005.

Singh, R., M. Kumar and N. R. Bishnoi. 2016. Development of biomaterial for chromium(VI) detoxification using *Aspergillus flavus* system supported with iron. *Ecological Engineering* 91: 31–40. doi: https://doi.org/10.1016/j.ecoleng.2016.01.060.

Srivastava, S. and I. S. Thakur. 2006. Isolation and process parameter optimization of *Aspergillus* sp. for removal of chromium from tannery effluent. *Bioresource Technology* 97 (10): 1167–1173. doi: https://doi.org/10.1016/j.biortech.2005.05.012.

Sumner, E. R., A. Shanmuganathan, T. C. Sideri, S. A. Willetts, J. E. Houghton and S. V. Avery. 2005. Oxidative protein damage causes chromium toxicity in yeast. *Microbiology* 151 (6): 1939–1948. doi: doi:10.1099/mic.0.27945-0.

Tamás, M. J. J. Labarre, M. B. Toledano and R. Wysocki. 2006. Mechanisms of toxic metal tolerance in yeast. In: *Molecular Biology of Metal Homeostasis and Detoxification: From Microbes to Man*, edited by Markus J. Tamas and Enrico Martinoia, 395–454. Berlin, Heidelberg: Springer Berlin Heidelberg.

Tekerlekopoulou, A. G., M. Tsiflikiotou, L. Akritidou, A. Viennas, G. Tsiamis, S. Pavlou, K. Bourtzis and D. V. Vayenas. 2013. Modelling of biological Cr(VI) removal in draw-fill reactors using microorganisms in suspended and attached growth systems. *Water Research* 47 (2): 623–636. doi: https://doi.org/10.1016/j.watres.2012.10.034.

Thatoi, H., S. Das, J. Mishra, B. P. Rath and N. Das. 2014. Bacterial chromate reductase, a potential enzyme for bioremediation of hexavalent chromium: A review. *Journal of Environmental Management* 146: 383–399. doi: https://doi.org/10.1016/j.jenvman.2014.07.014.

Viti, C., E. Marchi, F. Decorosi and L. Giovannetti. 2014. Molecular mechanisms of Cr(VI) resistance in bacteria and fungi. *FEMS Microbiology Reviews* 38 (4): 633–659. doi: doi:10.1111/1574-6976.12051.

Wang, J. and C. Chen. 2009. Biosorbents for heavy metals removal and their future. *Biotechnology Advances* 27 (2): 195–226. doi: https://doi.org/10.1016/j.biotechadv.2008.11.002.

Wrobel, K., A. R. Corrales Escobosa, A. A. Gonzalez Ibarra, M. Mendez Garcia, E. Yanez Barrientos and K. Wrobel. 2015. Mechanistic insight into chromium(VI) reduction by oxalic acid in the presence of manganese(II). *Journal of Hazardous Materials* 300: 144–152. doi: https://doi.org/10.1016/j.jhazmat.2015.06.066.

CHAPTER-7

Removal of Emerging Pollutants by Fungi
Elimination of Antibiotics

Carlos E. Rodríguez-Rodríguez,[1,*] *Juan Carlos Cambronero-Heinrichs,*[1,2] *Wilson Beita-Sandí*[1] and *J. Esteban Durán*[2]

1. Introduction

Antibiotics are a group of natural, synthetic, and semi-synthetic molecules therapeutically used in humans and animals to cure microbial infection diseases by inhibiting the growth of microorganisms. Natural antibiotics are produced by bacteria and fungi (e.g., benzylpenicillin and gentamicin) to inhibit or kill other competitor microorganisms through a bacteriostatic or bactericidal action. Semi-synthetic compounds are natural antibiotics chemically altered by inserting an additive within the drug formulation to improve its effectiveness. Also, antibiotics are used as food additives in animal feeding to improve production efficiency and to treat diseases (Grenni et al. 2018, Kümmerer 2009a, Sarmah et al. 2006, Yamaguchi et al. 2015).

These compounds have different functional groups within their structure and can be classified into different categories based on their chemical structure or mechanism of action: inhibition of cell wall synthesis, alteration of cell membranes, protein synthesis inhibition, synthesis of nucleic acids inhibition, and metabolic or anti-competitive antagonism (Grenni et al. 2018). Furthermore, antibiotics often possess different functionalities within the same molecule; hence antibiotics can be neutral, cationic, anionic, or zwitterionic under different pH conditions. This feature makes the physicochemical and biological properties of antibiotics, such as log P_{ow},

[1] Research Center of Environmental Pollution (CICA), Universidad de Costa Rica, 2060 San José, Costa Rica.
[2] School of Chemical Engineering, Universidad de Costa Rica, 2060 San José, Costa Rica.
* Corresponding author: carlos.rodriguezrodriguez@ucr.ac.cr

sorption behavior, photo reactivity, activity, and toxicity, change as a function of pH (Kümmerer 2009a, b).

There is a growing concern about the presence of antibiotics in the environment not only because the potential of polluting both the aquatic and soil environments, but also because their intensive use has dramatically increased the frequency of emergence, dissemination, and persistence of these compounds in the environment. Thus, these bioactive compounds may promote the development and spread of antibiotic-resistance among bacterial populations (Figueroa et al. 2004). Resistance dramatically reduces the possibility of treating infections effectively and increases the risk of clinical complications (Andersson and Hughes 2010, Zhang et al. 2011).

In polluted environments, antibiotic concentrations may range from a few nanograms to hundreds of nanograms per liter or kilogram of water/soil. Hospital effluents (Brown et al. 2006, Ory et al. 2016, Santos et al. 2013, Verlicchi et al. 2015), wastewater influents and effluents (Karthikeyan and Meyer 2006, Patrolecco et al. 2015, Watkinson et al. 2009), soils treated with manure or soils used for livestock (Hu et al. 2010, Kay et al. 2004, Martínez-Carballo et al. 2007) are prone to have higher concentrations of antibiotics due to their higher anthropogenic influence. Antibiotics can be released into water sources and are frequently detected in surface water, groundwater, and drinking water.

Many physicochemical processes have been applied to remove antibiotics from water; however, some antibiotics are often removed only partially during conventional wastewater treatment processes, it is consistent with the frequency of detection in effluents. Some conventional water treatment technologies to remove antibiotics from liquid media include adsorption, which is one of the most practical and economical methods (Acosta et al. 2016, Álvarez-Torrellas et al. 2016, Gao et al. 2012), membrane techniques (Dolar et al. 2012, Koyuncu et al. 2008, Shah et al. 2012), chemical oxidation (Ben et al. 2009, Carlesi Jara et al. 2007, Elmolla and Chaudhuri 2011), photodegradation (Norvill et al. 2017, Prados-Joya et al. 2011), and biodegradation (Polesel et al. 2016, Tran et al. 2016, Zhou et al. 2006).

Antibiotics enter wastewater treatment plants (WWTPs) through the sewage system and can be removed to a certain extent via several routes (e.g., biodegradation, adsorption, volatilization, or hydrolysis) (Li and Zhang 2010). Adsorption is an important process controlling the transport and fate of antibiotics in the environment. In WWTPs, the activated sludge may act as a reservoir that interacts with the compounds through sorption and biodegradation. While sorption removes some antibiotics, it also limits the availability of antibiotics to be degraded via biodegradation.

Engineered adsorbents (e.g., carbon nanotubes, activated carbons) have shown great potential in removing antibiotics from aqueous solutions due to their large surface areas and controllable pore size distributions (Gao et al. 2012, Ji et al. 2009, Ji et al. 2010). Adsorption is governed by electron donor/acceptor interactions and cation-π bonding mechanisms between the compounds and the surface of carbon.

However, bioremediation has gained significant attention for removing antibiotics. Among these eco-friendly treatment options is the use of white–rot fungi (WRF) (Gros et al. 2014, Prieto et al. 2011, Rodríguez-Rodríguez et al. 2012c, Wen et al. 2009). WRF have the potential to degrade an extremely diverse range of very persistent or toxic environmental pollutants due to their broad specificity to attack substrates, through the action of intracellular (i.e., cytochrome P450 system) and extracellular (i.e., laccases and peroxidases) enzymes (Barr and Aust 1994, Cerniglia 1997, Marco-Urrea et al. 2009).

The removal efficiency using WRF depends on several factors, including the physicochemical properties of the antibiotics. The molecular structure is a key factor rendering a wide range of antibiotics removal by WRF. Compounds with electron withdrawing functional groups (EWG) (e.g., amide ($-CONR_2$), carboxylic ($-COOH$), halogen ($-X$), and nitro ($-NO_2$) moieties) cause an electron deficiency, and thus render the compounds less susceptible to oxidative catabolism (Yang et al. 2013b). Instead, compounds with electron donating functional groups (EDG), e.g., $-NH_2$, hydroxyl ($-OH$), alkoxy ($-OR$), alkyl ($-R$), and acyl ($-COR$) moieties cause the molecules to be more prone to electrophilic attack by microbial oxygenases (Tadkaew et al. 2011, Yang et al. 2013b). We will try to exemplify how intrinsic characteristics of the antibiotics could play an important role in their fungal removal.

This chapter shows some of the latest research efforts at understanding the elimination of antibiotics using WRF whole cells and lignin-modifying enzymes. The first section of this chapter describes some general aspects of WRF and their enzymatic complexes; particular emphasis is placed on their role in the transformation of organic pollutants. The second section focuses on the strategies employed for the elimination of antibiotics by fungal-mediated processes with whole cells or enzymes, including the description of configurations and removal of data from synthetic wastewater and real matrices. Finally, ecotoxicological considerations regarding the potential detoxification of polluted matrices are included.

2. White-rot Fungi and their Potential in Biodegradation of Organic Pollutants

The physiological category of WRF groups, a wide variety of fungi with the shared capacity to degrade lignin, places them as the second most abundant raw material and the most abundant aromatic polymer in nature (Suhas et al. 2007). WRF are mainly represented by basidiomycetes and are the only organisms able to depolymerize and mineralize all the components of wood (i.e., cellulose, hemicellulose, and lignin) (Lundell et al. 2010); therefore, the decomposition of lignocellulose by WRF represents a key process in the carbon cycle, by releasing nutrients of easier assimilation by other organisms.

The breakdown of lignin by WRF is largely mediated by the production of a group of extracellular enzymes of low substrate specificity called lignin-modifying enzymes (LMEs) (Asgher et al. 2008). Particularly, the non-specificity of both LMEs and other enzymatic processes, such as those mediated by cytochrome P450, facilitates the transformation of diverse organic molecules and stands as the base for the potential use of WRF in the biodegradation of organic pollutants. In this respect, the ability of WRF to transform pollutants has been studied for a wide range of compounds, including pesticides, e.g., organophosphates (Bumpus et al. 1993), organochlorines (Quintero et al. 2007), carbamates (Mir-Tutusaus et al. 2014, Rodríguez-Rodríguez et al. 2017), triazoles (Woo et al. 2010), and triazines (Bending et al. 2002); textile dyes (Casas et al. 2013), polychlorinated biphenyls (PCBs) (Ruiz-Aguilar et al. 2002), polycyclic aromatic hydrocarbons (PAHs) (D'Annibale et al. 2005), and components of munitions wastes, i.e., TNT and its metabolites (Kim and Song 2000). More recently, the elimination of emerging pollutants, such as endocrine disrupting compounds (Cajthaml et al. 2009, Suzuki et al. 2003), brominated flame retardants (Rodríguez-Rodríguez et al. 2012a), pharmaceuticals (Cruz-Morató et al. 2013), and antibiotics (Prieto et al. 2011) has been also the target of fungal bioremediation.

The fungus *Phanerochaete chrysosporium* has been traditionally used as the model WRF. Recently, *Trametes versicolor* has also been utilized in the biodegradation of antibiotics and other pharmaceutical compounds (Mir-Tutusaus et al. 2018), and represents the main organism employed in the treatment of real matrices. Other relevant species of WRF employed in the removal of emerging pollutants include *Bjerkandera adusta* (Aydin 2016), *Dichomitus squalens* (Cajthaml et al. 2009), *Ganoderma lucidum* (Bilal et al. 2017), *Irpex lacteus* (Kum et al. 2009), *Lentinula edodes* (Dąbrowska et al. 2018), *Phlebia tremellosa* (Kum et al. 2011), *Pleurotus ostreatus* (Cajthaml et al. 2009), and *Stereum hirsutum* (Lee et al. 2005).

2.1 Enzymatic complexes of WRF: role in the removal of organic pollutants

LMEs play a key role in the regulation of the global C-cycle by WRF through their effect on the degradation of lignin in natural lignocellulosic substrates, but also in the transformation of organic pollutants. LMEs secreted by WRF are mostly oxidoreductases, roughly divided in peroxidases and phenoloxidases. Peroxidases comprise of mainly lignin peroxidase (LiP; E.C. 1.11.1.14), manganese peroxidase (MnP; E.C. 1.11.1.13), and the versatile peroxidase (VP; E.C. 1.11.1.16); whereas phenoloxidases are represented by laccases (E.C. 1.10.3.2) (Lundell et al. 2010, Wesenberg et al. 2003). The pattern of LMEs secretion seems to be species-dependent, thus resulting in a classification determined by the main enzymes produced by each species (Hatakka 1994). In general, the production of LMEs happens as part of the secondary metabolism of WRF, as no net energy is provided by the oxidation of lignin; similarly, induction occurs during limitation of nutrients, particularly nitrogen (Pointing 2001), and therefore, low C/N ratios are usually desired in a matrix to achieve high enzymatic production.

Peroxidases use H_2O_2 (or organic hydroperoxides) as electron acceptors in a catalytic cycle that is dependent on the oxidation of Mn^{2+} to Mn^{3+} in MnP (Hofrichter 2002), and the oxidation of the endogenously produced veratryl alcohol in LiP (Pointing 2001); VP exhibits a hybrid catalytic cycle between MnP and LiP (Hofrichter et al. 2010). In an H_2O_2-independent process, laccases employ molecular oxygen to oxidize diverse organic compounds through a catalytic cycle of electron transfer from the substrates to copper atoms; the formation of phenoxy-radicals produces further reactions that result in oxidized quinones or oligomeric products (Lundell et al. 2010, Nyanhongo et al. 2007). Details on the catalytic processes mediated by LMEs have been extensively reviewed (Hofrichter 2002, Hofrichter et al. 2010, Lundell et al. 2010, Martínez 2002, Wesenberg et al. 2003).

WRF endogenously produce low-molecular weight redox mediators during the transformation of organic molecules; these mediators are involved in the formation of radical moieties and reactive oxygen species, and consequently enhance the attack to lignin and organic pollutants (Asif et al. 2017, Wesenberg et al. 2003). The co-application of such mediators has improved the elimination of compounds, such as halogenated pesticides, textile dyes, and antibiotics in enzymatic-mediated processes (Chhabra et al. 2009, Torres-Duarte et al. 2009, Weng et al. 2013).

Besides the activity of extracellular enzymes, the ability of WRF to transform organic pollutants is also linked to the action of intracellular enzymatic complexes. Among them, one of the most studied is the cytochrome P450 complex, which comprises of heme *b* containing monooxygenases that catalyze very diverse reactions, such as aromatic hydroxylation, epoxidation of C=C double bonds, N-oxidation, dealkylation, deamination, dehalogenation, and cleavage of C-C bonds. The diversity of substrates for cytochrome P450 enzymes makes them particularly interesting in bioremediation, as they can act on antibiotics, pesticides, toxins, and other pollutants (Urlacher and Eiben 2006, Urlacher and Girhard 2012). The role of cytochrome P450 enzymes in the elimination of organopollutants by WRF is performed by means of their selective inhibition, which translates in a decrease or inhibition of their removal, as shown in the cases of the psychiatric drug carbamazepine (Golan et al. 2011) and the antibiotics ciprofloxacin and norfloxacin (Prieto et al. 2011).

2.2 *WRF as bioremediation agents*

As described above, the enzymatic machinery expressed by WRF proposes a broad spectrum of applications in bioremediation; moreover, these organisms exhibit several desirable characteristics that make them suitable candidates as bioaugmentation agents for the treatment of diverse polluted matrices. The highly non-specific nature of LMEs and other intracellular complexes in WRF represent an advantage in bioremediation processes, particularly for the broad spectrum of organic pollutants that they can degrade. This finding has been described as an advantage over bacterial-mediated processes, as WRF may potentially act on the simultaneous removal of pollutants from matrices containing complex mixtures of contaminants, such as those containing pharmaceuticals (Mir-Tutusaus et al. 2018). In this regard, there is no need for previous exposure or adaptation of WRF to the micropollutants (Rodríguez-Rodríguez et al. 2013), as usually required for the expression of induced degrading enzymes in bacteria. Furthermore, if the transformation were driven by extracellular LMEs, the process would not require the internalization of the pollutants to the cells-a convenient situation for the treatment of xenobiotics with low water solubility. Also, WRF are easily adapted to substrates or matrices; in this respect, the growth of fungi by hyphae permits their access to highly adsorbed contaminants, which is of particular interest in the treatment of solid effluents, such as dried sewage sludge (Rodríguez-Rodríguez et al. 2011). The use of WRF in the form of pellets, considered as a way of biomass immobilization (Espinosa-Ortiz et al. 2016), represents a popular approach employed for the treatment of effluents in liquid phase reactors. The versatility of WRF is also reflected in their pH requirements: even though the optimal pH for the growth and activity of several WRF is below 5, and some authors consider the pH adjustment as a disadvantage overcome by increasing costs (Mir-Tutusaus et al. 2018), several fungi have shown excellent removal capacity of micropollutants in close to neutral pH conditions of heavily buffered matrices, such as sewage sludge (Rodríguez-Rodríguez et al. 2011, Rodríguez-Rodríguez et al. 2014).

The transformation of organic pollutants by WRF takes place under aerobic conditions, through mechanisms of co-metabolic oxidation. This fact indicates that additional substrates are required as carbon sources (Asif et al. 2017), as in most cases, the pollutant is not completely assimilated by the fungi, even though mineralization has been reported (Badia-Fabregat et al. 2014). This condition might represent a disadvantage for the treatment of some effluents, such as those of hospital or veterinary origin, in which the concentration of micropollutants is not enough to support fungal growth and activity (Mir-Tutusaus et al. 2018), thus making the addition of other carbon sources necessary. In this respect, the search for low-cost residues to add in liquid phase systems still represents a topic to be further optimized. On the contrary, the same feature represents an advantage in solid phase systems, in which the addition of lignocellulosic substrates is employed not only as a component of the matrix (as a support and carrier for the WRF), but also to act as a specific substrate that promotes fungal growth and enzymatic activity (Rodríguez-Rodríguez et al. 2014); moreover, from the environmental perspective, this practice encourages the valorization of these residues.

3. Treatment of Antibiotics by WRF

Different WRF have been investigated for the removal of antibiotics, and most reports assayed the ability of whole cell systems in liquid media for the removal of spiked antibiotics. However, there are a few reports that describe the removal by fungi in real wastewater or solid residues (sludge) and other matrices using crude or purified enzyme extracts instead of living cells.

Whole cell approaches show many advantages for *in situ* application. First, the use of living fungi enables maintaining biological activity by the constant production of extracellular LMEs; moreover, some reports even describe an enhanced expression of these enzymes in the presence of antibiotics. For example, higher concentration of MnP was observed in fungi isolated from swine farms, when they were exposed to tiamulin (Nguyen et al. 2017). Also, WRF have shown higher activity of laccase in the presence of antibiotics, such as oxytetracycline (Migliore et al. 2012) and sulfamethoxazole (Guo et al. 2014). Living fungal cells also express other detoxification complexes; i.e., cytochrome P450 is constitutively expressed in eukaryotes and exhibits an important role in the removal of antibiotics, such as oxytetracycline (Mir-Tutusaus et al. 2014), sulfonamides (Rodríguez-Rodríguez et al. 2012b, Vasiliadou et al. 2016), and quinolones (Prieto et al. 2011).

3.1 Removal of antibiotics by WRF whole-cells

3.1.1 Fungal biomass: pelleted or free mycelial biomass and configurations

The use of whole cells can be roughly divided in liquid phase configurations, mostly employing fungi in the form of pellets, and in solid-phase approaches, which rely on the colonization of a substrate by fungal free mycelium. The use of liquid phase reactors with fungal pellets has attracted attention for environmental applications due to several advantages. Pellet production is considered a form of self-immobilization by fungi (Espinosa-Ortiz et al. 2016), which responds as a protection strategy against oxidative stress under oxygen rich conditions (Higashiyama et al. 2000). In this respect, immobilized biomass can be easily handled, in terms of settling and separation; moreover, pellet biomass can be potentially reused, for instance, fungal pellets have been successfully reused in several cycles of treatment of organic pollutants, such as textile dyes in repeated flask-scale batch (Sanghi et al. 2006, Yesilada et al. 2003, Yesilada et al. 2009), sequencing batch stirred tank (Borchert and Libra 2001), and repeated air-pulsed reactors (Baccar et al. 2011). On the other hand, the use of pellets permits long periods of treatment under continuous operation in packed bed (Mielgo et al. 2001) and air-pulsed reactors (Blánquez et al. 2006), or even in fungal membrane bioreactors (Yang et al. 2013a); nonetheless, neither long periods nor recycling of pellets have been explored in the treatment of antibiotics.

The use of mycelium in the form of pellets also reduces the risk of clogging and adherence to reactor surfaces; however, some mycelial growth can be observed at the upper level of the liquid matrix to be treated. Besides, operation with pellets results in low viscosity systems, contrary to mycelial growth that increases the viscosity in the medium, translating into higher power requirements to maintain aeration and agitation (Pazouki and Panda 2000). Immobilization also provides the fungus with protection against highly toxic pollutants, as the direct contact would occur on the pellet surface, thus, diffusion of the toxic compounds would result in lower concentrations inside the pellet. The population density in the pellet configuration represents a competing advantage against other microbial populations present in the polluted matrix.

Drawbacks related to the use of pelleted biomass are those corresponding to immobilized cells and include diffusional limitations of oxygen or nutrients (Espinosa-Ortiz et al. 2016) to the inner regions of the pellet, which can become anaerobic with limited growth (Pazouki and Panda 2000). Diffusion limitation could also hinder fungal activity by slowing the release of potentially toxic metabolites in the core of the pellet. These disadvantages will increase with the size of the pellet; consequently, smaller pellets of around 3 mm in diameter (Borràs et al. 2008) are usually preferable, as they provide a higher surface to volume ratio for oxygenation and interaction with the pollutants.

On the other hand, solid phase systems are regarded as highly suitable for the growth of fungi, as they reproduce their natural environmental conditions (Hölker et al. 2004). In these approaches, free mycelium has been employed for the treatment of solid polluted effluents, mostly sewage sludge (for antibiotic treatment) or soil. Such systems are based on the colonization of a mixture of a substrate and the sludge by the WRF mycelium. The substrates usually comprise of lignocellulosic wastes, which provide exclusive carbon sources for WRF; moreover, the substrate also plays the roles of support for the fungus and bulking material for the whole mixture. Several lignocellulosic wastes, such as wheat straw, coconut fiber, or wood chips have been used as substrates in fungal-mediated removal of pesticides, aromatic hydrocarbons, and pharmaceuticals (D'Annibale et al. 2005, Madrigal-Zúñiga et al. 2016, Rodríguez-Rodríguez et al. 2011).

3.1.2 Bioreactor configurations with WRF

Bioreactor configurations containing fungal pellets used for the removal of organic pollutants include stirred tanks (Libra et al. 2003), airlift columns (Rioja et al. 2008), bubble columns (Cerrone et al. 2011), and fluidized beds (Blánquez and Guieysse 2008). However, for the treatment of antibiotics, the stirred tank and fluidized bed configurations have been preferred. The stirred tank reactors (STR) constitute the most widespread configuration used with fungi (Moreira et al. 2003); they include air supply, usually from the bottom of the container, which is dispersed by mechanical agitation, thus resulting in the mixing of the system (Espinosa-Ortiz et al. 2016). Even though STR present the advantage to adjust the size of the pellets, operational limitations may occur during excess agitation, including the decrease in pellet size, the detachment of superficial hyphae fragments, or even the breakup of the pellets. In this regard, Kelly et al. (2006) propose three mechanisms for pellet breakup in STR: fluid induced shear stress by eddies, collisions between pellets, and collisions between pellets and impeller or baffles.

In fluidized bed reactors, the solid pellets are suspended in an upward-flowing stream of a fluid, usually the wastewater, in which the flow should be sufficient to maintain the pellets in suspension. This configuration creates a uniform dispersion in the system, which favors better mixing and oxygenation (Espinosa-Ortiz et al. 2016). Small pellets of a narrow distribution size are desirable for a proper fluidization. Despite being extensively used in the removal of organic pollutants, these systems present the limitation of potential pellet aggregation or overgrowth, which results in reduced fluidization (Moreira et al. 2003); partial purging of biomass has been used as a strategy to decrease these operational drawbacks (Rodríguez-Rodríguez et al. 2012b). A version used as an alternative for the traditional fluidized bed reactor, the "pulsed-reactor" has been widely employed in the removal of pharmaceuticals and textile dyes; this configuration includes a supply of air at the bottom of the system, which is applied at pulses not only to oxygenate and mix the content of the reactor, but also to maintain fluidization. The optimal pulsation frequency corresponds to the lowest pellet size and the narrowest dispersion in pellet diameter (Moreira et al. 2003).

As described above, the degradation by WRF mostly responds to a co-metabolic process; therefore the use of liquid phase configurations usually requires the addition of assimilable substrates, as the antibiotics or other micropollutants will not necessarily be used as carbon sources; moreover, whether the degradation is co-metabolic or not, their concentrations are not enough to support fungal growth (Badia-Fabregat et al. 2016). Consequently, experiments using synthetic wastewater usually contain carbon sources, such as glucose or malt extract (Mir-Tutusaus et al. 2018). In the case of real wastewater, several authors remark the need to supplement the system with glucose as carbon source and ammonium tartrate or ammonium chloride as nitrogen source to keep proper biological and enzymatic activity of *T. versicolor* in fluidized bed reactors and STR (Badia-Fabregat et al. 2015, Cruz-Morató et al. 2013, Mir-

Tutusaus et al. 2017, Rodríguez-Rodríguez et al. 2012a). Alternatively, some fungi can use the carbon sources dissolved in the wastewater (expressed as the chemical oxygen demand, COD), thus avoiding the need of additional carbon sources. Nowadays, the use of cheaper carbon sources based on wastes, such as molasses, should be a focal point in mycoremediation of wastewaters. On the other hand, this problem can be easily solved by the use of ligninolytic wastes as a mean of both fungal immobilization and nutrient supply, as described below for the solid-phase strategies.

Solid phase configurations using WRF include tray chambers, packed bed reactors, stirred bed reactors, and drum reactors (such as rotating, stirred, and rocking drum) (Mitchell et al. 2000); nonetheless, they have been mostly used for purposes different to bioremediation. For the treatment of pollutants (pharmaceuticals/ antibiotics), the configuration typically employed is the tray chamber, also referred to as biopiles. As described before, these systems contain a mixture of the polluted effluent (sewage sludge) with a lignocellulosic substrate, in which the addition of the fungal biomass is usually done by pre-colonization of the substrate (Rodríguez-Rodríguez et al. 2012c). From the ecological point of view, this configuration presents the advantages of low water demand, which translates to less wastewater production; and the revalorization of agricultural wastes used as substrates. From an operational perspective, their aeration and maintenance requirements are quite low (Hölker et al. 2004, Hölker and Lenz 2005). Nonetheless, biopiles present some drawbacks, mostly related to heat and mass limitations that result in the formation of moisture, pH, temperature, oxygen, and pollutant gradients (Mitchell et al. 2000), reflecting on operational problems, such as desiccation and overheating. Moreover, scaling-up has been difficult, and can only be achieved by increasing the area of the trays (Mitchell et al. 2000), given that increasing the bed height results in overheating problems. Besides these disadvantages, removal of organic pollutants in biopiles presents promising results for the treatment of pharmaceuticals from sludge. In this regard, the use of other solid phase configurations remains to be assayed for the elimination of pharmaceuticals, in particular, antibiotics.

3.1.3 Removal of antibiotics by WRF: studies under controlled conditions

Flask-scale assays using synthetic wastewater (aqueous solution of the antibiotics) have been made to demonstrate the transformation of antibiotics by pure WRF cultures (Table 7.1). The removal of sulfonamides has been studied with several WRF, usually applying biomass in the form of pellets or colonized plugs of agar. The adsorption of sulfonamides to fungal biomass in liquid media is limited, as demonstrated by the use of heat-killed controls; therefore, removal at flask scale has been ascribed to fungal activity. In the case of sulfamethoxazole, Guo et al. (2014) achieved an elimination of 74% after 10 days using *P. chrysosporium*; the presence of the antibiotic slightly affected fungal growth, but also increased laccase production. During the simultaneous treatment of sulfamethoxazole with other pharmaceutical compounds, Rodarte-Morales et al. (2011) achieved complete removal with *P. chrysosporium* and two strains of *Bjerkandera* sp. after 14 days; however, elimination was faster using *P. chrysosporium*. Less promising results were obtained with *T. versicolor* and *G. lucidum* during the treatment of sulfamethoxazole in a mixture of pharmaceutical agents: the former only eliminated around 10% after 7 days, whereas the removal by the latter was negligible. A slight improvement was observed when both fungi were simultaneously applied; however, in this case removal did not surpass 20% (Vasiliadou et al. 2016). The removal of sulfamethazine (> 95%) took place in 20 hours at flask scale using *T. versicolor* (García-Galán et al. 2011), whereas complete elimination required 48 hours for sulfapyridine and 168 hours for sulfathiazole. A potential role in the removal was ascribed to cytochrome P450 in the cases of sulfamethazine and sulfathiazole, according to cytochrome P450-inhibition assays; nonetheless, no clear conclusion could be drawn for sulfapyridine. A continuous fluidized bed reactor using fungal pellets was able to simultaneously remove the three sulfonamides (> 94%), employing a hydraulic residence time of 72 hours (Rodríguez-Rodríguez et al. 2012b). Elucidation of transformation products from the oxidation of these compounds revealed the formation of the desulfonated- and the formyl-intermediates in the three cases; other reactions included hydroxylations (at different positions), deaminations, and their combinations.

Similar to sulfonamides, the elimination of quinolones by fungi has been mainly ascribed to biological activity, as their adsorption to fungal biomass is negligible (Čvančarová et al. 2015, Prieto et al. 2011). A comparative study by Čvančarova et al. (2015) revealed a clearly more efficient capacity of *I. lacteus* and *T. versicolor* for the elimination of ciprofloxacin, norfloxacin, and ofloxacin, with respect to *Panus tigrinus*, *Dichomitus squalens*, and *Pleurotus ostreatus*; interestingly, the transformation was not enough to decrease the antimicrobial activity in most cases. The description of transformation products in liquid media suggests common degradation pathways for different quinolones, usually starting with the substitution or breakdown of the piperazine ring or the formation of hydroxylated congeners (Čvančarová et al. 2015, Prieto et al. 2011). A role of cytochrome-P450 in initial transformations has been described for ciprofloxacin and norfloxacin (Prieto et al. 2011), whereas mineralization of ^{14}C-radiolabeled compounds to $^{14}CO_2$ has been reported for ciprofloxacin and enrofloxacin (Martens et al. 1996, Wetzstein et al. 1997, Wetzstein et al. 1999). The breaking of active moieties in antibiotics has also been reported in the oxidation of the β-lactam ring in cefuroxime by *L. edodes* (Dąbrowska et al. 2018).

Although tetracyclines exhibit high adsorption to organic matter, the removal by fungal pellets or mycelium has shown an important adsorption component only in the elimination of oxytetracycline with *Pleurotus ostreatus*, which took place with a subsequent complete degradation after 14 days (Migliore et al. 2012). On the other hand, negligible adsorption was observed using *T. versicolor*; instead, this fungus was able to eliminate oxytetracycline, with a potential participation of cytochrome P450, in only 2 days (> 99%) at flask scale (Mir-Tutusaus et al. 2014), and after 2.5 days when the process was scaled-up in a fluidized bed reactor (Cambronero-Heinrichs et al. 2018).

Table 7.1 Removal of antibiotics from synthetic wastewater using WRF whole cells.

Antibiotic	WRF species	Initial concentration (mg/L)	Removal (%)	Incubation time (d)	References
Cefuroxime	Lentinula edodes	100–250	100	7	Dąbrowska et al. (2018)
	Trametes versicolor	2	>90	7	Prieto et al. (2011)
	Ganoderma lucidum	1000	0	4	Chakraborty and Abraham (2017)
	Irpex lacteus	0.01	100	10	
Ciprofloxacin	Panus tigrinus	0.01	60	14	
	Dichomitus squalens	0.01	23	14	Čvančarová et al. (2015)
	Trametes versicolor	0.01	100	14	
	Pleurotus ostreatus	0.01	47	14	
	Pleurotus ostreatus	500	95.1	14	Singh et al. (2017)
	Stropharia rugosoannulata	0.05	5.1	56	
	Phanerochaete chrysosporium	0.05	25.6	56	Martens et al. (1996)
Enrofloxacin (mineralization)	Irpex lacteus	0.05	13.7	56	
	Phellinus gilvus	0.05	0.19	56	
	Trametes versicolor	2	>90	7	Prieto et al. (2011)
	Ganoderma lucidum	2000	100	4	Chakraborty and Abraham (2017)
	Irpex lacteus	0.01	100	10	
Norfloxacin	Panus tigrinus	0.01	50	14	
	Dichomitus squalens	0.01	10	14	Čvančarová et al. (2015)
	Trametes versicolor	0.01	85	14	
	Pleurotus ostreatus	0.01	50	14	

Antibiotic	Fungus				Reference
Ofloxacin	Irpex lacteus	0.01	100	10	
	Panus tigrinus	0.01	28	14	
	Dichomitus squalens	0.01	44	14	Čvančarová et al. (2015)
	Trametes versicolor	0.01	100	14	
	Pleurotus ostreatus	0.01	56	14	
Oxytetracycline	Trametes versicolor	100	100	2,5	Cambronero-Heinrichs et al. 2018
	Trametes versicolor	20–25	100	2	Mir-Tutusaus et al. (2014)
	Pleurotus ostreatus	50–100	100	14	Migliore et al. (2012)
Sulfamethazine	Trametes versicolor	9	100	1	García-Galán et al. (2011)
Sulfamethoxazole	Phanerochaete chysosporium	10	74	10	Guo et al. (2014)
	Trametes versicolor	0.05	<20	7	
	Ganoderma lucidum	0.05	<20	7	Vasiliadou et al. (2016)
	Trametes versicolor + Ganoderma lucidum	0.05	<20	7	
	Bjerkandera sp.	1	100	14	
	Bjerkandera adusta	1	100	14	Rodarte -Morales et al. (2011)
	Phanerochaete chrysosporium	1	100	14	
Sulfapyridine	Trametes versicolor	9–11	100	2	Rodríguez-Rodríguez et al. (2012b)
Sulfathiazole	Trametes versicolor	9–11	80	3	Rodríguez-Rodríguez et al. (2012b)
Tiamulin	Trametes versicolor	10	86.1	12	Nguyen et al. (2017)
	Trametes hirsuta	10	66.8	12	

3.1.4 Removal from real matrices

When it comes to the treatment of antibiotics in real matrices, most of the processes have been applied to wastewater of veterinary or hospital origin, or reverse osmosis concentrate (ROC) wastewater, using liquid phase approaches, mainly in batch or continuous reactors or at flask scale. Solid matrices, such as sewage sludge have been treated in stirred tank bioslurry reactors, and dry sludge in solid phase biopiles, usually containing wheat straw or another lignocellulosic substrate. A compilation of antibiotic removal in real matrices is shown in Table 7.2.

Although few works describe the treatment in solid phase biopiles, these systems have shown consistently high efficiency in the elimination of antibiotics, as complete removal was determined for clarithromycin (Rodríguez-Rodríguez et al. 2011) and sulfonamides, sulfamethazine, sulfapyridine, and sulfathiazole (García-Galán et al. 2011) from dry sewage sludge after 42 days; since in these cases the treatment employed previously sterilized sludge, the elimination was mostly ascribed to the fungal effect of *T. versicolor*. High removal, though incomplete, has been observed for erythromycin, sulfamethoxazole, and tetracycline with *B. adusta* and *T. versicolor* in these configurations (Aydin 2016).

In the liquid phase approaches, systems using pre-sterilized or non-sterilized matrices have been used. In pre-sterilized matrices, the biological removal depends on fungal bioaugmentation and its effect on pollutants; on the contrary, in non-pre-sterilized matrices, the biological effect is given by a combination between the indigenous microbial community and the added fungus, which is desirable in those cases when bacterial-mediated degradation of the antibiotic also takes place. Nonetheless, potential antagonistic interactions between the WRF and autochthonous bacteria may lead to a decreased effect by bacteria or by the overall community. From reports in the literature, it is difficult to define which strategy results in better elimination for a given antibiotic, though in some cases an enhanced removal has been observed for compounds, such as ciprofloxacin (Cruz-Morató et al. 2013, Cruz-Morató et al. 2014, Llorens-Blanch et al. 2015) or ofloxacin (Llorens-Blanch et al. 2015) in pre-sterilized matrices. On the other hand, the antiprotozoan dimetridazole has shown negligible removal in pre-sterilized wastewater, but high elimination in non-sterilized systems (Badia-Fabregat et al. 2016, Cruz-Morató et al. 2014), whereas similar elimination rates have been determined for azithromycin or trimethoprim in both conditions (Cruz-Morató et al. 2014, Llorens-Blanch et al. 2015). Normally, flask-scale assays yield high elimination that decreases when the process is scaled-up to bioreactors; such behavior has been observed for compounds such as azithromycin and clarithromycin (Llorens-Blanch et al. 2015, Lucas et al. 2018, Mir-Tutusaus et al. 2017); nonetheless, enhanced elimination has increased with scaling up in some cases for ofloxacin (Cruz-Morató et al. 2014, Gros et al. 2014). The use of continuous systems has decreased the elimination performance with respect to batch reactors in the cases of ciprofloxacin and sulfamethoxazole (Badia-Fabregat et al. 2015, Badia-Fabregat et al. 2016, Badia-Fabregat et al. 2017, Mir-Tutusaus et al. 2017); however, this observation is not the norm. In general, the addition of external nutrients (glucose and ammonium tartrate) resulted in enhanced elimination of ciprofloxacin, ofloxacin, and other pharmaceuticals different from antibiotics

Table 7.2 Removal of antibiotics from real matrices using WRF whole cells.

Antibiotic	WRF species	Matrix	Reactor type	Initial concentration (μg/L)	Removal (%)	Incubation time (d)	References
		Sterile sludge from membrane biological reactor	Erlenmeyer flask	0.5947	98.6	7	Llorens-Blanch et al. (2015)
		Non-sterile sludge from membrane biological reactor			96.7	7	Llorens-Blanch et al. (2015)
Azithromycin	Trametes versicolor	Sterilized wastewater	Fluidized bed batch bioreactor	N.D.	–	7–8	Cruz-Morató et al. (2013, 2014)
		Non-sterilized wastewater		1.37–4.31	26–100		
		Non-sterile real veterinary hospital wastewater	Fluidized bed batch bioreactor	0.59	47	15	Badia-Fabregat et al. (2016)
			Fluidized bed fed-batch bioreactor	0.23	–34	15	Lucas et al. (2016)
			Continuous fluidized bed bioreactor	–	49	HRT: 3.3	Lucas et al. (2018)
		Non-sterile real hospital wastewater	Continuous fluidized bed bioreactor	0.0454	–95.5	HRT: 3	Mir-Tutusaus et al. (2017)
		Reverse osmosis concentrate from a WWTP		2.762	–60	HRT: 2	Badia-Fabregat et al. (2017)
Ampicillin	Verticillium leptobactrum	Defined medium	Erlenmayer flask	500–2000	100	14	Kumar et al. (2013)
	Trametes versicolor	Non-sterile real veterinary hospital wastewater	Fluidized bed fed-batch bioreactor	2.138	100	15	Lucas et al. (2016)

Table 7.2 contd. ...

...Table 7.2 contd.

Antibiotic	WRF species	Matrix	Reactor type	Initial concentration (μg/L)	Removal (%)	Incubation time (d)	References
Ampicillin B	*Trametes versicolor*	Non-sterile real veterinary hospital wastewater	Fluidized bed fed-batch bioreactor	1.965	94	15	Lucas et al. (2016)
Cefalexin	*Trametes versicolor*	Non-sterile real veterinary hospital wastewater	Fluidized bed batch bioreactor	15.941	100	15	Badia-Fabregat et al. (2016)
		Non-sterile real veterinary hospital wastewater	Fluidized bed fed-batch bioreactor	1.178	98	15	Lucas et al. (2016)
		Sterilized wastewater	Fluidized bed batch bioreactor	N.D.	–	8	Cruz-Morató et al. (2013)
		Non-sterilized wastewater		0.59	-56		
Cefazolin	*Trametes versicolor*	Non-sterile real veterinary hospital wastewater	Fluidized bed fed-batch bioreactor	4.81	100	15	Lucas et al. (2016)
Cefuroxime	*Trametes versicolor*	Non-sterile real veterinary hospital wastewater	Fluidized bed fed-batch bioreactor	0.162	64	15	Lucas et al. (2016)
Ciprofloxacin	*Trametes versicolor*	Sterile sludge from membrane biological reactor	Erlenmeyer flask	3.727	89.2	7	Llorens-Blanch et al. (2015)
		Non-sterile sludge from membrane biological reactor	Erlenmeyer flask		61	7	Llorens-Blanch et al. (2015)
		Sterilized wastewater	Fluidized bed batch bioreactor	ND-12.1	99	7–8	Cruz-Morató et al. (2013, 2014)
		Non-sterilized wastewater		13–84.71	35–69		

Organism	Water type	Reactor	Value	Removal (%)	Time/HRT	Reference
	Reverse osmosis concentrate from a WWTP, with nutrients	Fluidized bed batch bioreactor	0.631	100	6	Badia-Fabregat et al. (2015)
	Reverse osmosis concentrate from a WWTP, no nutrients		0.702	48	7	Badia-Fabregat et al. (2015)
	Reverse osmosis concentrate from a WWTP	Continuous fluidized bed bioreactor	0.361	43	HRT: 3	Badia-Fabregat et al. (2017)
	Non-sterile real hospital wastewater	Continuous fluidized bed bioreactor	0.3664 / 0.2669	47.1 / −7.2	HRT: 3	Mir-Tutusaus et al. (2017)
		Fluidized bed fed-batch bioreactor	13.881	91	15	Lucas et al. (2016)
	Non-sterile real veterinary hospital wastewater	Continuous fluidized bed bioreactor	–	88	HRT: 3.3	Lucas et al. (2018)
		Fluidized bed batch bioreactor	9.846	99	15	Badia-Fabregat et al. (2016)
	Non-sterilized wastewater	Erlenmayer flask and bioplastic entrapped fungus (fluidized bed)	N.R.	> 95	30	Accinelli et al. (2010)
	Activated sludge mixed liquor		N.R.	80		
Phanerochaete chrysosporium	Sterile sludge from membrane biological reactor	Erlenmeyer flask	0.0759	95.9	7	Llorens-Blanch et al. (2015)
	Non-sterile sludge from membrane biological reactor			100	7	Llorens-Blanch et al. (2015)

Table 7.2 contd. ...

...*Table 7.2 contd.*

Antibiotic	WRF species	Matrix	Reactor type	Initial concentration (µg/L)	Removal (%)	Incubation time (d)	References
Clarithromycin	*Trametes versicolor*	Non-sterile real veterinary hospital wastewater	Fluidized bed fed-batch bioreactor	0.013	–115	15	Lucas et al. (2016)
		Reverse osmosis concentrate from a WWTP	Continuous fluidized bed bioreactor	0.014	70	HRT: 2	Badia-Fabregat et al. (2017)
		Reverse osmosis concentrate from a WWTP, no nutrients	Fluidized bed batch bioreactor	0.07	75	7	Badia-Fabregat et al. (2015)
		Sterilized wastewater		N.D.	–		
		Non-sterilized wastewater		2.20	80	7	Cruz-Morató et al. (2014)
		Dry sewage sludge (sterile)	Biopile with wheat straw	21.0 (ng/g)	100	42	Rodríguez-Rodríguez et al. (2011)
Chlortetracycline	*Trametes versicolor*	Non-sterile real veterinary hospital wastewater	Fluidized bed fed-batch bioreactor	0.008	100	15	Lucas et al. (2016)
Danofloxacin	*Trametes versicolor*	Non-sterile real veterinary hospital wastewater	Fluidized bed fed-batch bioreactor	0.023	–25	15	Lucas et al. (2016)

Pollutant	Fungus	Matrix	Reactor	Concentration	Removal (%)	Time (d)	Reference
Dimetridazole	*Trametes versicolor*	Non-sterile real veterinary hospital wastewater	Fluidized bed batch bioreactor	0.192	70	15	Badia-Fabregat et al. (2016)
	Trametes versicolor	Reverse osmosis concentrate from a WWTP, with nutrients	Fluidized bed batch bioreactor	0.078	4	6	Badia-Fabregat et al. (2015)
		Sterilized wastewater		0.244	0		Cruz-Morató et al. (2014)
		Non-sterilized wastewater		0.068	100	7	
Doxycycline	*Trametes versicolor*	Non-sterile real veterinary hospital wastewater	Fluidized bed fed-batch bioreactor	4.697	29	15	Lucas et al. (2016)
Enoxacin	*Trametes versicolor*	Non-sterile real veterinary hospital wastewater	Fluidized bed fed-batch bioreactor	0.19	96	15	Lucas et al. (2016)
Enrofloxacin	*Trametes versicolor*	Non-sterile real veterinary hospital wastewater	Fluidized bed fed-batch bioreactor	0.982	76	15	Lucas et al. (2016)
	Gloeophyllum striatum	Spiked soil	Biopile with wheat straw	10 (µg/g)	53	56	Martens et al. (1996)
	Bjerkandera adusta	Sterile sludge	Biopile with wheat straw	100	< 60*	360	Aydin (2016)

Table 7.2 contd. ...

204 *Fungal Bioremediation: Fundamentals and Applications*

...Table 7.2 contd.

Antibiotic	WRF species	Matrix	Reactor type	Initial concentration (µg/L)	Removal (%)	Incubation time (d)	References
	Phanerochaete chrysosporium	Non-sterilized wastewater	Erlenmayer flask and bioplastic entrapped fungus (fluidized bed)	N.R.	> 95%	5	Accinelli et al. (2010)
		Activated sludge mixed liquor		N.R.	> 90%		
Erythromycin		Sterile sludge	Biopile with wheat straw	100	< 80*	360	Aydin (2016)
	Trametes versicolor	Sterilized wastewater	Fluidized bed batch bioreactor	N.D.	–	8	Cruz-Morató et al. (2013)
		Non-sterilized wastewater		0.008	100		
		Non-sterile real veterinary hospital wastewater	Fluidized bed fed-batch bioreactor	0.201	–33	15	Lucas et al. (2016)
Marbofloxacin	*Trametes versicolor*	Non-sterile real veterinary hospital wastewater	Fluidized bed fed-batch bioreactor	0.191	87	15	Lucas et al. (2016)
		Non-sterile real veterinary hospital wastewater	Fluidized bed batch bioreactor	4.387	76	15	Badia-Fabregat et al. (2016)
			Continuous fluidized bed bioreactor	–	64	HRT: 3.3	Lucas et al. (2018)
			Fluidized bed fed-batch bioreactor	4.588	17	15	Lucas et al. (2016)
Metronidazole	*Trametes versicolor*	Reverse osmosis concentrate from a WWTP, with nutrients	Fluidized bed batch bioreactor	BQL	0	6	Badia-Fabregat et al. (2015)
		Sterilized wastewater		0.05–2.13	31–100	7–8	Cruz-Morató et al. (2013, 2014)

Compound	Fungus	Water/Matrix	Reactor	Concentration	Removal	Time/HRT	Reference
Metronidazole OH		Non-sterilized wastewater		N.D.–0.912	85		
		Reverse osmosis concentrate from a WWTP, with nutrients	Fluidized bed batch bioreactor	0.195	–9	6	Badia-Fabregat et al. (2015)
	Trametes versicolor	Non-sterile real veterinary hospital wastewater	Fluidized bed batch bioreactor	0.612	60	15	Badia-Fabregat et al. (2016)
			Continuous fluidized bed bioreactor	–	73	HRT: 3.3	Lucas et al. (2018)
			Fluidized bed fed-batch bioreactor	0.186	–244	15	Lucas et al. (2016)
		Reverse osmosis concentrate from a WWTP	Continuous fluidized bed bioreactor	0.624	–37	HRT: 3	Badia-Fabregat et al. (2017)
Nalidixic Acid	*Trametes versicolor*	Non-sterile real veterinary hospital wastewater	Fluidized bed fed-batch bioreactor	0.016	–26	15	Lucas et al. (2016)
Norfloxacin	*Trametes versicolor*	Non-sterile real veterinary hospital wastewater	Fluidized bed fed-batch bioreactor	0.091	–55	15	Lucas et al. (2016)
Ofloxacin	*Trametes versicolor*	Sterile sludge from membrane biological reactor	Erlenmeyer flask	2.921	80.5	7	Llorens-Blanch et al. (2015)
		Non-sterile sludge from membrane biological reactor			64.5	7	Llorens-Blanch et al. (2015)

Table 7.2 contd. ...

...Table 7.2 contd.

Antibiotic	WRF species	Matrix	Reactor type	Initial concentration (μg/L)	Removal (%)	Incubation time (d)	References
		Non-sterile real veterinary hospital wastewater	Fluidized bed fed-batch bioreactor	0.22	46	15	Lucas et al. (2016)
			Fluidized bed batch bioreactor	0.03	100	15	Badia-Fabregat et al. (2016)
		Sterilized wastewater		32	98.5–99	7–8	Cruz-Morató et al. (2014); Gros et al. (2014)
		Non-sterilized wastewater		3.3–3.34	98–99		
		Reverse osmosis concentrate from a WWTP, no nutrients	Fluidized bed batch bioreactor	0.177	57	7	Badia-Fabregat et al. (2015)
		Reverse osmosis concentrate from a WWTP, with nutrients		0.067	100	6	Badia-Fabregat et al. (2015)
		Non-sterile real hospital wastewater	Continuous fluidized bed bioreactor	0.25371 / 0.4594	71.1 / −57.8	HRT: 3	Mir-Tutusaus et al. (2017)
		Reverse osmosis concentrate from a WWTP		0.11	44	HRT: 3	Badia-Fabregat et al. (2017)
Oxytetracyclin	*Trametes versicolor*	Non-sterile real veterinary hospital wastewater	Fluidized bed fed-batch bioreactor	0.038	57	15	Lucas et al. (2016)

Compound	Organism	Matrix	Reactor				Reference
Pipemidic Acid	*Trametes versicolor*	Non-sterile real veterinary hospital wastewater	Fluidized bed fed-batch bioreactor	0.088	37	15	Lucas et al. (2016)
		Non-sterile real veterinary hospital wastewater	Continuous fluidized bed bioreactor	–	82	HRT: 3.3	Lucas et al. (2018)
		Reverse osmosis concentrate from a WWTP, with nutrients	Fluidized bed batch bioreactor	0.123	23	15	Badia-Fabregat et al. (2016)
Ronidazole	*Trametes versicolor*		Fluidized bed batch bioreactor	0.049	24	6	Badia-Fabregat et al. (2015)
		Reverse osmosis concentrate from a WWTP	Continuous fluidized bed bioreactor	0.103	6	HRT: 3	Badia-Fabregat et al. (2017)
Spiramycin	*Trametes versicolor*	Non-sterile real veterinary hospital wastewater	Fluidized bed fed-batch bioreactor	0.009	100	15	Lucas et al. (2016)
Sulfadimethoxine	*Trametes versicolor*	Non-sterile real veterinary hospital wastewater	Fluidized bed fed-batch bioreactor	0.015	55	15	Lucas et al. (2016)
Sulfamethazine	*Trametes versicolor*	Dry sewage sludge (sterile)	Biopile with wheat straw	19.1–20.4 (ng/g)	100	42	García-Galán et al. (2011), Rodríguez-Rodríguez et al. (2011)
		Sewage sludge (sterile)	Bioslurry reactor	6.1	91		Rodríguez-Rodríguez et al. (2012a)
			Fluidized bed batch bioreactor	5.0	94	3	Rodríguez-Rodríguez et al. (2012b)

Table 7.2 contd. ...

...*Table 7.2 contd.*

Antibiotic	WRF species	Matrix	Reactor type	Initial concentration (µg/L)	Removal (%)	Incubation time (d)	References
	Bjerkandera adusta	Sterile sludge	Biopile with wheat straw	500	< 80*	300	Aydin (2016)
	Phanerochaete chrysosporium	Non-sterilized wastewater	Erlenmeyer flask and bioplastic entrapped fungus (fluidized bed)	N.R.	> 60%	5	Accinelli et al. (2010)
		Activated sludge mixed liquor		N.R.	> 80%		
		Sterile sludge	Biopile with wheat straw	500	< 90*	300	Aydin (2016)
		Sterile sludge from membrane biological reactor	Erlenmeyer flask		95	7	Llorens-Blanch et al. (2015)
Sulfamethoxazole		Non-sterile sludge from membrane biological reactor		0.158	95.7	7	Llorens-Blanch et al. (2015)
	Trametes versicolor	Non-sterile real veterinary hospital wastewater	Fluidized bed fed-batch bioreactor	0.016	47	15	Lucas et al. (2016)
		Reverse osmosis concentrate from a WWTP, no nutrients	Fluidized bed batch bioreactor	0.1	100	7	Badia-Fabregat et al. (2015)
		Reverse osmosis concentrate from a WWTP, with nutrients		BQL	100	6	Badia-Fabregat et al. (2015)

Compound	Fungus	Source	Reactor	Concentration	% Removal	Time/HRT	Reference
		Sterilized wastewater		5.29	100		Cruz-Morató et al. (2014)
		Non-sterilized wastewater		1.41	100	7	
		Non-sterile real hospital wastewater	Continuous fluidized bed bioreactor	1.130 / 0.0559	78.2 / 34.8	HRT: 3	Mir-Tuttusaus et al. (2017)
		Reverse osmosis concentrate from a WWTP		0.023	88	HRT: 2	Badia-Fabregat et al. (2017)
Sulfapyridine	*Trametes versicolor*	Dry sewage sludge	Biopile with wheat straw	29.4 (ng/g)	100	42	Garcia-Galán et al. (2011)
		Sewage sludge	Bioslurry reactor	21.5	100	26	Rodríguez-Rodríguez et al. (2012a)
		Sewage sludge	Fluidized bed batch bioreactor	5.0	94	3	Rodríguez-Rodríguez et al. (2012b)
Sulfathiazole	*Trametes versicolor*	Dry sewage sludge	Solid phase incubation in wheat-straw bed	71.1 (ng/g)	100	42	Garcia-Galán et al. (2011)
		Sewage sludge	Bioslurry reactor	143.0	91		Rodríguez-Rodríguez et al. (2012a)
		Sewage sludge	Fluidized bed batch bioreactor	5.0	94	3	Rodríguez-Rodríguez et al. (2012b)

Table 7.2 contd. ...

...*Table 7.2 contd.*

Antibiotic	WRF species	Matrix	Reactor type	Initial concentration (µg/L)	Removal (%)	Incubation time (d)	References
	Bjerkandera adusta	Sterile sludge	Biopile with wheat straw	100	<86*	300	Aydin (2016)
		Sterile sludge	Biopile with wheat straw	100	<77*	300	Aydin (2016)
		Sterilized wastewater		N.D.	–	7	Cruz-Morató et al. (2014)
		Non-sterilized wastewater	Fluidized bed batch bioreactor	0.01	0		
Tetracycline	*Trametes versicolor*	Reverse osmosis concentrate from a WWTP, with nutrients		BQL	–354	6	Badia-Fabregat et al. (2015)
		Non-sterile real veterinary hospital wastewater	Fluidized bed batch bioreactor	1.324	100	15	Badia-Fabregat et al. (2016)
		Non-sterile real veterinary hospital wastewater	Fluidized bed fed-batch bioreactor	0.063	26	15	Lucas et al. (2016)
		Reverse osmosis concentrate from a WWTP	Continuous fluidized bed bioreactor	0.092	90	HRT: 2	Badia-Fabregat et al. (2017)
Tilmicosin	*Trametes versicolor*	Non-sterile real veterinary hospital wastewater	Fluidized bed fed-batch bioreactor	0.057	–26	15	Lucas et al. (2016)

Antibiotic	Fungus	Matrix	Reactor	Concentration	Removal (%)	Time (days)	Reference
Trimethoprim	*Trametes versicolor*	Sterile sludge from membrane biological reactor	Erlenmeyer flask	0.0472	91.1	7	Llorens-Blanch et al. (2015)
		Non-sterilized sludge from membrane biological reactor			100	7	Llorens-Blanch et al. (2015)
		Sterilized wastewater	Fluidized bed batch bioreactor	2.57	75	7	Cruz-Morató et al. (2014)
		Non-sterilized wastewater		0.853	100	7	
		Reverse osmosis concentrate from a WWTP no nutrients		0.059	36	7	Badia-Fabregat et al. (2015)
		Non-sterile real veterinary hospital wastewater	Fluidized bed batch bioreactor	1.759	37	15	Badia-Fabregat et al. (2016)
			Fluidized bed fed-batch bioreactor	0.052	8	15	Lucas et al. (2016)
		Non-sterile real hospital wastewater	Continuous fluidized bed bioreactor	0.7483 / 0.0818	52.3 / −26.9	HRT: 3	Mir-Tuttusaus et al. (2017)
Tylosin	*Trametes versicolor*	Non-sterile real veterinary hospital wastewater	Fluidized bed fed-batch bioreactor	0.014	30	15	Lucas et al. (2016)

* Removal estimated based on the difference in the non-inoculated system versus the fungal inoculated system

N.D. - Not Detected

BQL - Below Quantifiable Levels

BLD - Below Limits of Detection

N.R. - Not reported

(Badia-Fabregat et al. 2015). In an interesting variation of the pellet use, Accinelli et al. (2010) employed *P. chrysosporium* entrapped in a granular bioplastic formulation and attained high elimination of erythromycin, sulfamethoxazole, and ciprofloxacin from activated sludge and wastewater.

High removal rates have been consistently reported for some chemical groups. This includes removals of > 64% for cephalosporines (cephalexin, cefazolin, and cefuroxime) (Badia-Fabregat et al. 2016, Cruz-Morató et al. 2013, Lucas et al. 2016) and complete elimination of tetracyclines (chlortetracycline, oxytetracycline, and tetracycline) (Aydin 2016, Badia-Fabregat et al. 2016, Badia-Fabregat et al. 2017). Likewise, sulfonamides were efficiently eliminated using pre-sterilized and non-sterile matrices: sulfadimethoxine (55%), sulfamethazine (91–100%), sulfamethoxazole (35–100%), sulfapyridine (94–100%), and sulfathiazole (91–100%) (García-Galán et al. 2011, Lucas et al. 2016, Rodríguez-Rodríguez et al. 2011, Rodríguez-Rodríguez et al. 2012b). However, removals of sulfonamides were usually higher in biopiles. On the contrary, the reported removal of quinolones ranged from no removal for danofloxacin, nalidixic acid, and norfloxacin (Lucas et al. 2016) to over 50% for ciprofloxacin, enoxacin, enrofloxacin, and marbofloxacin (Badia-Fabregat et al. 2015, Lucas et al. 2016, Martens et al. 1996).

The presence of antibiotic resistance genes in the environment is of great concern as they have also been considered as emerging pollutants (Pruden et al. 2006). The fungal treatment of veterinary hospital wastewater with *T. versicolor* was efficient to remove the antibiotic resistance genes *erm B*, *tetW*, *bla$_{TEM}$*, *sulI*, and *qnrS*, which provide resistance to macrolides, tetracyclines, β-lactams, sulfonamides, and fluoroquinolones, respectively, thus, expanding the potential applications of these processes (Lucas et al. 2016).

The differences in removal efficiencies reported in real matrices suggest that the elimination process is highly dependent on particular characteristics of the polluted effluent (including indigenous microbiota), and probably on the diversity and combination of antibiotic—and other micropollutants—present in the matrix.

3.1.5 *Removal of antibiotics by non-WRF*

Few reports have described the use of non-WRF for the removal of antibiotics; all of them employed whole-cell cultures in submerged synthetic media (Nguyen et al. 2017, Parshikov et al. 1999, 2000, I. A. Parshikov et al. 2001, I. Parshikov et al. 2001, Williams et al. 2007, Zhang et al. 2015). Although different fungi species have been examined, comparing their performance is challenging, as in most cases there is only one report for each antibiotic (Table 7.3). Remarkably, low removals have been obtained with non-WRF compared to WRF (e.g., *T. versicolor*), and even lower removal rates than those achieved with purified enzymatic assays (days vs. hours, respectively).

Elimination assays usually employed sugar-supplemented media and reported metabolites that were slightly modified in comparison to the precursor molecule, suggesting co-metabolic interactions and not the use of the antibiotics as carbon sources (Parshikov et al. 1999, 2000, I. A. Parshikov et al. 2001, I. Parshikov et al. 2001, Williams et al. 2007). As with WRF, the mechanism involved in the removal

Table 7.3 Removal of antibiotics using non-WRF.

Antibiotic	Non-WRF species	Initial concentration (mg/L)	Removal (%)	Incubation time (d)	References
Ampicillin	*Verticillium leptobactrum*	0.5–2.0	100	14	Kumar et al. (2013)
	Mucor ramannianus	1	89.1	14	Parshikov et al. (1999)
Ciprofloxacin	*Pestalotiopsis guepini*	1	68.7	18	Parshikov et al. (2001b)
	Gloeophyllum striatum	10	33	14	Wetztein et al. (1999)
	Mucor ramannianus	1	78	21	Parshikov et al. (2000)
Enrofloxacin	*Gloeophyllum striatum*	0.05	42.8–53.3	56	Martens et al. 1996
	Gloeophyllum striatum	10	27.3	56	Wetztein et al. (1997)
Flumequine	*Cunninghamella elegans*	1	66	7	Williams et al. (2007)
Gentamicin	*Aspergillus terreus*	100	95	7	Liu et al. (2016)
Lincomycin	*Galactomyces geotrichum*	Not reported	37	15	Zhang et al. (2015)
Norfloxacin	*Pestalotiopsis guepini*	1	69.9	18	Parshikov et al. (2001b)
Sarafloxacin	*Mucor ramannianus*	1	41	18	Parshikov et al. (2001a)
	Lasiodiplodia sp.	10	93.2	12	
Tiamulin	*Fusarium* sp.	10	82.4	12	Nguyen et al. (2017)
	Verticillium sp.	10	89.3	12	
	Galactomyces sp.	10	73.8	12	

of antibiotics by non-WRF appears to be an unspecific enzymatic transformation. The metabolites reported from quinolones are produced by unidentified enzymes, such as *N*-acetyl transferases, oxidases, and desethylenases (Parshikov et al. 1999, 2000, I. A. Parshikov et al. 2001, I. Parshikov et al. 2001), and their chemical structure is analogous to those produced in the treatment with WRF, as exemplified in the cases of ciprofloxacin and enrofloxacin using *Gleophyllum striatum* (Wetzstein et al. 1997, Wetzstein et al. 1999). Similarly, flumequine transformation products by *Cunninghamella elegans* are hydroxylated versions of the parental compound, perhaps produced by monooxygenases (Williams et al. 2007).

3.2 *Removal of antibiotics by WRF enzymes*

Removal of antibiotics has also been investigated using purified or semi-purified lignin oxidizing enzymes produced by WRF. Most of the scientific reports employed commercial laccase from *T. versicolor* (Migliore et al. 2012, Margot et al. 2015, Mir-Tutusaus et al. 2014, Prieto et al. 2011, Rodríguez-Rodríguez et al. 2012b), and just a few described biocatalysts of other origins, such as enzymes produced by *Perenniporia* sp. (Weng et al. 2012), or crude extracts of MnP and LiP prepared from *P. chrysosporium* cultures (Wen et al. 2009, 2010) (Table 7.4). These works employed batch reactors, even at flask scale, and free enzymes. Likewise, some reports using immobilized enzymes and more complex configurations such as membrane reactors are available (Becker et al. 2016, de Cazes et al. 2014, Nguyen et al. 2014).

Remarkably, enzymatic-reactor assays mostly reported incomplete removals of antibiotics in just a few hours, even using enzymatic mediators (Table 7.4). On the contrary, studies using whole-cell fungal reactors reported higher and even complete removals in larger incubation periods, reaching months in some cases. For example, Kumar et al. (2013) achieved complete elimination of ampicillin in a whole cell reactor over a period of 14 days, whereas Yang et al. (2017) attained a removal close to 60% in 48 hours, using 2,2-azino-bis(3-ethylbenzothiazoline-6-sulfonic acid) (ABTS) as a mediator in a reactor with immobilized laccase. Although whole-cell fungal assays need more time for removal, they combine other elimination strategies, such as biomass adsorption and intracellular enzymatic reactions, and not just an exoenzyme-mediated removal (Asif et al. 2017) (Yang et al. 2013b).

The performance of every enzyme could vary depending on the fungi that produced it, and from enzyme to enzyme. For instance, the removals of oxytetracycline and tetracycline using crude extracts of MnP and LiP from *P. chrysosporium* (Wen et al. 2009, 2010) are lower compared to assays using laccase coupled to mediators (Mir-Tutusaus et al. 2014, Migliore et al. 2012, Suda et al. 2012, Yang et al. 2017). Although Weng et al. (2012) reported the use of a laccase from *Perenniporia* sp. (and not *T. versicolor* as in other reports), this work used sulfadimethoxine and sulfamonomethoxine, sulfonamides, whose removals have not been studied with other fungal enzymes.

The work with enzymes clearly involves the control of media conditions that affect the biocatalytic removal capacity. Thus, optimal conditions such as pH must be evaluated to achieve the highest xenobiotic degradation. For instance, the optimal pH reported for laccases is slightly acidic (4,5–5) (Naghdi et al. 2018); however,

Table 7.4 Removal of antibiotics by fungal free-enzymes.

Enzyme	Antibiotic	Enzymatic mediator*	pH	Temperature (°C)	Initial concentration (mg/L)	Removal (%)	Incubation time (h)	Reference
Laccase	Ciprofloxacin	–	4.5	30	10	16	20	Prieto et al. (2011)
		ABTS	4.5	30	10	97.7	30	
	Norfloxacin	–	4.5	30	10	0	20	
		ABTS	4.5	30	10	33.7	30	
	Chlortetracycline	HBT	4.5	30	40–50	100	1	Suda et al. (2012)
	Doxycycline	HBT	4.5	30	40–50	100	0.25	
		–	4.5	25	10	0	96	Mir-Tutusaus et al. (2014)
		–	4.5	25	50–100	1.6–5	96	Migliore et al. (2012)
	Oxytetracycline	HBT	4.5	30	40–50	100	0.25	Suda et al. (2012)
		ABTS	4.5	25	10	0	96	Mir-Tutusaus et al. (2014)
		VA	4.5	25	10	0	96	
		HBT	4.5	25	10	0	96	
	Tetracycline	–	6	25	20	30	24	de Cazes et al. (2014)
		HBT	4.5	30	40–50	100	1	Suda et al. (2012)

Table 7.4 contd. ...

...Table 7.4 contd.

Enzyme	Antibiotic	Enzymatic mediator*	pH	Temperature (°C)	Initial concentration (mg/L)	Removal (%)	Incubation time (h)	Reference
	Sulfadimethoxime	-	4	30	50	8	5.5	
		HBA	4	30	50	67	5.5	Weng et al. (2012)
		HAP	4	30	50	0	5.5	
		SIR	4	30	50	90	5.5	
		CP	4	30	50	0	5.5	
		VA	4	30	50	100	5.5	
		ABTS	4	30	50	100	5.5	
	Sulfamethoxazole	ABTS	4–6	25	20–25	100	8	
		SIR	4–6	25	20–25	100	8	Margot et al. (2015)
		ASG	4–6	25	20–25	100	8	
	Sulfamonomethoxime	-	4	30	50	9	5.5	
		HBA	4	30	50	70	5.5	
		HAP	4	30	50	32	5.5	
		SIR	4	30	50	83	5.5	Weng et al. (2012)
		CP	4	30	50	19	5.5	
		VA	4	30	50	100	5.5	
		ABTS	4	30	50	100	5.5	

								Reference
	Sulfapyridine	–	4.5	25	20	0	50	Rodriguez-Rodriguez et al. (2012b)
		ABTS	4.5	25	20	100	50	
		VA	4.5	25	20	100	50	
		DMHAP	4.5	25	20	75	50	
	Sulfathiazole	–	4.5	25	16	0	50	
		ABTS	4.5	25	16	100	50	
		VA	4.5	25	16	100	50	
	DMHAP		4.5	25	16	82	50	
Manganese peroxidase	Tetracycline	H_2O_2	5	25	50	72.5	4	Wen et al. (2010)
	Oxytetracycline	H_2O_2	5	25	50	84.3	4	
Lignin peroxidase	Tetracycline	VRA	4.2	37	50	95	0.08	Wen et al. (2009)
	Oxytetracycline	VRA	4.2	37	50	95	0.08	

*2,2-azino-bis(3-ethylbenzothiazoline-6-sulfonic acid) (ABTS); 1-hydroxybenzotriazole hydrate (HBT); violuric acid (VA); 4-hydroxyacetophenone (HAP); 4-cyanophenol (CP); acetosyringone (ASG); 4-hydroxybenzil alcohol (HBA); syringaldehyde (SIR); 3,5-dimethoxy-4-hydroxyacetophenon (DMHAP); hydrogen peroxide (H_2O_2); veratryl alcohol (VRA); no mediator (–)

differences in activity have been observed towards different xenobiotics. Thus, it can be stated that optimal pH values depend on enzyme-antibiotic interaction and should be established for every case. Faster degradation of sulfamethoxazole was observed at lower pH values, near 5 (Margot et al. 2015), but a faster kinetics was achieved during the degradation of oxytetracycline at pH 7 (72% removal in 24 hours), compared to only 12% at pH 3 in a batch reactor with free enzyme (de Cazes et al. 2014).

Temperature also affects the removal capacity of enzymes since it is a parameter related to enzymatic optimal conditions. For example, a slow removal, i.e., 11% of tetracycline (0.01 g/L) with free laccase was reached at 15°C, whereas it was 35–36% at 25–35°C after 24 hours (de Cazes et al. 2014). Furthermore, dissolved constituents, such as sulfides and lipids, can inhibit the activity of laccase, an effect that is also observed with chloride and other halides; the last one is a parameter of concern due to the importance of chlorination in wastewater treatment (Naghdi et al. 2018). Finally, the presence of metals during the enzymatic degradation should be considered, for example, copper acts as a cofactor of laccase and its presence in the medium results in induction of enzyme production (Palmieri et al. 2000).

The use of laccase enzymatic mediators or enhancers must be acknowledged as well. These compounds work as electron shuttles that are easily oxidized by the laccase to free radicals, and diffuse to transport the electrons to the xenobiotics (Morozova et al. 2007, Yang et al. 2013b). These mediators can be classified as synthetic molecules such as ABTS, or lignin derived phenolic compounds (e.g., syringic acid) (Fabbrini et al. 2002, Martorana et al. 2013). The presence of laccase enzymatic mediators or enhancers on the removal system usually enhances the degradation of antibiotics or other pollutants; for instance, the use of ABTS-laccase systems improved the removal of antibiotics that are practically not oxidized by laccase alone, such as fluoroquinolones, e.g., ciprofloxacin and norfloxacin (Prieto et al. 2011), and sulfonamides, e.g., sulfadimethoxine, sulfamonomethoxine, sulfapiridine, and sulfathiazole (Weng et al. 2012).

In general, degradation assays demonstrated that laccase by itself cannot be considered responsible for the removal of antibiotics, such as norfloxacine (Prieto et al. 2011), oxytetracycline (Mir-Tutusaus et al. 2014), and some sulfonamides (Weng et al. 2012, Rodríguez-Rodríguez et al. 2012b). Furthermore, other antibiotics, such as oxytetracycline, are neither eliminated in laccase nor ABTS-laccase systems, but completely removed using HBT as enzymatic mediator (Mir-Tutusaus et al. 2014). Nonetheless, in other cases the use of enzymatic mediators resulted in negligible degradation of antibiotics. For instance, HAP or CP coupled to laccase achieved lower removals of sulfadimethoxine compared to the mediator-lacking assay (Weng et al. 2012). This casuistic efficiency seems to be related to the redox nature of each pollutant.

Transformation by free and immobilized laccase results in oxidation and hydrolysis reactions of parental compounds, as demonstrated in tetracyclines and fluoroquinolones (de Cazes et al. 2014, Migliore et al. 2012, Prieto et al. 2011, Yang et al. 2017). The transformation of tetracyclines and sulfonamides in the presence of mediators is also mediated by oligomerization of oxidized antibiotics via radical-radical coupling or cross-coupling with the radical form of the mediator

(Margot et al. 2015, Shi et al. 2014, Suda et al. 2012, Weng et al. 2012). Analogous transformation products have been described using either laccase-mediators systems or whole cells in the treatment of sulfonamides (García-Galán et al. 2011, Rodríguez-Rodríguez et al. 2012b).

3.2.1 Use of immobilized enzymes

The use of immobilized laccase for the removal of antibiotics has been scarcely reported in the literature. For example, laccases from *Echinodontium taxodii* linked to Fe_3O_4 nanoparticles (Shi et al. 2014), and magnetic cross-linked laccase aggregates from *Cerrena* sp. (Yang et al. 2017) were employed for the treatment of sulfonamides and tetracyclines, respectively (Table 7.5). Likewise, membrane reactors employing wet ceramic supports have been described for the elimination of penicillins, quinolones, tetracyclines, and sulfonamides (de Cazes et al. 2014, Becker et al. 2016) and polyacrylonitrile membranes for the removal of sulfamethoxazole (Nguyen et al. 2014).

The use of immobilized enzymes facilitates the application *in-situ* of the removal processes, due to the possibility of enzyme reuse (Yang et al. 2013b) and fluidization of the system. Despite not requiring the supplement of nutritional sources as in whole cell systems (Badia-Fabregat et al. 2014, Agrawal et al. 2017), the use of purified enzymes is more expensive, due to the downstream behind fungal enzymatic production (El-Batal et al. 2015, Chanudhary et al. 2016, Chigu and Ichinose 2017). The use of purified and immobilized enzymes must be based on other benefits, such as the higher apparent affinity of the enzyme-xenobiotic complex and higher removal efficiencies, which have been documented with antibiotics, such as tetracyclines (de Cazes et al. 2014) and sulfonamides (Shi et al. 2014). Moreover, immobilization enhances the enzyme stability, thus resulting in extended half-lives, as reported for the removal of tetracycline in a membrane reactor (constant degradation rate of 0.34 mg/h during 10 days) (de Cazes et al. 2014). On the contrary, low removals have been observed when using cross-linked laccase aggregates to remove chloramphenicol, trimethoprim, sulfamethoxazole, erythromycin, and ampicillin, even using ABTS as a mediator (Yang et al. 2017).

3.3 Fungal and physicochemical-coupled treatment

Combined physicochemical and fungal strategies have been scarcely used to obtain higher removal efficiencies. The main goal of these synergetic approaches is to minimize operational costs of effective technologies and improve the performance of pollutant removal. Strategies, such as wastewater coagulation/flocculation, addition of sorbents, and electrochemical-advanced oxidation have been successfully coupled to WRF treatments. These combined strategies will be briefly discussed, except for advanced oxidation processes that have been previously reviewed by Ganzenko et al. (2014).

Sorbent materials (e.g., activated carbon and soil) have been coupled to enzymatic batch assays to eliminate persistent pollutants with good removal rates (Ding et al. 2016, Nguyen et al. 2014). The removal mechanism in these absorbents is mainly produced by electrostatic sorption, and hence its success depends on the

Table 7.5 Use of immobilized enzymes for the removal of antibiotics.

Antibiotic family	Antibiotic	Enzymatic mediator*	pH	Temperature °C	Initial concentration (mg/L)	Removal (%)	Incubation time (h)	References
	Amoxicillin	–	3	25	0.01	96.6	24	Becker et al. (2016)
		SIR	3	25	0.01	94.7	24	
	Ampicillin	–	7	25	100	<40	48	Yang et al. (2017)
		ABTS	7	25	100	<60	48	
		–	3	25	0.01	88.6	24	Becker et al. (2016)
		SIR	3	25	0.01	99.9	24	
Penicillins	Cloxacillin	–	3	25	0.01	–4.6	24	
		SIR	3	25	0.01	54.3	24	
	Oxacillin	–	3	25	0.01	–23.1	24	
		SIR	3	25	0.01	53.5	24	Becker et al. (2016)
	Penicillin G	–	3	25	0.01	9.6	24	
		SIR	3	25	0.01	93.9	24	
	Penicillin V	–	3	25	0.01	–15.7	24	
		SIR	3	25	0.01	70.6	24	
	Sulfabenzamide	–	3	25	0.01	15.0	24	Becker et al. (2016)
		SIR	3	25	0.01	98.5	24	
Sulfonamides	Sulfadiazine	–	5	25	50	0	0.5	Shi et al. (2014)
		SIR	5	25	50	<100	0.5	
		SRA	5	25	50	<100	0.5	
		ASG	5	25	50	<100	0.5	

Compound							Reference
	VNL	5	25	50	< 20	0.5	
	AVNL	5	25	50	< 20	0.5	
	VNLA	5	25	50	< 20	0.5	
	FA	5	25	50	< 20	0.5	
	–	3	25	0.01	10.4	24	Becker et al. (2016)
	SIR	3	25	0.01	99.7	24	
Sulfadimethoxine	–	3	25	0.01	5.4	24	
	SIR	3	25	0.01	96.1	24	
Sulfamerazine	–	3	25	0.01	–1.3	24	Becker et al. (2016)
	SIR	3	25	0.01	100.0	24	
Sulfamethizole	–	3	25	0.01	8.54	24	
	SIR	3	25	0.01	96.4	24	
	-	7	25	100	< 20	48	Yang et al. (2017)
	ABTS	7	25	100	> 30	48	
	–	5	25	50	0	0.5	
	SIR	5	25	50	< 100	0.5	
	SRA	5	25	50	< 100	0.5	
Sulfamethoxazole	ASG	5	25	50	< 100	0.5	Shi et al. (2014)
	VNL	5	25	50	< 20	0.5	
	AVNL	5	25	50	< 20	0.5	
	VNLA	5	25	50	< 20	0.5	
	FA	5	25	50	< 20	0.5	

Table 7.5 contd. ...

... Table 7.5 contd.

Antibiotic family	Antibiotic	Enzymatic mediator*	pH	Temperature °C	Initial concentration (mg/L)	Removal (%)	Incubation time (h)	Reference
		–	5	25	50	0	0.5	
		SIR	5	25	50	<100	0.5	
		SRA	5	25	50	<100	0.5	
		ASG	5	25	50	<100	0.5	
		VNL	5	25	50	<20	0.5	
		AVNL	5	25	50	<20	0.5	
		VNLA	5	25	50	<20	0.5	
		FA	5	25	50	<20	0.5	Becker et al. (2016)
		–	3	25	0.01	14.2	24	
		SIR	3	25	0.01	97.2	24	
		SIR	6.8	28	1.66	97	24	Nguyen et al. (2014)
Sulfamethoxypyridazine		–	3	25	0.01	2.1	24	
		SIR	3	25	0.01	99.0	24	
Sulfanitran		–	3	25	0.01	5.9	24	
		SIR	3	25	0.01	49.5	24	
Sulfapyridine		–	3	25	0.01	–6.8	24	Becker et al. (2016)
		SIR	3	25	0.01	100.0	24	
Sulfathiazole		–	3	25	0.01	6.8	24	
		SIR	3	25	0.01	99.8	24	
Sulfisomidin		–	3	25	0.01	21.9	24	
		SIR	3	25	0.01	97.7	24	

Sulfisoxazole	–	3	25	0.01	12.3	24	
	SIR	3	25	0.01	100.0	24	Becker et al. (2016)
Chlorotetracycline	–	3	25	0.01	33.2	24	
	SIR	3	25	0.01	92.2	24	Becker et al. (2016)
Doxycycline	–	3	25	0.01	35.8	24	
	SIR	3	25	0.01	60.4	24	Becker et al. (2016)
	–	7	25	100	< 60	48	
	ABTS	7	25	100	< 20	48	Yang et al. (2017)
Tetracyclines							
Oxytetracycline	–	3	25	0.01	48.4	24	
	SIR	3	25	0.01	79.9	24	Becker et al. (2016)
	–	7	25	100	< 70	48	
	ABTS	7	25	100	< 40	48	Yang et al. (2017)
Tetracycline	–	3	25	0.01	26.0	24	
	SIR	3	25	0.01	69.7	24	Becker et al. (2016)
	–	6	25	20	56	24	De Cazes et al. (2014)
Quinolones							
Cinoxacin	–	3	25	0.01	12.3	24	
	SIR	3	25	0.01	14.6	24	Becker et al. (2016)
Nalidixic acid	–	3	25	0.01	–1.1	24	
	SIR	3	25	0.01	69.1	24	Becker et al. (2016)

Table 7.5 contd. ...

... *Table 7.5 contd.*

Antibiotic family	Antibiotic	Enzymatic mediator*	pH	Temperature °C	Initial concentration (mg/L)	Removal (%)	Incubation time (h)	Reference
	Pipemidic acid	–	3	25	0.01	54.6	24	
		SIR	3	25	0.01	85.5	24	
	Oxolinic acid	–	3	25	0.01	–4.5	24	
		SIR	3	25	0.01	73.5	24	
	Ciprofloxacin	–	3	25	0.01	59.4	24	
		SIR	3	25	0.01	93.0	24	
	Danofloxacin	–	3	25	0.01	59.6	24	
		SIR	3	25	0.01	75.8	24	
	Difloxacin	–	3	25	0.01	–4.8	24	Becker et al. (2016)
		SIR	3	25	0.01	48.8	24	
Fluoroquinolones	Enoxacin	–	3	25	0.01	–24.2	24	
		SIR	3	25	0.01	89.7	24	
	Enrofloxacin	–	3	25	0.01	50.1	24	
		SIR	3	25	0.01	76.6	24	
	Erythromycin	–	7	25	100	< 20	48	Yang et al. (2017)
		ABTS	7	25	100	> 20	48	
	Flumequine	–	3	25	0.01	–3.5	24	
		SIR	3	25	0.01	41.7	24	
	Marbofloxacin	–	3	25	0.01	56.1	24	Becker et al. (2016)
		SIR	3	25	0.01	73.1	24	
	Norfloxacin	–	3	25	0.01	58.1	24	
		SIR	3	25	0.01	82.4	24	

							Reference
Ofloxacin	-	3	25	0.01	54.9	24	Yang et al. (2017)
	SIR	3	25	0.01	77.7	24	
Orbifloxacin	-	3	25	0.01	7.4	24	Becker et al. (2016)
	SIR	3	25	0.01	33.0	24	
Trimethoprim	-	7	25	100	> 0	48	
Trimethoprim	ABTS	7	25	100	> 10	48	Becker et al. (2016)
	-	3	25	0.01	26.6	24	
	SIR	3	25	0.01	66.8	24	
Metronidazole	-	3	25	0.01	7.1	24	Becker et al. (2016)
Metronidazole	SIR	3	25	0.01	9.4	24	
Chloramphenicol	-	7	25	100	< 10	48	Yang et al. (2017)
Chloramphenicol	ABTS	7	25	100	< 10	48	

*Acronyms: syringaldehyde (SIR); 2,2-azino-bis(3-ethylbenzothiazoline-6-sulfonic acid) (ABTS); syringic acid (SRA); acetosyringone (ASG); vanillin (VNL); Acetovanillone (AVNL); vanillic acid (VNLA); ferulic acid (FA); no mediator (–)

antibiotic's electric charge. Antibiotics act as zwitterions, and depending on their pKa and the medium pH, they can be positively or negatively charged. For example, in neutral solution, Chinese clay soils (negatively charged) absorbed antibiotics with high pKa (Calisto et al. 2015), a property employed to remove quinolones and tetracyclines and improving removals of some antibiotics from negligible to almost 100% in 180 minutes (Ding et al. 2016). Granular activated carbon (GAC) improved the removal of sulfamethoxazole more than 50% in 35 days, compared to an enzymatic assay without GAC (Nguyen et al. 2014). The removals using enzymatic-sorptive methods are higher; however, since the pollutants are transferred only from the wastewater to the adsorptive material, in most cases a post-treatment of the matrix, such as incineration, is necessary.

A major difficulty in the fungal treatment of non-sterile wastewaters is to maintain the biological removal activity, due to the pressure of bacterial competition. Many studies have reported a pre-sterilization of wastewaters by heat treatment; however, the high-water influx in a wastewater treatment plant makes this strategy economically nonviable. Mir-Tutusaus et al. (2017) reported the maintenance of fungal activity for 28 days in non-sterile conditions, using a coagulation/flocculation procedure. This strategy is commonly used in treatment plants after the sedimentation process, to reduce the organic matter and phosphorous content of the wastewater. The coagulants and flocculants are polymers that act as binding particles, producing heavier and solid flakes that sink to the bottom of clarifier tanks, producing the matrix commonly known as sludge. Mir-Tutusaus et al. (2017) stated that fungal treatment could produce better removals after a coagulation/flocculation process, due to a reduction in the bacterial load.

3.4 *Ecotoxicological considerations of antibiotic transformation*

The objective of bioremediation processes is to eliminate the xenobiotic and the toxicity of the polluted matrix. Consequently, most antibiotic removal studies report the desired effect of toxicity decrease after fungal treatment. For example, a treated wastewater containing sulfamethoxazole/isoproturon was less toxic to the algae *Pseudokirchneriella subcapitata* than the untreated solution (Margot et al. 2015), and a mixture of oxytetracycline and agrochemicals showed a reduction of global toxicity against *Vibrio fisheri* (Microtox) after treatment with *T. versicolor* (Mir-Tutusaus et al. 2014). The solid phase treatment of sludge for the elimination of antibiotics and other pharmaceutical compounds revealed a marked detoxification in *Daphnia magna* and *V. fischeri* using *T. versicolor* (Rodríguez-Rodríguez et al. 2011) or the co-application of *T. versicolor* and *B. adusta* (Aydin 2016). Several tests have been performed after antibiotic treatment aimed at demonstrating the decrease in the antimicrobial activity. Suda et al. (2012) and Yang et al. (2017) reported reduced activity of a mixture of tetracyclines against *Escherichia coli, Bacillus subtilis, B. licheniformis,* and *Pseudokirchneriella subcapitata* after being treated by a laccase-mediator system; likewise, Shi et al. (2014) achieved the same effect on *E. coli* and *S. aureus* using laccase and mediators on sulfonamides. The use of whole cells also decreased antimicrobial activity in the case of ciprofloxacin transformation by *P.*

ostreatus (Singh et al. 2017); similarly, the transformation products reported for the transformation of flumequine are known to exhibit lower antimicrobial activity than the parental compound (Williams et al. 2007).

The degradation of organic molecules does not necessarily translate in the detoxification of the polluted matrix, since toxic compounds can be produced, even more toxic than parental compounds, and turn into dead-end products of the treatment process. This condition results in an increased toxicity of the treated effluent in some cases. For example, Čvančarová et al. (2015) reported the transformation of fluoroquinolones by WRF to toxic compounds that inhibit Gram-positive and Gram-negative bacteria; furthermore, carcinogenic N-nitrosamines were produced due to the treatment. Besides, although the fungal treatment at reactor scale with *T. versicolor* efficiently eliminated 100 mg/L oxytetracycline in only 60 hours, the removal did not correlate with a decrease in the toxicity against *Lactuca sativa* (Cambronero-Heinrichs et al. 2018). Furthermore, Rodríguez-Rodríguez et al. (2012a) found higher toxicity to *D. magna* and *V. fischeri* after the treatment of sludge containing antibiotics and other pollutants in a bioslurry reactor with *T. versicolor*, contrary to the lower toxicity observed after solid phase treatment with the same fungus and analogous sludge (Rodríguez-Rodríguez et al. 2011). Similar findings by Becker et al. (2016) describe the time-dependent increase of toxicity against *V. fischeri* and the growth of *B. subtilis* during the removal of antibiotics by immobilized laccase supplemented with the mediator syringaldehyde. Considering these reports, ecotoxicological assays should be carefully performed during the design of fungal treatments to avoid the production of more toxic effluents; in the particular case of antibiotics, a test aimed at demonstrating reduced antimicrobial activity should be considered in the battery of assays.

Concluding remarks

The versatility of WRF to be used on the removal of organic pollutants has been extensively demonstrated; nonetheless, their use to transform antibiotics and, in general, pharmaceutical compounds is a recently addressed issue. Transformation assays using whole cells or enzyme-mediator systems in synthetic wastewater usually show significant removals; however, whole cell processes may benefit of the complete metabolic machinery of the fungus. Fungal treatment of real antibiotic-polluted matrices has mostly been employed in fluidized beds (air pulsed) reactors to treat wastewater of hospital or veterinary origin, and bioslurry STR or solid phase biopiles to treat sewage sludge. The matrices have been either pre-sterilized or not; in the former case, removal is dependent on the fungal activity, whereas, in the latter, the elimination relies on the combined effects of the fungus and the indigenous microbiota (which could favor the process). The removal of antibiotics, so far described in real matrices, does not follow a clear pattern, suggesting that elimination is highly dependent on the specific features of the effluent, including physicochemical characteristics, diversity and combination of antibiotics (and other pollutants), and the indigenous microbial communities. Given that transformation of antibiotics does not necessarily correlate with the detoxification of the matrix or the decrease in antimicrobial activity due to the production of toxic metabolites, the use of ecotoxicological/activity assays should be implemented in the design of mycoremediation approaches to ensure the eco-friendliness and detoxification of the process. Future challenges in the development of enhanced fungal-mediated processes for antibiotic treatment should include: (i) The evaluation of other fungal species (as most reports, particularly with real matrices, employ *T. versicolor*), (ii) The use of different reactor configurations in liquid phase, and in particular in solid phase, where they have been minimally explored, (iii) The search of low cost carbon sources to promote co-metabolic elimination (especially in liquid phase systems), (iv) The use of immobilization strategies to increase enzyme stability during enzymatic degradation, (v) Design of approaches that couple physicochemical with biological treatments.

Acknowledgements

This work was supported by *Vicerrectoría de Investigación, Universidad de Costa Rica* (802-B6-137 and UCREA 802-B7-A09), and Ministry of Science Technology and Telecommunications (FI-197B-17). The authors thank Rebeca Tormo and Daniel Ulloa.

References

Accinelli, C., M. L. Saccà, I. Batisson, J. Fick, M. Mencarelli and R. Grabic. 2010. Removal of oseltamivir (Tamiflu) and other selected pharmaceuticals from wastewater using a granular bioplastic formulation entrapping propagules of *Phanerochaete chrysosporium*. *Chemosphere* 81 (3): 436–443. doi: https://doi.org/10.1016/j.chemosphere.2010.06.074.

Acosta, R., V. Fierro, A. Martinez de Yuso, D. Nabarlatz and A. Celzard. 2016. Tetracycline adsorption onto activated carbons produced by KOH activation of tyre pyrolysis char. *Chemosphere* 149: 168–176. doi: https://doi.org/10.1016/j.chemosphere.2016.01.093.

Agrawal, N., P. Verma, S. R. Singh and K. S. Shahi. 2017. Ligninolytic enzyme production by white rot fungi *Podoscypha elegans* strain FTG4. *International Journal of Current Microbiology and Applied Sciences* 6 (5): 2757–2764. doi: 10.20546/ijcmas.2017.605.309.

Álvarez-Torrellas, S., R. S. Ribeiro, H. T. Gomes, G. Ovejero, and J. García. 2016. Removal of antibiotic compounds by adsorption using glycerol-based carbon materials. *Chemical Engineering Journal* 296: 277–288. doi: https://doi.org/10.1016/j.cej.2016.03.112.

Andersson, D. I. and D. Hughes. 2010. Antibiotic resistance and its cost: is it possible to reverse resistance? *Nature Reviews Microbiology* 8: 260. doi: 10.1038/nrmicro2319.

Asgher, M., H. N. Bhatti, M. Ashraf and R. L. Legge. 2008. Recent developments in biodegradation of industrial pollutants by white rot fungi and their enzyme system. *Biodegradation* 19 (6): 771. doi: 10.1007/s10532-008-9185-3.

Asif, M. B., F. I. Hai, L. Singh, W. E. Price and L. D. Nghiem. 2017. Degradation of pharmaceuticals and personal care products by white-rot fungi—a critical review. *Current Pollution Reports* 3 (2): 88–103. doi: 10.1007/s40726-017-0049-5.

Aydin, S. 2016. Enhanced biodegradation of antibiotic combinations via the sequential treatment of the sludge resulting from pharmaceutical wastewater treatment using white-rot fungi *Trametes versicolor* and *Bjerkandera adusta*. *Applied Microbiology and Biotechnology* 100 (14): 6491–6499. doi: 10.1007/s00253-016-7473-0.

Baccar, R., P. Blánquez, J. Bouzid, M. Feki, H. Attiya and M. Sarrà. 2011. Decolorization of a tannery dye: From fungal screening to bioreactor application. *Biochemical Engineering Journal* 56 (3): 184–189. doi: https://doi.org/10.1016/j.bej.2011.06.006.

Badia-Fabregat, M., D. Lucas, M. Gros, S. Rodríguez-Mozaz, D. Barceló, G. Caminal and T. Vicent. 2015. Identification of some factors affecting pharmaceutical active compounds (PhACs) removal in real wastewater. Case study of fungal treatment of reverse osmosis concentrate. *Journal of Hazardous Materials* 283: 663–671. doi: https://doi.org/10.1016/j.jhazmat.2014.10.007.

Badia-Fabregat, M., M. Rosell, G. Caminal, T. Vicent and E. Marco-Urrea. 2014. Use of stable isotope probing to assess the fate of emerging contaminants degraded by white-rot fungus. *Chemosphere* 103: 336–342. doi: https://doi.org/10.1016/j.chemosphere.2013.12.029.

Badia-Fabregat, M., D. Lucas, M. A. Pereira, M. Alves, T. Pennanen, H. Fritze, S. Rodríguez-Mozaz, D. Barceló, T. Vicent and G. Caminal. 2016. Continuous fungal treatment of non-sterile veterinary hospital effluent: pharmaceuticals removal and microbial community assessment. *Applied Microbiology and Biotechnology* 100 (5): 2401–2415. doi: 10.1007/s00253-015-7105-0.

Badia-Fabregat, M., D. Lucas, T. Tuomivirta, H. Fritze, T. Pennanen, S. Rodríguez-Mozaz, D. Barceló, G. Caminal and T. Vicent. 2017. Study of the effect of the bacterial and fungal communities present in real wastewater effluents on the performance of fungal treatments. *Science of the Total Environment* 579: 366–377. doi: https://doi.org/10.1016/j.scitotenv.2016.11.088.

Barr, D. P. and S. D. Aust. 1994. Mechanisms white rot fungi use to degrade pollutants. *Environmental Science & Technology* 28 (2): 78A–87A. doi: 10.1021/es00051a724.

Becker, D., S. Varela Della Giustina, S. Rodriguez-Mozaz, R. Schoevaart, D. Barceló, M. de Cazes, M.-P. Belleville, J. Sanchez-Marcano, J. de Gunzburg, O. Couillerot, J. Völker, J. Oehlmann and M. Wagner. 2016. Removal of antibiotics in wastewater by enzymatic treatment with fungal laccase —Degradation of compounds does not always eliminate toxicity. *Bioresource Technology* 219: 500–509. doi: https://doi.org/10.1016/j.biortech.2016.08.004.

Ben, W., Z. Qiang, X. Pan and M. Chen. 2009. Removal of veterinary antibiotics from sequencing batch reactor (SBR) pretreated swine wastewater by Fenton's reagent. *Water Research* 43 (17): 4392–4402. doi: https://doi.org/10.1016/j.watres.2009.06.057.

Bending, G. D., M. Friloux and A. Walker. 2002. Degradation of contrasting pesticides by white rot fungi and its relationship with ligninolytic potential. *FEMS Microbiology Letters* 212 (1): 59–63. doi: doi:10.1111/j.1574-6968.2002.tb11245.x.

Bilal, M., M. Asgher, H. M. N. Iqbal, H. Hu and X. Zhang. 2017. Bio-based degradation of emerging endocrine-disrupting and dye-based pollutants using cross-linked enzyme aggregates. *Environmental Science and Pollution Research* 24 (8): 7035–7041. doi: 10.1007/s11356-017-8369-y.

Blánquez, P., M. Sarrà and M. T. Vicent. 2006. Study of the cellular retention time and the partial biomass renovation in a fungal decolourisation continuous process. *Water Research* 40 (8): 1650–1656. doi: https://doi.org/10.1016/j.watres.2006.02.010.

Blánquez, P. and B. Guieysse. 2008. Continuous biodegradation of 17β-estradiol and 17α-ethynylestradiol by *Trametes versicolor*. *Journal of Hazardous Materials* 150 (2): 459–462. doi: https://doi.org/10.1016/j.jhazmat.2007.09.085.

Borchert, M. and J. A. Libra. 2001. Decolorization of reactive dyes by the white rot fungus *Trametes versicolor* in sequencing batch reactors. *Biotechnology and Bioengineering* 75 (3): 313–321. doi: doi:10.1002/bit.10026.

Borràs, E., P. Blánquez, M. Sarrà, G. Caminal and T. Vicent. 2008. *Trametes versicolor* pellets production: Low-cost medium and scale-up. *Biochemical Engineering Journal* 42 (1): 61–66. doi: https://doi.org/10.1016/j.bej.2008.05.014.

Brown, K. D., J. Kulis, B. Thomson, T. H. Chapman and D. B. Mawhinney. 2006. Occurrence of antibiotics in hospital, residential, and dairy effluent, municipal wastewater, and the rio grande in new Mexico. *Science of the Total Environment* 366 (2): 772–783. doi: https://doi.org/10.1016/j.scitotenv.2005.10.007.

Bumpus, J. A., S. N. Kakar and R. D. Coleman. 1993. Fungal degradation of organophosphorous insecticides. *Applied Biochemistry and Biotechnology* 39 (1): 715–726. doi: 10.1007/bf02919030.

Cajthaml, T., Z. Křesinová, K. Svobodová and M. Möder. 2009. Biodegradation of endocrine-disrupting compounds and suppression of estrogenic activity by ligninolytic fungi. *Chemosphere* 75 (6): 745–750. doi: https://doi.org/10.1016/j.chemosphere.2009.01.034.

Calisto, V., C. I. A. Ferreira, J. A. B. P. Oliveira, M. Otero and V. I. Esteves. 2015. Adsorptive removal of pharmaceuticals from water by commercial and waste-based carbons. *Journal of Environmental Management* 152: 83–90. doi: https://doi.org/10.1016/j.jenvman.2015.01.019.

Cambronero-Heinrichs, J. C., M. Masís-Mora, J. Quirós-Fournier, V. Lizano-Fallas, I. Mata-Araya and C. E. Rodríguez-Rodríguez. 2018. Removal of herbicides in a biopurification system is not negatively affected by oxytetracycline or fungally pretreated oxytetracycline. *Chemosphere* 198: 198-203. doi:10.1016/j.chemosphere.2018.01.122.

Carlesi Jara, C., D. Fino, V. Specchia, G. Saracco and P. Spinelli. 2007. Electrochemical removal of antibiotics from wastewaters. *Applied Catalysis B: Environmental* 70 (1): 479–487. doi: https://doi.org/10.1016/j.apcatb.2005.11.035.

Casas, N., P. Blánquez, T. Vicent and M. Sarrà. 2013. Mathematical model for dye decoloration and laccase production by *Trametes versicolor* in fluidized bioreactor. *Biochemical Engineering Journal* 80: 45–52. doi: https://doi.org/10.1016/j.bej.2013.09.010.

Cerniglia, C. E. 1997. Fungal metabolism of polycyclic aromatic hydrocarbons: Past, present and future applications in bioremediation. *Journal of Industrial Microbiology & Biotechnology* 19: 324–33. doi: 10.1038/sj.jim.2900459.

Cerrone, F., P. Barghini, C. Pesciaroli and M. Fenice. 2011. Efficient removal of pollutants from olive washing wastewater in bubble-column bioreactor by *Trametes versicolor*. *Chemosphere* 84 (2): 254–259. doi: https://doi.org/10.1016/j.chemosphere.2011.03.066.

Cruz-Morató, C., L. Ferrando-Climent, S. Rodriguez-Mozaz, D. Barceló, E. Marco-Urrea, T. Vicent and M. Sarrà. 2013. Degradation of pharmaceuticals in non-sterile urban wastewater by *Trametes versicolor* in a fluidized bed bioreactor. *Water Research* 47 (14): 5200–5210. doi: https://doi.org/10.1016/j.watres.2013.06.007.

Cruz-Morató, C., D. Lucas, M. Llorca, S. Rodriguez-Mozaz, M. Gorga, M. Petrovic, D. Barceló, T. Vicent, M. Sarrà and E. Marco-Urrea. 2014. Hospital wastewater treatment by fungal bioreactor: Removal efficiency for pharmaceuticals and endocrine disruptor compounds. *Science of the Total Environment* 493: 365–376. doi: https://doi.org/10.1016/j.scitotenv.2014.05.117.

Čvančarová, M. M. Moeder, A. Filipová and T. Cajthaml. 2015. Biotransformation of fluoroquinolone antibiotics by ligninolytic fungi—Metabolites, enzymes and residual antibacterial activity. *Chemosphere* 136: 311–320. doi: https://doi.org/10.1016/j.chemosphere.2014.12.012.

Chakraborty, P. and J. Abraham. 2017. Comparative study on degradation of norfloxacin and ciprofloxacin by *Ganoderma lucidum* JAPC1. *Korean Journal of Chemical Engineering* 34 (4): 1122–1128. doi: 10.1007/s11814-016-0345-6.

Chanudhary, R., S. Hooda, S. Beniwal and A. Sindhu. 2016. Partial purification and immobilization of laccase isolated from medicinal mushroom (*Ganoderma lucidum*). *Indian Journal of Chemical Technology* 23 (4): 313–317.

Chhabra, M., S. Mishra and T. R. Sreekrishnan. 2009. Laccase/mediator assisted degradation of triarylmethane dyes in a continuous membrane reactor. *Journal of Biotechnology* 143 (1): 69–78. doi: https://doi.org/10.1016/j.jbiotec.2009.06.011.

Chigu, N. L. and H. Ichinose. 2017. Cellular uptake and metabolism of high molecular weight polycyclic aromatic hydrocarbons by the white-rot fungus *Phanerochaete chrysosporium*. *Resources and Environment* 7 (5): 138–144. doi: 10.5923/j.re.20170705.04.

D'Annibale, A., M. Ricci, V. Leonardi, D. Quaratino, E. Mincione and M. Petruccioli. 2005. Degradation of aromatic hydrocarbons by white-rot fungi in a historically contaminated soil. *Biotechnology and Bioengineering* 90 (6): 723–731. doi: 10.1002/bit.20461.

Dąbrowska, M., B. Muszyńska, M. Starek, P. Żmudzki and W. Opoka. 2018. Degradation pathway of cephalosporin antibiotics by *in vitro* cultures of *Lentinula edodes* and *Imleria badia*. *International Biodeterioration & Biodegradation* 127: 104–112. doi: https://doi.org/10.1016/j.ibiod.2017.11.014.

de Cazes, M. M. P. Belleville, E. Petit, M. Llorca, S. Rodríguez-Mozaz, J. de Gunzburg, D. Barceló and J. Sanchez-Marcano. 2014. Design and optimization of an enzymatic membrane reactor for tetracycline degradation. *Catalysis Today* 236: 146–152. doi: https://doi.org/10.1016/j.cattod.2014.02.051.

Ding, H., Y. Wu, B. Zou, Q. Lou, W. Zhang, J. Zhong, L. Lu and G. Dai. 2016. Simultaneous removal and degradation characteristics of sulfonamide, tetracycline, and quinolone antibiotics by laccase-mediated oxidation coupled with soil adsorption. *Journal of Hazardous Materials* 307: 350–358. doi: https://doi.org/10.1016/j.jhazmat.2015.12.062.

Dolar, D., A. Vuković, D. Asperger and K. Kosutic. 2012. Efficiency of RO/NF membranes at the removal of veterinary antibiotics. *Water Science and Technology* 65 (2): 317–323. doi: 10.2166/wst.2012.855.

El-Batal, A. I., N. M. ElKenawy, A. S. Yassin and M. A. Amin. 2015. Laccase production by pleurotus ostreatus and its application in synthesis of gold nanoparticles. *Biotechnology Reports* 5: 31–39. doi: https://doi.org/10.1016/j.btre.2014.11.001.

Elmolla, E. S. and M. Chaudhuri. 2011. Combined photo-fenton–SBR process for antibiotic wastewater treatment. *Journal of Hazardous Materials* 192 (3): 1418–1426. doi: https://doi.org/10.1016/j.jhazmat.2011.06.057.

Espinosa-Ortiz, E. J., E. R. Rene, K. Pakshirajan, E. D. van Hullebusch and P. N. L. Lens. 2016. Fungal pelleted reactors in wastewater treatment: Applications and perspectives. *Chemical Engineering Journal* 283: 553–571. doi: https://doi.org/10.1016/j.cej.2015.07.068.

Fabbrini, M., C. Galli and P. Gentili. 2002. Comparing the catalytic efficiency of some mediators of laccase. *Journal of Molecular Catalysis B: Enzymatic* 16 (5): 231–240. doi: https://doi.org/10.1016/ S1381-1177(01)00067-4.

Figueroa, R. A., A. Leonard and A. A. MacKay. 2004. Modeling tetracycline antibiotic sorption to clays. *Environmental Science & Technology* 38 (2): 476–483. doi: 10.1021/es0342087.

Ganzenko, O., D. Huguenot, E. D. van Hullebusch, G. Esposito and M. A. Oturan. 2014. Electrochemical advanced oxidation and biological processes for wastewater treatment: a review of the combined approaches. *Environmental Science and Pollution Research* 21 (14): 8493–8524. doi: 10.1007/ s11356-014-2770-6.

Gao, Y., Y. Li, L. Zhang, H. Huang, J. Hu, S. M. Shah and X. Su. 2012. Adsorption and removal of tetracycline antibiotics from aqueous solution by graphene oxide. *Journal of Colloid and Interface Science* 368 (1): 540–546. doi: https://doi.org/10.1016/j.jcis.2011.11.015.

García-Galán, M. J., C. E. Rodríguez-Rodríguez, T. Vicent, G. Caminal, M. S. Díaz-Cruz and D. Barceló. 2011. Biodegradation of sulfamethazine by *Trametes versicolor*: Removal from sewage sludge and identification of intermediate products by UPLC–QqTOF-MS. *Science of the Total Environment* 409 (24): 5505–5512. doi: https://doi.org/10.1016/j.scitotenv.2011.08.022.

Golan, N., B. Chefetz, J. Ben-Ari, J. Geva and Y. Hadar. 2011. *Transformation of the Recalcitrant Pharmaceutical Compound Carbamazepine by Pleurotus ostreatus: Role of Cytochrome P450 Monooxygenase and Manganese Peroxidase*. Vol. 45.

Grenni, P., V. Ancona and A. Barra Caracciolo. 2018. Ecological effects of antibiotics on natural ecosystems: A review. *Microchemical Journal* 136: 25–39. doi: https://doi.org/10.1016/j.microc.2017.02.006.

Gros, M., C. Cruz-Morato, E. Marco-Urrea, P. Longrée, H. Singer, M. Sarrà, J. Hollender, T. Vicent, S. Rodriguez-Mozaz and D. Barceló. 2014. Biodegradation of the X-ray contrast agent iopromide and the fluoroquinolone antibiotic ofloxacin by the white rot fungus *Trametes versicolor* in hospital wastewaters and identification of degradation products. *Water Research* 60: 228–241. doi: https:// doi.org/10.1016/j.watres.2014.04.042.

Guo, X.-l., Z.-w. Zhu and H.-l. Li. 2014. Biodegradation of sulfamethoxazole by *Phanerochaete chrysosporium*. *Journal of Molecular Liquids* 198: 169–172. doi: https://doi.org/10.1016/j. molliq.2014.06.017.

Hatakka, A. 1994. Lignin-modifying enzymes from selected white-rot fungi: production and role from in lignin degradation. *FEMS Microbiology Reviews* 13 (2): 125–135.

Higashiyama, K. K. Murakami, H. Tsujimura, N. Matsumoto and S. Fujikawa. 2000. Effects of dissolved oxygen on the morphology of an arachidonic acid production by *Mortierella alpina* 1S-4. *Biotechnology and Bioengineering* 63 (4): 442–448. doi: 10.1002/(sici)1097-0290(19990520)63:4<442::aid-bit7>3.0.co;2-9.

Hofrichter, M. 2002. Review: lignin conversion by manganese peroxidase (MnP). *Enzyme and Microbial Technology* 30 (4): 454–466. doi: https://doi.org/10.1016/S0141-0229(01)00528-2.

Hofrichter, M., R. Ullrich, M. J. Pecyna, C. Liers and T. Lundell. 2010. New and classic families of secreted fungal heme peroxidases. *Applied Microbiology and Biotechnology* 87 (3): 871–897. doi: 10.1007/s00253-010-2633-0.

Hölker, U., M. Höfer and J. Lenz. 2004. Biotechnological advantages of laboratory-scale solid-state fermentation with fungi. *Applied Microbiology and Biotechnology* 64 (2): 175–186. doi: 10.1007/ s00253-003-1504-3.

Hölker, U., and J. Lenz. 2005. Solid-state fermentation—are there any biotechnological advantages? *Current Opinion in Microbiology* 8 (3): 301–306. doi: https://doi.org/10.1016/j.mib.2005.04.006.

Hu, X., Q. Zhou and Y. Luo. 2010. *Occurrence and Source Analysis of Typical Veterinary Antibiotics in Manure, Soil, Vegetables and Groundwater from Organic Vegetable Bases, Northern China*. Vol. 158.

Ji, L., W. Chen, S. Zheng, Z. Xu and D. Zhu. 2009. Adsorption of sulfonamide antibiotics to multiwalled carbon nanotubes. *Langmuir* 25 (19): 11608–11613. doi: 10.1021/la9015838.

Ji, L., Y. Shao, Z. Xu, S. Zheng and D. Zhu. 2010. Adsorption of monoaromatic compounds and pharmaceutical antibiotics on carbon nanotubes activated by KOH etching. *Environmental Science & Technology* 44 (16): 6429–6436. doi: 10.1021/es1014828.

Karthikeyan, K. G. and M. T. Meyer. 2006. Occurrence of antibiotics in wastewater treatment facilities in Wisconsin, USA. *Science of the Total Environment* 361 (1): 196–207. doi: https://doi.org/10.1016/j. scitotenv.2005.06.030.

Kay, P., P. A. Blackwell and A. B. A. Boxall. 2004. Fate of veterinary antibiotics in a macroporous tile drained clay soil. *Environmental Toxicology and Chemistry* 23 (5): 1136–1144. doi: doi:10.1897/03-374.

Kelly, S., L. H. Grimm, C. Bendig, D. C. Hempel and R. Krull. 2006. Effects of fluid dynamic induced shear stress on fungal growth and morphology. *Process Biochemistry* 41 (10): 2113–2117. doi: https://doi.org/10.1016/j.procbio.2006.06.007.

Kim, H.-Y. and H.-G. Song. 2000. Comparison of 2,4,6-trinitrotoluene degradation by seven strains of white rot fungi. *Current Microbiology* 41 (5): 317–320. doi: 10.1007/s002840010142.

Koyuncu, I., O. A. Arikan, M. R. Wiesner and C. Rice. 2008. Removal of hormones and antibiotics by nanofiltration membranes. *Journal of Membrane Science* 309 (1): 94–101. doi: https://doi.org/10.1016/j.memsci.2007.10.010.

Kum, H., M. K. Kim and H. T. Choi. 2009. Degradation of endocrine disrupting chemicals by genetic transformants in *Irpex lacteus* with an inducible laccase gene of *Phlebia tremellosa*. *Biodegradation* 20 (5): 673. doi: 10.1007/s10532-009-9254-2.

Kum, H., S. Lee, S. Ryu and H. T. Choi. 2011. Degradation of endocrine disrupting chemicals by genetic transformants with two lignin degrading enzymes in *Phlebia tremellosa*. *The Journal of Microbiology* 49 (5): 824–827. doi: 10.1007/s12275-011-1230-y.

Kumar, R. R., B. J. Park, R. H. Jeong, T. J. Lee and Y. J. Cho. 2013. Biodegradation of β-lactam antibiotic 'ampicillin' by white rot fungi from aqueous solutions. *Journal of Pure and Applied Microbiology* 7 (4): 3163–3169.

Kümmerer, K. 2009a. Antibiotics in the aquatic environment—A review—Part I. *Chemosphere* 75 (4): 417–434. doi: https://doi.org/10.1016/j.chemosphere.2008.11.086.

Kümmerer, K. 2009b. Antibiotics in the aquatic environment—A review—Part II. *Chemosphere* 75 (4): 435–441. doi: https://doi.org/10.1016/j.chemosphere.2008.12.006.

Lee, S.-M., B.-W. Koo, J.-W. Choi, D. Choi, B.-S. An, E.-B. Jeung and I.-G. Choi. 2005. *Degradation of Bisphenol A by White Rot Fungi, Stereum hirsutum and Heterobasidium insulare, and Reduction of Its Estrogenic Activity*. Vol. 28.

Li, B. and T. Zhang. 2010. Biodegradation and adsorption of antibiotics in the activated sludge process. *Environmental Science & Technology* 44 (9): 3468–3473. doi: 10.1021/es903490h.

Libra, J. A., M. Borchert and S. Banit. 2003. Competition strategies for the decolorization of a textile-reactive dye with the white-rot fungi *Trametes versicolor* under non-sterile conditions. *Biotechnology and Bioengineering* 82 (6): 736–744. doi: doi:10.1002/bit.10623.

Liu, Y., H. Chang, Z. Li, C. Zhang, Y. Feng and D. Cheng. 2016. Gentamicin removal in submerged fermentation using the novel fungal strain Aspergillus terreus FZC3. Scientific reports 6: 35856. doi:10.1038/srep35856.

Lucas, D., M. Badia-Fabregat, T. Vicent, G. Caminal, S. Rodríguez-Mozaz, J. L. Balcázar and D. Barceló. 2016. Fungal treatment for the removal of antibiotics and antibiotic resistance genes in veterinary hospital wastewater. *Chemosphere* 152: 301–308. doi: https://doi.org/10.1016/j.chemosphere.2016.02.113.

Lucas, D., F. Castellet-Rovira, M. Villagrasa, M. Badia-Fabregat, D. Barceló, T. Vicent, G. Caminal, M. Sarrà and S. Rodríguez-Mozaz. 2018. The role of sorption processes in the removal of pharmaceuticals by fungal treatment of wastewater. *Science of the Total Environment* 610–611: 1147–1153. doi: https://doi.org/10.1016/j.scitotenv.2017.08.118.

Lundell, T., K., M. Mäkelä, R. and K. Hildén. 2010. Lignin-modifying enzymes in filamentous basidiomycetes—ecological, functional and phylogenetic review. *Journal of Basic Microbiology* 50 (1): 5–20. doi: 10.1002/jobm.200900338.

Llorens-Blanch, G., M. Badia-Fabregat, D. Lucas, S. Rodriguez-Mozaz, D. Barceló, T. Pennanen, G. Caminal and P. Blánquez. 2015. Degradation of pharmaceuticals from membrane biological reactor sludge with *Trametes versicolor*. *Environmental Science: Processes & Impacts* 17 (2): 429–440. doi: 10.1039/c4em00579a.

Madrigal-Zúñiga, K., K. Ruiz-Hidalgo, J. S. Chin-Pampillo, M. Masís-Mora, V. Castro-Gutiérrez and C. E. Rodríguez-Rodríguez. 2016. Fungal bioaugmentation of two rice husk-based biomixtures for the removal of carbofuran in on-farm biopurification systems. *Biology and Fertility of Soils* 52 (2): 243–250. doi: 10.1007/s00374-015-1071-7.

Marco-Urrea, E., M. Pérez-Trujillo, T. Vicent and G. Caminal. 2009. Ability of white-rot fungi to remove selected pharmaceuticals and identification of degradation products of ibuprofen by *Trametes versicolor*. *Chemosphere* 74 (6): 765–772. doi: https://doi.org/10.1016/j.chemosphere.2008.10.040.

Margot, J., P.-J. Copin, U. von Gunten, D. A. Barry and C. Holliger. 2015. Sulfamethoxazole and isoproturon degradation and detoxification by a laccase-mediator system: Influence of treatment conditions and mechanistic aspects. *Biochemical Engineering Journal* 103: 47–59. doi: https://doi.org/10.1016/j.bej.2015.06.008.

Martens, R., H. G. Wetzstein, F. Zadrazil, M. Capelari, P. Hoffmann and N. Schmeer. 1996. Degradation of the fluoroquinolone enrofloxacin by wood-rotting fungi. *Applied and Environmental Microbiology* 62 (11): 4206–4209.

Martínez-Carballo, E., C. González-Barreiro, S. Scharf and O. Gans. 2007. Environmental monitoring study of selected veterinary antibiotics in animal manure and soils in Austria. *Environmental Pollution* 148 (2): 570–579. doi: https://doi.org/10.1016/j.envpol.2006.11.035.

Martínez, A. T. 2002. Molecular biology and structure-function of lignin-degrading heme peroxidases. *Enzyme and Microbial Technology* 30 (4): 425–444. doi: https://doi.org/10.1016/S0141-0229(01)00521-X.

Martorana, A., L. Sorace, H. Boer, R. Vazquez-Duhalt, R. Basosi and M. C. Baratto. 2013. A spectroscopic characterization of a phenolic natural mediator in the laccase biocatalytic reaction. *Journal of Molecular Catalysis B: Enzymatic* 97: 203–208. doi: https://doi.org/10.1016/j.molcatb.2013.08.013.

Mielgo, I., M. T. Moreira, G. Feijoo and J. M. Lema. 2001. A packed-bed fungal bioreactor for the continuous decolourisation of azo-dyes (orange II). *Journal of Biotechnology* 89 (2): 99–106. doi: https://doi.org/10.1016/S0168-1656(01)00319-4.

Migliore, L., M. Fiori, A. Spadoni and E. Galli. 2012. Biodegradation of oxytetracycline by *Pleurotus ostreatus* mycelium: a mycoremediation technique. *Journal of Hazardous Materials* 215–216:227–232. doi: https://doi.org/10.1016/j.jhazmat.2012.02.056.

Mir-Tutusaus, J. A., M. Masís-Mora, C. Corcellas, E. Eljarrat, D. Barceló, M. Sarrà, G. Caminal, T. Vicent and C. E. Rodríguez-Rodríguez. 2014. Degradation of selected agrochemicals by the white rot fungus *Trametes versicolor*. *Science of the Total Environment* 500–501: 235–242. doi: https://doi.org/10.1016/j.scitotenv.2014.08.116.

Mir-Tutusaus, J. A., E. Parladé, M. Llorca, M. Villagrasa, D. Barceló, S. Rodriguez-Mozaz, M. Martinez-Alonso, N. Gaju, G. Caminal and M. Sarrà. 2017. Pharmaceuticals removal and microbial community assessment in a continuous fungal treatment of non-sterile real hospital wastewater after a coagulation-flocculation pretreatment. *Water Research* 116: 65–75. doi: https://doi.org/10.1016/j.watres.2017.03.005.

Mir-Tutusaus, J. A., R. Baccar, G. Caminal and M. Sarrà. 2018. Can white-rot fungi be a real wastewater treatment alternative for organic micropollutants removal? A review. *Water Research* 138: 137–151. doi: https://doi.org/10.1016/j.watres.2018.02.056.

Mitchell, D. A., N. Krieger, D. M. Stuart and A. Pandey. 2000. New developments in solid-state fermentation: II. Rational approaches to the design, operation and scale-up of bioreactors. *Process Biochemistry* 35 (10): 1211–1225. doi: https://doi.org/10.1016/S0032-9592(00)00157-6.

Moreira, M. T., G. Feijoo and J. M. Lema. 2003. Fungal bioreactors: Applications to white-rot fungi. *Reviews in Environmental Science and Biotechnology* 2 (2): 247–259. doi: 10.1023/B:RESB.0000040463.80808.dc.

Morozova, O. V., G. P. Shumakovich, S. V. Shleev and Y. I. Yaropolov. 2007. Laccase-mediator systems and their applications: A review. *Applied Biochemistry and Microbiology* 43 (5): 523–535. doi: 10.1134/s0003683807050055.

Naghdi, M., M. Taheran, S. K. Brar, A. Kermanshahi-pour, M. Verma and R. Y. Surampalli. 2018. Removal of pharmaceutical compounds in water and wastewater using fungal oxido reductase enzymes. *Environmental Pollution* 234: 190–213. doi: https://doi.org/10.1016/j.envpol.2017.11.060.

Nguyen, L. N., F. I. Hai, W. E. Price, F. D. L. Leusch, F. Roddick, H. H. Ngo, W. Guo, S. F. Magram and L. D. Nghiem. 2014. The effects of mediator and granular activated carbon addition on degradation of trace organic contaminants by an enzymatic membrane reactor. *Bioresource Technology* 167: 169–177. doi: https://doi.org/10.1016/j.biortech.2014.05.125.

Nguyen, T. K. X., P. Thayanukul, O. Pinyakong and O. Suttinun. 2017. Tiamulin removal by wood-rot fungi isolated from swine farms and role of ligninolytic enzymes. *International Biodeterioration & Biodegradation* 116: 147–154. doi: https://doi.org/10.1016/j.ibiod.2016.10.010.

Norvill, Z. N., A. Toledo-Cervantes, S. Blanco, A. Shilton, B. Guieysse and R. Muñoz. 2017. Photodegradation and sorption govern tetracycline removal during wastewater treatment in algal ponds. *Bioresource Technology* 232: 35–43. doi: https://doi.org/10.1016/j.biortech.2017.02.011.

Nyanhongo, G. S., G. Gübitz, P. Sukyai, C. Leitner, D. Haltrich and R. Ludwig. 2007. Oxido reductases from *Trametes* spp. in biotechnology: A wealth of catalytic activity. *Food Technology and Biotechnology* 45 (3): 250–268.

Ory, J., G. Bricheux, A. Togola, J. L. Bonnet, F. Donnadieu-Bernard, L. Nakusi, C. Forestier and O. Traore. 2016. Ciprofloxacin residue and antibiotic-resistant biofilm bacteria in hospital effluent. *Environmental Pollution* 214: 635–645. doi: https://doi.org/10.1016/j.envpol.2016.04.033.

Palmieri, G., P. Giardina, C. Bianco, B. Fontanella and G. Sannia. 2000. Copper induction of laccase isoenzymes in the ligninolytic fungus *Pleurotus ostreatus*. *Applied and Environmental Microbiology* 66 (3): 920–924.

Parshikov, I. A., J. P. Freeman, J. O. Lay, R. D. Beger, A. J. Williams and J. B. Sutherland. 1999. Regioselective transformation of ciprofloxacin to N-acetylciprofloxacin by the fungus *Mucor ramannianus*. *FEMS Microbiology Letters* 177 (1): 131–135. doi: 10.1111/j.1574-6968.1999.tb13723.x.

Parshikov, I. A., J. P. Freeman, J. O. Lay, R. D. Beger, A. J. Williams and J. B. Sutherland. 2000. Microbiological transformation of enrofloxacin by the fungus *Mucor ramannianus*. *Applied and Environmental Microbiology* 66 (6): 2664–2667.

Parshikov, I. A., J. P. Freeman, J. J. O. Lay, J. D. Moody, A. J. Williams, R. D. Beger and J. B. Sutherland. 2001a. Metabolism of the veterinary fluoroquinolone sarafloxacin by the fungus *Mucor ramannianus*. *Journal of Industrial Microbiology and Biotechnology* 26 (3): 140–144. doi: 10.1038/sj.jim.7000077.

Parshikov, I., T. Heinze, J. Moody, J. Freeman, A. Williams and J. Sutherland. 2001b. The fungus *Pestalotiopsis guepini* as a model for biotransformation of ciprofloxacin and norfloxacin. *Applied Microbiology and Biotechnology* 56 (3): 474–477. doi: 10.1007/s002530100672.

Patrolecco, L., S. Capri and N. Ademollo. 2015. Occurrence of selected pharmaceuticals in the principal sewage treatment plants in Rome (Italy) and in the receiving surface waters. *Environmental Science and Pollution Research* 22 (8): 5864–5876. doi: 10.1007/s11356-014-3765-z.

Pazouki, M. and T. Panda. 2000. Understanding the morphology of fungi. *Bioprocess Engineering* 22 (2): 127–143. doi: 10.1007/s004490050022.

Pointing, S. 2001. Feasibility of bioremediation by white-rot fungi. *Applied Microbiology and Biotechnology* 57 (1): 20–33. doi: 10.1007/s002530100745.

Polesel, F., H. R. Andersen, S. Trapp and B. G. Plósz. 2016. Removal of antibiotics in biological wastewater treatment systems—A critical assessment using the activated sludge modeling framework for xenobiotics (ASM-X). *Environmental Science & Technology* 50 (19): 10316–10334. doi: 10.1021/acs.est.6b01899.

Prados-Joya, G., M. Sánchez-Polo, J. Rivera-Utrilla and M. Ferro-garcía. 2011. Photodegradation of the antibiotics nitroimidazoles in aqueous solution by ultraviolet radiation. *Water Research* 45 (1): 393–403. doi: https://doi.org/10.1016/j.watres.2010.08.015.

Prieto, A., M. Möder, R. Rodil, L. Adrian and E. Marco-Urrea. 2011. Degradation of the antibiotics norfloxacin and ciprofloxacin by a white-rot fungus and identification of degradation products. *Bioresource Technology* 102 (23): 10987–10995. doi: https://doi.org/10.1016/j.biortech.2011.08.055.

Pruden, A., R. Pei, H. Storteboom and K. H. Carlson. 2006. Antibiotic resistance genes as emerging contaminants: Studies in Northern Colorado. *Environmental Science & Technology* 40 (23): 7445–7450. doi: 10.1021/es060413l.

Quintero, J. C., T. A. Lú-Chau, M. T. Moreira, G. Feijoo and J. M. Lema. 2007. Bioremediation of HCH present in soil by the white-rot fungus *Bjerkandera adusta* in a slurry batch bioreactor. *International Biodeterioration & Biodegradation* 60 (4): 319–326. doi: https://doi.org/10.1016/j.ibiod.2007.05.005.

Rioja, R., M. T. García, M. Peña and G. González. 2008. Biological decolourisation of wastewater from molasses fermentation by *Trametes versicolor* in an airlift reactor. *Journal of Environmental Science and Health, Part A* 43 (7): 772–778. doi: 10.1080/10934520801960102.

Rodarte-Morales, A. I., G. Feijoo, M. T. Moreira and J. M. Lema. 2011. Degradation of selected pharmaceutical and personal care products (PPCPs) by white-rot fungi. *World Journal of Microbiology and Biotechnology* 27 (8): 1839–1846. doi: 10.1007/s11274-010-0642-x.

Rodríguez-Rodríguez, C. E., A. Jelić, M. Llorca, M. Farré, G. Caminal, M. Petrović, D. Barceló and T. Vicent. 2011. Solid-phase treatment with the fungus *Trametes versicolor* substantially reduces pharmaceutical concentrations and toxicity from sewage sludge. *Bioresource Technology* 102 (10): 5602–5608. doi: https://doi.org/10.1016/j.biortech.2011.02.029.

Rodríguez-Rodríguez, C. E., E. Barón, P. Gago-Ferrero, A. Jelić, M. Llorca, M. Farré, M. S. Díaz-Cruz, E. Eljarrat, M. Petrović, G. Caminal, D. Barceló and T. Vicent. 2012a. Removal of pharmaceuticals, polybrominated flame retardants and UV-filters from sludge by the fungus *Trametes versicolor* in bioslurry reactor. *Journal of Hazardous Materials* 233–234: 235–243. doi: https://doi.org/10.1016/j.jhazmat.2012.07.024.

Rodríguez-Rodríguez, C. E., J. M. García-Galán, P. Blánquez, M. S. Díaz-Cruz, D. Barceló, G. Caminal and T. Vicent. 2012b. Continuous degradation of a mixture of sulfonamides by *Trametes versicolor* and identification of metabolites from sulfapyridine and sulfathiazole. *Journal of Hazardous Materials* 213–214: 347–354. doi: https://doi.org/10.1016/j.jhazmat.2012.02.008.

Rodríguez-Rodríguez, C. E., A. Jelić, M. A. Pereira, D. Z. Sousa, M. Petrović, M. M. Alves, D. Barceló, G. Caminal and T. Vicent. 2012c. Bioaugmentation of sewage sludge with *Trametes versicolor* in solid-phase biopiles produces degradation of pharmaceuticals and affects microbial communities. *Environmental Science & Technology* 46 (21): 12012–12020. doi: 10.1021/es301788n.

Rodríguez-Rodríguez, C. E., V. Castro-Gutiérrez, J. S. Chin-Pampillo and K. Ruiz-Hidalgo. 2013. On-farm biopurification systems: role of white rot fungi in depuration of pesticide-containing wastewaters. *FEMS Microbiology Letters* 345 (1): 1–12. doi: 10.1111/1574-6968.12161.

Rodríguez-Rodríguez, C. E., D. Lucas, E. Barón, P. Gago-Ferrero, D. Molins-Delgado, S. Rodríguez-Mozaz, E. Eljarrat, M. Silvia Díaz-Cruz, D. Barceló, G. Caminal and T. Vicent. 2014. Re-inoculation strategies enhance the degradation of emerging pollutants in fungal bioaugmentation of sewage sludge. *Bioresource Technology* 168: 180–189. doi: https://doi.org/10.1016/j.biortech.2014.01.124.

Rodríguez-Rodríguez, C. E., K. Madrigal-León, M. Masís-Mora, M. Pérez-Villanueva and J. S. Chin-Pampillo. 2017. Removal of carbamates and detoxification potential in a biomixture: Fungal bioaugmentation versus traditional use. *Ecotoxicology and Environmental Safety* 135: 252–258. doi: https://doi.org/10.1016/j.ecoenv.2016.10.011.

Ruiz-Aguilar, G. M. L., J. M. Fernández-Sánchez, R. Rodríguez-Vázquez and H. Poggi-Varaldo. 2002. Degradation by white-rot fungi of high concentrations of PCB extracted from a contaminated soil. *Advances in Environmental Research* 6 (4): 559–568. doi: https://doi.org/10.1016/S1093-0191(01)00102-2.

Sanghi, R., A. Dixit and S. Guha. 2006. Sequential batch culture studies for the decolourization of reactive dye by coriolus versicolor. *Bioresource Technology* 97 (3): 396–400. doi: 10.1016/j.biortech.2005.03.010.

Santos, L. H. M. L. M. M. Gros, S. Rodriguez-Mozaz, C. Delerue-Matos, A. Pena, D. Barceló and M. C. B. S. M. Montenegro. 2013. Contribution of hospital effluents to the load of pharmaceuticals in urban wastewaters: Identification of ecologically relevant pharmaceuticals. *Science of the Total Environment* 461–462: 302–316. doi: https://doi.org/10.1016/j.scitotenv.2013.04.077.

Sarmah, A. K., M. T. Meyer and A. B. A. Boxall. 2006. A global perspective on the use, sales, exposure pathways, occurrence, fate and effects of veterinary antibiotics (VAs) in the environment. *Chemosphere* 65 (5): 725–759. doi: https://doi.org/10.1016/j.chemosphere.2006.03.026.

Shah, A. D., C.-H. Huang and J.-H. Kim. 2012. Mechanisms of antibiotic removal by nanofiltration membranes: Model development and application. *Journal of Membrane Science* 389: 234–244. doi: https://doi.org/10.1016/j.memsci.2011.10.034.

Shi, L., F. Ma, Y. Han, X. Zhang and H. Yu. 2014. Removal of sulfonamide antibiotics by oriented immobilized laccase on Fe_3O_4 nanoparticles with natural mediators. *Journal of Hazardous Materials* 279: 203–211. doi: https://doi.org/10.1016/j.jhazmat.2014.06.070.

Singh, S. K., R. Khajuria and L. Kaur. 2017. Biodegradation of ciprofloxacin by white rot fungus *Pleurotus ostreatus*. *3 Biotech* 7 (1): 69. doi: 10.1007/s13205-017-0684-y.

Suda, T., T. Hata, S. Kawai, H. Okamura and T. Nishida. 2012. Treatment of tetracycline antibiotics by laccase in the presence of 1-hydroxybenzotriazole. *Bioresource Technology* 103 (1): 498–501. doi: https://doi.org/10.1016/j.biortech.2011.10.041.

Suhas, P. J. M. Carrott and M. M. L. Ribeiro Carrott. 2007. Lignin—from natural adsorbent to activated carbon: A review. *Bioresource Technology* 98 (12): 2301–2312. doi: https://doi.org/10.1016/j.biortech.2006.08.008.

Suzuki, K., H. Hirai, H. Murata and T. Nishida. 2003. Removal of estrogenic activities of 17β-estradiol and ethinylestradiol by ligninolytic enzymes from white rot fungi. *Water Research* 37 (8): 1972–1975. doi: https://doi.org/10.1016/S0043-1354(02)00533-X.

Tadkaew, N., F. I. Hai, J. A. McDonald, S. J. Khan and L. D. Nghiem. 2011. Removal of trace organics by MBR treatment: The role of molecular properties. *Water Research* 45 (8): 2439–2451. doi: https://doi.org/10.1016/j.watres.2011.01.023.

Torres-Duarte, C., R. Roman, R. Tinoco and R. Vazquez-Duhalt. 2009. Halogenated pesticide transformation by a laccase–mediator system. *Chemosphere* 77 (5): 687–692. doi: https://doi.org/10.1016/j.chemosphere.2009.07.039.

Tran, N., H. Chen, M. Reinhard, F. Mao and K. Gin. 2016. Occurrence and removal of multiple classes of antibiotics and antimicrobial agents in biological wastewater treatment processes. *Water Research* 104: 461–472. doi: 10.1016/j.watres.2016.08.040.

Urlacher, V. B. and S. Eiben. 2006. Cytochrome P450 monooxygenases: perspectives for synthetic application. *Trends in Biotechnology* 24 (7): 324–330. doi: 10.1016/j.tibtech.2006.05.002.

Urlacher, V. B. and M. Girhard. 2012. Cytochrome P450 monooxygenases: an update on perspectives for synthetic application. *Trends in Biotechnology* 30 (1): 26–36. doi: https://doi.org/10.1016/j.tibtech.2011.06.012.

Vasiliadou, I. A., R. Sánchez-Vázquez, R. Molina, F. Martínez, J. A. Melero, L. F. Bautista, J. Iglesias and G. Morales. 2016. Biological removal of pharmaceutical compounds using white-rot fungi with concomitant FAME production of the residual biomass. *Journal of Environmental Management* 180: 228–237. doi: https://doi.org/10.1016/j.jenvman.2016.05.035.

Verlicchi, P., M. Al Aukidy and E. Zambello. 2015. What have we learned from worldwide experiences on the management and treatment of hospital effluent— An overview and a discussion on perspectives. *Science of the Total Environment* 514: 467–491. doi: https://doi.org/10.1016/j.scitotenv.2015.02.020.

Watkinson, A. J., E. J. Murby, D. W. Kolpin and S. D. Costanzo. 2009. The occurrence of antibiotics in an urban watershed: From wastewater to drinking water. *Science of the Total Environment* 407 (8): 2711–2723. doi: https://doi.org/10.1016/j.scitotenv.2008.11.059.

Wen, X., Y. Jia and J. Li. 2009. Degradation of tetracycline and oxytetracycline by crude lignin peroxidase prepared from *Phanerochaete chrysosporium*—A white rot fungus. *Chemosphere* 75 (8): 1003–1007. doi: https://doi.org/10.1016/j.chemosphere.2009.01.052.

Wen, X., Y. Jia and J. Li. 2010. Enzymatic degradation of tetracycline and oxytetracycline by crude manganese peroxidase prepared from *Phanerochaete chrysosporium*. *Journal of Hazardous Materials* 177 (1): 924–928. doi: https://doi.org/10.1016/j.jhazmat.2010.01.005.

Weng, S.-S., K.-L. Ku and H.-T. Lai. 2012. The implication of mediators for enhancement of laccase oxidation of sulfonamide antibiotics. *Bioresource Technology* 113: 259–264. doi: https://doi.org/10.1016/j.biortech.2011.12.111.

Weng, S.-S. S.-M. Liu and H.-T. Lai. 2013. Application parameters of laccase–mediator systems for treatment of sulfonamide antibiotics. *Bioresource Technology* 141: 152–159. doi: https://doi.org/10.1016/j.biortech.2013.02.093.

Wesenberg, D., I. Kyriakides and S. N. Agathos. 2003. White-rot fungi and their enzymes for the treatment of industrial dye effluents. *Biotechnology Advances* 22 (1): 161–187. doi: https://doi.org/10.1016/j.biotechadv.2003.08.011.

Wetzstein, H.-G., M. Stadler, H.-V. Tichy, A. Dalhoff and W. Karl. 1999. Degradation of ciprofloxacin by basidiomycetes and identification of metabolites generated by the brown rot fungus *Gloeophyllum striatum*. *Applied and Environmental Microbiology* 65 (4): 1556–1563.

Wetzstein, H. G., N. Schmeer and W. Karl. 1997. Degradation of the fluoroquinolone enrofloxacin by the brown rot fungus *Gloeophyllum striatum*: identification of metabolites. *Applied and Environmental Microbiology* 63 (11): 4272–4281.

Williams, A. J., J. Deck, J. P. Freeman, M. Paul Chiarelli, M. D. Adjei, T. M. Heinze and J. B. Sutherland. 2007. Biotransformation of flumequine by the fungus *Cunninghamella elegans*. *Chemosphere* 67 (2): 240–243. doi: https://doi.org/10.1016/j.chemosphere.2006.10.016.

Woo, C., B. Daniels, R. Stirling and P. Morris. 2010. Tebuconazole and propiconazole tolerance and possible degradation by basidiomycetes: A wood-based bioassay. *International Biodeterioration & Biodegradation* 64 (5): 403–408. doi: https://doi.org/10.1016/j.ibiod.2010.01.009.

Yamaguchi, T., M. Okihashi, K. Harada, Y. Konishi, K. Uchida, M. Hong Ngoc Do, H. Dang Thien Bui, T. Duc Nguyen, P. Do Nguyen, V. Van Chau, K. Thi Van Dao, H. Thi Ngoc Nguyen, K. Kajimura,

Y. Kumeda, C. Trong Bui, M. Quang Vien, N. Le, K. Hirata and Y. Yamamoto. 2015. *Antibiotic Residue Monitoring Results for Pork, Chicken, and Beef Samples in Vietnam in 2012–2013.* Vol. 63.

Yang, J., Y. Lin, X. Yang, T. B. Ng, X. Ye and J. Lin. 2017. Degradation of tetracycline by immobilized laccase and the proposed transformation pathway. *Journal of Hazardous Materials* 322: 525–531. doi: https://doi.org/10.1016/j.jhazmat.2016.10.019.

Yang, S., F. I. Hai, L. D. Nghiem, L. N. Nguyen, F. Roddick, and W. E. Price. 2013a. Removal of bisphenol A and diclofenac by a novel fungal membrane bioreactor operated under non-sterile conditions. *International Biodeterioration & Biodegradation* 85: 483–490. doi: https://doi.org/10.1016/j.ibiod.2013.03.012.

Yang, S., F. I. Hai, L. D. Nghiem, W. E. Price, F. Roddick, M. T. Moreira and S. F. Magram. 2013b. Understanding the factors controlling the removal of trace organic contaminants by white-rot fungi and their lignin modifying enzymes: A critical review. *Bioresource Technology* 141: 97–108. doi: https://doi.org/10.1016/j.biortech.2013.01.173.

Yesilada, O., D. Asma and S. Cing. 2003. Decolorization of textile dyes by fungal pellets. *Process Biochemistry* 38 (6): 933–938. doi: https://doi.org/10.1016/S0032-9592(02)00197-8.

Yesilada, O., S. C. Yildirim, E. Birhanli, E. Apohan, D. Asma and F. Kuru. 2009. The evaluation of pre-grown mycelial pellets in decolorization of textile dyes during repeated batch process. *World Journal of Microbiology and Biotechnology* 26 (1): 33. doi: 10.1007/s11274-009-0138-8.

Zhang, L., Y. Shen, F. Hui and Q. Niu. 2015. Degradation of residual lincomycin in fermentation dregs by yeast strain S9 identified as *Galactomyces geotrichum*. *Annals of Microbiology* 65 (3): 1333–1340. doi: 10.1007/s13213-014-0971-3.

Zhang, Q., G. Lambert, D. Liao, H. Kim, K. Robin, C.-k. Tung, N. Pourmand and R. H. Austin. 2011. Acceleration of emergence of bacterial antibiotic resistance in connected microenvironments. *Science* 333 (6050): 1764.

Zhou, P., C. Su, B. Li and Y. Qian. 2006. Treatment of high-strength pharmaceutical wastewater and removal of antibiotics in anaerobic and aerobic biological treatment processes. *Journal of Environmental Engineering* 132 (1): 129–136. doi: 10.1061/(asce)0733-9372(2006)132:1(129).

CHAPTER-8

Fungi and Remediation of Radionuclide Pollution

John Dighton

1. Introduction

The behavior of radionuclides in the terrestrial ecosystem was initially evaluated during the testing of atomic weapons in the 1970s amid concerns around nuclear facilities, such as the Savannah River Nuclear Weapons Complex (Williams et al. 1999). More recently, accidents have occurred in the nuclear power industry, involving Three Mile Island (1979), Chernobyl (1986), both caused by human error, and Fukushima (2011), caused by an earthquake and tsunami. These accidents have caused great concern about human safety. They have stimulated a need to investigate, not only the effects of the ensuing radionuclide fallout on the health of humans, but also a desire to find remediation techniques that can minimize the effect of these chemicals. The ecological aspects of radionuclide release are discussed by Coughtree et al. (1983) in relation to recommendations for remediation made by The International Commission on Radiological Protection. In this book, studies of the element transfers within terrestrial ecosystems highlight the importance of organic soil horizons and the microbial communities within as potential accumulators of both nutrient elements and radionuclides (Heal and Horrill 1983). In a model of radiocesium migration in forest ecosystems, fungal activity was placed in a pivotal point of importance in the biotic regulators of radionuclide movement in soils (Avila and Moberg 1999).

Radionuclides cannot be destroyed by biological or any other activity and can only be reduced in activity by natural decay of the isotope. However, biological entities in ecosystems can change the availability of radionuclides to other organisms by sequestration within themselves or alter retention on soil particles. To this end, fungi are probably the most important organisms in terrestrial ecosystems as these

Rutgers Pinelands Field Station, PO Box 206, 501 Four Mile Road, New Lisbon, NJ 08064, USA.
Email: dighton@camden.rutgers.edu

'microorganisms' have a large and persistent biomass for adsorption and absorption, along with the ability to translocate materials over distances of meters. Additionally, fungal association with plants, as mycorrhizae, make their interaction with phytoextraction of pollutants more direct than the effects of bacteria.

Both the source and nature of the radioisotopes in the environment are likely to affect their impact on the organisms and the ability of fungi to access them and effect any transformations. The Chernobyl accident released ^{134}Cs, ^{137}Cs, ^{131}I, ^{90}Sr, ^{132}Te, and ^{132}I into the atmosphere (Nesterenko and Yablokov 2009), much of which was deposited into local ecosystems or dispersed in the atmosphere to be deposited at more remote sites throughout Europe and Russia. The Fukushima Daiichi accident released ^{129}Te, ^{131}I, ^{134}Cs, ^{136}Cs, and ^{137}Cs, and there is some evidence to suggest deposition of ^{241}Pu into soil (Zheng et al. 2012). In addition, trace amounts of ^{95}Nb, ^{110}Ag, and ^{140}La were found in soil, though not ^{95}Zr, $^{103, \ 106}$Ru, or ^{140}Ba, which were found in Chernobyl (Tagami et al. 2011). From the Fukushima nuclear power plant some 10^{12} Bq h^{-1} of ^{137}Cs and 10^{14} Bq h^{-1} of ^{131}I was released in the first month after the accident (Chino et al. 2011).

A 30 km radius (2,600 km^2) area of land was demarcated around Chernobyl, as an area of high contamination and designated as an exclusion zone. Much of this area was agricultural land and there was concern for not only direct effects of exposure of people to the radiation sources, but also indirectly through the uptake of radionuclides into food crops and transfer through the food chain. The availability of radionuclides to roots is strongly influenced by agricultural practices. In wheat cultivation on an andosol soil, the concentration of ^{131}I, ^{134}Cs, and ^{137}Cs was highest in the top 2.5 cm of soil, but with tillage a more even distribution occurred to a depth of 15 cm (Yamaguchi et al. 2012) (Table 8.1). Tillage of dry soil risks aeolean loss of small soil particles, hence between 7 and 10 Bq kg^{-1} of ^{134}Cs and ^{137}Cs were lost as dust ($< 10 \ \mu m$ and $< 2.5 \ \mu m$) as a result of tillage. In addition to the gaseous emissions from the Chernobyl accident, numerous 'hot particles' were produced. These materials, contaminated particles of building material, and reactor core elements containing transuranium elements of ^{238}Pu, ^{239}Pu, ^{240}Pu, and ^{242}Pu (Boulyga et al. 1997) represent materials from which the radionuclide may be disassociated and released into the environment as soluble and bioavailable ions.

Table 8.1 Radioactivity in upper layers of an andosol soil near Fukushima under wheat cultivation prior to and after tillage. Loss of activity due to more even distribution down the soil profile and aeolean dust loss (after Yamaguchi et al. (2012)).

Soil depth	Isotope	Before tillage	After tillage
		Bq kg^{-1}	
0–0.5 cm	^{131}I	6320	206
	^{134}Cs	4300	358
	^{137}Cs	4270	377
0–5 cm	^{131}I	866	163
	^{134}Cs	438	227
	^{137}Cs	405	231

Fungi are an important group of organisms involved in most essential functions of ecosystem development and maintenance, including essential beneficial effects of their interactions with pollutants (Dighton 2016). Indeed, Steiner et al. (2002) say that 'Fungi are one of the most important components of forest ecosystems, since they determine, to a large extent, the fate and transport processes of radionuclides...' (Steiner et al. 2002). Fungi are also potentially important agents for bioremediation (Gray 1998). The important role of fungi in terrestrial and aquatic systems, and the fact that they are 'good candidates for remediation of radionuclide waste' is acknowledged in the Environmental Protection Agency (EPA) resource guide for radionuclide biological remediation (Ibeanusi and Grab 2004).

In this Chapter 8, the discussion will be restricted mainly to the role of fungi in terrestrial ecosystems in sequestering and retaining radionuclides. The potential role of fungi in radionuclide remediation by this process is important, as '...degradation of metals, metalloids, and radionuclides is impossible, thus leaving risk reduction by stabilization as the only option' (Harms et al. 2011). Deposition of radionuclides onto the soil surface from gaseous sources and the availability of surface contaminated particles from the reactor of buildings (hot particles) leads to the idea that fungi may play roles in: (i) stabilization of radionuclides to restrict mobility in the ecosystem, (ii) as mycorrhizae, modify accumulation rates in plants, which could be harvested (phytoremediation), (iii) accumulate radionuclides in fungal biomass that could be removed from a location (myco-remediation), and (iv) extraction of radionuclides bound to the surface of materials. These four roles are the foci of this chapter (Fig. 8.1).

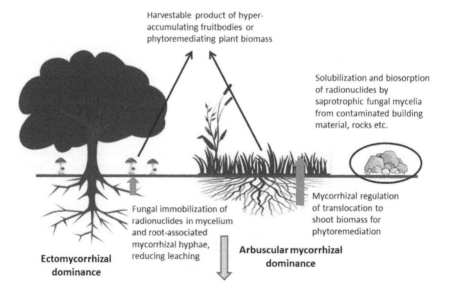

Figure 8.1 Schematic representation of the four major areas of fungal interactions with radionuclides.

2. Fungi—Properties and Functions

The kingdom Fungi consists of a very diverse collection of organisms that are largely filamentous in their hyphal structure. They are non-discrete organisms in which the hyphae continue to grow and branch and consist of spatially distinct age classes (Andrews 1992, Rayner 1991). Three attributes of fungi make them important components of ecosystems in relation to retention and movement of heavy metals and radionuclides: (i) the connection between adjacent cells of the hyphal filament through septal pores allows for translocation of material within the mycelium (Gray et al. 1996, Gray et al. 1995), (ii) the longevity of the mycelium and the area exploited by it (Smith et al. 1992), and (iii) cell wall components, such as chitin and melanin, that bind metals and radionuclides (Gadd 1993, Gadd 2004).

Within terrestrial ecosystems, the saprotrophic fungal community largely occupies the litter layer of the soil, forming a large surface area for interaction with the organic resources it is utilizing for nutrition, along with metals and radionuclides that may also be in the soil (Dighton et al. 1991, Dighton et al. 2008). In association with plants, mycorrhizal fungi are important in ecosystems for the promotion of plant growth and defense against environmental stressors (Dighton 2009) and are important in radionuclide retention, phytoremediaton or immobilization (Leyval et al. 2002, Vinichuk et al. 2005). Fungi are potentially useful as bioremediators of a number of pollutant chemical groups (Goltapeh et al. 2013).

Above-ground, fungi in the form of lichens are good at intercepting atmospherically deposited materials (Crittenden 1995, Eckl et al. 1986, Gaare 1990), including radionuclides. Foliose and terricolous lichens have a biological half-life of 1 to 2.5 years for ^{123}Cs and ^{90}Sr (Guillette et al. 1990), suggesting long-term accumulation and high risk of transfer to grazing animals, where lichens are dominant in the vegetation (Gaare 1990). Indeed, in a comparison between concentrations of radioisotopes of Np and Pu, the ratio of isotopes appearing in lichens indicated long term accumulation from nuclear weapons testing compared to more recently deposited isotopes from Chernobyl (Lindahl et al. 2004).

Due to the fungal ability to produce exoenzymes and organic acid secretions, along with the unique characteristics of cell wall chemistry, fungi are able to both solubilize complexed ions and then absorb or adsorb them into their own biomass allowing the potential for decontamination of surfaces of structures (Dhami et al. 1998, Fomina et al. 1999, Gadd 2009). We can consider fungi to have three primary roles in remediation of radionuclide contamination: (i) Hyphal immobilization of radionuclides by metabolic uptake and passive biosorption to reduce radionuclide mobility in soil, leaching, and availability to other ecosystem components (plant above-ground biomass and transfer to grazing fauna), (ii) Fungal accumulation and hyper-accumulation of radionuclides in above-ground and harvestable fruiting structures that could be removed from the ecosystem, (iii) Regulating adsorption and desorption characteristics of radionuclides on the surface of structures and in liquids, allowing biosorption onto harvestable live or dead mycelia in the form of films or filters.

3. Fungi as Remediators

3.1 Fungi and stability of radionuclides in soil

Fungal biomass in soil is considerable and can provide an absorptive and adsorptive surface for retention of radionuclides. In forest ecosystems fungal biomass has been estimated at 0.66 to 6.24 mg g^{-1} in leaf litter and 0.22 to 0.68 mg g^{-1} in soil (Baldrian et al. 2013). Where these fungi associate with tree roots in the form of ectomycorrhizae the extraradical hyphal mass (fungal hyphae extending from the root surface into soil) can be 125–299 kg ha^{-1} and when the mass of fungus associated with the mantle or sheath on the root surface is added, these numbers increase to 100–900 kg ha^{-1} (Wallander et al. 2001). With hyphal absorption values for Th of about 120 mg g^{-1} of fungal mycelium (White and Gadd 1990), an average of 200 kg ha^{-1} of extraradical mycelium could be involved in retaining some 24 kg Th ha^{-1}. Uptake of ^{137}Cs into mycelia of a range of, largely basidiomycete, fungi in culture conditions showed values ranging between 100 and 300 nmol g^{-1} with highest uptake by saprotrophic species and a range of values for ectomycorrhizal fungal species (Clint et al. 1991). In this study, pre-treatment of the fungi with cold Cs had little effect on the uptake of the radionuclide, suggesting minimal saturation.

Immobilization of radionuclides by fungi reduces their downward movement through soil due to leaching (Rommelt et al. 1990, Guillette et al. 1990). It has been shown that most ^{137}Cs in spruce forest soil is in the Ah layer (just under the leaf litter), retained in the high abundance of fungal mycelia in the surficial organic horizons of forest soils (Baeza et al. 2005, Drissner et al. 1998). This surface accumulation prevents leaching loss of radionuclides and enhances uptake by plants through mycorrhizal associations. Fungi compete more strongly than plants for Cs^{+}, accumulating 10 to 150 times more of this element than plants (Avery 1996). As a result, retention of ^{137}Cs and ^{90}Sr has been shown to be increased by 30 and 20%, respectively, in organic soil colonized by a single isolate of a variety of ascomycete and basidiomycete fungi (Parekh et al. 2008), and this retention is largely stable to any subsequent extraction.

The physico-chemical properties of fungal cell walls may be important for immobilization of radionuclides. Bacterial biosorbents have been made with dead bacterial cells. Pellets of Pseudomonas cells have been shown to sorb up to 541 µg U g^{-1} bacteria and 430 µg Th g^{-1} at pH 4–5 and with 90% recovery using sodium or calcium carbonate (Sar et al. 2004). In experimental systems, the incorporation of Fe$_3$O$_4$ nanoparticles on the surface of Penicillum mycelia has been shown to increased sorption of Sr, U, and Th at 110, 224, and 281 mg g^{-1} respectively for aqueous media at pH 5.0 (Ding et al. 2015). This may be a way to enhance retention of radionuclides from soil pore water. However, this enhanced adsorption is only useful if the fungal mycelium can be recovered, which is unlikely in a natural ecosystem. However, this novel idea may be of interest to consider in the removal of radionuclides from contaminated water or in effluent systems if batch production of the modified fungal mycelium can be attained.

After 6 to 7 years following the Chernobyl accident, most of the ^{137}Cs and ^{90}Sr was confined to the topmost (10–15 cm deep) layer of soil, with ^{90}Sr showing greater mobility (Ivanov et al. 1997). Their diffusion models indicate that greater downward flow of ^{137}Cs will occur in wetter and peaty soils compared to podsolic sandy soils (Table 8.2). However, given these rates of downward migration of radionuclides through soil, some 25 years after the Chernobyl accident only 0.01–0.1% of the deposited ^{137}Cs and ^{90}Sr has moved to depth in forest soils adjacent to the reactor, and is much less than the biological uptake and retention in roots and associated mycorrhizae (Shcheglov et al. 2014). These authors also report the different behaviors of contrasting isotopes, in that 15–30% of ^{137}Cs appeared in arboreal biomass and 65–80% in soil microbial biomass. In contrast, approximately 80–95% of ^{90}Sr was present in arboreal biomass, but negligible amounts in soil microbial biomass (Shcheglov et al. 2014). Hence the importance of bacteria and fungi in radionuclide recycling varies with isotope.

Uptake of radionuclides into plant roots is dependent on numerous factors. In simple hydroponic and soil systems, the rate of uptake of Cs and Sr is dependent on the extra-radical availability of K and Ca, respectively, in a competitive manner (Casadesus et al. 2008). Their findings suggested that as near root K and Ca depletions are established due to influx into roots, so too the uptake of Cs and Sr would increase. However, these findings did not include mycorrhizal fungal interactions, nor did they consider radionuclide binding onto soil mineral or organic components. In a study of the saprotrophic basidiomycete, *Pleurotus eryngii*, and the ectomycorrhizal fungus *Hebeloma cylindrocarpon*, the interaction between K and Ca as analogues for ^{137}Cs and ^{90}Sr, respectively showed there was a positive correlation of ^{90}Sr uptake with increasing K concentration (Baeza et al. 2005). The uptake of the radionuclides ^{90}Sr, [239+240]Pu and [241]Am by six ectomycorrhizal fungal fruitbodies had ranges of 1–6, 0.03–0.24 and 0.009–0.4 Bq kg[-1], respectively (Baeza et al. 2006), but the available transfer factors (ATF) from the radionuclide available pool in soil to the fruitbody varies much less between radionuclides (0.7–4.8 for ^{90}Sr; 0.6–2.3 for [239+240]Pu and 0.4–1.7 for [241]Am). These data suggest a much closer relationship in efficiency of the accumulation of contrasting radioisotopes than can be gained by just measuring radionuclide content in the fruitbodies. Thus, if mushroom removal is recommended

Table 8.2 Convective-diffusion model vertical migration rates of ^{137}Cs and ^{90}Sr in contrasting soils (after Ivanov et al. (1997)).

Site	Soil type	[137]Cs Darcy velocity (mm y[-1])	[90]Sr Darcy velocity (mm y[-1])
BP1	Loamy sand	17.0	14.5
UPP15	Soddy podsolic sand	15.5	23.7
RKP1	Soddy podsolic loamy sand	8.5	12.3
RKP2	Soddy podsolic loamy sand	6.6	6.0
RKP3	Soddy podsolic sand	14.4	14.1
RKP4	Humic gley sandy loam	31.4	28.2
RKP5	Soddy podsolic loamy sand	13.5	21.5
RKP6	Humic gley sandy loam	57.5	66.6
RKP7	Lowland peat	89.6	91.9

as a remediation process, the transfer factor is a more important metric than absolute isotope concentration in the fruitbody.

Soil saprotrophic fungi have great potential for uptake and immobilization of radiocesium fallout (Dighton et al. 1991, Olsen et al. 1990). With laboratory measures of Cs influx into mycelia of 134 nmol Cs g^{-1} dry weight of mycelium along with an estimate of hyphal biomass of 6 g dry weight cm^{-3} of soil (determined by hyphal length measurements of field collected samples), it is estimated that the fungal biomass of upland grass ecosystems in northern England could immobilize 350–800 nmol Cs m^{-2} h^{-1} (Dighton et al. 1991). Estimates of radiocesium concentration in pore water in these soils are at the micromolar level (Oughton 1989). Thus, these fungi have the potential to accumulate a large percentage of the total fallout from the Chernobyl accident. Compared to sterilized soils, soils with active bacterial and fungal activity can retain some 40–80% of radionuclide (Cs), which would otherwise be lost to groundwater, due to leaching (Guillette et al. 1990, Sanchez et al. 2000).

Bacterial and fungal surface chemistry, excretions and production of oxalates, are cited as chemicals that complexed with radionuclides, particularly uranium (Arwidsson and Allard 2009, Birch and Bachofen 1990). Fungal excretions, such as glycoproteins, are important in creating soil aggregates whose presence, abundance, and stability significantly stabilize soil and prevent erosion (Tisdall 1994, Tisdall et al. 2012). These soil aggregates also increase stability of soil and the radionuclides associated with the fungi, preventing loss through leaching and erosion. The abundance of arbuscular mycorrhizal spores in upper soil layers may also influence radionuclide retention as they have been found to adsorb radionuclides and heavy metals to spore walls (Baghvardani and Zare-Maivan 2000).

Relative plant availability of ^{137}Cs and ^{90}Sr was studied in intact soil lysimeters, where sequential extractions of the elements were performed by increasingly aggressive extractants (Forsberg et al. 2001). For ^{137}Cs, only 11–17% was extracted in labile fractions, compared to 31–53% for ^{90}Sr, suggesting that Sr was more plant available than Cs. Similar results were found for ^{144}Ce, ^{137}Cs, ^{106}Ru, and ^{90}Sr in sod podzolic soils from within the Chernobyl exclusion zone (Krouglov et al. 1998). The inference here, with respect to fungi, is that fungal activity, by the production of enzymes or organic acids, needs to be adequate to disassociate the radionuclides from binding sites on soil particle surfaces to enable uptake and immobilization into fungal tissue (Gadd 2004), or release from soil into soil pore water for plant uptake.

3.2 *Arbuscular mycorrhizae assist phytoremediation*

Arbuscular mycorrhizae have fungal biomass in soil as extraradical hyphae and in the host plant roots. Their function could be regarded both as immobilizing radionuclides in soil and mediating below-ground to above-ground movement of radionuclides within the plant. In the latter capacity, these fungi could be important in the selection of and efficiency of plant species as accumulators of radionuclides and used for phytoremediation. There are some 440 species of plants that have been identified as potential hyperaccumulators of heavy metals and suitable for phytoremediation (Leung et al. 2013, Prasad and Freitas 2003, Sharma et al. 2015) and some constraints on the phytoextraction capabilities are discussed by Zhu and Shaw (2000).

The potential role of phytoremediation of radionuclide contaminated soil came from studies of Alamo switchgrass and uptake of ^{137}Cs and ^{90}Sr (Entry and Watrud 1998). In a review of phytoremediation of radionuclides (Dushenkov 2003), four main methods of plant interaction are highlighted. Phyto-extraction, where plants take up radionuclides from soil and accumulate them in biomass; rhizo-filtration, where the retention of radionuclides in the rhizosphere limit leaching and lateral flow; phyto-volatilization, whereby foliar volatilization of volatile radionuclides can increase transfer from soil to air; and phyto-stabilization, where plant roots enhance retention of radionuclides in soil. Despite the fact that all mechanisms discussed in this review rely on the ability of the plant to take up radionuclides via their root systems, there was only one mention of mycorrhizal associations in the whole review. Similarly, in the selection of plant species to limit transfer of radionuclides up the food chain (White et al. 2003) there was no mention of the influence of mycorrhizal associations, probably because of the plant taxa discussed, where both Brassicacea and Caryophyllaceae are largely non mycorrhizal, although the Poaceae are mycorrhizal. From their data most of the high foliar Cs accumulators are in the Brassicacea and Caryophyllacae, where low foliar accumulation of radionuclides in the Poaceae may be as a result of mycorrhizal retention of radionuclides in the root system (Clint and Dighton 1992, White et al. 2003). In a similar search for efficient phyto-accumulators of radionuclides, dicotyledons (Magnoliopsidae) had three times higher accumulation than monocotyledons (Lilopsida), with Caryophyllidae (beets, buckwheats, quinoas, and amaranths) having the highest accumulation rates (Broadley et al. 1999).

In a review of the role of arbuscular mycorrhizae in plant uptake of radionuclides, arbuscular mycorrhizal inoculation of sweet clover and Sudan grass showed slight increases in uptake of ^{137}Cs and ^{60}Co, compared to non-mycorrhizal plants (Haselwandter and Berreck 1994). However, in a later study it was shown that mycorrhizal colonization of the grass *Agrostis tenuis* by *Glomus mosseae* significantly decreased shoot Cs content (Berreck and Haselwandter 2001). In Chinese cabbage and tomato seedlings in culture, the addition of dark septate fungi increased Cs uptake into above-ground parts of cabbage plants, but suppressed uptake in tomato (Diene et al. 2014). Dark septate fungi are a relatively understudied group of endophytic and mycorrhizal fungi whose abundance has been positively correlated to environmental stress (Jumpponen and Trappe 1998, Jumpponen and Egerton-Warburton 2005, Zhang et al. 2008).

With respect to other toxic elements, arsenic concentration in shoots of *Cynodon dactylon* (Bermudagrass) increased from 2 to 88 mg kg^{-1} in above-ground plant parts (and 2–99 mg kg^{-1} in the root) due to colonization by arbuscular mycorrhizae (Leung et al. 2006). However, there are studies that show mycorrhizae increase root concentration of metals and reduce translocation to above-ground plant parts (e.g., Leung et al. (2013)). The ability of mycorrhizae to increase translocation of radionuclides to above-ground plant parts is essential if phytoremediation is be successful, as it is the plant biomass that is most easily harvested. This point is alluded to in the review of Leyval et al. (2002), where they cite examples of foliar tissue having up to twice the ^{137}Cs concentrations in mycorrhizal plants compared to non-mycorrhizal plants of the same species, whereas in other examples, ^{137}Cs is concentrated in roots and prevented from accumulating in leaves of mycorrhizal plants, thus reducing transfer to grazing animals (Berreck and Haselwandter 2001, Haselwandter and Berreck 1994). However, where above-ground accumulation does occur, mycorrhizal-assisted phytoremediation is reported to remove some 75% of ^{137}Cs in about 15 years (Ebbs et al. 2008). Management of the system may also be important in modifying the concentration of radionuclides in shoots of plants. The time between mowing of a mycorrhizal perennial ryegrass system altered shoot ^{137}Cs concentration by a factor of 10 (Rosén et al. 2005), however the concentration of ^{137}Cs in roots was 10 times than in the shoot.

Uptake of ^{109}Cd was enhanced by the arbuscular mycorrhization of subterranean clover with a large proportion of accumulation in the root system of the plants (Table 8.3) (Joner and Leyval 1997). In a study of radionuclide accumulation by three grass species, recovery of ^{137}Cs and ^{90}Sr was a factor of 1.6 and 1.7 times, respectively, greater in arbuscular mycorrhizal than non-mycorrhizal plants after three harvests of above-ground biomass (Entry et al. 1999). Influx of radiocesium into mycorrhizal heather plants (*Calluna vulgaris*) with ericoid mycorrhizae was, overall, lower than that into non-mycorrhizal plants. However, the internal redistribution of Cs within mycorrhizal plants allowed a greater proportion of the Cs taken up to be translocated to shoots in mycorrhizal plants than in non-mycorrhizal plants, especially when incubated in a high potassium environment prior to radiocesium exposure (Clint and Dighton 1992). Mycorrhizal heather plants (*Calluna vulgaris*) were shown to have 18% higher shoot accumulation of ^{137}Cs than non-mycorrhizal plants (Strandberg and Johansson 1998).

Similar enhanced translocation of radiocesium into shoots of arbuscular mycorrhizal *Festuca ovina* has been reported but, in the same experiment, there was no observed enhanced shoot translocation in clover (*Trifolium repens*) (Dighton and Terry 1996). In contrast, a decrease in the Cs translocated to the shoots of *Agrostis tenuis* in the presence of arbuscular mycorrhizae has been observed (Berreck and Haselwandter 2001). ^{137}Cs can be retained and recirculated from senescing leaves to new growth with minimal loss in successive leaf generations (Jones et al. 1998b), although leaching loss from leaves can be important. Loss of ^{137}Cs from *Eriophorum vaginatum* leaves at senescence was 43%, compared to only 22% loss during late growing season (Jones et al. 1998a).

Table 8.3 Uptake of ^{109}Cd into mycorrhizal or non-mycorrhizal subterranean clover, where ^{109}Cd was placed in a hyphal ingrowth part of the experimental system (after Joner and Leyval 1997).

^{109}Cd added (mg kg^{-1})	Mycorrhizal status	Shoot (Bq)	Root (Bq)	Root:Shoot
1	NM	1269	663	0.52
	M	1339	1203	0.90
10	NM	1392	987	0.74
	M	1478	2285	1.52
100	NM	1612	3731	2.48
	M	2379	8907	3.83

In general, arbuscular mycorrhizae are important in regulating plant growth and the plant uptake and internal distribution of heavy metals radionuclides. Where their activity both increase plant biomass and stimulate radionuclide incorporation into shoot tissue, an advantage is made for the use of that plant/fungal combination as a radionuclide phytoremediator. The radionuclides extracted from soil by these plants can be harvested and the radionuclide subsequently retained.

3.3 *Ectomycorrhizae as above-ground accumulators: mycoremediation*

Ectomycorrhizae are mainly associations between a limited number of tree species and a large diversity of basidiomycete and ascomycete fungal symbionts. With a large biomass of fungal mycelium in soil as extraradical hyphae and in the root as sheath and Hartig net, these fungi are important in retaining radionuclides within their biomass in soil. Along with saprotrophic forms, the production of above-ground fruiting bodies (mushrooms), and a degree of preferential translocation of radionuclides to these structures, they may be considered harvestable products for ecosystem decontamination (Gray 1998). However, the accumulation of trace metals and radionuclides by fungi and their potential transfer up the food chain is a concern for human health (Falandysz and Borovička 2013).

The literature has many examples of radionuclide accumulation in mycorrhizal fungal fruitbodies following the Chernobyl accident (Dighton and Horrill 1988, Eckl et al. 1986, Haselwandter 1978, Vinichuk et al. 2013).

Following the Chernobyl accident there was renewed interest in fungi as radionuclide accumulators were based on earlier works of Witkamp (1968) and Witkamp and Barzansky (1968), who demonstrated that fungi were accumulated radionuclides in mushrooms Haselwandter (1978), Eckl et al. (1986), Haselwandter et al. (1988), and Byrne (1988). Byrne (1988) paid special attention to members of the Cortinaraiacea, which are known to be Cs accumulators. In addition to radiocesium, fungi have been shown to take up ^{7}Be, ^{60}Co, ^{90}Sr, ^{95}Zr, ^{95}Nb, ^{100}Ag, ^{125}Sb, ^{144}Ce, ^{226}Ra, and ^{238}U (Haselwandter and Berreck 1994). Radiocesium accumulation into 41 species of mushrooms from sites of contrasting surface contamination in the Ukraine showed that fruitbody concentrations from low level contaminated regions ($< 3.7 \times 10^{10}$ Bq km^{-2}) varied from zero to 33 kBq kg^{-1} dry weight to values between 1.4 and 3.7 MBq kg^{-1} dry weight in heavily contaminated regions (148×10^{10} Bq km^{-1} around Chernobyl and Prypyat) (Grodzinsky 1989).

Radiocesium concentration factors of between 0.4 and 99 (ratio of radiocesium content of mushroom to substrate) into basidiomycete fungal fruit bodies have been reported (Horyna and Žanda 1988). The highest radiocesium concentrations were found in the ectomycorrhizal fungal fruitbodies (*Boletus, Paxillus, Tylopilus, Lactarius, Leccinum, Amanita, Cortinarius, and Suillus*) and lowest in saprotrophs (*Scleroderma, Lepista*, and *Agaricus*) (Horyna and Žanda 1988, Randa and Benada 1990), a trend also seen by Yoshida and Muramatsu (1994) and Yoshida et al. (1994). Mycorrhizal fungal fruitbodies accumulated more radioactivity than most other organisms in terrestrial ecosystems (Fraiture et al. 1990, Giovani et al. 1990).

Translocation of [85]Sr and [134]Cs from substrate to mycelium, and then to fruitbody in *Pleurotus eringii* was documented by Baeza et al. (2002). These observations support findings of radiocesium accumulation in basidiomycete fungi that could be both high and long-lived (Dighton and Horrill 1988, Yoshida and Muramatsu 1994). For example, up to 92% of the radiocesium in mycorrhizal basidiocarps in the UK may have been derived from pre-Chernobyl sources of fallout (Dighton and Horrill 1988); resulting from accumulation over the prior 30 years or so.

Transfer ratios (TR) of radionuclides from soil to fruitbodies of fungi measure the soil surface area of radionuclide deposition per kg of fungus fruitbody. From a review of published literature the TR of ^{137}Cs in ectomycorrhizal fungi was 1.0, compared to 0.36–0.38 in saprotrophic and parasitic fungal species (Gillett and Crout 2000). Within fungal species there was considerable variation in TR with spatial location relative to Chernobyl and, in most of the recorded species (*Paxillus involutus*, *Cantharellus cibarius*, *Lactarius rufus*, and *Lactarius necator*), the TR increased during the first 6 years following the Chernobyl accident. This suggests that during the first 6 years there was little leaching of ^{137}Cs and cumulative immobilization into fungal tissue.

Transfer ratios of ^{137}Cs into ectomycorrhizal fruiting bodies was shown to be dependent on soil characteristics, with ^{137}Cs accumulation decreasing with the following ranked factors: (i) increasing clay content and exchangeable potassium or exchangeable bases, (ii) pH, (iii) cation exchange capacity and organic matter content (Kaduka et al. 2006). The authors also showed significant differences in transfer ratios for ^{137}Cs between ectomycorrhizal fungal species within the same site with a potential 35-fold difference in transfer ratio (Table 8.4). The addition of stable potassium, cesium, calcium, or strontium has effects on the accumulation of radionuclides into fruitbodies of *Pleurotus eryngii* in straw/wheat grain and soil culture (Guillén et al. 2012). Potassium, cesium, calcium, and strontium all suppressed ^{90}Sr uptake into fruitbodies. Potassium and strontium increased accumulation of [134]Cs. Additions of these alkaline and alkaline earth elements had little effect on the uptake of [60]Co. Thus, edaphic conditions play a role in basidiomycete fruitbody accumulation of radionuclides. It is interesting in this study that addition of potassium did not suppress [134]Cs accumulation in this fungal species, as potassium has been applied to compete with Cs in soil to limit phytoaccumulation of [134]Cs (Guillén et al. 2012).

The important role of mycorrhizal fungi in accumulation of radionuclides in natural terrestrial ecosystems is mentioned in Smith and Beresford (2005), where orders of magnitude differences in accumulation of ^{137}Cs occur between the fruitbodies of different fungal species. Recently the importance of ecosystem type on the degree of radiocesium accumulation into ectomycorrhizal fruitbodies has been reported from Sweden, following the Chernobyl accident. A comparison between contrasting forested systems showed that the percentage of deposited radiocesium appearing in above-ground fruiting structures varied from 0.1% in *Sphagnum* peat bogs, through 2% in a pine swamp with ericaceous understory, to 11% in a Scots pine forest in mineral soil (Vinichuk et al. 2013). The greater accumulation in the pine forest was partly due to the soil type, but mainly the greater proportion of ectomycorrhizal fungi in the forest ecosystem. The concept of supply and demand in the accumulation and translocation of radiocesium in mycorrhizal fungi shows that the concentration of

Table 8.4 Transfer ratios (10^{-3} m^2 kg^{-1} dry weight) of ^{137}Cs from soil into fruitbodies of mycorrhizal fungi from three contrasting soil types. Major edaphic factors regulating transfer factors in rank order are (i) clay content, exchangeable bases (S), exchangeable K (K$_2$O); (ii) pH; (iii) cation exchange capacity (CEC), organic matter content (C %). After Kaduka et al (2006).

	Bryansk	Murmansk/Arkhangelsk	Orel/Tula
	Turf pozol	**Pozol/peaty gley**	**Forest grey and black**
Clay %	5.8	6.1	28
S mEq 100 g^{-1}	1.1	1.3	9.8
K$_2$O mg 100 g^{-1}	2.1	2.2	13
pH	4	4	4.5
CEC mEq 100 g^{-1}	13	15	22
C%	1.4	1.5	2.3
Lactarius rufus	1390	230	
Suillus luteus	1340	250	50
Xercomus sp.	900	400	2.6
Paxillus involutus	870		8.3
Lactarius tormentosus	400	190	7.2
Leccinum scabrum	340	210	5.7
Lactarius necator	330		1.5
Russula sp.	320	160	3.8
Armillaria mellea	310		8.7
Boletus edulis	250	42	15
Tricholoma auratum	140		300
Leccinum aurantiacum	60	50	1.9
Lycoperdon pyriforme	40		2.2

^{137}Cs in the mycorrhizae before and during fruitbody formation alters and can promote preferential translocation of ^{137}Cs to the fruitbody (Nikolova et al. 1997). In *Suillus variagatus* ^{137}Cs in the mycorrhizae declined from 68 kBq kg^{-1} dry mass, prior to fruiting, to 35 kBq kg^{-1} during fruiting with 68 kBq kg^{-1} in mushrooms. In *Lactarius rufus* the values were 12, 46, and 52 respectively. It may be assumed from these data that much of the ^{137}Cs pool resides in the extensive extraradical hyphae and is re-translocated to 'sinks' of increased metabolic activity in highly active mycorrhizal roots and the production of fruitbody tissue and spores. Indeed, the importance of mycorrhizal mycelium in soil is suggested to be due to its ability to exploit a greater volume of soil than roots alone, increase bioavailability of metals and radionuclides, and transport material more readily and over longer pathways than root systems (Steiner et al. 2002).

The impact of accumulation and hyper-accumulation of radionuclides into ectomycorrhizal fungal fruiting structures leads to the concept of mycoremediation, where fruitbodies could be harvested, in comparison to plant material in phytoremediation, and the radionuclides contained in the fungal biomass safely contained (Gray 1998).

In the same way that herbaceous plants and their arbuscular mycorrhizal associations may enhance phytoremediation, ectomycorrhizae also regulate radionuclide incorporation in host plant biomass. Seedlings of *Pinus radiata* and *Pinus ponderosa* showed approximately tenfold increase in ^{90}Sr uptake into shoot tissue when associated with five ectomycorrhizal fungi compared to uninoculated tree seedlings, accounting a 6-fold (*P. radiata*) or 3-fold (*P. ponderosa*) percentage uptake from the Sr pool in soil accounting for bioaccumulation ratios of 120–138 in mycorrhizal plants compared to 27 from non-mycorrhizal (Entry et al. 1994). These authors speculated that this mycorrhizal induced uptake of radionuclides could be used for the reforestation and remediation of radionuclide contaminated soil (Entry et al. 1993).

In a comparison between pine trees colonized by the ectomycorrhizal fungus *Rhizopogon roseolus* and control plants for uptake and internal plant distribution of ^{137}Cs, ^{85}Sr, and ^{95m}Tc, there was greater percentage root retention of ^{85}Sr and ^{95}Tc in mycorrhizal roots than non- mycorrhizal plants, whereas for ^{137}Cs mycorrhizal colonization of roots did not alter the above-ground: below-ground distribution of the isotope (Ladeyn et al. 2008) (Table 8.5).

It seems possible that tree plantations could play a dual role in radionuclide remediation in (i) increasing radionuclide incorporation into above-ground tree biomass, and (ii) accumulation into fruitbodies. However, harvest of tree biomass and recovery of radionuclides is probably less feasible than for herbaceous vegetation. Nevertheless, with long biological half-lives of radionuclides in both plant and ectomycorrhizal fungal biomass (Dighton and Horrill 1988, Yoshida and Muramatsu 1994, Yoshida et al. 1994), tree plantations may be a good method of stabilizing radionuclides in the ecosystem.

Table 8.5 Percent allocation of ^{137}Cs, ^{85}Sr and ^{95m}Tc to above-ground needs and roots of *Pinus pinaster* growing conjunction with the ectomycorrhizal fungus *Rhizopogon roseolus* or non-mycorrhizal. Data reworked from Ladeyn et al. (2008).

		^{137}Cs	^{85}Sr	^{95m}Tc
Non-mycorrhizal	Needles	10.1	29.1	36.6
	Roots	89.9	70.9	63.4
Mycorrhizal	Needles	10.6	25.5	21.4
	Roots	89.4	74.5	78.6

3.4 *Fungi as extractors, accumulators, decontamination of surfaces and bio-recovery*

Radionuclides do not always occur as soluble ions in the ecosystem. Following the Chernobyl accident particularly, large amounts of contaminated debris (building material and reactor core components) were distributed in the environment. Fungi, acting as saprotrophs, have the capacity to produce extracellular enzymes and organic acids, both of which play a part in solubilizing and desorbing bound elements from the resources they colonize. As such, they may play a major role in accessing radionuclides from solid contaminated material.

Soil saprotrophic microfungi are known to grow into and decompose carbon-based radioactive debris from the Chernobyl reactor (Zhdanova et al. 1991). *Cladosporium cladosporioides* and *Penicillium roseopurpureum* were shown to overgrow 'hot particles' with less than 1147 Bq of β-activity and destroy them within 50 to 150 days. Fungal induced rate of release of ^{152}Eu from intact 'hot particles' is greater than for ^{137}Cs (Zhdanova et al. 2000), but rates are similar if the particles were ground into small particles. Thus, capacity to acquire radionuclides in these fungi is determined by an interaction between the physical nature of the radioactive source, fungal species and, presumably, enzymatic potential.

In order to prevent mobilization of radionuclides following disassociation from contaminated surfaces, fungi need to be able to sequester the released isotopes into fungal biomass. A two-step process of adsorption and metabolic incorporation of radionuclides into cells was suggested in a review of biosorption of radionuclides (Das 2012). Cell surface adsorption capacity is reached in approximately 30–40 mins, and following cell membrane passage to the cytoplasm depends on the radionuclide in question and the group pf microorganism (alga, fungus, bacteria, etc.). All of these processes are dependent on pH, extracellular concentration of the radionuclide (Dhami et al. 1998, Li et al. 2014).

Clay minerals in soil are also likely to affect the degree to which radionuclides are sorbed onto fungal/clay complexes. In a study of various clay materials (bentonite and kaolinite), the growth of Cladosporium fungal biomass changed with various minerals, and ^{137}Cs sorption was increased between 18 and 22 times, compared to fungus grown alone and increased up to 1.8 times for ^{90}Sr (Fomina et al. 1999). This shows that for true soils, rather than experimental situations, the nature of the mineral matrix of soil could play a major part in determining fungal retention of radionuclides. Biomass of *Rhizopus* species were shown to be efficient in adsorbing trivalent actinides and fission products (^{241}Am, ^{144}Ce, ^{147}Pm, $^{152+154}$Eu) from solution at pH 2, whereas ^{233}U and ^{230}Pu was maximally adsorbed at pH 6–7 (Dhami et al. 1998, Dhami et al. 2002).

Within these recovery systems, redox reactions are important for metal transformations to enhance immobilization (Gadd 2009, Gadd and Pan 2016). These reactions alter availability of radionuclides in the environment, allowing opportunities for biosorption, which is a physico-chemical process that binds material to both live and dead microbial (fungal) biomass. Within fungal cell walls, chitosan and other chitin derivatives are important components which bind metals and radionuclides (Gadd 2004, Gadd 2009), especially by precipitation, where the bound metal acts a focus for greater sorption. Additionally, pigments, such as melanin, polysaccharides, proteins (González-Chávez et al. 2004), and siderophores may be important components of fungal immobilization of radionuclides (Gadd 1993). Biosorption accounts for large amounts of immobilization of U and Th in dead fungal biomass, where desorption from these fungal hyphal filters can be achieved with acids, alkalis, or complexing agents (Gadd 2009). The biological processes related to decontamination of metal contaminated systems are discussed by Eccles (1999). Changes in redox potential can be influenced by soil microbial communities, including fungi, which influence the solubility of metal ions. For example, oxidized forms of Tc(VII), U(VI), and Pu(V) are much more soluble than the reduced forms (Tc(IV), U(IV), and Pu(IV)) (van Hullebusch et al. 2005).

Using the solubilizing and complexing activities of fungi (production of enzymes and acids and oxalates), a patent has been published to decontaminate surface, particularly porous rock substrates using a wrap or bandage of material containing fungal mycelium (Fomina and Gadd 2011). Solubilization of the pollutant with either decontamination fluids or the activities of fungal mycelia, the released radionuclides are sorbed into the fungi and can then be removed. This method is of particular interest where rough or porous materials, such as concrete, are polluted and standard cleaning methods are less effective.

Thus, fungi are important agents in the release of bound radionuclide ions and their subsequent adsorption or absorption into fungal mycelial tissue. Combined with the knowledge that mycelia are efficient biosorbants of radionuclides, there is great potential for them to be used as filters (Akhtar et al. 2009, Das 2012, Mishra et al. 2009) and decontaminators of contaminated solid surfaces.

4. Uranium—A Case Study

^{238}U series of radionuclides are an important type of pollutant in that this is the raw mined material used for nuclear energy production and for weapons. Mine tailings represent an important ecological hazard and deep soil repositories for waste U are perceived as possible sources of surface contamination if not well constructed and maintained (Mitchell et al. 2013). In most aerobic systems U exists as U(VI), which is readily soluble and mobile in ecosystems, although it is readily adsorbed onto clay at pH above 7 and, due to strong binding to plant cell walls and plant uptake is low (Malaviya and Singh 2012, Mitchell et al. 2013). Some 25 plus plant species have been identified as possible phytoaccumuators of U with uptake values of between 1 and 1000 mg kg^{-1} plant material (Malaviya and Singh 2012), despite the fact that U is highly toxic and can cause up to 60% reduction in maize seed germination.

In reviewing some of the literature on the interactions of arbuscular mycorrhizal fungi and plant uptake of the non-essential element U (Davies et al. 2015), state that in many studies that show enhanced U uptake in the mycorrhizal condition with reduced translocation to above-ground plant parts, studies have either used single mycorrhizal fungi or hydroponics. Phyto-stabilization of U occurs in plant roots when associated with arbuscular mycorrhizae, where translocation to shoots is reduced (Chen et al. 2005a, Chen et al. 2008). Differential U accumulation into plant parts may be dependent on the degree of mycorrhizal colonization. A greater suppression of U translocation to shoots was seen in the highly mycorrhizal *Medicago truncatula* compared to ryegrass (*Loilum perenne*) with lower root mycorrhizal colonization (Chen et al. 2008). The Chinese brake fern (*Pteris vittata*) shows good promise for bioremediation, where colonization by the arbuscular mycorrhizal fungus *Glomus mosseae* results in transfer factors from soil of 7 to 14 with up to 1.6 g kg^{-1} plant biomass U content (Chen et al. 2006). Thus, the author's results indicate mycorrhizal colonization of root systems of plants contributes significantly to phytostabilization of U in mine tailings (Chen et al. 2008).

In a review of the role of arbuscular mycorrhiza in U uptake (de Boulois et al. 2008), three U species exist in soil depending on the soil pH. Uranyl sulfate dominates at pH 4, uranyl phosphate at pH 5.5, and uranyl carbonate at pH 8. Each of these states provide contrasting availabilities of U to mycorrhizae. To enhance plant growth for phytoremediation, the application of phosphate fertilizer may enhance plant biomass, but reduce shoot U concentration (Chen et al. 2005b), as uranyl phosphate is less available than other forms. However, P fertilizer may be a useful soil amendment to enhance colonization of U contaminated soils by plants, such as subterranean clover and barley (Rufyikiri et al. 2006).

Using carrot in a laboratory experiment, arbuscular mycorrhizal fungi increase both hyphal and root accumulation of U (Rufyikiri et al. 2002), with concentrations of U between 5.5 to 9.6 times higher in hyphae than in either mycorrhizal or non-mycorrhizal roots (Rufyikiri et al. 2003). This suggests that arbuscular mycorrhizae may be more important in reducing movement of U in soil than regulating uptake into host plants. Indeed, at low pH (3.5) the presence of fungal mycelium immobilized some 97% of the 10 mg L^{-1} of U in U bearing shale due to immobilization onto fungal metabolites (Ogar et al. 2014). Uptake of U into shoots of subterreanean clover was actually reduced by the presence of the arbuscular mycorrhiza *Glomus interradices* at high soil concentrations of U (Rufyikiri et al. 2004).

In their review of the interactions between fungi and uranium (Gadd and Fomina 2011), suggest the retention and accumulation of U by fungi is largely due to adsorption of U onto fungal cell walls, where chitin has been suggested as the main cell wall constituent involved (Gadd and Fomina 2011, Tsezos 1983). Arbuscular mycorrhizal fungi also produce glycoprotein exudates, such as glomalin, which have been shown to complex heavy metals (González-Chávez et al. 2004). It is also likely that they are involved in the sequestration and immobilization of radionuclides also. Synergistic effects between bacteria and mycorrhizae are reported for some functions (Berta et al. 2014, Castro-Sowinski et al. 2007, Garbaye 1994), but the addition of arbuscular mycorrhizae and *Streptomyces* bacteria to soil only slightly increased phytoaccumulation of U by *Helianthus* but suppressed uptake in *Triticale* and had no effect in *Brassica* (Willscher et al. 2013). Fungi, such as *Beauvaria*, *Hymenoscyphus*, and *Rhizopogon* can accumulate 300 400 mg g^{-1} of U from depleted uranium, where U is chelated with oxalic acid produced by the fungal hyphae (Fomina et al. 2008) and from uranium oxides, where there is direct correlation between U accumulation in fungal biomass and the concentration of excreted oxalic acid (Fomina et al. 2007). Within arbuscular mycorrhizal roots, increased accumulation of U was found in the storage structures (vesicles), along with elevated concentrations of Mn, Cu, and Ni (Weiersbye et al. 1999).

The soil pore water and the leachate from uranium rich deposits and mine spoils can be a source of contamination and spread of the radionuclide in the environment. Studies have been made to establish the role of fungi to immobilize U from aquatic systems. Between 200 to 300 mg g^{-1} of uranium can be adsorbed by pellets of dead *Rhizopus arrhizus* mycelium (Tsezos and Deutschmann 1990). Seven fungal strains were shown to increase removal of U from water significantly more than a non-

Table 8.6 Bioleaching of uranium from low grade ore by different local fungal isolates. Data from Mishra et al. (2009).

Fungal strain	U in leachate (ppm)	% U recovery
Penicillium decumbens	1.9	5
Phoma tropica	1.8	5
Penicillium citrinum NM1	2.1	6
Aspergillus flavus UC1F4	2.5	7
Scytalidium lignicola	2.5	7
Penicillium citrinum UC1F1	3	8
Penicillium citrinum UC2F5	4.3	12
Curvularia clavata	18.6	50
Aspergillus flavus BM1	21.7	59
Cladosporium oxsporum	26	71

fungal control due to hyphal low molecular weight organic secretions (Grandin et al. 2015). Uranium recovery from low grade ore by local fungal strains was shown to be 71% by *Cladosporium oxysporum*, 59% from *Aspergillus flavus*, and 50% from *Curvularia clavata* isolates (Mishra et al. 2009), showing that these fungi were important factors influencing leaching of U from the ore (Table 8.6). Removal of uranium from water by adsorption onto mycelium of *Tricholoma harzianum* decreased from 75% U extraction by approximately 50%, as the U concentration increased from 100 to 400 mg L^{-1}. However, when the mycelium was incorporated into calcium alginate granules, the extraction efficiency rose to > 90% with reduction with increased U concentration (Akhtar et al. 2009), suggesting that bioremediation could be enhanced with modification of the fungal matrix.

Uranium can be considered a special case for fungal remediation, as this radionuclide is actively being mined, and contamination of the environment is due more to this activity than from accidental release. Fungi can play an important role in retaining U within the site of production, immobilize into both fungal and plant associated tissues, and recover U from effluent aquatic systems.

5. Future Directions

Due to their extensive and indeterminate hyphal growth, enzyme, and acid secretions, and high absorptive and adsorptive capacities, fungi have been observed to be important agents of regulating desorption of radionuclides from surfaces, adsorption and absorption of radionuclides, leading to immobilization and translocation into fruiting structures for hyperacumulation. As such they have great potential for remediation of radionuclide contaminated terrestrial systems. However, as far as is known, there have been no demonstrations of field trials to assess the overall effect of fungi in this context. This would certainly be an area of study for the future, in order to see if the hypothetical arguments above could be put into practice.

In addition to the potentially beneficial effects of fungi, they are affected by the ionizing radiation emitted from radionuclides. Due to differential susceptibility of contrasting fungal species, the fungal composition of the ecosystem can be changed by both the elemental concentration of a radionuclide and as a result of the ionizing radiation they produce. Root colonization by arbuscular mycorrhizae and the degree to which arbuscules are formed decreases with increasing Cs concentration in soil, with mycorrhizal colonization dropping from 45% to 15%, and arbuscule abundance from 80 to 20%, with Cs concentrations increasing from 3 to 78 µg kg^{-1} soil (Wiesel et al. 2015). It is known that the intensity of ionizing radiation causes changes in the fungal community in soil (Zhdanova et al. 2005). These possible changes in fungal communities need to be taken into account when considering the potential of fungi for remediation of radionuclide contaminated areas, as these organisms will be exposed to longer term exposure to both the radiation and the chemical toxicity of the elements in natural conditions compered to short term experimental conditions.

As Bradshaw et al. (2014) point out, the risk of exposure to radiation varies by organism, so the notion of using 'Reference Animals and Plants' (ICRP 2007) may be better replaced by ecological risk based on the whole ecosystem response (Bradshaw et al. 2014, Bréchignac et al. 2016). By placing the risk of exposure in a food web context, they place fungi as both a food source for small mammals and as mycorrhizal symbionts with trees. Although we are aware that mushroom fruitbodies are hyper-accumulators of radionuclides (Bakken and Olsen 1990, Byrne 1988, Eckl et al. 1986), it must be appreciated that the bulk of the fungal biomass is in soil or other resources (Baldrian et al. 2013). Retention of radionuclides by this below-ground fungal biomass and the translocation of radionuclides through extensive mycelia, is likely to affect transfer to other organisms on the ecosystem scale.

Using the unique fingerprint of a 2:1 ratio of ^{137}Cs and ^{134}Cs emitted from Chernobyl, Dighton and Horrill (1988) showed that a large proportion (25 to 92%) of ^{137}Cs that accumulated in ectomycorrhizal fungal fruitbodies originated from sources prior to the Chernobyl accident. Similar figures (13 to 69%) of pre-Chernobyl accumulation of radiocesium are given by Byrne (1988), and Giovani et al. (1990) commented on the deviance from the 2:1 ratio and fungal accumulation of radiocesium from atmospheric nuclear tests. Thus, fungi can be long-term accumulators and retainers of radionuclides in the environment. Together with an ecological half-life of 8–13 years (Vinichuk et al. 2013), it is anticipated that fungi will continue to be important in the retention of radiocesium in the ecosystem for the foreseeable future.

Emergent properties of fungi growing in areas of high radiation lead to behavioral adaptations of fungi to evolve enhanced spore germination, elevated hyphal extension rates, and radiotropism hyphal growth towards a source of ionizing radiation (Tugay 2006, Dighton et al. 2008). The role of melanin mediated sequestration of ionizing radiation (Bryan et al. 2011, Dadachova et al. 2007) has important ramifications for fungal physiology suggesting a photosynthesis-like attribute of melanized fungal species, which may be able to grow as autotrophs. Melanin has been identified to protect lichens from UV light damage (Solhaug et al. 2003). Melanin in the apothecial wall of the lichen *Trapelia involuta* has been shown to accumulate U in the lichen growing on U contaminated soil and is likely an adaptation to these stressed conditions allowing protection of the sexual reproductive structures from U toxicity (McLean et al. 1998).

In the context of the potential role of fungi in remediation of radionuclide contaminated environments, the role of melanin and other pigments in the protection of fungal hyphae from ionizing radiation and the potential of melanin to sequester energy for fungal function requires more attention.

References

Akhtar, K., A. M. Khalid, M. W. Akhtar and M. A. Ghauri. 2009. Removal and recovery of uranium from aqueous solutions by ca-alginate immobilized *Trichoderma harzianum*. *Bioresource Technology* 100 (20): 4551–4558. doi: https://doi.org/10.1016/j.biortech.2009.03.073.

Andrews, J. H. 1992. Fungal Life-history strategies. In: *The Fungal Community: Its Organization and Role in the Ecosystem*, edited by G. C. Carrol and D. T. Wicklow, 119–145. New York: Marcel Dekker.

Arwidsson, Z. and B. Allard. 2009. Remediation of metal-contaminated soil by organic metabolites from fungi II—metal redistribution. *Water, Air, and Soil Pollution* 207 (1): 5–18. doi: 10.1007/s11270-009-0222-6.

Avery, S. V. 1996. Fate of caesium in the environment: Distribution between the abiotic and biotic components of aquatic and terrestrial ecosystems. *Journal of Environmental Radioactivity* 30 (2): 139–171. doi: https://doi.org/10.1016/0265-931X(96)89276-9.

Avila, R. and L. Moberg. 1999. A systematic approach to the migration of ^{137}Cs in forest ecosystems using interaction matrices. *Journal of Environmental Radioactivity* 45 (3): 271–282. doi: https://doi.org/10.1016/S0265-931X(98)00111-8.

Baeza, A., F. J. Guillén and S. Hernández. 2002. Transfer of ^{134}Cs and ^{85}Sr to *Pleurotus eryngii* fruiting bodies under laboratory conditions: A compartmental model approach. *Bulletin of Environmental Contamination and Toxicology* 69 (6): 0817–0828. doi: 10.1007/s00128-002-0133-3.

Baeza, A., J. Guillén, S. Hernández, A. Salas, M. Bernedo, J. L. Manjón and G. Moreno. 2005. Influence of the nutritional mechanism of fungi (mycorrhize/saprophyte) on the uptake of radionuclides by mycelium. *Radiochimica Acta* 93 (4): 233–238. doi: 10.1524/ract.93.4.233.64074.

Baeza, A., J. Guillén, W. Mietelski Jerzy and P. Gaca. 2006. Soil-to-fungi transfer of ^{90}Sr, $^{239+240}$Pu, and ^{241}Am. *Radiochimica Acta* 94 (2): 75. doi: 10.1524/ract.2006.94.2.75.

Baghvardani, M. and H. Zare-Maivan. 2000. Evaluation of adsorption capacity of heavy and radioactive metals by mycorrhizal fungi. *Iranian Journal of Plant Pathology* 36 (1/2):Pe1-Pe15.

Bakken, L. R. and R. A. Olsen. 1990. Accumulation of radiocaesium in fruit bodies of fungi. In: *Transfer of Radionuclides in Natural and Semi-natural Environments*, edited by G. Desmet, P. Nassimbeni and M. Belli, 664–668. United Kingdom, London: Elsevier Applied Science.

Baldrian, P., T. Větrovský, T. Cajthaml, P. Dobiášová, M. Petránková, J. Šnajdr and I. Eichlerová. 2013. Estimation of fungal biomass in forest litter and soil. *Fungal Ecology* 6 (1):1-11. doi: https://doi.org/10.1016/j.funeco.2012.10.002.

Berreck, M. and K. Haselwandter. 2001. Effect of the arbuscular mycorrhizal symbiosis upon uptake of cesium and other cations by plants. *Mycorrhiza* 10 (6): 275–280. doi: 10.1007/s005720000089.

Berta, G., A. Copetta, E. Gamalero, E. Bona, P. Cesaro, A. Scarafoni and G. D'Agostino. 2014. Maize development and grain quality are differentially affected by mycorrhizal fungi and a growth-promoting pseudomonad in the field. *Mycorrhiza* 24 (3): 161–170. doi: 10.1007/s00572-013-0523-x.

Birch, L. and R. Bachofen. 1990. Complexing agents from microorganisms. *Experientia* 46 (8): 827–834. doi: 10.1007/bf01935533.

Boulyga, S. F., N. Erdmann, H. Funk, M. K. Kievets, E. M. Lomonosova, A. Mansel, N. Trautmann, O. I. Yaroshevich and I. V. Zhuk. 1997. Determination of isotopic composition of plutonium in hot particles of the chernobyl area. *Radiation Measurements* 28 (1): 349–352. doi: https://doi.org/10.1016/S1350-4487(97)00098-X.

Bradshaw, C., L. Kapustka, L. Barnthouse, J. Brown, P. Ciffroy, V. Forbes, S. Geras'kin, U. Kautsky and F. Bréchignac. 2014. Using an ecosystem approach to complement protection schemes based on organism-level endpoints. *Journal of Environmental Radioactivity* 136: 98–104. doi: https://doi.org/10.1016/j.jenvrad.2014.05.017.

Bréchignac, F., D. Oughton, C. Mays, L. Barnthouse, J. C. Beasley, A. Bonisoli-Alquati, C. Bradshaw, J. Brown, S. Dray, S. Geras'kin, T. Glenn, K. Higley, K. Ishida, L. Kapustka, U. Kautsky, W. Kuhne, M. Lynch, T. Mappes, S. Mihok, A. P. Møller, C. Mothersill, T. A. Mousseau, J. M. Otaki, E. Pryakhin,

O. E. Rhodes, B. Salbu, P. Strand and H. Tsukada. 2016. Addressing ecological effects of radiation on populations and ecosystems to improve protection of the environment against radiation: Agreed statements from a consensus symposium. *Journal of Environmental Radioactivity* 158–159: 21–29. doi: https://doi.org/10.1016/j.jenvrad.2016.03.021.

Broadley, M. R., N. J. Willey and A. Mead. 1999. A method to assess taxonomic variation in shoot caesium concentration among flowering plants. *Environmental Pollution* 106 (3): 341–349. doi: https://doi.org/10.1016/S0269-7491(99)00105-0.

Bryan, R., Z. Jiang, M. Friedman and E. Dadachova. 2011. The effects of gamma radiation, UV and visible light on ATP levels in yeast cells depend on cellular melanization. *Fungal Biology* 115 (10): 945–949. doi: https://doi.org/10.1016/j.funbio.2011.04.003.

Byrne, A. R. 1988. Radioactivity in fungi in slovenia, yugoslavia, following the chernobyl accident. *Journal of Environmental Radioactivity* 6 (2): 177–183. doi: https://doi.org/10.1016/0265-931X(88)90060-4.

Casadesus, J., T. Sauras-Yera and V. R. Vallejo. 2008. Predicting soil-to-plant transfer of radionuclides with a mechanistic model (BioRUR). *Journal of Environmental Radioactivity* 99 (5): 864–871. doi: https://doi.org/10.1016/j.jenvrad.2007.10.013.

Castro-Sowinski, S., Y. Herschkovitz, Y. Okon and E. Jurkevitch. 2007. Effects of inoculation with plant growth-promoting rhizobacteria on resident rhizosphere microorganisms. *FEMS Microbiology Letters* 276 (1): 1–11. doi: 10.1111/j.1574-6968.2007.00878.x.

Clint, G. M., J. Dighton and S. Rees. 1991. Influx of ^{137}Cs into hyphae of basidiomycete fungi. *Mycological Research* 95 (9): 1047–1051. doi: https://doi.org/10.1016/S0953-7562(09)80544-3.

Clint, G. M. and J. Dighton. 1992. Uptake and accumulation of radiocaesium by mycorrhizal and non-mycorrhizal heather plants. *New Phytologist* 121 (4): 555–561. doi: doi:10.1111/j.1469-8137.1992.tb01125.x.

Coughtrey, P. J., J. N. B. Bell, T. M. Roberts and B. E. S. I. E. Group. 1983. *Ecological Aspects of Radionuclide Release*: Blackwell Scientific.

Crittenden, P. 1995. Lichens and radioactive fallout over north west Russia. *British Lichen Society Bulletin* 77: 34–35.

Chen, B., I. Jakobsen, P. Roos and Y.-G. Zhu. 2005a. Effects of the mycorrhizal fungus *Glomus intraradices* on uranium uptake and accumulation by *Medicago truncatula L.* from uranium-contaminated soil. *Plant and Soil* 275 (1): 349–359. doi: 10.1007/s11104-005-2888-x.

Chen, B., Y. Zhu, X. Zhang and I. Jakobsen. 2005b. The influence of mycorrhiza on uranium and phosphorus uptake by barley plants from a field-contaminated soil. *Environmental Science and Pollution Research* 12: 325–331. doi: 10.1065/espr2005.06.267.

Chen, B. D., Y. G. Zhu and F. A. Smith. 2006. Effects of arbuscular mycorrhizal inoculation on uranium and arsenic accumulation by Chinese brake fern (*Pteris vittata L.*) from a uranium mining-impacted soil. *Chemosphere* 62 (9): 1464–1473. doi: https://doi.org/10.1016/j.chemosphere.2005.06.008.

Chen, B., P. Roos, Y.-G. Zhu and I. Jakobsen. 2008. Arbuscular mycorrhizas contribute to phytostabilization of uranium in uranium mining tailings. *Journal of Environmental Radioactivity* 99 (5): 801–810. doi: https://doi.org/10.1016/j.jenvrad.2007.10.007.

Chino, M., H. Nakayama, H. Nagai, H. Terada, G. Katata and H. Yamazawa. 2011. Preliminary estimation of release amounts of ^{131}I and ^{137}Cs accidentally discharged from the fukushima daiichi nuclear power plant into the atmosphere. *Journal of Nuclear Science and Technology* 48 (7):1129-1134. doi: 10.1080/18811248.2011.9711799.

Dadachova, E., R. A. Bryan, X. Huang, T. Moadel, A. D. Schweitzer, P. Aisen, J. D. Nosanchuk and A. Casadevall. 2007. Ionizing radiation changes the electronic properties of melanin and enhances the growth of melanized fungi. *PLOS ONE* 2 (5):e457. doi: 10.1371/journal.pone.0000457.

Das, N. 2012. Remediation of radionuclide pollutants through biosorption—an overview. *CLEAN – Soil, Air, Water* 40 (1): 16–23. doi: doi:10.1002/clen.201000522.

Davies, H. S., F. Cox, C. H. Robinson and J. K. Pittman. 2015. Radioactivity and the environment: technical approaches to understand the role of arbuscular mycorrhizal plants in radionuclide bioaccumulation. *Frontiers in Plant Science* 6: 580. doi: 10.3389/fpls.2015.00580.

de Boulois, H. D., E. J. Joner, C. Leyval, I. Jakobsen, B. D. Chen, P. Roos, Y. Thiry, G. Rufyikiri, B. Delvaux and S. Declerck. 2008. Impact of arbuscular mycorrhizal fungi on uranium accumulation by plants. *Journal of Environmental Radioactivity* 99 (5): 775–784. doi: https://doi.org/10.1016/j.jenvrad.2007.10.009.

Dhami, P. S., V. Gopalakrishnan, R. Kannan, A. Ramanujam, N. Salvi and S. R. Udupa. 1998. Biosorption of radionuclides by *Rhizopus arrhizus*. *Biotechnology Letters* 20 (3): 225–228. doi: 10.1023/a:1005313532334.

Dhami, P. S., R. Kannan, P. W. Naik, V. Gopalakrishnan, A. Ramanujam, N. A. Salvi and S. Chattopadhyay. 2002. Biosorption of americium using biomasses of various *Rhizopus* species. *Biotechnology Letters* 24 (11): 885–889. doi: 10.1023/a:1015533129642.

Diene, O., N. Sakagami and K. Narisawa. 2014. The role of dark septate endophytic fungal isolates in the accumulation of cesium by chinese cabbage and tomato plants under contaminated environments. *PLOS ONE* 9 (10): e109233. doi: 10.1371/journal.pone.0109233.

Dighton, J., G. M. Clint and J. Poskitt. 1991. Uptake and accumulation of ^{137}Cs by upland grassland soil fungi: a potential pool of Cs immobilization. *Mycological Research* 95 (9): 1052–1056. doi: https://doi.org/10.1016/S0953-7562(09)80545-5.

Dighton, J. and G. M. Terry. 1996. Uptake and immobilization of caesium in UK grassland and forest soils by fungi, following the chernobyl accident. In: *Fungi and Environmental Change*, edited by N. Magan F. J. C. and G. M. Gadd, 184–200. Cambridge, UK: Cambridge University Press.

Dighton, J. and A. D. Horrill. 1988. Radiocaesium accumulation in the mycorrhizal fungi *Lactarius rufus* and *Inocybe longicystis*, in upland Britain, following the chernobyl accident. *Transactions of the British Mycological Society* 91 (2): 335–337. doi: https://doi.org/10.1016/S0007-1536(88)80223-7.

Dighton, J. 2009. Mycorrhizae. In: *Encyclopedia of Microbiology*, edited by M. Schaechter, 153–162. Elsevier.

Dighton, J., T. Tugay and N. Zhdanova. 2008. Interactions of fungi and radionuclides in soil. In: *Microbiology of Extreme Soils*, edited by Patrice Dion and Chandra Shekhar Nautiyal, 333–355. Berlin, Heidelberg: Springer Berlin Heidelberg.

Dighton, J. 2016. *Fungi in Ecosystem Processes* Second Edition ed. Boca Raton: CRC Taylor & Francis.

Ding, C., W. Cheng, Y. Sun and X. Wang. 2015. Novel fungus-Fe$_3$O$_4$ bio-nanocomposites as high performance adsorbents for the removal of radionuclides. *Journal of Hazardous Materials* 295: 127–137. doi: https://doi.org/10.1016/j.jhazmat.2015.04.032.

Drissner, J., W. Bürmann, F. Enslin, R. Heider, E. Klemt, R. Miller, G. Schick and G. Zibold. 1998. Availability of caesium radionuclides to plants—classification of soils and role of mycorrhiza. *Journal of Environmental Radioactivity* 41 (1): 19–32. doi: https://doi.org/10.1016/S0265-931X(97)00092-1.

Dushenkov, S. 2003. Trends in phytoremediation of radionuclides. *Plant and Soil* 249 (1): 167–175. doi: 10.1023/a:1022527207359.

Ebbs, S., L. Kochian, M. Lasat, N. Pence and T. Jiang. 2008. An integrated investigation of the phytoremediation of heavy metal and radionuclide contaminated soils: from laboratory to the field. In: *Bioremediation of Contaminated Soils*, edited by D. L. Wise, 745–769. New York: Marcel Dekker.

Eccles, H. 1999. Treatment of metal-contaminated wastes: why select a biological process? *Trends in Biotechnology* 17 (12): 462–465. doi: 10.1016/S0167-7799(99)01381-5.

Eckl, P., W. Hofmann and R. Türk. 1986. Uptake of natural and man-made radionuclides by lichens and mushrooms. *Radiation and Environmental Biophysics* 25 (1): 43–54. doi: 10.1007/bf01209684.

Entry, J. A., P. T. Rygiewicz and W. H. Emmingham. 1993. Accumulation of cesium-137 and strontium-90 in ponderosa pine and monterey pine seedlings. *Journal of Environmental Quality* 22 (4): 742–746. doi: 10.2134/jeq1993.00472425002200040016x.

Entry, J. A., P. T. Rygiewicz and W. H. Emmingham. 1994. [90]Sr uptake by *Pinus ponderosa* and *Pinus radiata* seedlings inoculated with ectomycorrhizal fungi. *Environmental Pollution* 86 (2): 201–206. doi: https://doi.org/10.1016/0269-7491(94)90191-0.

Entry, J. A. and L. S. Watrud. 1998. Potential remediation of ^{137}Cs and ^{90}Sr contaminated soil by accumulation in alamo switch grass. *Water, Air, and Soil Pollution* 104 (3): 339–352. doi: 10.1023/a:1004994123880.

Entry, J. A., L. S. Watrud and M. Reeves. 1999. Accumulation of ^{137}Cs and ^{90}Sr from contaminated soil by three grass species inoculated with mycorrhizal fungi. *Environmental Pollution* 104 (3): 449–457. doi: https://doi.org/10.1016/S0269-7491(98)00163-8.

Falandysz, J. and J. Borovička. 2013. Macro and trace mineral constituents and radionuclides in mushrooms: health benefits and risks. *Applied Microbiology and Biotechnology* 97 (2): 477–501. doi: 10.1007/s00253-012-4552-8.

Fomina, M. A., V. M. Kadoshnikov and B. P. Zlobenko. 1999. Fungal biomass grown on media containing clay as a sorbent of radionuclides. *Process Metallurgy* 9: 245–254. doi: https://doi.org/10.1016/S1572-4409(99)80114-X.

Fomina, M., J. M. Charnock, S. Hillier, R. Alvarez and G. M. Gadd. 2007. Fungal transformations of uranium oxides. *Environmental Microbiology* 9 (7): 1696–1710. doi: doi:10.1111/j.1462-2920.2007.01288.x.

Fomina, M., J. M. Charnock, S. Hillier, R. Alvarez, F. Livens and G. M. Gadd. 2008. Role of fungi in the biogeochemical fate of depleted uranium. *Current Biology* 18 (9): R375–R377. doi: https://doi.org/10.1016/j.cub.2008.03.011.

Fomina, M. and G. M. Gadd. 2011. Decontamination method patent WO 2011121291 A1.

Forsberg, S., K. Rosén and F. Bréchignac. 2001. Chemical availability of ^{137}Cs and ^{90}Sr in undisturbed lysimeter soils maintained under controlled and close-to-real conditions. *Journal of Environmental Radioactivity* 54 (2): 253–265. doi: https://doi.org/10.1016/S0265-931X(00)00150-8.

Fraiture, A., O. Guillette and J. Lambinon. 1990. Interest of fungi as bioindicators of the radio contamination in forest ecosystems. In: *Transfer of Radionuclides in Natural and Semi-natural Environments*, edited by G. Desmet, P. Nassimbeni and M. Belli, 477–484. London: Elsevier Applied Science.

Gaare, E. 1990. Lichen content of radio cesium after the Chernobyl accident in mountains in southern Norway. In: *Transfer of Radionuclides in Natural and Semi-natural Environments*, edited by G. Desmet, P. Nassimbeni and M. Belli, 492–501. London: Elsevier Applied Science.

Gadd, G. M. 1993. Interactions of fungi with toxic metals. *New Phytologist* 124 (1): 25–60. doi: doi:10.1111/j.1469-8137.1993.tb03796.x.

Gadd, G. M. 2004. Microbial influence on metal mobility and application for bioremediation. *Geoderma* 122 (2): 109–119. doi: https://doi.org/10.1016/j.geoderma.2004.01.002.

Gadd, G. M. 2009. Biosorption: critical review of scientific rationale, environmental importance and significance for pollution treatment. *Journal of Chemical Technology & Biotechnology* 84 (1): 13–28. doi: 10.1002/jctb.1999.

Gadd, G. M. and M. Fomina. 2011. Uranium and fungi. *Geomicrobiology Journal* 28 (5–6): 471–482. doi: 10.1080/01490451.2010.508019.

Gadd, G. M. and X. Pan. 2016. Biomineralization, bioremediation and biorecovery of toxic metals and radionuclides. *Geomicrobiology Journal* 33 (3–4): 175–178. doi: 10.1080/01490451.2015.1087603.

Garbaye, J. 1994. Helper bacteria: a new dimension to the mycorrhizal symbiosis. *New Phytologist* 128: 197–210. doi: 10.1111/j.1469-8137.1994.tb04003.x.

Gillett, A. G. and N. M. J. Crout. 2000. A review of ^{137}Cs transfer to fungi and consequences for modelling environmental transfer. *Journal of Environmental Radioactivity* 48 (1): 95–121. doi: https://doi.org/10.1016/S0265-931X(99)00060-0.

Giovani, C., P. L. Nimis, P. Land, and R. Padovani. 1990. Investigation on the performance of macromycetes as bioindicators of radioactive contamination. In: *Transfer of Radionuclides in Natural and Semi-natural Environments*, edited by G. Desmet, P. Nassimbeni and M. Belli, 485–491. London: Elsevier Applied Science.

Goltapeh, E. M., R. Danesh and A. Varma. 2013. *Fungi as Bioremediators*. Heidelberg: Springer.

González-Chávez, M. C., R. Carrillo-González, S. F. Wright and K. A. Nichols. 2004. The role of glomalin, a protein produced by arbuscular mycorrhizal fungi, in sequestering potentially toxic elements. *Environmental Pollution* 130 (3): 317–323. doi: https://doi.org/10.1016/j.envpol.2004.01.004.

Grandin, A., A. Ogar, V. Sjöberg, B. Allard and S. Karlsson. 2015. Potential use of native fungal strains for assisted uranium retention. *Minerals Engineering* 81: 173–178. doi: https://doi.org/10.1016/j.mineng.2015.04.003.

Gray, S. N., J. Dighton, S. Olsson and D. H. Jennings. 1995. Real-time measurement of uptake and translocation of ^{137}Cs within mycelium of *Schizophyllum commune* Fr. by autoradiography followed by quantitative image analysis. *New Phytologist* 129 (3): 449–465. doi: doi:10.1111/j.1469-8137.1995.tb04316.x.

Gray, S. N., J. Dighton and D. H. Jennings. 1996. The physiology of basidiomycete linear organs. *New Phytologist* 132 (3): 471–482. doi: doi:10.1111/j.1469-8137.1996.tb01867.x.

Gray, S. N. 1998. Fungi as potential bioremediation agents in soil contaminated with heavy or radioactive metals. *Biochemical Society Transactions* 26 (4): 666.

Grodzinsky, D. M. 1989. *Radiobiology of plants*. K: Nauk.dumka, 384.

Guillén, J., A. Baeza and A. Salas. 2012. Influence of alkali and alkaline earth elements on the uptake of radionuclides by *Pleurotus eryngii* fruit bodies. *Applied Radiation and Isotopes* 70 (4): 650–655. doi: https://doi.org/10.1016/j.apradiso.2012.01.009.

Guillette, O., A. Fraiture and J. Lambinon. 1990. Soil-fungi radiocaesium transfers in forest ecosystems. In: *Transfer of Radionuclides in Natural and Semi-natural Environments*, edited by G. Desmet, P. Nassimbeni and M. Belli, 468–476. London: Elsevier Applied Science.

Harms, H., D. Schlosser and L. Y. Wick. 2011. Untapped potential: exploiting fungi in bioremediation of hazardous chemicals. *Nature Reviews Microbiology* 9: 177. doi: 10.1038/nrmicro2519 https://www.nature.com/articles/nrmicro2519#supplementary-information.

Haselwandter, K. 1978. Accumulation of the radioactive nuclide ^{137}Cs in fruitbodies of basidiomycetes. *Health Physics* 34 (6): 713–715.

Haselwandter, K., and M. Berreck. 1994. Accumulation of radionuclides in fungi. In: *Metal Ions in Fungi*, edited by G. Winkelmann and D. R. Winge, 259–277. New York: Marcel Dekker.

Haselwandter, K., M. Berreck and P. Brunner. 1988. Fungi as bioindicators of radiocaesium contamination: Pre- and post-chernobyl activities. *Transactions of the British Mycological Society* 90 (2): 171–174. doi: https://doi.org/10.1016/S0007-1536(88)80085-8.

Heal, O. W. and A. D. Horrill. 1983. Terrestrial ecosystems: an ecological context for radionuclide research. In: *Ecological Aspects of Radionuclide Release*, edited by P. J. Coughtree, 31–46. Oxford: Blackwell Scientific Publications.

Horyna, J. and Z. Žanda. 1988. Uptake of radiocesium and alkali metals by mushrooms. *Journal of Radioanalytical and Nuclear Chemistry* 127 (2): 107–120. doi: 10.1007/bf02164600.

Ibeanusi, V. M. and D. A. Grab. 2004. EPA Radionuclide Biological Remediation Resource Guide. Chicago, IL, USA: EPA Region 5 Superfund Division.

Ivanov, Y. A., N. Lewyckyj, S. E. Levchuk, B. S. Prister, S. K. Firsakova, N. P. Arkhipov, A. N. Arkhipov, S. V. Kruglov, R. M. Alexakhin, J. Sandalls and S. Askbrant. 1997. Migration of ^{137}Cs and ^{90}Sr from chernobyl fallout in ukrainian, belarussian and Russian soils. *Journal of Environmental Radioactivity* 35 (1): 1–21. doi: https://doi.org/10.1016/S0265-931X(96)00036-7.

Joner, E. J. and C. Leyval. 1997. Uptake of [109]Cd by roots and hyphae of a *Glomus mosseae/Trifolium subterraneum* mycorrhiza from soil amended with high and low concentrations of cadmium. *New Phytologist* 135 (2): 353–360. doi: undefined.

Jones, D. R., W. R. Eason and J. Dighton. 1998a. Foliar leaching of ^{137}Cs from *Eriophorum vaginatum* L., *Scirpus caespitosus* L. and *Erica tetralix* L. *Environmental Pollution* 99 (2): 247–254. doi: https://doi.org/10.1016/S0269-7491(97)00171-1.

Jones, D. R., W. R. Eason and J. Dighton. 1998b. Investigation of spatial and temporal patterns of ^{137}Cs partitioning in *Eriophorum Vaginatum* L. in relation to its nutrient retrieval and storage strategy. *Journal of Environmental Radioactivity* 40 (3): 271–288. doi: https://doi.org/10.1016/S0265-931X(97)00089-1.

Jumpponen, A. R. I. and J. M. Trappe. 1998. Dark septate endophytes: a review of facultative biotrophic root-colonizing fungi. *New Phytologist* 140 (2): 295–310. doi: 10.1046/j.1469-8137.1998.00265.x.

Jumpponen, A. and L. Egerton-Warburton. 2005. Mycorrhizal fungi in successional environments: A community assembly model incorporating host plant, environmental, and biotic filters. In: *The fungal community: its organization and role in the ecosystem*, edited by J. Dighton, J. White and P. Oudemans, 139–168. New York: CRC Press.

Kaduka, M. V., V. N. Shutov, G. Y. Bruk, M. I. Balonov, J. E. Brown and P. Strand. 2006. Soil-dependent uptake of ^{137}Cs by mushrooms: experimental study in the chernobyl accident areas. *Journal of Environmental Radioactivity* 89 (3): 199–211. doi: https://doi.org/10.1016/j.jenvrad.2006.05.001.

Krouglov, S. V., A. D. Kurinov and R. M. Alexakhin. 1998. Chemical fractionation of ^{90}Sr, [106]Ru, ^{137}Cs and [144]Ce in chernobyl-contaminated soils: an evolution in the course of time. *Journal of Environmental Radioactivity* 38 (1): 59–76. doi: https://doi.org/10.1016/S0265-931X(97)00022-2.

Ladeyn, I., C. Plassard and S. Staunton. 2008. Mycorrhizal association of maritime pine, *Pinus pinaster*, with *Rhizopogon roseolus* has contrasting effects on the uptake from soil and root-to-shoot transfer of ^{137}Cs, [85]Sr and [95m]Tc. *Journal of Environmental Radioactivity* 99 (5): 853–863. doi: https://doi.org/10.1016/j.jenvrad.2007.10.012.

Leung, H. M., Z. H. Ye and M. H. Wong. 2006. Interactions of mycorrhizal fungi with Pteris vittata (As hyperaccumulator) in As-contaminated soils. *Environmental Pollution* 139 (1): 1–8. doi: https://doi.org/10.1016/j.envpol.2005.05.009.

Leung, H.-M., Z.-W. Wang, Z.-H. Ye, K.-L. Yung, X.-L. Peng and K.-C. Cheung. 2013. Interactions between arbuscular mycorrhizae and plants in phytoremediation of metal-contaminated soils: A review. *Pedosphere* 23 (5): 549–563. doi: https://doi.org/10.1016/S1002-0160(13)60049-1.

Leyval, C., E. J. Joner, C. del Val and K. Haselwandter. 2002. Potential of arbuscular mycorrhizal fungi for bioremediation. In: *Mycorrhizal Technology in Agriculture: From Genes to Bioproducts*, edited by Silvio Gianinazzi, Hannes Schüepp, José Miguel Barea and Kurt Haselwandter, 175–186. Basel: Birkhäuser Basel.

Li, F., Z. Gao, X. Li and L. Fang. 2014. The effect of environmental factors on the uptake of ^{60}Co by *Paecilomyces catenlannulatus*. *Journal of Radioanalytical and Nuclear Chemistry* 299 (3): 1281–1286. doi: 10.1007/s10967-013-2827-x.

Lindahl, P., P. Roos, M. Eriksson and E. Holm. 2004. Distribution of Np and Pu in Swedish lichen samples (*Cladonia stellaris*) contaminated by atmospheric fallout. *Journal of Environmental Radioactivity* 73 (1): 73–85. doi: https://doi.org/10.1016/j.jenvrad.2003.08.003.

Malaviya, P. and A. Singh. 2012. Phytoremediation strategies for remediation of uranium-contaminated environments: A review. *Critical Reviews in Environmental Science and Technology* 42 (24): 2575–2647. doi: 10.1080/10643389.2011.592761.

McLean, J., O. W. Purvis, B. J. Williamson and E. H. Bailey. 1998. Role for lichen melanins in uranium remediation. *Nature* 391:649. doi: 10.1038/35533.

Mishra, A., N. Pradhan, R. N. Kar, L. B. Sukla and B. K. Mishra. 2009. Microbial recovery of uranium using native fungal strains. *Hydrometallurgy* 95 (1): 175–177. doi: https://doi.org/10.1016/j.hydromet.2008.04.005.

Mitchell, N., D. Pérez-Sánchez and M. C. Thorne. 2013. A review of the behaviour of U-238 series radionuclides in soils and plants. *Journal of Radiological Protection* 33 (2): R17.

Nesterenko, V. B. and A. V. Yablokov. 2009. Chapter I. Chernobyl contamination: An overview. *Annals of the New York Academy of Sciences* 1181 (1): 4–30. doi: doi:10.1111/j.1749-6632.2009.04820.x.

Nikolova, I., K. J. Johanson and A. Dahlberg. 1997. Radiocaesium in fruitbodies and mycorrhizae in ectomycorrhizal fungi. *Journal of Environmental Radioactivity* 37 (1): 115–125. doi: https://doi.org/10.1016/S0265-931X(96)00038-0.

Ogar, A., A. Grandin, V. Sjöberg, K. Turnau and S. Karlsson. 2014. Stabilization of uranium(VI) at low pH by fungal metabolites: Applications in environmental biotechnology. *APCBEE Procedia* 10: 142–148. doi: https://doi.org/10.1016/j.apcbee.2014.10.032.

Olsen, R. A., E. Joner and L. R. Bakken. 1990. Soil fungi and the fate of radioceasium in the soil ecosystem—a discussion of possible mechanisms involved in the radiocaesium accumulation in fungi, and the role of fungi as a Cs-sink in the soil. In: *Transfer of Radionuclides in Natural and Semi-natural Environments*, edited by G. Desmet, P. Nassimbeni and M. Belli, 657–663. London: Elsevier Applied Science.

Oughton, D. H. 1989. The Environmental Chemistry of Radiocaesium and Other Nuclides. PhD thesis, Univerity of Manchester, U.K.

Parekh, N. R., J. M. Poskitt, B. A. Dodd, E. D. Potter and A. Sanchez. 2008. Soil microorganisms determine the sorption of radionuclides within organic soil systems. *Journal of Environmental Radioactivity* 99 (5): 841–852. doi: https://doi.org/10.1016/j.jenvrad.2007.10.017.

Prasad, M. N. V. and H. M. d. O. Freitas. 2003. Metal hyperaccumulation in plants—Biodiversity prospecting for phytoremediation technology. *Electronic Journal of Biotechnology* 6 (3).

Randa, Z. and J. Benada. 1990. Mushrooms-significant source of internal contamination by radiocesium. In: *Transfer of Radionuclides in Natural and Semi Natural Environments*, edited by G. Desmet, P. Nassimbeni and M. Belli, 169–178. London: Elsevier Applied Science.

Rayner, A. D. M. 1991. The challenge of the individualistic mycelium. *Mycologia* 83 (1): 48–71. doi: 10.2307/3759832.

Rommelt, R., L. Hiersche, G. Schaller and E. Wirth. 1990. Influence of soil fungi (basidiomycetes) on the migration of Cs 134 + 137 and SR 90 in coniferous forest soils. In: *Transfer of Radionuclides in Natural and Semi-natural Environments*, edited by G. Desmet, P. Nassimbeni and M. Belli, 152–160. London: Elsevier Applied Science.

Rosén, K., Z. Weiliang and A. Mårtensson. 2005. Arbuscular mycorrhizal fungi mediated uptake of ^{137}Cs in leek and ryegrass. *Science of the Total Environment* 338 (3): 283–290. doi: https://doi.org/10.1016/j.scitotenv.2004.07.015.

Rufyikiri, G., Y. Thiry, L. Wang, B. Delvaux and S. Declerck. 2002. Uranium uptake and translocation by the arbuscular mycorrhizal fungus, *Glomus intraradices*, under root-organ culture conditions. *New Phytologist* 156 (2): 275–281. doi: 10.1046/j.1469-8137.2002.00520.x.

Rufyikiri, G., Y. Thiry and S. Declerck. 2003. Contribution of hyphae and roots to uranium uptake and translocation by arbuscular mycorrhizal carrot roots under root-organ culture conditions. *New Phytologist* 158 (2): 391–399. doi: 10.1046/j.1469-8137.2003.00747.x.

Rufyikiri, G., L. Huysmans, J. Wannijn, M. Van Hees, C. Leyval and I. Jakobsen. 2004. Arbuscular mycorrhizal fungi can decrease the uptake of uranium by subterranean clover grown at high levels of uranium in soil. *Environmental Pollution* 130 (3): 427–436. doi: https://doi.org/10.1016/j.envpol.2003.12.021.

Rufyikiri, G., J. Wannijn, L. Wang and Y. Thiry. 2006. Effects of phosphorus fertilization on the availability and uptake of uranium and nutrients by plants grown on soil derived from uranium mining debris. *Environmental Pollution* 141 (3): 420–427. doi: https://doi.org/10.1016/j.envpol.2005.08.072.

Sanchez, A. L., N. R. Parekh, B. A. Dodd and P. Ineson. 2000. Microbial component of radiocaesium retention in highly organic soils. *Soil Biology and Biochemistry* 32 (14): 2091–2094. doi: https://doi.org/10.1016/S0038-0717(00)00099-7.

Sar, P., S. K. Kazy and S. F. D'Souza. 2004. Radionuclide remediation using a bacterial biosorbent. *International Biodeterioration & Biodegradation* 54 (2): 193–202. doi: https://doi.org/10.1016/j.ibiod.2004.05.004.

Sharma, S., B. Singh and V. K. Manchanda. 2015. Phytoremediation: role of terrestrial plants and aquatic macrophytes in the remediation of radionuclides and heavy metal contaminated soil and water. *Environmental Science and Pollution Research* 22 (2): 946–962. doi: 10.1007/s11356-014-3635-8.

Shcheglov, A., O. g. Tsvetnova and A. Klyashtorin. 2014. Biogeochemical cycles of chernobyl-born radionuclides in the contaminated forest ecosystems. Long-term dynamics of the migration processes. *Journal of Geochemical Exploration* 144: 260–266. doi: https://doi.org/10.1016/j.gexplo.2014.05.026.

Smith, J. and N. Beresford (eds.). 2005. *Chernobyl—Catastrophe and Consequences*. Berlin: Springer-Verlag.

Smith, M. L., J. N. Bruhn and J. B. Anderson. 1992. The fungus *Armillaria bulbosa* is among the largest and oldest living organisms. *Nature* 356: 428. doi: 10.1038/356428a0.

Solhaug, K. A., Y. Gauslaa, L. Nybakken and W. Bilger. 2003. UV-induction of sun-screening pigments in lichens. *New Phytologist* 158 (1): 91–100. doi: doi:10.1046/j.1469-8137.2003.00708.x.

Steiner, M., I. Linkov and S. Yoshida. 2002. The role of fungi in the transfer and cycling of radionuclides in forest ecosystems. *Journal of Environmental Radioactivity* 58 (2): 217–241. doi: https://doi.org/10.1016/S0265-931X(01)00067-4.

Strandberg, M. and M. Johansson. 1998. ^{134}Cs in heather seed plants grown with and without mycorrhiza. *Journal of Environmental Radioactivity* 40 (2): 175–184. doi: https://doi.org/10.1016/S0265-931X(97)00074-X.

Tagami, K., S. Uchida, Y. Uchihori, N. Ishii, H. Kitamura and Y. Shirakawa. 2011. Specific activity and activity ratios of radionuclides in soil collected about 20 km from the fukushima daiichi nuclear power plant: Radionuclide release to the south and southwest. *Science of the Total Environment* 409 (22): 4885–4888. doi: https://doi.org/10.1016/j.scitotenv.2011.07.067.

Tisdall, J. M. 1994. Possible role of soil microorganisms in aggregation in soils. *Plant and Soil* 159 (1): 115–121. doi: 10.1007/bf00000100.

Tisdall, J. M., S. E. Nelson, K. G. Wilkinson, S. E. Smith and B. M. McKenzie. 2012. Stabilisation of soil against wind erosion by six saprotrophic fungi. *Soil Biology and Biochemistry* 50: 134–141. doi: https://doi.org/10.1016/j.soilbio.2012.02.035.

Tsezos, M. and A. A. Deutschmann. 1990. An investigation of engineering parameters for the use of immobilized biomass particles in biosorption. *Journal of Chemical Technology & Biotechnology* 48 (1): 29–39. doi: doi:10.1002/jctb.280480104.

Tsezos, M. 1983. Recovery of uranium from biological adsorbents—desorption equilibrium. *Biotechnology and Bioengineering* 26 (8): 973–981. doi: doi:10.1002/bit.260260823.

Tugay, T. I. 2006. Features of evince of adaptable reactions at micromycetes, isolated from radioactively polluted territories. Abstract of XII meeting of Ukrainian Botany Society, 15–18 May, Odessa, p 26.

van Hullebusch, E. D., P. N. L. Lens and H. H. Tabak. 2005. Developments in bioremediation of soils and sediments polluted with metals and radionuclides. 3. influence of chemical speciation and bioavailability on contaminants immobilization/mobilization bio-processes. *Reviews in Environmental Science and Bio/Technology* 4 (3): 185–212. doi: 10.1007/s11157-005-2948-y.

Vinichuk, M. M., K. J. Johanson K. Rosén and I. Nilsson. 2005. Role of the fungal mycelium in the retention of radiocaesium in forest soils. *Journal of Environmental Radioactivity* 78 (1): 77–92. doi: https://doi.org/10.1016/j.jenvrad.2004.02.008.

Vinichuk, M., K. Rosén and A. Dahlberg. 2013. ¹³⁷Cs in fungal sporocarps in relation to vegetation in a bog, pine swamp and forest along a transect. *Chemosphere* 90 (2): 713–720. doi: https://doi.org/10.1016/j.chemosphere.2012.09.054.

Wallander, H., L. O. Nilsson, D. Hagerberg and E. Bååth. 2001. Estimation of the biomass and seasonal growth of external mycelium of ectomycorrhizal fungi in the field. *New Phytologist* 151 (3): 753–760. doi: doi:10.1046/j.0028-646x.2001.00199.x.

Weiersbye, I. M., C. J. Straker and W. J. Przybylowicz. 1999. Micro-PIXE mapping of elemental distribution in arbuscular mycorrhizal roots of the grass, *Cynodon dactylon*, from gold and uranium mine tailings. *Nuclear Instruments and Methods in Physics Research Section B: Beam Interactions with Materials and Atoms* 158 (1): 335–343. doi: https://doi.org/10.1016/S0168-583X(99)00467-X.

White, C. and G. M. Gadd. 1990. Biosorption of radionuclides by fungal biomass. *Journal of Chemical Technology & Biotechnology* 49 (4): 331–343. doi: doi:10.1002/jctb.280490406.

White, P. J., K. Swarup, A. J. Escobar-Gutiérrez, H. C. Bowen, N. J. Willey and M. R. Broadley. 2003. Selecting plants to minimise radiocaesium in the food chain. *Plant and Soil* 249 (1): 177–186. doi: 10.1023/a:1022593307224.

Wiesel, L., S. Dubchak, K. Turnau, M. R. Broadley and P. J. White. 2015. Caesium inhibits the colonization of *Medicago truncatula* by arbuscular mycorrhizal fungi. *Journal of Environmental Radioactivity* 141: 57–61. doi: https://doi.org/10.1016/j.jenvrad.2014.12.001.

Williams, B. L., S. Brown and M. Greenberg. 1999. Determinants of trust perceptions among residents surrounding the savannah river nuclear weapons site. *Environment and Behavior* 31 (3): 354–371. doi: 10.1177/00139169921972146.

Willscher, S., D. Mirgorodsky, L. Jablonski, D. Ollivier, D. Merten, G. Büchel, J. Wittig and P. Werner. 2013. Field scale phytoremediation experiments on a heavy metal and uranium contaminated site, and further utilization of the plant residues. *Hydrometallurgy* 131–132: 46–53. doi: https://doi.org/10.1016/j.hydromet.2012.08.012.

Witkamp, M. 1968. Accumulation of ¹³⁷Cs by *Trichoderma viride* relative to ¹³⁷Cs in soil organic matter and soil solution. *Soil Science* 106 (4): 309–311.

Witkamp, M. and B. Barzansky. 1968. Microbial immobilization of ¹³⁷Cs in forest litter. *Oikos* 19 (2): 392–395. doi: 10.2307/3565024.

Yamaguchi, N., S. Eguchi, H. Fujiwara, K. Hayashi and H. Tsukada. 2012. Radiocesium and radioiodine in soil particles agitated by agricultural practices: Field observation after the fukushima nuclear accident. *Science of the Total Environment* 425: 128–134. doi: https://doi.org/10.1016/j.scitotenv.2012.02.037.

Yoshida, S. and Y. Muramatsu. 1994. Accumulation of radiocesium in basidiomycetes collected from Japanese forests. *Science of the Total Environment* 157: 197–205. doi: https://doi.org/10.1016/0048-9697(94)90580-0.

Yoshida, S., Y. Muramatsu and M. Ogawa. 1994. Radiocesium concentrations in mushrooms collected in Japan. *Journal of Environmental Radioactivity* 22 (2): 141–154. doi: https://doi.org/10.1016/0265-931X(94)90019-1.

Zhang, Y., Y. Zhang, M. Liu, X. Shi and Z. Zhao. 2008. Dark septate endophyte (DSE) fungi isolated from metal polluted soils: Their taxonomic position, tolerance, and accumulation of heavy metals *in vitro*. *The Journal of Microbiology* 46 (6): 624–632. doi: 10.1007/s12275-008-0163-6.

Zhdanova, N. N., T. N. Lashko, T. I. Redchits, A. I. Vasilevskaia, L. G. Borisiuk, O. I. Siniavskaia, V. I. Gavriliuk and P. N. Muzalev. 1991. The interaction of soil micromycetes with "hot" particles in a model system. *Microbiologichny Zhurnal* 53: 9–17.

Zhdanova, N. N., V. A. Zakharchenko, and K. Haselwandter. 2005. Radionuclides and Fungal Communities. In *The Fungal Community: Its Organization and Role in the Ecosystem*, edited by J. Dighton, J. W. White and P. Oudemans, 759–7685. Boca Raton, Fl, USA: Taylor & Francis.

Zhdanova, N. N., V. A. Zakharchenko, V. V. Vember and L. T. Nakonechnaya. 2000. Fungi from chernobyl: mycobiota of the inner regions of the containment structures of the damaged nuclear reactor. *Mycological Research* 104 (12): 1421–1426. doi: https://doi.org/10.1017/S0953756200002756.

Zheng, J., K. Tagami, Y. Watanabe, S. Uchida, T. Aono, N. Ishii, S. Yoshida, Y. Kubota, S. Fuma and S. Ihara. 2012. Isotopic evidence of plutonium release into the environment from the fukushima DNPP accident. *Scientific Reports* 2: 304. doi: 10.1038/srep00304.

Zhu, Y. G. and G. Shaw. 2000. Soil contamination with radionuclides and potential remediation. *Chemosphere* 41 (1): 121–128. doi: https://doi.org/10.1016/S0045-6535(99)00398-7.

CHAPTER-9

Removal of Gaseous Pollutants from Air by Fungi

Alberto Vergara-Fernández,[1] *Felipe Scott,*[1]
Patricio Moreno-Casas[1] **and** *Sergio Revah*[2,*]

1. Introduction

Air quality represents a worldwide concern, with consequences not only for the health of those exposed to the polluted environments, but also to the future of global economic development. Actual trends, showing a mounting population and increased urbanization, have led governments to enact and enforce legislation that is more comprehensive and stringent. Air quality is a very relevant issue, both outdoor and indoor, where progressively more time is spent at home, transportation, work, and leisure activities. The sources and types of air pollution vary widely for each location, and the air quality surveillance and enforcement may be under the responsibility of different agencies, including those in charge of the environment, health, occupational and labor, etc., leading to very diverse guidelines and regulations (Colome et al. 1994, World Health Organization 2006, 2010).

Establishing the most efficient and cost effective technologies for indoor and outdoor air purification requires considering on the one hand, the physical, thermodynamic, and reaction properties of the pollutants involved, and on the other hand, their impact on health and the environment. The existent technologies to improve air quality can be classified as physical, chemical, or biological. While physical processes, such as filters or centrifugal cyclones, are applied for waste gas streams with particulate matter, the removal of gases and vapors from volatile inorganic or organic compounds (VICs and VOCs) require further treatment. Incineration, thermal oxidation, catalytic oxidation, absorption, adsorption, and condensation are widely

[1] Green Technology Research Group, Facultad de Ingeniería y Ciencias Aplicadas, Universidad de los Andes. Mons. Alvaro del Portillo 12455, Las Condes. Santiago, 7620001 Chile.
[2] Departamento de Procesos y Tecnología, Universidad Autónoma Metropolitana-Cuajimalpa, México. Avenida Vasco de Quiroga 4871, Delegación Cuajimalpa. Cd. de México, 05348; México.
* Corresponding author: srevah@correo.cua.uam.mx

utilized among the traditional physicochemical technologies (Revah and Morgan-Sagastume 2005). The group of technologies relying on biological processes for the abatement and control of air pollution has been in constant evolution for the past 40 years, and nowadays represent an economic and environmentally sound alternative to limit the emission of noxious pollutants that are increasingly used in industry and explored for other non-industrial applications (Guieysse et al. 2008, Mudliar et al. 2010, Ralebitso-Senior et al. 2012, Estrada et al. 2012, Barbusinski et al. 2017).

The core of the biological air treatment systems is the capacity of microbes to transform certain organic and inorganic volatile pollutants in air streams to lower impact compounds, such as carbon dioxide, water, sulfate, and nitrate. As a first step to the microbial transformation of the gaseous pollutants, the compounds need to be transferred from the air stream to the biologically active phase. These molecules, along with oxygen, are then utilized by the microorganisms, which may include bacteria, yeast, and fungi, as carbon and energy sources to produce more biomass. In the catabolic process, carbon dioxide, water, sulfate, and nitrate are produced and may leave with the air stream or remain in the aqueous phase. The overall pollutant degradation is determined by the relative rates of the physical, chemical, and biological processes.

Compared to traditional technologies, biological systems are regarded as effective, robust, with simpler operation, and ecologically friendly, especially for airstreams with low pollutant concentration (< 5 g m^{-3}). While investment expenditures can be, in general, similar or lower to physicochemical systems, the operating costs are lower than the traditional technologies (Estrada et al. 2012). Different reactor configurations exist, and several recent reviews present detailed description of the systems (Barbusinski et al. 2017, Mudliar et al. 2010). The main configurations are the biofilters (BF) and the biotrickling filters (BTF) (Fig. 9.1).

Biofilters consist of columns packed with a porous solid supporting the growth of microorganisms adapted to the degradation of the gaseous contaminants. The polluted gaseous stream is generally pre-humidified to prevent desiccation of the packed bed and fed to the biofilter. Water, nutrients, or buffer are often intermittently

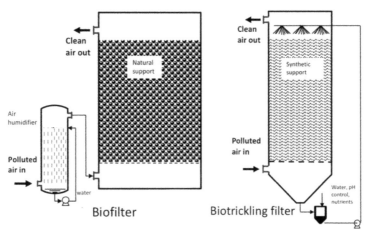

Figure 9.1 Biofilters and biotrickling filter.

sprayed to control humidity, microbial growth, and pH. The supports must be capable of retaining water (or a nutrient solution) and may be organic, such as compost, wood shavings, peat, etc. or inert inorganic, such as perlite, vermiculite, activated carbon, etc. (Barbusinski et al. 2017). The low water content in BF facilitates the treatment of hydrophobic contaminants favoring the direct mass transport between the gas stream and the biomass. Furthermore, a large void volume (50–80%) is available for biomass growth before observing loss of pressure effects and the formation of preferential channels due to excessive growth. The non-polar character of the packaging material in biofilters promotes the adsorption and subsequent degradation of hydrophobic compounds (Acuña et al. 1999, Y. Cheng et al. 2016). Water content is a critical issue as overwatering promotes excessive microbial growth and anaerobic zones. On the other hand, low water content affects the microbial activity and hence the biofilter performance (Cox et al. 1996, Gostomski et al. 1997, Morales et al. 2003). While intermittent water addition may restore activity, repeated drying, accumulation of by-products, and lack of pH control may lead to a short working life of the support (Cercado et al. 2012).

Biotrickling filters are packed with an inert support, having high specific area and empty fraction. The packing allows microbial cell attachment and the flow of both the gaseous and liquid streams. In contrast with the BF, a liquid is recirculated continuously facilitating the solubilization of the gaseous pollutants, the availability of nutrients to the biofilm, purging of reaction products, and pH control (Revah and Morgan-Sagastume 2005, Barbusinski et al. 2017).

The terms Elimination Capacity ($EC = (C_{G_{in}} - C_{G_{out}}) \cdot Q \cdot V_r^{-1}$); Load ($L = C_{G_{in}} \cdot Q \cdot V_r^{-1}$), and Removal Efficiency ($\%RE = 100 \cdot \{(C_{G_{in}} - C_{G_{out}})/C_{G_{in}}\}$) are generally used to characterize and compare the performance of these reactors. In these equations, $C_{G_{in}}$ and $C_{G_{out}}$ are the inlet and outlet concentrations (gm^{-3}), Q is the gas volumetric flow ($m^3\ h^{-1}$), and V_r is the packed volume (m^3). The EC and L units are usually expressed as $g\ m^{-3}\ h^{-1}$ and are referred to as the reactor volume.

Biological air waste treatment has been successfully applied with a wide variety of emissions, including industrial (chemical, food, pharmaceutical), odors from waste processing plants (wastewater, compost, solid wasters), odors from animal husbandry, indoor air, etc. (Guieysse et al. 2008, Mudliar et al. 2010, Ralebitso-Senior et al. 2012, Ferdowsi et al. 2017).

Biofiltration systems are generally inoculated with microbial sources that allow an initial rich microbial diversity, such as activated sludge from water treatment plants, compost, peat, etc. The characteristics of the incoming polluted stream and reactor operation conditions favor the selection of specialized microbial population adapted to degrade the incoming volatile compounds. The relevant aspects determining the development of a specific microbial consortium include: (A) the composition of the polluted air to be treated (chemical species present and their concentration), (B) Operating conditions, such as temperature, air feed flow, and type (continuous or intermittent), (C) For the BTF, liquid medium nutrient composition, pH, packing type, (D) For the BF, packing type (nutrient rich natural supports, such as compost, wood chips, peat; or inert nutrient free as activated carbon, perlite, sponges); support water content; pH, and nutrient composition and availability.

In general, biological systems treating highly water soluble and biodegradable compounds under conditions of high water activity (A_w) tend to develop bacterial rich consortia. On the other hand, reactor environments with reduced A_w, low pH, and the presence of hydrophobic (slightly soluble in water) biodegradable molecules tend to also favor the development of fungi, which may increase the reactor performance.

The following chapter presents the state of the art and the technical advances in the research of biofiltration of gaseous contaminants using filamentous fungi. It is organized as follows: first, an outline of the basic and technological aspects of biofiltration is given in section 2, next the research topics in fungal biofiltration are reviewed, and finally, in section 4, the most important operational considerations in fungal biofilters are analyzed.

2. Fungal Biotechnologies for air Treatment

Most of the literature regarding biofiltration of VOCs considers a global microbial catalysis regardless of the population structure. Recent studies have shown that highly diverse consortia develop, including bacteria, yeast, and filamentous fungi (Xie et al. 2009, Cabrol and Malhautier 2011, Prenafeta-Boldú et al. 2012, Portune et al. 2015, Z. Cheng et al. 2016). Growth in BTF and BF occurs generally as a biofilm on the support and may increase in depth with time if sufficient nutrients (nitrogen, phosphates, sulfates, etc.) and water are available. Flat biofilms are preferentially established in the BTF due to the effect of the descending liquid but in BF aerial structures may form. The presence of fungi was initially recognized as part of the normal development in traditional biofilters based on organic supports (Acuña et al. 1999, van Groenestijn et al. 2001), but has merited further research due to the increase in elimination capacity obtained with certain pollutants (van Groenestijn et al. 2001, Jorio et al. 2009, Revah et al. 2011).

To favor the development of a fungal population in BF at least two strategies have been implemented. The first is to foster the growth of a non-axenic fungal consortium by establishing appropriate conditions in the support, such as mild acidity and low humidity (van Groenestijn et al. 2001, Arriaga and Revah 2005a, Zhai et al. 2017). Fungi in these systems may have been in the support, seeded with a non-axenic diverse inoculum, such as activated sludge, or developed from spores dragged along with the incoming air. A second approach is to inoculate axenic cultures selected for their ability to degrade particular pollutants (García-Peña et al. 2001, Cheng et al. 2016, Lebrero et al. 2016, Vergara-Fernández et al. 2016, Morales et al. 2017). In several of the previous reports, the selected culture was isolated from BF inoculated with a rich microbial source operating with the same pollutants. It is important to consider that the normal operation of BF and BTF excludes, for economical and practical reasons, the filtration of the incoming air, so the reactors are continuously being exposed to microbes. Consequently, it should be assumed that pollutants are degraded by a mixed and diverse microbial population.

While fungi can degrade many VOCs, the reports show fungal biofiltration preferably to improve the elimination capacity of hydrophobic VOCs. The low performance in bacterial biofilters shown with these molecules has been chiefly attributed to their low bioavailability in the biofilm due to limitations in the mass transfer rates observed for slightly water-soluble molecules at low concentrations in the gas phase (Ferdowsi et al. 2017).

When compared to bacterial biofiltration systems, the use of fungi may present several advantages for the abatement of hydrophobic volatile organic compounds (VOCs) present in a gas phase (Kennes and Veiga 2004, Estévez et al. 2005, Arriaga and Revah 2005b, Vergara-Fernández et al. 2006, Ralebitso-Senior et al. 2012, Y. Cheng et al. 2016, Botros et al. 2017):

(a) Fungi degrade a large number of different organic compounds due to their rich and complex catabolic enzymatic system,

(b) They have a higher tolerance to low humidity conditions (low A_w) and acid pH,

(c) They have a capacity to increase the area for pollutants mass transfer due to their aerial hyphal growth,

(d) They colonize the void spaces within the biofilter through the formation of aerial hyphae (Fig. 9.2),

(e) They have higher affinity to hydrophobic compounds due to their cell composition, lower water content, and presence of hydrophobins on their surface,

(f) They can penetrate into the solid support, thus increasing nutrient availability from organic supports.

On the other hand, compared to bacteria, fungal growth in biofilters shows some disadvantages which include (Vergara-Fernández et al. 2012a, Vergara-Fernández et al. 2012b, Miller and McMullin 2014):

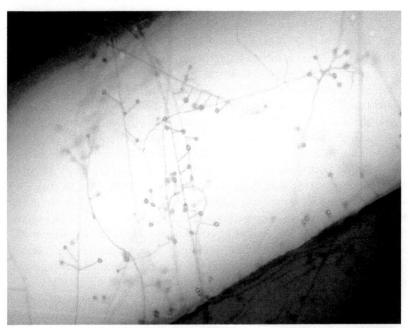

Figure 9.2 *Fusarium solani* hypha in the void space of the support material (perlite) in a biofilter treating *n*-hexane contaminated air.

(a) They have lower growth and catabolic rates,

(b) They have low activity toward inorganic pollutants, such as H_2S and NH_3 that may be present in mixed streams,

(c) They may sharply increase the pressure drop when filamentous growth is not controlled,

(d) They may release harmful volatiles, and under certain circumstances spores, that might represent a concern for health if uncontained.

The fungal biomass in BF can improve the mass transfer rate and the solubility of hydrophobic molecules through several mechanisms associated to their growth and structure. The presence of aerial mycelia (Vergara-Fernández et al. 2006, Vergara-Fernández et al. 2008) (see Figs. 9.2 and 9.3), which are in contact with the gas phase, promote direct absorption of the VOCs from the gas. Mass transfer increases substantially due to the extended exchange surface produced by the filamentous nature of fungal growth in contrast to the flat biofilm formation by bacteria (Vergara-Fernández et al. 2016).

Furthermore, the increase in the solubility of hydrophobic VOCs is facilitated by the hydrophobic nature of the aerial hypha. This has been shown to be favored by the cell accumulation of fatty molecules and to the presence of a class of hydrophobic proteins known as hydrophobins that will be discussed later (Vergara-Fernández et al. 2006, Vigueras et al. 2009). Table 9.1 shows some of the most relevant reports on fungal biofiltration, highlighting the obtained elimination capacity, the contaminant abated, and the species of fungi used.

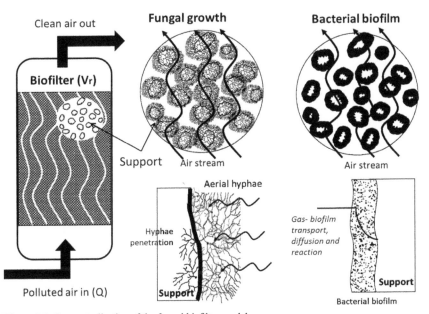

Figure 9.3 Conceptualization of the fungal biofilter model.

Table 9.1 Species of fungi reported that degrade VOCs in biofilter.

Fungi	Compounds/ support	Elimination capacity (g m^{-3} h^{-1})	References
Exophiala jeanselmei	Styrene/perlite	79	Cox et al. (1996)
Sporothrix variecibatus	Styrene/perlite	336	Rene et al. (2010)
E. lecanii corni	Toluene/silica pellets	270	Woertz et al. (2001a)
Paecilomyces variotii	Toluene/vermiculite Toluene/ceramic rings	258	García-Peña et al. (2001)
		290	Aizpuru et al. (2005)
Exophiala oligosperma *Paecilomyces variotii*	Toluene/perlite	55–77	Estévez et al. (2005)
Paecilomyces lilacinus	Toluene/perlite	80	Vigueras et al. (2008)
Cladophialophora sp.	Toluene/perlite and polyurethane foam cubes	50	Prenafeta-Boldú et al. (2008)
Trichoderma viride	Toluene/lava rock	26	Z. Cheng (2016)
Exophiala oligosperma	Toluene, Ethyl benzene, *p*-xylene/organic pellet	Tol: 39; EB 33; Xyl: 20	Prenafeta-Boldú et al. (2012)
Aspergillus niger	*n*-hexane/expanded clay	200	Spigno and de De Faveri (2005)
Ophiostoma species	α-pinene/perlite and Pall rings	143	Jin et al. (2007)
Fusarium solani	*n*-hexane/perlite	248	Vergara-Fernández et al. (2008)
			Vergara-Fernández et al. (2012a)
	n-pentane/vermiculite	65	Vergara-Fernández et al. (2012b)
Rhinocladiella similis	*n*-hexane; toluene; ethanol; phenol/perlite and wheat bran	Hex: 74; Tol: 85; EtOH: 230; Phe: 30	Vigueras et al. (2009)
Candida utilis	Ethanol/cane bagasse	108	Christen et al. (2002)
Graphium sp.	Methane/compost	38.5	Lebrero et al. (2016)
Non axenic mixed fungi	Toluene/perlite	60–100	van Groenestijn et al. (2001)
Non axenic mixed fungi	Toluene/coconut fiber, peat, pine leaves, compost	75–95	Maestre et al. (2007)
Non axenic mixed fungi	*n*-hexane/perlite	150	Arriaga and Revah (2005a)

Table 9.1 shows wide differences in EC, which does not necessarily reflect intrinsic fungal degradation rates. One of the main reasons is that the EC depends on the inlet load, which varies in the different reports. For instance, the first two entries show that *Sporothrix variecibatus* has a four-fold styrene EC compared to *Exophiala jeanselmei* (Cox et al. 1996, Rene et al. 2010). These results were obtained with inlet loads of around 500 g m^{-3} h^{-1} and 120 g m^{-3} h^{-1}, respectively, yielding similar removal efficiencies for both systems of around 60–70%. Other relevant aspects associated with performance include biomass content and operational conditions, such as pH, temperature, and water content, that will be revised in the next sections.

Figure 9.3 shows a conceptual model of a fungal biofilter, highlighting the differences with a bacterial one. A column, generally cylindrical, is packed with a solid support, either organic or inorganic, where the fungi grow. The stream of contaminated air enters the biofilter through a distributor located on the top or at the bottom of the biofilter, flowing through the void spaces of the solid support. The gaseous molecules, including both oxygen and the contaminants, are transported to the biological active phase, where they are solubilized and biodegraded. The solubility of these molecules in the humid biomass can be described at equilibrium using the contaminant—biofilm partition coefficient, as will be described later. The rate of transport, i.e., the amount of contaminants and oxygen attaining the biofilm, depends also on the available contact transfer area of the biomass and on the turbulence of the gas phase (Vergara-Fernández et al. 2008). For bacterial biofilms, the contact area corresponds to the flat surface where oxygen and pollutants solubilize, diffuse, and react through the biofilm depth. On the other hand, for fungal biofilters, the area corresponds both to the surface of the biofilm adhered to the surface of the support and to the area generated by the aerial hyphae (Arriaga and Revah 2009, Vergara-Fernández et al. 2016). Transport is enhanced by both the extremely high surface increase due to the aerial hyphae and to the reduced diffusion due to the small mycelial diameter (Vergara-Fernández et al. 2016). Recently, Vergara-Fernández et al. (2016) used a biofilter packed with vermiculite and inoculated with *Fusarium solani* to establish that although only 26% of the total biomass corresponded to aerial biomass, this fraction was responsible for 72% of the *n*-pentane elimination capacity. In this study, authors also determined the average specific superficial area of the aerial biomass as 5.5×10^6 m^2 m^{-3}, a value 18 times higher than the one obtained by Vergara-Fernández et al. (2008) using a fixed-bed biofilter for *n*-hexane abatement inoculated with the same fungus. The increased surface area was attributed to the bioreactor configuration used by Vergara-Fernández et al. (2016), where a perforated box filled with vermiculite was utilized instead of the traditional biofilter. The vermiculite box allows the development of the aerial hypha without any physical constraint. These results highlight the relevance of the aerial growth of the filamentous fungi in the biodegradation of hydrophobic VOCs.

3. Research in Fungal Biofilters

Among the first studies reporting the use of fungi in a biofiltration system, Cox et al. (1993) reported that styrene was degraded predominantly by fungi in a mixed bacterial-fungal consortium that developed in biofilters packed with perlite or polyurethane. These BFs were initially inoculated with a suspension of garden soil. Later, Cox et al. (1993), reported the elimination of styrene (EC = 79 g m^{-3} h^{-1}) with a mixed population enriched with *Exophiala jeanselmei*, that had been isolated from the previous work. Toluene vapors abatement has been performed with mixed fungal populations developed under appropriate conditions (low pH) or by inoculating the BF with fungi, such as *Paecilomyces variotii, Exophiala lecanii-corni, and Exophiala oligosperma*. Most of the strains inoculated were previously isolated from operating biofilters and had achieved elimination capacities that were much higher than those obtained with bacterial systems (García-Peña et al. 2001, Woertz et al. 2001a, Aizpuru et al. 2005, Estévez et al. 2005, Prenafeta-Boldú et al. 2012). For toluene, García Peña et al. (2005) showed that *Paecilomyces variotii* degradation involved hydroxylation occurring both on the methyl group and through the *p*-cresol pathway, and that low concentrations of the intermediates benzyl alcohol, benzaldehyde, and *p*-cresol may be present in the exit gas only with high initial toluene concentrations.

It has been reported that BF that were started under normal conditions (i.e., favoring bacterial development by high humidity and neutral pH) may develop fungal growth when operational conditions induce both lower humidity and pH. Dorado et al. (2008) described a dynamic mathematical model for the abatement of toluene in a mixed bacterial-fungal biofilter. This system was initially inoculated with a bacterial consortium, and after 60 days of operation it was colonized by fungi. There was an increase in the removal efficiency from 20%, when a predominant bacterial population was present, to 80% when fungi colonized the support was observed. Zhai et al. (2017) reported that changes in pH, from 7.0 to 4.0, and in nitrogen source, from ammonia to nitrate, favored the development of a fungal population in a polyurethane packed biofilter, improving the RE by 20%.

In contrast to the previous results, Z. Cheng et al. (2016) compared the operation of a bacterial biofilter (inoculated with activated sludge), a fungal biofilter (*Trichoderma viride*), and a fungal-bacterial biofilter for the elimination of toluene. The lowest removal efficiency was achieved by the fungal biofilter (20%), followed by the bacterial biofilter (60%). Notably, the fungal-bacterial biofilter achieved a removal efficiency of 90%. The elimination capacities followed the same trend, reaching 140, 90, and 26 g m^{-3} h^{-1} for the fungal-bacterial, bacterial, and fungal biofilter respectively. Similarly, Cheng et al. (2017) reported for the first time the elimination of chlorobenzene in a fungal-bacterial biofilter. Results indicate that the performance, in term of biodegradation rate, of the fungal-bacterial biofilter was 1.5 times higher than either the bacterial or the fungal one.

There has been an interest in studying the abatement of highly hydrophobic VOCs, including α-pinene and alkanes (C5–C6) (Hernández-Meléndez et al. 2008, Saucedo-Lucero et al. 2014, Lebrero et al. 2016, Jin et al. 2007). Spigno and coworkers (Spigno et al. 2003, Spigno and De Faveri 2005), addressed the abatement of *n*-hexane in a biofilter inoculated with *Aspergillus niger*, reaching an average EC of 150 g m^{-3} h^{-1}. These results are comparable to those reported by Arriaga and Revah (2005a), who obtained sustained hexane EC of 100 g m^{-3} h^{-1} and maximum EC of 150 g m^{-3} h^{-1} with perlite packed BF that developed a fungal population. Similar results were obtained when the BF was inoculated with *Fusarium solani* that had been isolated from the previous BF (Arriaga and Revah 2005b). Jin et al. (2007)

studied the biofiltration of α-pinene using a species from the genus *Ophiostoma*, obtaining an elimination capacity of 143 g m^{-3} h^{-1} with a removal efficiency of 89%. Temperatures of 15–40°C, relative humidity of 16–95%, and transient loads were also studied to assess the performance and resilience of the biofilter. The results showed that the fungal biofilter could recover completely after a starvation period of 3 to 7 days.

Fungal biofiltration has been tested also with VOCs mixtures. Woertz et al. (2001b), demonstrated that a fungal biofilter degrading toluene at high rate efficiently consumed nitric oxide (NO). Qi et al. (2002) tested five fungal species from microbial collections for their ability to utilize nine different VOCs commonly found in industrial off-gas emissions, and found that two strains (*Cladosporium sphaerospermum* and *Exophiala lecanii-corni*) could degrade all of the molecules. Moe and Qi (2004) studied the elimination of a *n*-butyl-acetate, methyl ethyl ketone, methyl ethyl propyl ketone, and toluene, using a culture of several species of fungi, including *Cladosporium sphaerospermun*, *Exophilia jenselmei*, *Fusarium oxysporum*, and *Fusarium nygamai*, reaching an elimination capacity of 92.1 g m^{-3} h^{-1} (as total VOCs). García-Peña et al. (2008) showed that *Paecilomyces variotii* degraded a BTEX mixture with a preference for toluene > ethyl benzene > benzene > xylenes; with o-xylene being the most recalcitrant of the isomers. Gutiérrez-Acosta et al. (2012) demonstrated that high EC can be obtained with *Fusarium solani* in mixtures of variable hydrophobicity, such as hexane, toluene, and methyl-ethyl-ketone, methyl ethyl ketone. In contrast to the previous result, Estrada et al. (2013), performed a comparative assessment of the bacterial and fungal biofiltration of a mixture of propanal, methyl isobutyl ketone (MIBK), toluene, and hexanol. The fungal biofilters showed lower elimination capacities compared to the bacterial biofilters (27.7 vs 40.2 gC m^{-3} h^{-1}, respectively). Moreover, the fungal biofilters showed a 60% higher pressure loss compared to its bacterial counterpart. In this work, it was shown that the presence of more soluble molecules may offset the advantages demonstrated by fungi with hydrophobic molecules.

The use of filamentous fungi for the abatement of methane has been recently explored. The report by Lebrero et al. (2016) shows that a fungal (*Graphium* sp.)-bacterial biofilter packed with composts fed with a stream containing 2% methane and an empty bed residence time of 20 and 40 minutes attained EC of 36.6 g m^{-3} h^{-1}. The specific role of fungi in these systems, either as catabolic agents or simply as bacterial support, remains to be elucidated.

4. Operational Considerations in Fungal Biofilters

The biological abatement of organic compounds is based on the capacity of microorganisms to use these molecules as carbon and energy sources. Degradation in biofilters depends mainly on the bioavailability and biodegradability of the compounds involved. As mentioned previously, bioavailability refers to how much pollutant present in the gas phase can reach the biomass and is determined by its concentration in the gas, the solubility in the biotic phase, the exchange surface, and mass transfer attributes in the reactor, such as gas velocity, mixing properties, etc. (Revah and Morgan-Sagastume 2005). Biodegradation requires the fungus having the necessary catabolic apparatus and being able to express it under conditions allowing fungal development, such as available growth nutrients and appropriate pH, temperature, and moisture. These aspects will be detailed in the following sections.

4.1 *Nature and concentration of the gaseous pollutants*

The physicochemical properties of the gaseous pollutants, their concentration, and biodegradability determine the overall biofilter performance by impacting the mass transfer rates, the development of the biomass layer, and the biodegradation rates.

Hydrophobic VOCs including, among others, alkanes, alkenes, and aromatics, show a lower mass transfer rate in biofiltration systems due to their low solubility in water (Y. Cheng et al. 2016, Ferdowsi et al. 2017). This is in contrast to molecules that can attain higher transfer rates, such as alcohols, fatty acids, ketones, aldehydes, etc., where the limiting step is generally the biodegradation rate. With highly soluble molecules, the limiting step may be the availability of oxygen, even at low pollutant concentration (Christen et al. 2002). The liquid layer surrounding the biofilm in bacterial biofilters hinders the possibility of increasing the solubility of hydrophobic VOCs, thus constraining the attainment of higher elimination capacities (Vergara-Fernández et al. 2008).

The amount of gaseous pollutant that can be solubilized in the biofilm at equilibrium for gas-biomass systems at low concentrations can be represented by the partition coefficient, which corresponds to an apparent Henry's constant, m, and defined by $m = C_g/C_l$. Here, m is the partition coefficient, C_g is the concentration of pollutant in the gas phase, and C_l is the concentration of the pollutant in the biofilm. The value of m increases as the molecule is less soluble in the liquid/biotic phase. The non-dimensional Henry's constant (mg_{gas} L^{-1})/(mg_{liq} L^{-1}), values for some often reported hydrophobic pollutants in pure water, at 25°C, are hexane: 70; oxygen: 29; methane: 24; α-pinene: 18; toluene: 0.272, and styrene: 0.113. These values contrast to soluble pollutants, such as ethanol: 0.002; methyl ethyl ketone: 0.002, or acetic acid: 0.0000122; other equilibrium values may be found in Revah et al. (2011) and Ferdowsi et al. (2017).

Literature shows that microbial biomass, due to its organic composition, favors the solubility of hydrophobic molecules. Davison et al. (2000) reported that a 100 g L^{-1} yeast suspension had a Henry's constant value that was about one fifth of that of pure water. Furthermore, hydrophobic carbon sources have been shown to induce a response in the morphology and superficial properties of fungi, favoring an increase in the solubility of the contaminant and in the superficial area for contaminants exchange. Vergara-Fernández et al. (2006), analyzed the effect of the carbon source on the characteristics of *Fusarium solani*. Results showed an increase of up to 200 times of the solubility of *n*-hexane in wet biomass compared to the control in water. Recently, Morales et al. (2017) studied the partition coefficient of toluene, formaldehyde, and benzo(α)pyrene in biomass of the bacteria *Rhodococcus erythropolis*, the fungus *Fusarium solani*, and a consortium of both organisms at temperatures ranging from 17 to 25°C. The results showed that the partition coefficients of toluene and benzo(α) pyrene in *Fusarium solani* biomass were 38 times lower than those in water, while this decrease was only 1.5 times for the bacteria.

Vergara-Fernández et al. (2011b) studied the effect of glycerol, 1-hexanol, and *n*-hexane as carbon sources on the morphology of the aerial hypha of the filamentous fungi *Fusarium solani* as a model system of hydrophobic VOCs biofiltration. Results showed that the morphology of the fungus adapted to the different substrates leading to improved consumption. There was a decrease in the diameter of the hypha from 2.99 to 2.01 µm and increased ramification (average hyphal lengths were 603 µm and 280 µm when grown in glycerol and *n*-hexane vapors, respectively). When combined with a mathematical model, these changes indicate that an increase in the transport area for the same amount of biomass may be an efficient adaptive response to increase the uptake of volatile hydrophobic molecules.

Fungi development under the conditions found in biofilters treating hydrophobic pollutants is favored also by the production of a certain type of hydrophobic proteins on the fungal surface called hydrophobins. These are a class of small amphipathic proteins (~ 100 amino acids) with surface activity and are produced exclusively by filamentous fungi (ascomycetes and basidiomycetes) (Linder et al. 2005, Vigueras et al. 2008). These proteins are secreted by fungi to decrease the surface tension of water by self-assembling in the fungi's surface allowing for the aerial hyphae to break the surface of the water and to emerge to the gas phase, thus creating a direct contact between the VOCs and the hydrophobic surface of the fungi (Vergara-Fernández et al. 2016). The increased superficial area due to the aerial growth and the decrease in the water surface tension promote the solubilization and availability of hydrophobic VOCs (Arriaga and Revah 2009, Y. Cheng et al. 2016, Vergara-Fernández et al. 2016). The presence of hydrophobins was confirmed in the surface of *Paecilomyces lilacinus* and *Rhinocladiella similis* in biofilters treating hydrophobic VOCs (Vigueras et al. 2008, Vigueras et al. 2009).

Vergara-Fernández et al. (2006) showed that the production of hydrophobins increased the surface hydrophobicity of *Fusarium solani* as demonstrated by measuring the contact angle of a water drop deposited on the surface of the biomass. The contact angle changed from 75°, when the fungus was cultivated on glucose, to 113° when it was grown in an environment containing *n*-hexane vapors. This angle was close to 115° that was reported for the highly hydrophobic surface of Teflon. This effect was confirmed by Vigueras et al. (2009) with a strain of *Rhinocladiella similis*, showing that the contact angle was 105° when the biomass was grown on *n*-hexane, but only 84° when grown on ethanol. Similar results were reported in the same study for the fungus *Cladosporium* sp. grown on different carbon sources (contact angles between 107° and 100°).

During the biodegradation process, the concentration of the contaminants in the microenvironment surrounding the microorganisms exerts a profound impact on the biological activity, and thereby the elimination rate, of the pollutants. At high substrates concentrations, contaminants are metabolized and used to synthesize new biomass. When the concentration decreases below a critical level (related to cell maintenance), the substrate no longer supports the production of new cells but it is still biodegraded if enough active biomass is present and the gene or genes responsible for the production of the required enzyme are active or can be expressed (Kovárová-Kovar and Egli 1998). The simultaneous presence of several contaminants can aid in sustaining the microbial growth, or at least, to induce the mineralization of the contaminants (Pahm and Alexander 1993). Moreover, under limitation conditions, some microorganisms are able to increase their affinity for the limiting substrates or adopt a strategy of simultaneous consumption of several carbon sources (Tros et al. 1996, Kovárová-Kovar and Egli 1998). In this sense, Vergara-Fernández et al. (2014) showed in *Fusarium solani* that the ratio of available carbon source to biomass determined the substrate to biomass yield. Values ranging from 0.3 $g_{biomass}$ $g^{-1}_{n-pentane}$ to 0.9 $g_{biomass}$ $g^{-1}_{n-pentane}$ were found at variable substrate to biomass ratios.

4.2 *Packing media*

Several packing materials have been used in BF to support fungal growth for VOCs treatment including peat, perlite, wood chips, straw, cane bagasse, compost, vermiculite, activated carbon, and polymers (Cox et al. 1996, Estévez et al. 2005, Maestre et al. 2007, García-Peña et al. 2008, Xie et al. 2009, Hernández-Meléndez et al. 2008, Gutiérrez-Acosta et al. 2010, Gutiérrez-Acosta et al. 2012, Vergara-Fernández et al. 2016).

The choice of packing material is an important decision during the biofilter design, since it should provide a proper environment for fungal growth and biological activity, facilitate the gas and liquid flows, enhance the gas/liquid contact, sustain long term operation, and resist compaction (Hernández-Meléndez et al. 2008, Ferdowsi et al. 2017). It is well established that the nature of the support influences the degradation kinetic of the carbon source by fungi (Vergara-Fernández et al. 2011c). These authors assessed the effect of the packing material in the *n*-pentane/biomass partition coefficient with *Fusarium solani* using two inorganic supports (perlite and vermiculite) and two organic supports (peat and compost). Results showed that the use of inorganic packing materials increase the solubility of *n*-pentane in the biomass, as demonstrated by a decrease of the partition coefficient from 0.21 when the biomass growth in an organic material, versus 0.05 when the fungi developed on the surface of an inorganic support. Similarly, Prenafeta-Boldú et al. (2008) compared a synthetic packing material (polyurethane foam cubes) to granular perlite, focusing in the dispersion and biodegradation of gaseous toluene in a fungal biofilter, showing lower mass transfer rate in the perlite supported system as a consequence of the reduced transport area compared to the polyurethane support. Interestingly, Cox et al. (1993) found that activated carbon did not support a strong fungal styrene biofiltration as compared to perlite or polyurethane, despite having a high superficial area and being a strong adsorbant.

Studies have also been made to create or adapt supports for fungal biofilters. Hernández-Meléndez et al. (2008) improved the structural resistance of the supports by grafting polymers to peat and pine sawdust. The composites had defined shapes, high hexane sorption, and allowed sustained hexane biofiltration after inoculation with the filamentous fungus *Fusarium solani*. Gutiérrez-Acosta et al. (2012) studied the effect of different packing materials in the elimination of a mixture of VOCs (hexane, toluene, and methyl-ethyl-ketone, MEK) in a biofilter inoculated with *Fusarium solani*. The EC of the biofilters was compared when perlite or polyurethane that was chemically modified with starch or agave were used. Results showed that the maximum EC recorded at varying inlet loads were 160 and 350 gC m^{-3} h^{-1} for the biofilters packed with perlite and modified polyurethane, respectively. The elimination capacity was also determined for each contaminant in the polyurethane packed biofilter, obtaining elimination capacities of 300, 360, and 400 g m^{-3} h^{-1} for hexane, toluene and MEK vapors, respectively.

4.3 *Moisture content*

The moisture content of the packing support is a critical factor to sustain the metabolic activity of the microorganisms and hence the performance of fungal biofilters (Cox et al. 1996, Estévez et al. 2005, Vergara-Fernández et al. 2012b). Moreover, it also affects the mass transfer of pollutants in the gas phase to the biofilm and the onset of sporulation processes in fungi (Gervats et al. 1988). It has been demonstrated that BF tend to dry, even if incoming air is close to 100% humidity, from water evaporation that is favored by small temperature increases due to metabolic heat and other thermal perturbations (Morales et al. 2003).

Maintaining the appropriate water content in BF is generally done both by pre-humidifying the stream of inlet air and by intermittent water or nutrient addition directly to the support (Barbusinski et al. 2017). Typical moisture content for biofilters packed with organic supports is around 60%. Low moisture contents (below 40% w/w) are responsible for more than 75% of the problems in a biofilter, causing drying of the packed support, canalization, and the decrease in the biological activity (Sun et al. 2002, Cercado et al. 2012). On the other hand, an excess humidity (> 80% w/w) promotes the formation of anaerobic zones and high pressure drops, hindering the solubilization of hydrophobic molecules and oxygen transfer (Morgan-Sagastume et al. 2001).

Although in fungal BF it has been reported that relative humidity in the inlet air stream as low as 70% are sufficient to support fungal growth (Deacon 1997), this has not been the case when continuous long-term operation is desired. Cox et al. (1996), reported that biofilters packed with the xerotolerant *Exophiala jeanselmei* degrading styrene on perlite required pre-humidification of the incoming gas to 90%–95% RH and intermittent weekly water addition for sustained biodegradation. Similar inlet humidity requirements were reported by Jin et al. (2007), who reported higher elimination capacities of α-pinene using intermittent loads when the inlet air was at 90% RH. In fungal biofilters, Maestre et al. (2007) showed that the water holding capacity of certain packings, such as compost, pine leaves, or peat, allows longer operation time without watering than others, such as coconut fiber.

The risks associated with spore emissions from BF treating aromatic compounds were studied by Prenafreta-Boldú et al. (2006). In this study, it was reported that some fungal genera found in BF, such as *Cladophialophora* and *Exophiala*, have species that have been associated with pathogenic species, although the virulence of BF grown strains has to be demonstrated. Therefore, proper control of the biofilter operation should be performed, especially if biological indoor air is envisaged (Guieysse et al. 2008). To address this problem, post treatment alternatives, such as filtration, UV, photocatalysis, etc., may be considered (Ralebitso-Senior et al. 2012, Saucedo-Lucero et al. 2014). Vergara-Fernández et al. (2011b) studied the effect of temperature and moisture content of the packing material on the elimination ability and spore emission of the fungi *Fusarium solani* in a biofilter treating *n*-pentane. Results indicated that the highest elimination capacity (65 g m^{-3} h^{-1}) was obtained at a moisture content of the support of 50%. However, the lowest rate of spore emission was attained at 80% moisture, which may be related to both lower sporulation and higher spore retention by the humid support. Moreover, it was reported that a decrease in nitrogen availability increases sporulation and reduces the elimination capacity.

4.4 Temperature

Fungi are mostly mesophilic organisms, growing in the 25–35°C range, being the optimal temperature for growth characteristic of each species (Deacon 1997), and thereby, fungal biofilters are normally operated in that range. Temperature can also have a major influence in the solubility of gaseous pollutants in the biofilm (Vergara-Fernández et al. 2006, Vergara-Fernández et al. 2008). This effect can play an important role in biofiltration systems where the low concentration of contaminants imposes a reduced gradient force for mass transfer to the biofilm. In this sense, Morales et al. (2017) reported a decrease in the solubility of toluene and benzo(α)pyrene as temperature increases from 17°C to 25°C. The balance between temperature effects on growth and solubility was reported by Vergara-Fernández et al. (2012a) in a system treating *n*-pentane using *Fusarium solani* and temperatures of 15, 25, and 30°C. A maximum elimination capacity of 64 g m^{-3} h^{-1} and a low spore emission (2.0 × 10^3 CFU m^{-3} h^{-1}) was achieved at 25°C. This value compares with the report of Saucedo-Lucero et al. (2014) with a spore emission of 2.4 × 10^3 a 9.0 × 10^4 CFU m^{-3} h^{-1} in a biofilter inoculated with a fungal consortium treating *n*-hexane at 25°C.

Biological activity sharply decreases at temperatures above the optimal growth temperature. In this regard, Salehahmadi et al. (2012) studied the effect of temperature, in a 30 to 45°C range, on the hexane elimination capacity in a fungal biofilter using a fungal consortium. A maximal elimination capacity of 400 g m^{-3} h^{-1} at 35°C, decreasing to 80 g m^{-3} h^{-1} at 45°C.

4.5 pH

Values of pH from 6.0 to 7.0 are commonly used in biofiltration systems to promote the growth and activity of heterotrophic bacteria (Revah and Morgan-Sagastume 2005, Ralebitso-Senior et al. 2012). It has been widely reported that the operation of biofilters under mild-acidic conditions, such as pH of 3.0–4.5, promote fungal growth, thus increasing the elimination ability of some VOCs (García-Peña et al. 2001, Moe and Qi 2004, Arriaga and Revah 2005a, b, Estévez et al. 2005). Furthermore, van Groenestijn et al. (2001) showed that low pH (2.5–4) was a prerequisite for stable fungal biofiltration.

The pH can vary in the BF when treating certain compounds, such as chlorinated organic molecules, hydrogen sulfide, methyl sulfide, and ammonia (Ergas et al. 1994, Chung et al. 1997, Kennes and Veiga 2004). The solubilization of the contaminants or the generated byproducts change the pH and the composition of the liquid media surrounding the biofilm and might alter the microbial activity in the biofilter and induce medium and long-term variations in the population. Controlling the pH in BF can be made by adding intermittently a soluble buffer calculated from pH data from the lixiviates (water percolating after intermittent addition) or from extracted biofilter samples (Kennes and Veiga 2004, Arriaga and Revah 2005a). An interesting alternative, reported by Morales et al. (1998), is the addition of gaseous molecules, such as ammonia, which may induce changes in the pH support without changing the packing water content or structure.

4.6 Reactor operation

The aspects previously mentioned need to be controlled within close range to maintain the expected performance. Other relevant issues to the operation of the fungal BF are related to the duration of the start-up phase required for the system to attain the maximum elimination capacity and efficiency and the control of the accumulation of excessive biomass that reduces the gas retention time, provokes increased pressure drop, and eventually reduces performance.

It has been reported that the start-up time following reactor inoculation may be a slow process, requiring up to a few weeks (Ralebitso-Senior et al. 2012). This is due to several factors, including the intrinsic slow fungal growth, the reduced availability of substrate, and the environmental conditions, such as low water content. The start-up time can be further extended if inoculation was made from spores or from complex mixed microbial consortia, such as compost, activated sludge, etc. where the degrading fungi may be in very low initial concentrations (Cox et al. 1993, Arriaga and Revah 2005a, Xie et al. 2009). Several reports address the possibility of shortening start-up by inoculating active vegetative cells and by adding an initial selective substrate (Vergara-Fernández et al. 2011a).

Filamentous fungi in biofilters might increase the pressure drop (compared to a bacterial biofilter) due to biomass accumulation and the development of an aerial mycelium that colonizes the void spaces of the support where the air flows. Pressure drop, which depends on the permeability of the bed and the gas velocity, determines the energy required to force the desired flow of polluted air through the reactor (Auria et al. 1993). The increase in the BF pressure drop depends, among others, on the pollutant load, the characteristics of the support, and the addition of nutrients (García-Peña et al. 2001, Maestre et al. 2007, Vergara-Fernández et al. 2008). Several reports regarding the pressure drop of fungal biofilters (*Fusarium solani*, perlite packed) treating *n*-pentane and *n*-hexane, obtained a maximum pressure drop of 55 mmH_2O m^{-1}_{bed} (Arriaga and Revah 2005b, Vergara-Fernández et al. 2011b, Vergara-Fernández et al. 2012b). This value was nearly three times higher than a similar system inoculated with a bacterial consortium for the abatement of *n*-hexane. Alternatives to control excessive biomass include nutrient control, backwashing, addition of chemicals, and possibly the use of mites (Cox et al. 1996, Mendoza et al. 2004, Estrada et al. 2013).

Conclusions

The biofiltration of gases contaminated with organic volatile compounds has emerged as a viable technological and economic alternative to the physicochemical treatments, especially for high-volume and low concentration streams. Although biofilters inoculated with bacterial consortia and pure bacterial cultures have been used in the industry, the abatement of hydrophobic VOCs at low concentrations is still a challenging task. In this context, results obtained with fungal biofilters show superior mass transfer rates of the contaminants, stability, resilience, tolerance to pH and moisture variations, and enhanced elimination capacities. The challenge remains to scale-up the laboratory results to systems that allow the high performance found in bench scale while controlling the critical aspects, such as pressure drop increase and spore emission.

Acknowledgement

The present work has been sponsored by CONICYT – Chile (National Commission for Scientific and Technological Research) (Fondecyt 1160220).

References

Acuña, M. E., F. Pérez, R. Auria and S. Revah. 1999. Microbiological and kinetic aspects of a biofilter for the removal of toluene from waste gases. *Biotechnology and Bioengineering* 63 (2): 175–184. doi: doi:10.1002/(SICI)1097-0290(19990420)63:2<175::AID-BIT6>3.0.CO;2-G.

Aizpuru, A., B. Dunat, P. Christen, R. Auria, I. García-Peña and S. Revah. 2005. Fungal biofiltration of toluene on ceramic rings. *Journal of Environmental Engineering* 131 (3): 396–402. doi: doi:10.1061/(ASCE)0733-9372(2005)131:3(396).

Arriaga, S. and S. Revah. 2005a. Improving hexane removal by enhancing fungal development in a microbial consortium biofilter. *Biotechnology and Bioengineering* 90 (1): 107–115. doi: doi:10.1002/bit.20424.

Arriaga, S. and S. Revah. 2005b. Removal of n-hexane by *Fusarium solani* with a gas-phase biofilter. *Journal of Industrial Microbiology and Biotechnology* 32 (11): 548–553. doi: 10.1007/s10295-005-0247-9.

Arriaga, S. and S. Revah. 2009. Mathematical modeling and simulation of hexane degradation in fungal and bacterial biofilters: effective diffusivity and partition aspects. *Canadian Journal of Civil Engineering* 36 (12): 1919–1925. doi: 10.1139/L09-090.

Auria, R., M. Morales, E. Villegas and S. Revah. 1993. Influence of mold growth on the pressure drop in aerated solid state fermentors. *Biotechnology and Bioengineering* 41 (11): 1007–1013. doi: doi:10.1002/bit.260411102.

Barbusinski, K. K. Kalemba, D. Kasperczyk, K. Urbaniec and V. Kozik. 2017. Biological methods for odor treatment—A review. *Journal of Cleaner Production* 152: 223–241. doi: https://doi.org/10.1016/j.jclepro.2017.03.093.

Botros, M. M., A. A. Hassan and G. A. Sorial. 2017. Role of fungal biomass in N-hexane biofiltration. *Advances in Microbiology* 7 (10): 673. doi: 10.4236/aim.2017.710053.

Cabrol, L. and L. Malhautier. 2011. Integrating microbial ecology in bioprocess understanding: the case of gas biofiltration. *Applied Microbiology and Biotechnology* 90 (3): 837–849. doi: 10.1007/s00253-011-3191-9.

Cercado, B., R. Auria, B. Cardenas and S. Revah. 2012. Characterization of artificially dried biofilms for air biofiltration studies. *Journal of Environmental Science and Health, Part A* 47 (7): 940–948. doi: 10.1080/10934529.2012.667292.

Colome, S., R. J. McCunney and J. M. Samet. 1994. Indoor air pollution: an introduction for health professionals. In: Indoor air pollution: an introduction for health professionals. *EPA*.

Cox, H. H. J., J. H. M. Houtman, H. J. Doddema and W. Harder. 1993. Enrichment of fungi and degradation of styrene in biofilters. *Biotechnology Letters* 15 (7): 737–742. doi: 10.1007/bf01080148.

Cox, H. H. J., F. J. Magielsen, H. J. Doddema and W. Harder. 1996. Influence of the water content and water activity on styrene degradation by *Exophiala jeanselmei* in biofilters. *Applied Microbiology and Biotechnology* 45 (6): 851–856. doi: 10.1007/s002530050773.

Cheng, Y., H. He, C. Yang, G. Zeng, X. Li, H. Chen and G. Yu. 2016. Challenges and solutions for biofiltration of hydrophobic volatile organic compounds. *Biotechnology Advances* 34 (6): 1091–1102. doi: https://doi.org/10.1016/j.biotechadv.2016.06.007.

Cheng, Z., C. Li, C. Kennes, J. Ye, D. Chen, S. Zhang, J. Chen and J. Yu. 2017. Improved biodegradation potential of chlorobenzene by a mixed fungal-bacterial consortium. *International Biodeterioration & Biodegradation* 123: 276–285. doi: https://doi.org/10.1016/j.ibiod.2017.07.008.

Cheng, Z., L. Lu, C. Kennes, J. Yu and J. Chen. 2016. Treatment of gaseous toluene in three biofilters inoculated with fungi/bacteria: Microbial analysis, performance and starvation response. *Journal of Hazardous Materials* 303: 83–93. doi: https://doi.org/10.1016/j.jhazmat.2015.10.017.

Christen, P., F. Domenech, G. Michelena, R. Auria and S. Revah. 2002. Biofiltration of volatile ethanol using sugar cane bagasse inoculated with *Candida utilis*. *Journal of Hazardous Materials* 89 (2): 253–265. doi: https://doi.org/10.1016/S0304-3894(01)00314-4.

Chung, Y.-C., C. Huang and C.-P. Tseng. 1997. Removal of hydrogen sulphide by immobilized *Thiobacillus* sp. strain CH11 in a biofilter. *Journal of Chemical Technology & Biotechnology* 69 (1): 58–62. doi: doi:10.1002/(SICI)1097-4660(199705)69:1<58::AID-JCTB660>3.0.CO;2-H.

Davison, B. H., J. W. Barton, K. T. Klasson and A. B. Francisco. 2000. Influence of high biomass concentrations on alkane solubilities. *Biotechnology and Bioengineering* 68 (3): 279–284. doi: doi:10.1002/(SICI)1097-0290(20000505)68:3<279::AID-BIT6>3.0.CO;2-P.

Deacon, J. W. 1997. *Introduction to Modern Mycology*. 3 ed. Doncaster, United Kingdom: Wiley-Blackwell Science.

Dorado, A. D., G. Baquerizo, J. P. Maestre, X. Gamisans, D. Gabriel and J. Lafuente. 2008. Modeling of a bacterial and fungal biofilter applied to toluene abatement: Kinetic parameters estimation and model validation. *Chemical Engineering Journal* 140 (1): 52–61. doi: https://doi.org/10.1016/j.cej.2007.09.004.

Ergas, S. J., K. Kinney, M. E. Fuller and K. M. Scow. 1994. Characterization of compost biofiltration system degrading dichloromethane. *Biotechnology and Bioengineering* 44 (9): 1048–1054. doi: doi:10.1002/bit.260440905.

Estévez, E., M. C. Veiga and C. Kennes. 2005. Biofiltration of waste gases with the fungi *Exophiala oligosperma* and *Paecilomyces variotii*. *Applied Microbiology and Biotechnology* 67 (4): 563–568. doi: 10.1007/s00253-004-1786-0.

Estrada, J. M., N. J. R. Kraakman, R. Lebrero and R. Muñoz. 2012. A sensitivity analysis of process design parameters, commodity prices and robustness on the economics of odour abatement technologies. *Biotechnology Advances* 30 (6): 1354–1363. doi: https://doi.org/10.1016/j.biotechadv.2012.02.010.

Estrada, J. M., S. Hernández, R. Muñoz and S. Revah. 2013. A comparative study of fungal and bacterial biofiltration treating a VOC mixture. *Journal of Hazardous Materials* 250–251: 190–197. doi: https://doi.org/10.1016/j.jhazmat.2013.01.064.

Ferdowsi, M., A. A. Ramirez, J. P. Jones and M. Heitz. 2017. Elimination of mass transfer and kinetic limited organic pollutants in biofilters: A review. *International Biodeterioration & Biodegradation* 119: 336–348. doi: https://doi.org/10.1016/j.ibiod.2016.10.015.

García-Peña, E. I., S. Hernández, E. Favela-Torres, R. Auria and S. Revah. 2001. Toluene biofiltration by the fungus *Scedosporium apiospermum* TB1. *Biotechnology and Bioengineering* 76 (1): 61–69. doi: doi:10.1002/bit.1026.

García-Peña, I., S. Hernández, R. Auria and S. Revah. 2005. Correlation of biological activity and reactor performance in biofiltration of toluene with the fungus *Paecilomyces variotii* CBS115145. *Applied and Environmental Microbiology* 71 (8): 4280–4285. doi: 10.1128/AEM.71.8.4280-4285.2005.

García-Peña, I., I. Ortiz, S. Hernández and S. Revah. 2008. Biofiltration of BTEX by the fungus *Paecilomyces variotii*. *International Biodeterioration & Biodegradation* 62 (4): 442–447. doi: https://doi.org/10.1016/j.ibiod.2008.03.012.

Gervats, P. P. Molin, W. Grajek and M. Bensoussan. 1988. Influence of the water activity of a solid substrate on the growth rate and sporogenesis of filamentous fungi. *Biotechnology and Bioengineering* 31 (5): 457–463. doi: doi:10.1002/bit.260310510.

Gostomski, P. A., J. B. Sisson and R. S. Cherry. 1997. Water content dynamics in biofiltration: The role of humidity and microbial heat generation. *Journal of the Air & Waste Management Association* 47 (9): 936–944. doi: 10.1080/10473289.1997.10463952.

Guieysse, B., C. Hort, V. Platel, R. Munoz, M. Ondarts and S. Revah. 2008. Biological treatment of indoor air for VOC removal: Potential and challenges. *Biotechnology Advances* 26 (5): 398–410. doi: https://doi.org/10.1016/j.biotechadv.2008.03.005.

Gutiérrez-Acosta, O. B., V. A. Escobar-Barrios and S. Arriaga. 2010. Batch biodegradation of hydrocarbon vapors using a modified polymeric support. *Journal of Chemical Technology & Biotechnology* 85 (3): 410–415. doi: doi:10.1002/jctb.2329.

Gutiérrez-Acosta, O. B., S. Arriaga, V. A. Escobar-Barrios, S. Casas-Flores and A. Almendarez-Camarillo. 2012. Performance of innovative PU-foam and natural fiber-based composites for the biofiltration of a mixture of volatile organic compounds by a fungal biofilm. *Journal of Hazardous Materials* 201–202: 202–208. doi: https://doi.org/10.1016/j.jhazmat.2011.11.068.

Hernández-Meléndez, O., E. Bárzana, S. Arriaga, M. Hernández-Luna and S. Revah. 2008. Fungal removal of gaseous hexane in biofilters packed with poly(ethylene carbonate) pine sawdust or peat composites. *Biotechnology and Bioengineering* 100 (5): 864–871. doi: doi:10.1002/bit.21825.

Jin, Y., L. Guo, M. C. Veiga and C. Kennes. 2007. Fungal biofiltration of α-pinene: Effects of temperature, relative humidity, and transient loads. *Biotechnology and Bioengineering* 96 (3): 433–443. doi: doi:10.1002/bit.21123.

Jorio, H., Y. Jin, H. Elmrini, J. Nikiema, R. Brzezinski and M. Heitz. 2009. Treatment of VOCs in biofilters inoculated with fungi and microbial consortium. *Environmental Technology* 30 (5): 477–485. doi: 10.1080/09593330902778849.

Kennes, C. and M. C. Veiga. 2004. Fungal biocatalysts in the biofiltration of VOC-polluted air. *Journal of Biotechnology* 113 (1): 305–319. doi: https://doi.org/10.1016/j.jbiotec.2004.04.037.

Kovárová-Kovar, K. and T. Egli. 1998. Growth kinetics of suspended microbial cells: From single-substrate-controlled growth to mixed-substrate kinetics. *Microbiology and Molecular Biology Reviews* 62 (3): 646–666.

Lebrero, R., J. C. López, I. Lehtinen, R. Pérez, G. Quijano and R. Muñoz. 2016. Exploring the potential of fungi for methane abatement: Performance evaluation of a fungal-bacterial biofilter. *Chemosphere* 144: 97–106. doi: https://doi.org/10.1016/j.chemosphere.2015.08.017.

Linder, M. B., G. R. Szilvay, T. Nakari-Setälä and M. E. Penttilä. 2005. Hydrophobins: the protein-amphiphiles of filamentous fungi. *FEMS Microbiology Reviews* 29 (5): 877–896. doi: https://doi.org/10.1016/j.femsre.2005.01.004.

Maestre, J. P., X. Gamisans, D. Gabriel and J. Lafuente. 2007. Fungal biofilters for toluene biofiltration: Evaluation of the performance with four packing materials under different operating conditions. *Chemosphere* 67 (4): 684–692. doi: https://doi.org/10.1016/j.chemosphere.2006.11.004.

Mendoza, J. A., Ó. J. Prado, M. C. Veiga and C. Kennes. 2004. Hydrodynamic behaviour and comparison of technologies for the removal of excess biomass in gas-phase biofilters. *Water Research* 38 (2): 404–413. doi: https://doi.org/10.1016/j.watres.2003.09.014.

Miller, J. D. and D. R. McMullin. 2014. Fungal secondary metabolites as harmful indoor air contaminants: 10 years on. *Applied Microbiology and Biotechnology* 98 (24): 9953–9966. doi: 10.1007/s00253-014-6178-5.

Moe, W. M. and B. Qi. 2004. Performance of a fungal biofilter treating gas-phase solvent mixtures during intermittent loading. *Water Research* 38 (9): 2259–2268. doi: https://doi.org/10.1016/j.watres.2004.02.017.

Morales, M., S. Hernández, T. Cornabé, S. Revah and R. Auria. 2003. Effect of drying on biofilter performance: Modeling and experimental approach. *Environmental Science & Technology* 37 (5): 985–992. doi: 10.1021/es025970w.

Morales, M., S. Revah and R. Auria. 1998. Start-up and the effect of gaseous ammonia additions on a biofilter for the elimination of toluene vapors. *Biotechnology and Bioengineering* 60 (4): 483–491. doi: doi:10.1002/(SICI)1097-0290(19981120)60:4<483::AID-BIT10>3.0.CO;2-J.

Morales, P., M. Cáceres, F. Scott, L. Díaz-Robles, G. Aroca and A. Vergara-Fernández. 2017. Biodegradation of benzo[α]pyrene, toluene, and formaldehyde from the gas phase by a consortium of *Rhodococcus erythropolis* and *Fusarium solani*. *Applied Microbiology and Biotechnology* 101 (17): 6765–6777. doi: 10.1007/s00253-017-8400-8.

Morgan-Sagastume, F., B. E. Sleep and D. G. Allen. 2001. Effects of biomass growth on gas pressure drop in biofilters. *Journal of Environmental Engineering* 127 (5): 388–396. doi: doi:10.1061/(ASCE)0733-9372(2001)127:5(388).

Mudliar, S., B. Giri, K. Padoley, D. Satpute, R. Dixit, P. Bhatt, R. Pandey, A. Juwarkar and A. Vaidya. 2010. Bioreactors for treatment of VOCs and odours—A review. *Journal of Environmental Management* 91 (5): 1039–1054. doi: https://doi.org/10.1016/j.jenvman.2010.01.006.

Pahm, M. A. and M. Alexander. 1993. Selecting inocula for the biodegradation of organic compounds at low concentrations. *Microbial Ecology* 25 (3): 275–286. doi: 10.1007/bf00171893.

Portune, K. J., M. C. Pérez, F. J. Álvarez-Hornos and C. Gabaldón. 2015. Investigating bacterial populations in styrene-degrading biofilters by 16S rDNA tag pyrosequencing. *Applied Microbiology and Biotechnology* 99 (1): 3–18. doi: 10.1007/s00253-014-5868-3.

Prenafeta-Boldú, F. X., R. Summerbell and G. Sybren de Hoog. 2006. Fungi growing on aromatic hydrocarbons: biotechnology's unexpected encounter with biohazard? *FEMS Microbiology Reviews* 30 (1): 109–130. doi: 10.1111/j.1574-6976.2005.00007.x.

Prenafeta-Boldú, F. X., J. Illa, J. W. van Groenestijn and X. Flotats. 2008. Influence of synthetic packing materials on the gas dispersion and biodegradation kinetics in fungal air biofilters. *Applied Microbiology and Biotechnology* 79 (2): 319–327. doi: 10.1007/s00253-008-1433-2.

Prenafeta-Boldú, F. X., M. Guivernau, G. Gallastegui, M. Viñas, G. S. de Hoog and A. Elías. 2012. Fungal/bacterial interactions during the biodegradation of TEX hydrocarbons (toluene, ethylbenzene and p-xylene) in gas biofilters operated under xerophilic conditions. *FEMS Microbiology Ecology* 80 (3): 722–734. doi: 10.1111/j.1574-6941.2012.01344.x.

Qi, B., W. Moe and K. Kinney. 2002. Biodegradation of volatile organic compounds by five fungal species. *Applied Microbiology and Biotechnology* 58 (5): 684–689. doi: 10.1007/s00253-002-0938-3.

Ralebitso-Senior, T. K., E. Senior, R. Di Felice and K. Jarvis. 2012. Waste gas biofiltration: Advances and limitations of current approaches in microbiology. *Environmental Science & Technology* 46 (16): 8542–8573. doi: 10.1021/es203906c.

Rene, E. R., M. C. Veiga and C. Kennes. 2010. Biodegradation of gas-phase styrene using the fungus *Sporothrix variecibatus*: Impact of pollutant load and transient operation. *Chemosphere* 79 (2): 221–227. doi: https://doi.org/10.1016/j.chemosphere.2010.01.036.

Revah, S. and J. M. Morgan-Sagastume. 2005. Methods of odor and VOC control. In: *Biotechnology for Odor and Air Pollution Control*, edited by Zarook Shareefdeen and Ajay Singh, 29–63. Berlin, Heidelberg: Springer Berlin Heidelberg.

Revah, S., A. Vergara-Fernández and S. Hernández. 2011. Fungal biofiltration for the elimination of gaseous pollutants from air. In: *Mycofactories*, edited by Ana Lúcia Leitão, 109–120. Bentham Science Publishers.

Salehahmadi, R., R. Halladj and S. M. Zamir. 2012. Unsteady-state mathematical modeling of a fungal biofilter treating hexane vapor at different operating temperatures. *Industrial & Engineering Chemistry Research* 51 (5): 2388–2396. doi: 10.1021/ie2014718.

Saucedo-Lucero, J. O., G. Quijano, S. Arriaga and R. Muñoz. 2014. Hexane abatement and spore emission control in a fungal biofilter-photoreactor hybrid unit. *Journal of Hazardous Materials* 276: 287–294. doi: https://doi.org/10.1016/j.jhazmat.2014.05.040.

Spigno, G., C. Pagella, M. D. Fumi, R. Molteni and D. M. De Faveri. 2003. VOCs removal from waste gases: gas-phase bioreactor for the abatement of hexane by *Aspergillus niger*. *Chemical Engineering Science* 58 (3): 739–746. doi: https://doi.org/10.1016/S0009-2509(02)00603-6.

Spigno, G. and D. M. De Faveri. 2005. Modeling of a vapor-phase fungi bioreactor for the abatement of hexane: Fluid dynamics and kinetic aspects. *Biotechnology and Bioengineering* 89 (3): 319–328. doi: doi:10.1002/bit.20336.

Sun, Y., X. Quan, J. Chen, F. Yang, D. Xue, Y. Liu and Z. Yang. 2002. Toluene vapour degradation and microbial community in biofilter at various moisture content. *Process Biochemistry* 38 (1): 109–113. doi: https://doi.org/10.1016/S0032-9592(02)00056-0.

Tros, M. E., T. N. Bosma, G. Schraa and A. J. Zehnder. 1996. Measurement of minimum substrate concentration (smin) in a recycling fermentor and its prediction from the kinetic parameters of *Pseudomonas* strain B13 from batch and chemostats cultures. *Applied and Environmental Microbiology* 62 (10): 3655–3661.

van Groenestijn, J. W., W. N. M. van Heiningen and N. J. R. Kraakman. 2001. Biofilters based on the action of fungi. *Water Science and Technology* 44 (9): 227–232. doi: 10.2166/wst.2001.0546.

Vergara-Fernández, A., B. Van Haaren and S. Revah. 2006. Phase partition of gaseous hexane and surface hydrophobicity of *Fusarium solani* when grown in liquid and solid media with hexanol and hexane. *Biotechnology Letters* 28 (24): 2011–2017. doi: 10.1007/s10529-006-9186-4.

Vergara-Fernández, A., S. Hernández and S. Revah. 2008. Phenomenological model of fungal biofilters for the abatement of hydrophobic VOCs. *Biotechnology and Bioengineering* 101 (6): 1182–1192. doi: doi:10.1002/bit.21989.

Vergara-Fernández, A., S. Hernández and S. Revah. 2011a. Elimination of hydrophobic volatile organic compounds in fungal biofilters: Reducing start-up time using different carbon sources. *Biotechnology and Bioengineering* 108 (4): 758–765. doi: doi:10.1002/bit.23003.

Vergara-Fernández, A., S. Hernández, J. San Martin-Davison and S. Revah. 2011b. Morphological characterization of aerial hyphae and simulation growth of *Fusarium solani* under different carbon source for application in the hydrophobic VOCs biofiltration. *Revista Mexicana de ingenieria Quimica* 10: 225–233.

Vergara-Fernández, A., O. Soto-Sanchez and J. Vasquez. 2011c. Effects of packing material type on n-pentane/biomass partition coefficient for use in fungal biofilters. *Chemical and Biochemical Engineering Quarterly* 25 (4): 439–444.

Vergara-Fernández, A., S. Hernández, R. Muñoz and S. Revah. 2012a. Influence of the inlet load, EBRT and mineral medium addition on spore emission by *Fusarium solani* in the fungal biofiltration of hydrophobic VOCs. *Journal of Chemical Technology & Biotechnology* 87 (6): 778–784. doi: doi:10.1002/jctb.3762.

Vergara-Fernández, A., V. Salgado-Ísmodes, M. Pino, S. Hernández and S. Revah. 2012b. Temperature and moisture effect on spore emission in the fungal biofiltration of hydrophobic VOCs. *Journal of Environmental Science and Health, Part A* 47 (4): 605–613. doi: 10.1080/10934529.2012.650581.

Vergara-Fernández, A., J. San Martín-Davison, L. A. Díaz-Robles and O. Soto-Sánchez. 2014. Effect of initial substrate/inoculum ratio on cell yield in the removal of hydrophobic VOCs in fungal biofilters. *Revista Mexicana de Ingeniería Química* 13: 749–755.

Vergara-Fernández, A., F. Scott, P. Moreno-Casas, L. Díaz-Robles and R. Muñoz. 2016. Elucidating the key role of the fungal mycelium on the biodegradation of n-pentane as a model hydrophobic VOC. *Chemosphere* 157: 89–96. doi: https://doi.org/10.1016/j.chemosphere.2016.05.034.

Vigueras, G., K. Shirai, D. Martins, T. T. Franco, L. F. Fleuri and S. Revah. 2008. Toluene gas phase biofiltration by *Paecilomyces lilacinus* and isolation and identification of a hydrophobin protein produced there of. *Applied Microbiology and Biotechnology* 80 (1): 147. doi: 10.1007/s00253-008-1490-6.

Vigueras, G., S. Arriaga, K. Shirai, M. Morales and S. Revah. 2009. Hydrophobic response of the fungus *Rhinocladiella similis* in the biofiltration with volatile organic compounds with different polarity. *Biotechnology Letters* 31 (8): 1203–1209. doi: 10.1007/s10529-009-9987-3.

Woertz, J. R., K. A. Kinney, N. D. P. McIntosh and P. J. Szaniszlo. 2001a. Removal of toluene in a vapor-phase bioreactor containing a strain of the dimorphic black yeast *Exophiala lecanii*-corni. *Biotechnology and Bioengineering* 75 (5): 550–558. doi: doi:10.1002/bit.10066.

Woertz, J. R., K. A. Kinney and P. J. Szaniszlo. 2001b. A fungal vapor-phase bioreactor for the removal of nitric oxide from waste gas streams. *Journal of the Air & Waste Management Association* 51 (6): 895–902. doi: 10.1080/10473289.2001.10464321.

World Health Organization. 2006. Air quality guidelines for particulate matter, ozone, nitrogen dioxide and sulfur dioxide—Global update 2005—Summary of risk assessment. *WHO*.

World Health Organization. 2010. WHO guidelines for indoor air quality: selected pollutants. *WHO*.

Xie, B., S. B. Liang, Y. Tang, W. X. Mi and Y. Xu. 2009. Petrochemical wastewater odor treatment by biofiltration. *Bioresource Technology* 100 (7): 2204–2209. doi: https://doi.org/10.1016/j.biortech.2008.10.035.

Zhai, J., P. Shi, Z. Wang, H. Jiang and C. Long. 2017. A comparative study of bacterial and fungal-bacterial steady-state stages of a biofilter in gaseous toluene removal: performance and microbial community. *Journal of Chemical Technology & Biotechnology* 92 (11): 2853–2861. doi: doi:10.1002/jctb.5302.

PART III
USEFUL TOOLS

CHAPTER-10

A Learning Journey on Toxico-Proteomics

The Neglected Role of Filamentous Fungi in the Environmental Mitigation of Pentachlorophenol

Celso Martins,[1,#] *Isabel Martins,*[1,#] *Tiago Martins,*[1,#]
Adélia Varela[1,2,#] **and** *Cristina Silva Pereira*[1,*]

1. Introduction

Proteomics stands for the *"protein complement of the expressed genome by an organism in a particular biological state"* (Wilkins et al. 1996). Proteins are key multifunctional operators in cells, acting as essential structural components, catalysts, and major regulatory elements. Therefore, proteomic-based tools can potentially unravel metabolic reactions and regulatory cascades taking place in an organism and/or cellular compartment at a specific condition, as well as their relative abundance, stability, and post-translational modifications (Bianco and Perrotta 2015). When "proteomics" is headed by "meta-", "environmental-" or "community-", it moves beyond that of a single species or cellular compartment to reach a microbial population: *"the large-scale characterization of the entire protein complement of environmental microbiota at a given point in time"* (Wilmes et al. 2015). It goes without saying that as such, the contribution of uncultured microorganisms is also accounted for, offering means to resolve major catalytic units of microbial populations (Wilmes et al. 2015) and to understand microbial networks in an effective way

---

[1] Instituto de Tecnologia Química e Biológica António Xavier, Universidade Nova de Lisboa (ITQB NOVA), Av. da República, 2780–157, Oeiras, Portugal.

[2] Instituto Nacional Investigação Agrária e Veterinária, Av. da República, 2780–157, Oeiras, Portugal

[#] Equally contributing authors listed alphabetically.

[*] Corresponding Author: spereira@itqb.unl.pt

(Gotelli et al. 2012). Scientific progresses are continuously increasing the resolution of proteomics and facilitating its integration in systems approaches for modelling complex phenomena. However, the major bottleneck in proteomics research is still that *"our ability to generate (proteomic) data now outstrips our ability to analyse it"* (Patterson 2003).

This chapter discloses our learning journey in toxicoproteomics, aiming to understand how pollution affects the functioning of below-ground fungi, from a single species to communities. It covers also fundamentals of environmental pollution, with emphasis on mitigation processes mediated by filamentous fungi. A historical perspective of the use and hazards of pentachlorophenol (PCP)—our archetypal man-made halogenated pollutant; is also enclosed. The grounds and major developments of proteomics are comprehensively analysed, including our toxicoproteomics-led studies in filamentous fungi. Our studies have contributed to highlight the widespread remedial potential of Ascomycota and Zygomycota, and to disclose central pathways for the catabolism of aromatics in a model fungus. Collectively our journey has been setting the knowledge foundations to move beyond that of a single fungal species; we are now undertaking complex metaproteomic analyses on the specialisation of fungal communities during the mitigation of PCP.

2. Environmental Pollution is Globally Dispersed

The global dispersion of pollutants constitutes a critical and extremely alarming problem; especially as many undergo transport through middle and long range distances, from Persistent Organic Pollutants (POPs), to microplastics and nanoparticles (Fenger 2009). Numerous environmental processes are contributing to their global dispersion, yet the atmosphere plays the most central role in scattering pollution with limitless impacts on public health. Volatile compounds are generally regarded as the major contributors of atmospheric pollution, yet aggregation of compounds (including poor or non-volatile) to dust or particulate air matter ensures their long range atmospheric transport (LRAT) (Muir and Howard 2006, United Nations 2001). It has been shown that LRAT can be responsible for *ca.* 70–90% of the POPs levels found in EU countries (United Nations 2001, Kallenborn et al. 2007).

POPs persist for long in the environment, may reach regions far away from their application source, and are dispersed throughout the environment as a result of natural processes (Fenger 2009). They are detected all over the planet, including remote locations where no significant local reservoirs prevail (United Nations 2015). Their global dispersion is known as the "Grasshopper Effect" or "Global Distillation"—a geochemical process that mediates the transport of chemicals from warmer to colder regions of the Earth (Semeena and Lammel 2005). This phenomenon justifies the detection of POPs in the Arctic environment and in the corpses of local animals and people, especially in fatty tissues (Li and Macdonald 2005); compounds not used locally.

Reports on acute exposure to toxic pollutants (short-term exposure to a high concentration) are becoming increasingly scarce, possibly due to increase liability issues in most developed countries (Briggs 2003). On the contrary, the continuous discharge of various toxic chemicals at (very) low doses is increasingly imposing a major (often inaudible) threat to public health and environmental stability (Muir and Howard 2006). The impact of both acute and chronic pollution has been widely investigated using model organisms and laboratory simulations, as well as incidence reports and/or theoretical models (Briggs 2003).

3. Pentachlorophenol: from a Reputable Pesticide to an Obsolete Chemical

Pentachlorophenol (PCP), which was first described in 1841 (Erdmann 1841), started to be commercialised in 1936 essentially as pesticide, antifungal, and antimicrobial due to its *"high degree of effectiveness in biological control, combined with its desirable physical properties"* (Carswell and Nason 1938) (Fig. 10.1). PCP (as many other pesticides) was initially seen as a useful chemical to solve severe contamination problems in agriculture and industry (Varela et al. 2017). Soon the perception of safe use of pesticides conflicted with scientific knowledge on their poisoning effects. Until 1940, applications and properties of PCP were often reported, e.g., Carswell and Nason 1938, Carswell 1939, whereas its toxicity was analysed only twice, specifically in rabbits (Kehoe et al. 1939) and in humans (skin irritation) (Boyd et al. 1940). During the following decades, new reports on PCP toxicity came to light (Nomura 1953, Sugiura et al. 1956), including the first report on poisoning in humans (Baader and Bauer 1951). Its mechanism of action was also disclosed, *viz.* uncouples mitochondrial oxidative phosphorylation (Weinbach 1957) (Fig. 10.1). Throughout the 1960s, PCP poisoning effects in humans were more often reported, associated to direct industrial handling (Chapman and Rbson 1965) or use of daily products washed with PCP based products, such as diapers (Brown 1970). Progresses in analytical techniques allowed higher accuracy in the detection of PCP, either in human tissues/fluids or environmental samples (Chapman and Rbson 1965, Bevenue et al. 1966). The Stockholm Declaration was adopted by the United Nations (UN)

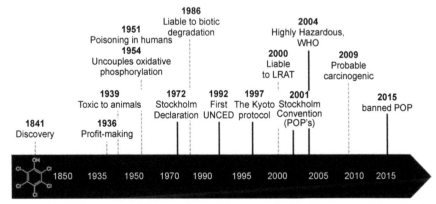

Figure 10.1 Schematic representation of the historical path of pentachlorophenol (PCP), since its discovery to its classification as banned persistent organic pollutant (POP). Continuous and dashed lines mark key landmarks on PCP liability and scientific discoveries, respectively.

in 1972, acknowledging the right to a healthy environment (United Nations 1972), and giving rise to the first restrictions on PCP production and uses (Hattemer-Frey and Travis 1989); finally the era on the liability of PCP commenced (Fig. 10.1). Not surprisingly also PCP biotic degradation started to be scrutinised (Lamar and Dietrich 1990, Mikesell and Boyd 1986). The recognition that human activities were seriously affecting the balance of the Earth's ecosystems- UN Conference on Environment and Development (UNCED, Rio de Janeiro, 1992) (United Nations 1992)—boosted the importance of environmental pollution; naturally PCP was not an exception (McAllister et al. 1996, Reddy and Gold 2000). In the Stockholm Convention on POPs (United Nations 2001), unexpected PCP was not classified as POP, and was left out from the original list of the "Dirty Dozen" (United Nations 2001). Only in 2015, PCP was recognised as a banned POP (United Nations 2015), obviously supported by numerous scientific evidences on its far-reaching hazards (Borysiewicz 2008, Zheng et al. 2011, Czaplicka 2004).

The global dispersion of PCP is a consequence of its long history of use, persistence, and transboundary nature (Hoferkamp et al. 2010, Borysiewicz 2008, Czaplicka 2004), as well as its current usage in many locations worldwide, e.g., in China (Zheng et al. 2012), and its formation during the degradation of more volatile pollutants, e.g., pentachlorobenzene and hexachlorobenzene (Sandau et al. 2000, Kovacevic and Sabljic 2016). Nowadays, PCP is detected in human fluids and tissues, worldwide, due to exposure in both indoor and outdoor environments, e.g. (Morgan 2015, Piskorska-Pliszczynska et al. 2016). One of the main reasons behind the unwillingness of classifying PCP as a POP was its low vapor pressure, mostly advertised in the early years of usage (Carswell and Nason 1938), initially understood as a poor ability for undergoing LRAT. Conversely, in the last decade, PCP's capability to undergo long-range transport has been well established (Borysiewicz 2008, United Nations 2015), similar to that reported to other POPs.

Since the 1990s, scientific and anecdotal evidence of contamination of the bark of *Quercus suber* (cork oak) with PCP and its derivatives exists (Silva Pereira et al. 2000a, McLellan et al. 2007). These forests cover *ca.* 1.5 million ha in Europe and 700,000 ha in North Africa (Bugalho et al. 2011); it is presently a source of income for thousands of people, especially due to its most added-value product, the cork bottle-stoppers. The cork taint defect in bottled-wines (responsible for major losses in the cork industry in the 1990s) is largely associated with the presence of chloroanisoles (Silva Pereira et al. 2000a). Their most direct precursors are PCP, 2,4,6-trichlorophenol (2,4,6-TCP) and 2,3,4,6-tetrachlorophenol (2,3,4,6-TeCP), originating through microbial O-methylation, the corresponding anisole. Chloroanisoles are still occasionally identified in cork and its derivatives, regardless of current industrial best practice designed to eradicate any chlorinated phenol and/ or anisole from processed cork. The cork bark behaves as a sampler, accumulating both gaseous and particulate pollutants (Zhao et al. 2008), yet PCP partition to the soil is likely to be significant, estimated to be as high as 95% (Borysiewicz 2008). This has inspired us to seek the prevalence of active sources of PCP pollution in soils from Tunisian cork oak forests (Tabarka, N.W. Tunisia (McLellan et al. 2014, Varela et al. 2015)). Notably, soils collected along 100 km^2 within these forests contained PCP levels ranging from 2 to 30 µg·kg⁻¹ (Varela et al. 2015), similar to levels detected at locations now using PCP as antiparasitic (Zheng et al. 2012). The atmospheric deposition of PCP (or its precursors) is likely contributing to the prevalence of the biocide in these forests; its half-life in air and environment has been estimated to be 7.4 days and 1.5 months, respectively, with the potential to be transported over 1500 to 3000 km (Borysiewicz 2008). The contribution of regional forest management practices still relying on PCP usage cannot be disregarded, especially since the Tunisian legislation on PCP is not very prohibitive (Varela et al. 2017). PCP was also found prevalent in soils of European cork oak forests under qualified management (*unpublished data*), where the most probable source is the atmospheric deposition of the biocide. It remains poorly understood how chronic exposure to this biocide is affecting the functional diversity of below-ground microbes—in particular, of fungi that contribute greatly to preserve the functioning and ecological balance of forest soils (Harms et al. 2011, Žifčáková et al. 2016, Baldrian et al. 2012).

4. Microbial Decay of Critical Pollutants: Fungi as Main Protagonists

All the polluted Earth ecosystems are colonised by microorganism that display matchless capabilities to degrade and/or transform a wide diversity of POPs (El-Naas et al. 2017). Soil is likely the most critical environmental compartment when accounting for the reception, transference, and dispersion of pollution, particularly through microbial biotic processes. Many of these processes are potentially mediated by fungi in virtue of their high abundance (nearly ¾ of the soil microbial biomass), broad enzymatic capacities, extensive hyphae reach, and high surface-to-cell ratio (Van Der Heijden et al. 2008, Harms et al. 2011).

Fungi are a highly diverse and heterogeneous Kingdom of organisms, and most likely the second most common on Earth. They are widely distributed across all biomes (in mutualistic, pathogenic, and commensal relationships) and can grow in adverse conditions of low nutrient availability, low water activity, and low pH. Till date, over 120,000 fungi species are described, yet 2.2 to 3.8 million species have been estimated to exist (Hawksworth and Lücking 2017).

Fungi (heterotrophs) act as the major recyclers of organic matter, secreting enzymes that ensure degradation of complex and recalcitrant natural polymers, such as keratin, chitin, and lignin. Taken as an example, plant litter decomposition is initially accomplished mostly by Ascomycota, which are largely replaced at latter stages and at lower soil depths by Basidiomycota (Voříšková and Baldrian 2013, Voříšková et al. 2014). The soil mycobiota ensures even the degradation of lignin; its amount decreases in lower soil horizons reaching vestigial levels in the stable clay fractions (Schöning et al. 2005). Lignin degradation mainly involves lignin-modifying class II heme peroxidases (*viz.* lignin peroxidase – LiP, and manganese peroxidase – MnP), of which the origin has been linked to an ancestor of wood white rot Basidiomycota (Floudas et al. 2012). Other types of fungal peroxidases have been identified, namely unspecific or aromatic peroxygenases, dye peroxidase, and haloperoxidase, which are capable of undertaking diverse reactions, e.g., aromatic peroxygenation, double-bond epoxidation, and hydroxylation of aliphatic compounds, or to transform compounds poorly accepted by other peroxidases (e.g., textile dyes) (Hofrichter et al. 2010, Kües 2015). The lignolytic mechanisms used by Ascomycota are less understood, notwithstanding they are efficient wood degraders in high moisture conditions (Fukasawa et al. 2011), possibly assuming a major role in aquatic habitats where Basidiomycota are rare (Harms et al. 2011). Fungi can also use non-enzymatic Fenton chemistry to generate highly reactive hydroxyl radicals that mediate depolymerization of cellulose, as well as degradation of some pollutants (Harms et al. 2011). In addition, fungi own numerous and diverse cytochrome P450 enzymes (nearly 150 in some genomes) that are involved in secondary metabolism and detoxification of xenobiotics (Lah et al. 2011).

The non-specificity and versatility of fungal degradative enzymes has been the source of inspiration for research on the degradation of aromatic and aliphatic xenobiotics by fungi (Pinedo-Rivilla et al. 2009). A parallel between the ligninolytic metabolism in fungi and the degradation of pollutants had been initially considered; it is well-illustrated in a first report on POPs oxidation by *Phanerochaete chrysosporium* (Basidiomycota) (Bumpus et al. 1985). Nonetheless, in many subsequent studies, the capability of non-ligninolytic Ascomycota and Zygomycota to degrade diverse pollutants become apparent (Harms et al. 2011, Pinedo-Rivilla et al. 2009). Currently, there are no doubts that fungi from every phylum can effectively degrade an extraordinary diversity of structurally dissimilar xenobiotics, including dichlorodiphenyltrichloroethane (DDT), polychlorinated biphenyls (PCBs), polycyclic aromatic hydrocarbons (PAHs), benzene/toluene/ethyl benzene/xylene (BTEX), and dioxins, among others (Pinedo-Rivilla et al. 2009, Harms et al. 2011, Marco-Urrea et al. 2015), as well as complex mixtures, e.g., creosotes or kerosene (Lamar et al. 1994, Saratale et al. 2007). It is also known that the high tolerance to xenobiotics in fungi is associated with pleiotropic drug resistance mechanisms (Monk and Goffeau 2008).

5. Proteomics: Matchless Means for Pushing Discovery on Fungal Processes

Proteomic tools aim to separate, identify, and quantify the polypeptides (i.e., protein species) present in a biological sample. Their identification by mass spectrometry (MS) may be physically detached from their electrophoretic separation (gel dependent) or coupled using liquid chromatography (LC) separation in tandem (gel independent) (Aebersold and Mann 2016, Westermeier 2016). Gel dependent methods, which have been the work-horse of proteomics (Westermeier 2016), usually use high resolution two-dimensional gel electrophoresis (2DE), i.e., isoelectric focussing (IEF) of polypeptides, followed by electrophoretic orthogonal separation by molecular weight (MW) (i.e., sodium dodecyl sulfate-polyacrylamide gel electrophoresis, SDS-PAGE). Another example is the Blue-native (BN) PAGE that separates protein complexes according to their size/shape in a native run, subsequently splitting their individual components using SDS-PAGE (Schägger and von Jagow 1991). Relevant polypeptides can then be excised from the gels and their identification retrieved using MS-based techniques. Progressively more powerful MS-based technologies are becoming available, pushing development of gel independent proteomics (Aebersold and Mann 2016, Domon and Aebersold 2010), especially shotgun proteomics. In the end, proteins (that may have previously undergone a SDS-PAGE separation) are submitted to proteolytic digestion, then the polypeptides are analysed using LC-MS/MS.

Quantification of the polypeptides may be "label free" requiring calculation of the number of MS/MS spectra assigned to each peptide/protein or measuring the MS-signal intensity by direct integration of the chromatographic peak area of the peptide precursor ion (Lindemann et al. 2017). Before the electrophoresis run, the polypeptides can be also covalently labelled (either *in vivo* or *in vitro*) to pairs of chemically identical molecules of different stable-isotope composition (containing ^{13}C, ^{15}N, ^{18}O, or ^{36}S); heavy and light samples are then pooled and analysed together by MS for relative protein quantification with high accuracy (Aebersold and Mann 2016, Domon and Aebersold 2010).

Currently, key improvements in MS instrumental resolution and cost reduction as well as the high availability of sequenced genomes, favor the use of gel independent methods in detriment of the gel dependent ones (Lindemann et al. 2017). The lower amount of protein needed for a gel independent analysis compared to gel dependent is likely another key causal factor. Moreover, the last presents some limitations, such as weak-resolution of polypeptides with "extreme" IE and/or MW and highly hydrophobic, presence of highly abundant proteins masking the low copy number ones, and unprecise identification/quantification of polypeptides overlapping in the same protein-spot (Westermeier 2016). Nonetheless, still today, one remarkable advantage of the gel based methods over the gel free ones is the possibility of *de novo* sequencing of individual intact proteins identified by quantitation of their levels in two or more biologically relevant states. This possibility is particularly relevant in metaproteomics, of which the information derives largely from microorganisms with non-sequenced genomes (Speda et al. 2017, Wang et al. 2016).

A search in the web of knowledge (https://apps.webofknowledge.com) using 'Proteom*' and 'Fungi' as keywords revealed nearly 1,500 publications (in July 2017), illustrating that proteomics has been widely used in research on fungi. Four to five times more studies are found when "Fungi" is replaced by either "Bacteria" or "Plant", possibly reflecting the size of the associated disciplines. Historically, proteomic analyses in fungi have been restrained by poor efficiency in protein extraction (especially due to the rigid cell wall) and reduced availability of sequenced genomes. Protein extraction/enrichment is generally hampered by their delicateness, existence of multiple isoforms, and limited abundance of a diversity of polypeptides (Patterson 2003). Nonetheless standard methods for extracting proteins from distinct fungal cellular compartments have been established whenever cultures are grown axenically (Bianco and Perrotta 2015). When moving towards environmental samples, protein extraction faces unforeseen challenges, especially from soils where many interfering compounds are particularly abundant, e.g., humic colloids that strongly bind to proteins (Keiblinger et al. 2012). Soils usually display low abundance and high heterogeneity of proteins; taken as an example, 0.1–2 µg of protein per g of semiarid soil can be directly extracted, estimated to correspond to *ca.* 7% of the total protein (Bastida et al. 2014), similar to levels typically achieved by us (*unpublished results*). Several methods have been described to overcome these constrains, yet no definitive standard protocol has been established, and some may lead to bias recovery (Speda et al. 2017, Bastida et al. 2014, Keiblinger et al. 2012).

The limited availability of sequenced genomes from filamentous fungi has initially reduced the pace of proteomics research; the genome of *Neurospora crassa*—the first filamentous fungus to be fully sequenced—was released in 2003, seven years after the completion of the yeast *Saccharomyceas cerevisiae* genome. Luckily, this paradigm has been already altered; currently there are > 800 fungal genomes sequenced. Moreover, "The 1,000 Fungal Genomes Project" aims to sequence two species for every family-level clade of Fungi (http://1000.fungalgenomes. org), producing genomic data fully descriptive of the phylogenetic diversity of the Kingdom of Fungi.

The first 2DE study on a filamentous fungus was reported more than twenty five years ago: discovery of differentially accumulated polypeptides in distinct infection structures in the phytopathogenic fungus *Uromyces viciae-fabae* (Deising et al. 1991). In the following years, gel dependent proteomics has been extensively applied to investigate the response of fungi to either xenobiotics (Carvalho et al. 2013, Martins et al. 2013, Szewczyk et al. 2014, Tang et al. 2010), antifungal drugs (Gautam et al. 2016), or various stresses (Albrecht et al. 2010, Hu et al. 2016, Ilyas et al. 2016), fungal secondary metabolism (Jami et al. 2010), development (Gonzalez-Rodriguez et al. 2015) and pathogenesis (Asif et al. 2010, Li et al. 2015), and interactions of fungi with bacteria (Moretti et al. 2010), plants (Marra et al. 2006) or insects (Khan et al. 2003). Taken as an example, extracellular multi-enzymatic complexes of hydrolytic enzymes that mediate the degradation of biomass in *Trichoderma harzianum* were disclosed using BN-PAGE (Silva et al. 2012). Presently, gel independent proteomics is providing means for in-depth characterisation of membrane (Zhao et al. 2017), mitochondrial (Casaletti et al. 2017), and extracellular (Xie et al. 2016) sub-proteomes. Moreover, it is supporting discovery in fungi biotyping (Del Chierico et al. 2012, Barker et al. 2014) and interaction with a host (Kim et al. 2013).

Representative studies on the metaproteome of soils displaying very distinct geochemical properties are depicted in Table 10.1, highlighting the major research questions and findings of each study. Metaproteomics (2DE) was first applied to unravel the functioning of a bacterial community during optimal phosphorus removal from sludge (Wilmes and Bond 2004). This has inspired further studies on metaproteomes using either gel dependent methods, e.g., soil rhizosphere (Wang et al. 2011) (Table 10.1), or gel independent ones, e.g., biofilm formation (Mueller et al. 2011, Pan et al. 2011) and litter decomposition (Schneider et al. 2012) (Table 10.1). Taken as an example, long-term deforestation was demonstrated to foster the diversity of below-ground bacteria, particularly of *Cyanobacteria* mediating carbon-fixation processes (Bastida et al. 2015a). Most metaproteomic studies on the functioning of the soil microbiota still highlight the role of bacteria in detriment of fungi, regardless that the contribution of fungal proteins may be significant (Wang et

Table 10.1 Selected metaproteomics studies on microbial populations associated to different soil habitats, in reverse chronological order. The central research question is marked in bold.

Sample	Methods	Research question and major findings	Work
Forest soils (0–5 cm)	LC-MS/MS	**Microbial functioning of the soil community during short-term and long-term warming.** Bacterial proteins >> fungal proteins (linked to high pH). Soil warming altered taxonomic diversity: ↓ [Ascomycota/Basidiomycota] ratio, ↑ Actinobacteria and Cyanobacteria; Long-term soil warming: enhanced soil respiration, increased CO_2 efflux consequently altering the functioning of the community.	(Liu et al. 2017)
Soil with tobacco litter	Protein-SIP $^{15}NO_3^-$-labeled LC-MS/MS	**Analysis of the assimilation flux of plant-derived N by the soil microbiota.** The most abundant ^{15}N-utilizer are Rhizobiales. Short-term assimilation of N: dominance of bacteria over fungi; later stages of short-term ^{15}N-leaf litter degradation: Saccharomycetales and Hypocreales are more active. Fungi abundance increases with the decreasing of water soluble C.	(Starke et al. 2016)
Dryland soils, predesertic, marshes and forests	LC-MS/MS	**Microbial functioning of the soil community in response to the C load.** Ascomycota and Basidiomycota occupy different nutritional niches: Ascomycota expressed proteins in soils of moderate C content, while Basidiomycota protein levels were higher in soils with high DOC content.	(Felipe Bastida et al. 2016b)
Permafrost (30–35 cm) and soil (65–75 cm)	LC-MS/MS	**Microbial functioning in permafrost and adjacent soil layer.** Fungi and fungal processes are poorly represented in the permafrost active microbial population. Fungal proteins (genes and transcripts) corresponded only to a small fraction in the analysed samples.	(Hultman et al. 2015)
10 years old amended soils with sewage sludge or compost	LC-MS/MS	**Functioning of the soil microbial community in soils undergoing different amendments.** Fungal proteins (of which 95% are *Ascomycota*) were 20-times < than bacterial ones, decreasing further in the amended soils. The phylogeny of the mycobiota is highly influenced by the amendment type, with the decrease and/or loss of some families (e.g., *Glomerellales* are virtually absent in soils undergoing amendment with sludge but increase for compost).	(Bastida et al. 2015b)

Table 10.1 contd. ...

... *Table 10.1 contd.*

Sample	Methods	Research question and major findings	Work
Cryosol,[1] 0–5-cm	LC-MS/MS	**Functioning of the microbial community in cryosol.** Several methanotrophic proteins were identified, constituting the first evidence of active atmospheric CH_4-oxidizing bacteria in permafrost-affected cryosols, which may help to explain the atmospheric CH_4 uptake in the polar region.	(Lau et al. 2015)
Semiarid soils	LC/MS/MS	**Functioning of the microbial community in semiarid soils.** Proteomic data are consistent with an ecological adaptation for carbon and nitrogen fixation. The amount of protein extracted from soils is scarce and largely influence by the extraction method which may be bias, e.g., Chourey and Singleton methods favor extraction of bacterial or fungal proteins with 1048 or 238 total identified proteins, respectively.	(Bastida et al. 2014)
Forest soils (0–10 cm) and commercial potting soil	LC-MS/MS	**Functioning of the soil microbial community & comparison of 4 ≠ protein extraction protocols.** The amount of protein extracted from soil is scarce and largely influence by the extraction method and soil type; methods retrieved very few identical unique spectra (0.9% and 2.9% for potting soil and forest soil, respectively) and should be optimized for particular soil types and/or research questions.	(Keiblinger et al. 2012)
Leaf litter from forests	1D-GE LC-MS/MS	**Functioning of the microbial community in leaf litter.** The season and litter nutrient content influence the community structure and function; yet Ascomycota dominated the litter decomposer community and cellulases are the most representative class of degradative enzymes. A total of 1724 unique protein clusters were identified.	(Schneider et al. 2012)
Rhizospheric soil Monocultures (1–2 years)	2DE	**Functioning of rizhospheric microbial communities (monoculture).** Proteomic data consistent with the existence of interaction between plants and microorganisms. The large majority of the identified proteins were derived from plants (75.73%), with only *ca.* 12% associated to either bacteria or fungi. Total of 103 protein spots were identified, categorized into 14 groups.	(Wu et al. 2011)

Rhizospheric soil	2DE LC-MS/MS	**Functioning of rhizospheric microbial communities (crops).** (Wang et al. 2011) Proteomic data consistent with the existence of interaction between plants and microorganisms. The large majority of the identified proteins (122 from 189 spots) were derived from plants (107), with 72 associated to either bacteria of fungi of which only 29 were associated to fungi (comprising functional classes of energy metabolism, protein metabolism, secondary metabolism, nucleotide metabolism and signal transduction).

[1] Mineral soils formed in a permafrost environment; C, N, and DOC stands for Carbon, nitrogen, and Dissolved organic carbon.

al. 2016, Arsene-Ploetze et al. 2015). For example, in the rhizosphere fungal proteins account for 12–30% of the total identified proteins (Wu et al. 2011, Wang et al. 2011) (Table 10.1).

6. Toxicoproteomics on Fungi, from Single species to Communities

Representative toxicoproteomics-based studies on the degradation of xenobiotics by filamentous fungi are depicted in Table 10.2. Most studies focussed Ascomycota (Wu et al. 2010, Yang et al. 2012, Szewczyk et al. 2014, Szewczyk et al. 2015) and Zygomycota (Sobon et al. 2016, Carvalho et al. 2013), covering xenobiotic degradation yields from 60% (1 day) (Szewczyk et al. 2014) to ~ 99% (3 days) (Carvalho et al. 2013), regardless of major increase in stress responsive processes and proteins.

In Table 10.3, selected metaproteomics studies on the influence/degradation of several xenobiotics in the functioning of microbial communities are shown, stressing their major observations. Taken as an example, the decay of 2,4-dichlorophenoxy acetic acid in soils and groundwater, was analysed using both 2DE and shot gun proteomics (Benndorf et al. 2007) (Table 10.3). Data made apparent the involvement of bacterial enzymes belonging to the chlorobenzene degradation pathway via 3-chlorocatechol to 3-oxoadipate. Direct analysis of the soil microbiota led to discovery of a bacterial enoyl-CoA that possibly mediates the degradation of benzene in an anoxic aquifer (Benndorf et al. 2009). Comparative shotgun proteomics was used instead to study the compost-assisted bioremediation of crude oils; efficient remediation was driven by *Sphingomonadales* and uncultured bacteria, with increased accumulation of catechol 2,3-dioxygenases, *cis*-dihydrodiol dehydrogenase, and 2-hydroxymuconic semialdehyde dehydrogenase (Bastida et al. 2016a) (Table 10.3).

7. Lessons from a Learning Journey on Toxicoproteomics

7.1 Past—Deciphering how fungi mitigate PCP

The first study on the degradation of PCP by fungi was undertaken in *P. chrysosporium*. PCP degradation yielded tetrachlorobenzoquinone (TeCBQ) and tetrachlorodihydroquinone (TeCHQ) (Reddy and Gold 2000). These intermediates were further dechlorinated through a reductive dehalogenase system involving a glutathione conjugate reductase that ultimately leads to full mineralisation (Reddy and Gold 2001). Subsequent studies demonstrated the potential of numerous Ascomycota and Zygomycota to degrade PCP, e.g., *T. harzianum* (Van Leeuwen et al. 1997), *Penicillium camemberti* (Taseli and Gokcay 2005), and several strains isolated by us either from cork slabs (Carvalho et al. 2011, Silva Pereira et al. 2000b) or from PCP contaminated forest soils (Varela et al. 2015). TeCHQ was also formed during the degradation of PCP by non-lignolytic Zygomycota, namely *Mucor ramosissimusis* (Szewczyk and Długoński 2009) and *M. plumbeus* (Carvalho et al. 2011), yet the degradation pathways differ as only in the first cytochrome activity it was apparently involved.

Table 10.2 Overview of major functional categories of proteins identified in representative toxicoproteomics studies on different fungal systems grown axenically in media supplemented with a xenobiotic compound.

Xenobiotic	Fungus	Method	Major classes of protein species identified	Ref.
Tributyltin	*Cunninghamella echinulata* (Z)	LC-MS/MS	Carbohydrate & Energy metabolism (malate dehydrogenase, enolase and ATP synthase) Amino acid metabolism Cell wall remodelling (chitin deacetylase) Detoxification of ROS (peroxiredoxin, nuclease C1)	(Sobon et al. 2016)
Alachlor	*Paecilomyces marquandii (A)*	2DE	Carbohydrate & Energy metabolism (malate dehydrogenase, enolase) Detoxification of ROS (SOD, catalase) Stress (HSP70) **Xenobiotic transformation** (nitrilase)	(Szewczyk et al. 2015)
4-*n*-Nonylphenol	*Metarhizium robertsii (A)*	2DE	Carbohydrate & Energy metabolism (malate dehydrogenase, pyruvate dehydrogenase) Detoxification of ROS Species (peroxiredoxin, SOD)	(Szewczyk et al. 2014)
Pentachlorophenol	*Mucor plumbeus (Z)*	2DE	Carbohydrate & Energy metabolism (enolase, glyceraldehyde 3P dehydrogenase) Cell wall remodelling (chitin deacetylase) Detoxification of ROS (cytochrome c peroxidase, thiamine biosynthesis) Stress (HSP70) **Xenobiotic transformation** (alcohol dehydrogenase)	(Carvalho et al. 2013)
Gossypol	*Aspergillus niger (A)*	2DE	Carbohydrate & Energy metabolism (malate dehydrogenase, citrate synthase) Detoxification of ROS (thiamine biosynthesis) Stress (HEX1)	(Yang et al. 2012)
Anthracene	*Fusarium solani (A)*	BN-PAGE	**Xenobiotic transformation** (Laccase)	(Wu et al. 2010)

A and Z stand for Ascomycota and Zygomycota, respectively; ROS stands for Reactive Oxygen Species.

Table 10.3 Overview of major functional categories of proteins identified in representative toxicoproteomics studies on different fungal systems grown axenically in media supplemented with a xenobiotic compound.

Sample	Methods	Major findings	Work
Microcosms exposed to **diesel** with or without compost	SDS-PAGE	Diesel increases the abundance of proteobacterial proteins yet decreasing Rhizobiales proteins. Compost addition stimulated diesel degradation; the compost-assisted bioremediation was mainly driven by Sphingomonadales; the abundance of catechol 2,3-dioxygenases, *cis*-dihydrodiol dehydrogenase and 2-hydroxymuconic semialdehyde dehydrogenase increased. Several of the identified proteins (total of 2883) are involved in the biotransformation of byphenyls.	(Bastida et al. 2016a)
Bed reactor of a rhizospheric community during **toluene** degradation	LC-MS/MS	A total of 553 proteins were identified in day and night samples, of which 32 were differential. Burkholderiales proteins constituted 40% of the total, including catabolic enzymes involved in aerobic toluene degradation. The Rhizospheric community exhibited stable protein profiles during day and night; with a stable aerobic toluene turnover by Burkholderiales. PHA synthesis was upregulated in these bacteria during day, suggesting feeding on organic root exudates, while re-utilizing the stored carbon compounds during night via the glyoxylate cycle.	(Lünsmann et al. 2016b)
Bed reactor during **toluene** biodegradation	Protein-SIP ^{13}C-labelled toluene	Burkholderiales proteins increased during toluene degradation; the microbiota apparently ensured the anaerobic toluene degradation *via* benzylsuccinate and benzoyl-CoA. Several proteins were involved in the metabolism of PHA, yet a correlation between toluene degradation and carbon storage could not be established.	(Lünsmann et al. 2016a)
Microcosms exposed to **naphthalene** or **fluorene**	Protein-SIP ^{13}C-labeled naphthalene or fluorene	The naphthalene degrading microbiota comprised essentially members of the orders Burkholderiales, Actinomycetales and Rhizobiales. The fluorene degrading community could not be disclosed. In total 847 proteins have been identified.	(Herbst et al. 2013)

An enriched **toluene** degrading community under submerged cultivation	2DE LC-MS/MS	The proteins involved in anaerobic toluene activation, dissimilatory sulphate reduction, H$_2$ production/consumption and autotrophic C fixation were associated to Desulfobulbaceae. In total 202 protein were identified (out of 236 protein spots) comprising the key enzymes for toluene degradation and sulfate reduction.	(Jehmlich et al. 2010)
Microcosms contaminated or not with **toluene**	1-DE MALDI-TOF/TOF	No proteins known to be consistently involved in toluene degradation were identified. The identified glutamine synthetase, ABC transporters, extracellular solute-binding proteins, and outer membrane proteins, possibly play a role in the detoxification of toluene. In total 47 proteins were identified.	(Williams et al. 2010)
Groundwater and microcosms contaminated with **2,4-D**	2DE 1DE, LC-MS/MS	The enzymes identified in groundwater reflected the metabolism of chlorobenzene that should represent a part of the functional metaproteome of this environment. The proteins extracted from the microcosms—autochthonous community established on 2,4-D—were similar to those of the community established after bio-stimulation (i.e., inoculation with a 2,4-D degrading bacterial community isolated from a contaminated aquifer). The optimised protocol (allows the metaproteome analysis in soils and groundwater) led to identification of 29 proteins (out of 19 in 1DE bands) and 26 proteins (out of 50 in 2DE spots), including enzymes, such as chlorocatechol dioxygenases, likely participating in the degradation of the xenobiotic.	(Benndorf et al. 2007)

2,4-D, 2,4-dichlorophenoxy acetic acid; PHA, polyhydroxyalkanoate; C, carbon

The hypothesis that the prevalence of chloroanisoles in cork is linked to PCP pollution (Silva Pereira et al. 2000a, Silva Pereira et al. 2000b) has inspired our opening studies on the PCP degradation capacity of fungal strains capable to completely perforate the cork cell walls (Silva Pereira et al. 2006). The cork cell wall comprises an inner layer of suberin (Ferreira et al. 2013), hence its degradation also suggests a potential for the degradation of both aromatic and aliphatic polymers. Most strains found prevalent in cork were observed to efficiently degrade chlorophenols (Silva Pereira et al. 2000a), including PCP, namely *P. glandicola, P. jancjewskii, T. longibrachiatum, Chrysonilia sitophila,* and *M. plumbeus* (Carvalho et al. 2009). The PCP degradation pathway of *M. plumbeus,* which could deplete virtually all the PCP in the medium, was disclosed using a metabolomics-based study (Carvalho et al. 2011). The identified intermediates included tetra- and tri-CHQ and phase II-conjugated metabolites resulting from the conjugation of sulphate, glucose, or ribose with PCP, TeCHQ, and TCHQ; the sulphate glucose conjugates were reported for the first time in fungi.

These early observations inspired our first toxicoproteomics-based study (2DE): to disclose the molecular events associated with PCP exposure in *M. plumbeus* (Carvalho et al. 2013) (Table 10.2). The identified differential polypeptides confirmed PCPs capacity to uncouple oxidative-phosphorylation in mitochondria; PCP induced stress responses and led to alterations in cell wall architecture and cytoskeleton. However, the PCP degradation pathway in this fungus remained largely concealed, similar to that seen in subsequent 2DE studies on the degradation of alachlor (Szewczyk et al. 2015) or 4-*n*-nonylphenol (Szewczyk et al. 2014) by Ascomycota (Table 10.2). In *M. plumbeus,* PCP exposure increased an alcohol dehydrogenase which may be involved in the last steps of the degradation of PCP (Carvalho et al. 2013). Instead, alachlor increased a cyanide hydratase (nitrilase) in *Paecilomyces marquandii,* which is involved in the regular cyanoamino acid metabolism (Szewczyk et al. 2015). Nitrilase putative role in the direct metabolism of alachlor cannot be disregarded since functional redundancy is expected in the co-metabolic transformation of xenobiotics. 4-*n*-Nonylphenol metabolism in *Metarhizium robertsii* proceeds by consecutive oxidations of the alkyl chain with accumulation of 4-hydroxybenzaldehyde as major intermediate, followed by aromatic ring oxidation, presumably through the protocatechuate branch of the 3-oxoadipate pathway (Szewczyk et al. 2014). The differential proteome induced by 4-*n*-nonylphenol in this fungus failed to disclose any enzyme directly involved in its metabolism (including those of the 3-oxoadipate pathway).

Recently, we have also used a 2DE differentially analysis to investigate how labdanolic acid—terpenoid found abundantly in *Cistus ladanifer* (Frija et al. 2011)—impacts *P. jancjewskii* metabolism during its stereo-selective hydroxylation to 3β-hydroxy-labdanolic acid (Frija et al. 2013). The plant terpenoid increased one putative P450 enzyme as well as stress responses, especially against oxidative stress (e.g., accumulation of superoxide dismutase) and apparently altered mitochondria functioning (Martins et al. 2017). One P450 enzyme likely hydroxylates the terpenoid, yet its unequivocal identification is yet to be attained. Disappointing results were also attained by us when using a 2DE approach to disclose the degradation pathway of suberin in *Aspergillus nidulans* (cork media versus wood media), especially as suberin degradation was negligible compared to that of cork polysaccharides (Martins et al. 2014a). The successful identification of extracellular enzymes associated with suberin degradation (shot gun proteomics) required supplementation of the growth medium with suberin, though most of the pathway was revealed by transcriptomics (Martins et al. 2014b). The last study reinforces that one critical aspect for the success of any proteomic study is the experimental design.

Similar to fungi, numerous and diverse bacteria are also capable of degrading PCP, usually relying on successive reductive dechlorination reactions that yield non-chlorinated or chlorinated phenol derivatives that are often highly toxic and recalcitrant (Bosso et al. 2015, Zhang et al. 2012, McAllister et al. 1996). Using (toxic) metabolomics we further unravelled the uniqueness of the biochemical reactions used by filamentous fungi for the degradation of chlorophenols, specifically in *A. nidulans* (Martins et al. 2014). Monochlorocatechols are recognised as key degradation intermediates of numerous chlorinated aromatic hydrocarbons, including monochlorophenols, yet their degradation in fungi was largely unknown. Two novel degradation paths were described in *A. nidulans*, namely for 4-chlorocatechol and 3-chlorocatechol, yielding 3-chlorodienelactone and catechol, respectively. Our results reinforced previous findings that enzymes mediating lactonisation of chloromuconates in fungi (i.e., 1,4-cycloisomerisation) differ from their bacterial counterparts (i.e., 3,6-cycloisomerisation). However, once more disappointing results were attained in a complementary 2DE analysis; the enzymes directly involved in the metabolism of the monochlorophenols could not be detected in the proteomic gels (*unpublished data*). To overpass these limitations, we decided to focus our toxicoproteomics-based studies on model aromatic compounds instead of chlorophenols (*see below*).

7.2 *Present: knowledge gaps on the catabolism of aromatic compounds by fungi*

The aerobic catabolism of aromatic hydrocarbons in fungi occurs ultimately via catechol, protocatechuate, hydroxyquinol, homogentisate, or gentisate, which are channelled to central pathways (Fig. 10.2) (Harwood and Parales 1996, Rieble et al. 1994). Transcriptomic-based studies have ensured discovery of the homogentisate pathway (involved, e.g., in the phenylalanine metabolism) in *P. chrysogenum* (Veiga et al. 2011), and of both the gentisate and the hydroxyquinol variant of the 3-oxoadipate pathway in *Candida parapsilosis* (Holesova et al. 2011). Recently, we have also disclosed the 3-oxoadipate pathway in *A. nidulans* by relying essentially on 2DE and gene expression assays (Martins et al. 2015). In our study, benzoate and salicylate were used as upper precursors for the protocatechuate branch and catechol branch of the 3-oxoadipate pathway, respectively (Fig. 10.2). Instead, benzoate or vanillin, have been used as precursors of the hydroxyquinol variant of the 3-oxoadipate pathway in *P. chrysosporium* (Fig. 10.2) (Shimizu et al. 2005, Matsuzaki et al. 2008, Nakamura

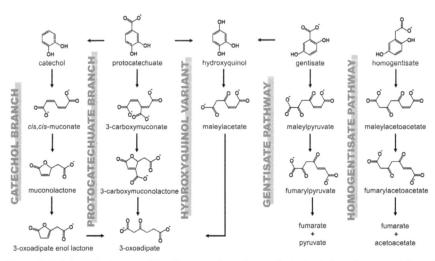

Figure 10.2 Schematic representation of the catabolism of aromatics in fungi; the pathways used for each central intermediate are shown.

et al. 2012). Up-accumulation of arylaldehyde and arylalcohol dehydrogenases, glutathione S-transferase, and flavonol reductase/cinnamoyl-CoA reductase, was observed in response to both compounds in *P. chrysosporium* (Matsuzaki et al. 2008, Shimizu et al. 2005). The first dehydrogenases possibly convert vanillin to vanillate or benzaldehyde to benzoate. In addition, a metabolic switch from the glyoxylate cycle to the TCA was observed, likely to increase the pool of succinyl-CoA required as co-substrate by 3-oxoadipate CoA-transferase (Matsuzaki et al. 2008, Shimizu et al. 2005). Two P450s were found only to increase in *P. chrysosporium* exposed to benzoate (Matsuzaki et al. 2008): PcCYP1f and CYP63A1. The first was previously proposed as a functional benzoate 4-monooxygenase and is highly homologous to P450s of the CYP53 family hypothetically displaying similar activity (van Gorcom et al. 1990, Fujii et al. 1997, Matsuzaki and Wariishi 2005), whereas the second, a putative 4-hydroxybenzoate 2-monooxygenase (Matsuzaki et al. 2008) is also found expressed in ligninolytic, nutrient limited, conditions (Yadav et al. 2003). Two flavin-containing monooxygenases (PcFMO1 and PcFMO2) were expressed in response to vanillin in *P. chrysosporium* (Nakamura et al. 2012). PcFMO1 displayed activity against monocyclic phenols, e.g., phenol, HQ and 4-CP, and is homologous to other phenol 2-monooxygenases (Eppink et al. 2000, Martins et al. 2015, Nakamura et al. 2012). PcFMO2 was not yet functionally characterized as it did not show activity against the tested monocyclic phenols, including vanillin, vanillyl alcohol, or vanillate (Nakamura et al. 2012). Vanillin addition increased an extradiol homogentisate dioxygenase family protein of 53 kDa (Shimizu et al. 2005), yet a previously biochemically characterized intradiol hydroxyquinol 1,2-dioxygenase has only 45 kDa (Rieble et al. 1994). A maleylacetate reductase—required in the hydroxyquinol variant of the 3-oxoadipate pathway—remains to be identified, though different alcohol dehydrogenases increased upon addition of either benzoate or vanillin (Matsuzaki et al. 2008, Shimizu et al. 2005).

Aspergillus nidulans and other Ascomycota have both the protocatechuate and catechol branches of the 3-oxoadipate pathway, whereas *P. chrysosporium* (and many other) have the hydroxyquinol variant of the 3-oxoadipate pathway or only have one of the branches (Boschloo et al. 1990, Gérecová et al. 2015, Kuswandi and Roberts 1992, Harwood and Parales 1996). Benzoate and salicylate as sole carbon and energy sources are directly channelled to the TCA cycle as sources of acetyl-CoA and succinate; consequently they have up-regulated gluconeogenesis and pentose phosphate pathway in *A. nidulans* (Martins et al. 2015). The aromatics led to the differential up-accumulation of several short chain dehydrogenase/reductase family proteins and stress response proteins (Martins et al. 2015), similar to that seen in *P. chrysosporium* (Matsuzaki et al. 2008). The benzoate catabolism in *A. nidulans* is mediated by five specific enzymes (Fig. 10.2), of which only the benzoate 4-monooxygenase (BzuA) has been previously described (Fraser et al. 2002). In fungi, the intermediate 3-carboxymuconolactone is converted by a single enzyme to 3-oxoadipate (Fig. 10.2), in opposition of formation of 4-carboxymuconolactone and a two steps conversion in bacteria (Harwood and Parales 1996, Martins et al. 2015). The salicylate catabolism involves eight specific enzymes (Fig. 10.2) and has two catabolic shunts: salicylate is converted to catechol directly or in two steps; *cis,cis*-muconate is converted to muconolactone either directly or indirectly through its isomer *cis,trans*-muconate (Martins et al. 2015). The genes codifying for muconate cycloisomerase and muconolactone isomerase were also identified for the first time in *A. nidulans* (Martins et al. 2015). These genes are present in a catechol gene cluster in several fungi, but not in *A. nidulans*, as subsequently verified in *C. albicans* and other phylogenetically related species (Gérecová et al. 2015). The muconolactone isomerase has no known specific sequence domains, and homologous proteins were only found in Dykaria fungi, mostly Ascomycota, indicating the uniqueness of the catechol branch in fungi. Of foremost importance is the functional redundancy of the salicylate and catechol branches owing fungi different tools to achieve degradation of an aromatic hydrocarbon (Martins et al. 2015).

The *A. nidulans* mutant carrying deletion of the 3-oxoadipic enol-lactone hydrolase gene, showed a phenotype able to utilise salicylate as carbon source during growth in solid media (Martins et al. 2015). The hypothesised activation of the hydroxyquinol variant of the 3-oxoadipate pathway, is now refuted since the expression levels of hydroxyquinol 1,2-dioxygenase and maleylacetate reductase genes were unaltered in the mutant (*unpublished results*). Additional unpublished observations support that a redundant and constitutive mechanism may exist for the degradation of either muconolactone or 3-oxoadipate enol-lactone, possibly related to the degradation pathway of the lactone protoanemonin previously reported for *A. nidulans* (Martins et al. 2014). Further knowledge on the degradation of aromatic compounds in fungi (and their complex regulatory mechanisms) is essential to support further developments in the degradation of xenobiotics by fungi.

7.3 *Future: frontier studies on the degradation of PCP by belowground fungi*

The impact of PCP pollution in the functioning of below-ground fungal communities is poorly understood. PCP (a)biotic transformation processes may led to formation of degradation intermediates that are usually more toxic to both bacteria and fungi than PCP, for example 2,3,4,5-TeCP or 3,4,5-TCP (McGrath and Singleton 2000). The fungal below-ground community responds rapidly to PCP pollution during composting with diversity shifts that are apparently associated with PCP removal (Zeng et al. 2011). Ascomycota strains isolated from soils, namely *Byssochlamys nivea* and *Scopulariopsis brumptii*, were also demonstrated to degrade PCP efficiently; interestingly, more efficiently when co-cultivated due to synergist effects (Bosso et al. 2015).

We have recently disclosed the PCP degradation pathway used by fungal communities from soils chronically affected by the biocide, through integration of the PCP-derived metabolomes (LC-ESI-HRMS) of each fungal strain (grown axenically) within the community (Fig. 10.3) (Varela et al. 2015). The cultivable

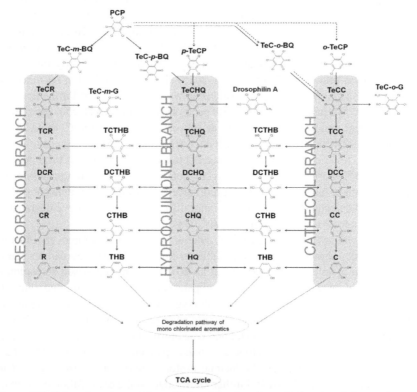

Figure 10.3 Schematic representation of the three branches of the degradation pathway of pentachlorophenol by filamentous fungi, either grown axenically or as a community (submerged cultivation). Continuous and dashed lines mark biotic and abiotic steps, respectively. Conjugated intermediates (apart from those involving tetra-chlorinated species) are not shown in the pathway for matters of simplicity. *Adapted from* (Varela et al. 2015).

below-ground community comprised seventy-seven isolates covering thirty-three species, with a clear dominance by *Penicillium* species (Varela et al. 2015). Remarkably, fifty-three out of the seventy-seven fungal strains within the community, could significantly deplete PCP from the media (concentrations ranging from 19 to 56 µM), further highlighting fungi's widespread capacity to degrade the biocide. Most of the fungal PCP-degradation intermediates identified by us, have been observed before in pure cultures of fungi (Bosso and Cristinzio 2014) or bacteria (Copley et al. 2011). The exceptions consist of compounds never linked before to PCP degradation, namely those in the resorcinol branch and the tetrachloroguaiacol (TeC-G) isomers (Fig. 10.3). In more detail, PCP initial dehalogenation to TeCP may involve its reductive dechlorination, either via biotic or abiotic steps (yielding the *meta* and the p*ara/ortho* isomers, respectively) or its peroxidative dechlorination (forming transient benzoquinone immediately followed by H^+ mediated reductions) (Varela et al. 2015, Gadd 2001, Reddy and Gold 2000). Both reactions are consistent with the subsequent formation of TeCHQ and tetrachlorocatechol (TeCC), as well as tetrachlororesorcinol (TeCR). The toxicity of the TeCBQ is extremely high, however *Sphingobium chlorophenolicum* is able to circumvent its toxicity, using a TeCBQ reductase that catalyses its reduction to TeCHQ (Yadid et al. 2013). These tetrachlorinated PCP-derivatives, can undergo successive biotic reductive dechlorinations through the resorcinol (R), hydroquinone (HQ), or catechol (C) branches (Fig. 10.3), which intersect through additional hydroxylation of their derivatives, yielding the corresponding trihydroxybenzenes (THB) (Varela et al. 2015, Gadd 2001). As often reported, fungi can also form several conjugated compounds, e.g., through the *O*-methylation and/or sulphation of PCP and/or its derivatives, e.g., sulphate trichlorometoxyphenol (S-TCMP) (Carvalho et al. 2011, Varela et al. 2015). In particular, the *O*-methylation of chlorophenols leads to formation of chloroanisoles as abovementioned, e.g. (Callejón et al. 2016). The PCP degradation pathway observed in community-based cultures perfectly matched that inferred by integration of the pathways observed in the axenic cultures, regardless of much higher diversity of PCP-derivatives in the last (Varela et al. 2015). The metabolites formed by the community, including those internalised by the mycelia, further support the superior use of the C and HQ branches over the R branch by fungi (Varela et al. 2017).

The PCP-derivatives so far identified may be also used as footprints of PCP environmental pollution, some of which, e.g., DCTHB and TeC-G (Fig. 10.3), can directly link the mitigation of the biocide with fungal activity (Varela et al. 2015). Once forest soils of high geochemical homogeneity may display very dissimilar fungal diversity (both taxonomic and functional) (Stursova et al. 2016), community-based proteomic studies may provide complementary means to reveal key functional trends during the decay of the pollutant. Our initial observations on the degradation of PCP by fungal communities isolated from chronically stressed forest soils show that most alterations are related to stress responsive proteins and carbohydrate metabolism; the major up- and down-proteins so far identified are depicted in Table 10.4 (Martins et al. 2018). Metagenomics was successfully used to assign specific fungi species to specific functions.

Table 10.4 List of the major up- and down-regulated proteins identified during the degradation of pentachlorophenol by a fungal community isolated from polluted forest soils compared to control conditions.

	Accession	$\log_2$(FC)	Protein name	Organism	Biological function
up regulated	Q5B2V1	9,201	Heat shock 70 kDa protein	*Aspergillus nidulans*	Stress response
	P07038	8,452	Plasma membrane ATPase (Proton pump)	*Neurospora crassa*	ATP biosynthesis
	P24487	8,452	ATP synthase subunit alpha, mitochondrial	*Schizosaccharomyces pombe*	ATP biosynthesis
	P34825	8,304	Elongation factor 1-alpha	*Trichoderma reesei*	Protein biosynthesis
	Q00640	8,224	Glyceraldehyde-3-phosphate dehydrogenase	*Blumeria graminis*	Glycolysis
	Q8X097	7,904	Probable ATP-citrate synthase subunit 1	*Neurospora crassa*	Fatty acid biosynthesis
	P34085	7,621	Citrate synthase, mitochondrial	*Neurospora crassa*	TCA cycle
	P41797	7,621	Heat shock protein SSA1	*Candida albicans*	Stress response
	C7Z8P6	7,491	Thiamine thiazole synthase	*Fusarium solani*	Stress response
	P28876	7,491	Plasma membrane ATPase 2 (Proton pump 2)	*Schizosaccharomyces pombe*	ATP biosynthesis
down regulated	Q7RV85	-9,672	Enolase	*Neurospora crassa*	Glycolysis
	P54118	-8,443	Glyceraldehyde-3-phosphate dehydrogenase	*Neurospora crassa*	Glycolysis
	J9N5G7	-8,346	Thiamine thiazole synthase	*Fusarium oxysporum*	Stress response
	P87197	-7,921	Glyceraldehyde-3-phosphate dehydrogenase	*Trichoderma atroviride*	Glycolysis
	P14228	-7,875	Phosphoglycerate kinase	*Trichoderma reesei*	Glycolysis
	Q01291	-7,828	40S ribosomal protein S0	*Neurospora crassa*	Protein biosynthesis
	Q9HGY7	-7,729	Glyceraldehyde-3-phosphate dehydrogenase	*Aspergillus oryzae*	Glycolysis
	P23617	-7,677	Thiamine thiazole synthase	*Fusarium solani*	Stress response
	A4QVP2	-7,508	ATP-dependent RNA helicase	*Magnaporthe oryzae*	Protein biosynthesis
	O00098	-7,096	Citrate synthase, mitochondrial	*Aspergillus nidulans*	TCA cycle

Conclusions

The comprehension of *in situ* microbial interactions and their response to environmental changes has been recognized as a major challenge for science (Abram 2015). Fungi own an impressive diverse array of mechanisms to tackle the stress imposed by xenobiotics, many of which are not yet properly characterized or remain unidentified. In this context, proteomic analyses may provide matchless means to discover new functional genes, proteins, and metabolic pathways, which can be considered as functional bio-indicators for assessing the sustainability of ecosystems. We will further use proteomic-based tools (complemented by additional methods whenever necessary) for solving major knowledge omissions on the degradation of xenobiotics by filamentous fungi, including in the functioning of contaminated soils. The success of any experimental proteomic-based approach relies intrinsically in the experimental design (linked to the biological question under study), the protein extraction method, and the downstream MS-based analysis. In each of the past studies (see examples listed in Tables 10.1 to 10.3), either in axenic fungal cultures or communities (enriched or naturally existing), several fungal proteins with unknown function were found to differentially accumulate upon exposure to a xenobiotic. The scientific community should periodically revisit and upgrade such old toxicoproteomics datasets to seek new protein identifications, including some potentially involved in the degradation pathways of the targeted xenobiotics.

Acknowledgements

C.M., I.M. and T.M. are grateful to Fundação para a Ciência e a Tecnologia (FCT), Portugal, for the fellowships SFRH/BD/118377/2016, SFRH/BPD/110841/2015 and SFRH/BPD/70064/2010. The work has been partially supported by grants from FCT through grants UID/Multi/04551/2013 (Research unit GREEN-it "Bioresources for Sustainability"). The authors are extremely thankful to scientific collaborators from the University of the West of Scotland (Andrew Hursthouse & Iain McLellan), the University of Barcelona (Oscar Núñez and co-workers) and the Luxembourg Institute of Science and Technology (Jenny Renaut and co-workers) for helpful discussions.

References

1000 Fungal Genomes Project, Joint Genome Institute. Lawrence Berkeley National Laboratory. http://1000.fungalgenomes.org, accessed in July 2017.

Abram, F. 2015. Systems-based approaches to unravel multi-species microbial community functioning. *Computational and Structural Biotechnology Journal* 13: 24–32. doi: 10.1016/j.csbj.2014.11.009.

Aebersold, R. and M. Mann. 2016. Mass-spectrometric exploration of proteome structure and function. *Nature* 537 (7620): 347–355.

Albrecht, D., R. Guthke, A. A. Brakhage and O. Kniemeyer. 2010. Integrative analysis of the heat shock response in *Aspergillus fumigatus*. *BMC Genomics* 11. doi: 10.1186/1471-2164-11-32.

Arsene-Ploetze, F., P. N. Bertin and C. Carapito. 2015. Proteomic tools to decipher microbial community structure and functioning. *Environmental Science and Pollution Research* 22 (18): 13599–13612. doi: 10.1007/s11356-014-3898-0.

Asif, A. R., M. Oellerich, V. W. Amstrong, U. Gross and U. Reichard. 2010. Analysis of the cellular *Aspergillus fumigatus* proteome that reacts with sera from rabbits developing an acquired immunity after experimental aspergillosis. *Electrophoresis* 31 (12): 1947–1958. doi: 10.1002/elps.201000015.

Baader, E. W. and H. J. Bauer. 1951. Industrial intoxication due to pentachlorophenol. *Industrial Medicine & Surgery* 20: 286–290.

Baldrian, P., M. Kolařík, M. Štursová, J. Kopecký, V. Valášková, T. Větrovský, L. Žifčáková, J. Šnajdr, J. Rídl and Č. Vlček. 2012. Active and total microbial communities in forest soil are largely different and highly stratified during decomposition. *ISME Journal* 6 (2):248.

Barker, A. P., J. L. Horan, E. S. Slechta, B. D. Alexander and K. E. Hanson. 2014. Complexities associated with the molecular and proteomic identification of *Paecilomyces* species in the clinical mycology laboratory. *Medical Mycology* 52 (5): 535–543. doi: 10.1093/mmy/myu001.

Bastida, F., T. Hernandez and C. Garcia. 2014. Metaproteomics of soils from semiarid environment: functional and phylogenetic information obtained with different protein extraction methods. *Journal of Proteomics* 101: 31–42. doi: 10.1016/j.jprot.2014.02.006.

Bastida, F., C. Garcia, M. von Bergen, J. L. Moreno, H. H. Richnow and N. Jehmlich. 2015a. Deforestation fosters bacterial diversity and the cyanobacterial community responsible for carbon fixation processes under semiarid climate: A metaproteomics study. *Applied Soil Ecology* 93: 65–67. doi: 10.1016/j.apsoil.2015.04.006.

Bastida, F., N. Selevsek, I. Torres, T. Hernández and C. García. 2015b. Soil restoration with organic amendments: Linking cellular functionality and ecosystem processes. *Scientific Reports* 5.

Bastida, F., N. Jehmlich, K. Lima, B. E. L. Morris, H. H. Richnow, T. Hernandez, M. von Bergen and C. Garcia. 2016a. The ecological and physiological responses of the microbial community from a semiarid soil to hydrocarbon contamination and its bioremediation using compost amendment. *Journal of Proteomics* 135: 162–169. doi: 10.1016/j.jprot.2015.07.023.

Bastida, F., I. F. Torres, J. L. Moreno, P. Baldrian, S. Ondoño, A. Ruiz-Navarro, T. Hernández, H. H. Richnow, R. Starke and C. García. 2016b. The active microbial diversity drives ecosystem multifunctionality and is physiologically related to carbon availability in mediterranean semi-arid soils. *Molecular Ecology* 25 (18): 4660–4673.

Benndorf, D., G. U. Balcke, H. Harms and M. von Bergen. 2007. Functional metaproteome analysis of protein extracts from contaminated soil and groundwater. *ISME Journal* 1 (3): 224–234. doi: 10.1038/ismej.2007.39.

Benndorf, D., C. Vogt, N. Jehmlich, Y. Schmidt, H. Thomas, G. Woffendin, A. Shevchenko, H. H. Richnow and M. von Bergen. 2009. Improving protein extraction and separation methods for investigating the metaproteome of anaerobic benzene communities within sediments. *Biodegradation* 20 (6): 737–750. doi: 10.1007/s10532-009-9261-3.

Bevenue, A., J. Wilson, E. Potter, M. K. Song, H. Beckman and G. Mallett. 1966. A method for the determination of pentachlorophenol in human urine in picogram quantities. In: *Bulletin of Environmental Contamination and Toxicology*, 257–266. Springer.

Bianco, L. and G. Perrotta. 2015. Methodologies and perspectives of proteomics applied to filamentous fungi: From sample preparation to secretome analysis. *International Journal of Molecular Sciences* 16 (3): 5803–5829. doi: 10.3390/ijms16035803.

Borysiewicz, M. 2008. Risk profile of pentachlorophenol. In: *Institute of Environmental Protection*. Institute of Environmental Protection.

Boschloo, J. G., A. Paffen, T. Koot, W. J. J. Vandentweel, R. F. M. Vangorcom, J. H. G. Cordewener and C. J. Bos. 1990. Genetic analysis of benzoate metabolism in *Aspergillus niger*. *Applied Microbiology and Biotechnology* 34 (2): 225–228.

Bosso, L. and G. Cristinzio. 2014. A comprehensive overview of bacteria and fungi used for pentachlorophenol biodegradation. *Reviews in Environmental Science and Bio/Technology* 13 (4): 387–427.

Bosso, L., R. Scelza, A. Testa, G. Cristinzio and M. A. Rao. 2015. Depletion of pentachlorophenol contamination in an agricultural soil treated with *Byssochlamys nivea*, *Scopulariopsis brumptii* and urban waste compost: A laboratory microcosm study. *Water, Air, and Soil Pollution* 226 (6): 1.

Boyd, L. J., T. H. McGavack, R. Terranova and F. V. Piccione. 1940. Toxic effects following the cutaneous administration of sodium pentachlorophenate. *Bulletin New York Medical College, Flower and Fifth Avenue Hospitals* 3 (323).

Briggs, D. 2003. Environmental pollution and the global burden of disease. *British Medical Bulletin* 68 (1): 1–24.

Brown, B. 1970. Fatal phenol poisoning from improperly laundered diapers. *American Journal of Public Health and the Nation's Health* 60 (5): 901–902.

Bugalho, M. N., M. C. Caldeira, J. S. Pereira, J. Aronson and J. G. Pausas. 2011. Mediterranean cork oak savannas require human use to sustain biodiversity and ecosystem services. *Frontiers in Ecology and the Environment* 9 (5): 278–286.

Bumpus, J. A., M. Tien, D. Wright and S. D. Aust. 1985. Oxidation of persistent environmental pollutants by a white rot fungus. *Science* 228: 1434–1437.

Callejón, R., C. Ubeda, R. Ríos-Reina, M. Morales and A. Troncoso. 2016. Recent developments in the analysis of must odour compounds in water and wine: A review. *Journal of Chromatography A* 1428: 72–85.

Carswell, T. and H. Nason. 1938. Properties and uses of pentachlorophenol. *Industrial & Engineering Chemistry* 30 (6): 622–626.

Carswell, T. 1939. Pentachlorophenol for wood preservation. *Industrial & Engineering Chemistry* 31 (11): 1431–1435.

Carvalho, M. B., I. Martins, M. C. Leitão, H. Garcia, C. Rodrigues, V. San Romão, I. McLellan, A. Hursthouse and C. Silva Pereira. 2009. Screening pentachlorophenol degradation ability by environmental fungal strains belonging to the phyla Ascomycota and Zygomycota. *Journal of Industrial Microbiology & Biotechnology* 36 (10): 1249–1256. doi: 10.1007/s10295-009-0603-2.

Carvalho, M. B., S. Tavares, J. Medeiros, O. Núñez, H. Gallart-Ayalla, M. C. Leitão, M. T. Galceran, A. Hursthouse and C. Silva Pereira. 2011. Degradation pathway of pentachlorophenol by *Mucor plumbeus* involves phase II conjugation and oxidation-reduction reactions. *Journal of Hazardous Materials* 198: 133–142.

Carvalho, M. B., I. Martins, J. Medeiros, S. Tavares, S. Planchon, J. Renaut, O. Nunez, H. Gallart-Ayala, M. T. Galceran, A. Hursthouse and C. Silva Pereira. 2013. The response of *Mucor plumbeus* to pentachlorophenol: A toxicoproteomics study. *Journal of Proteomics* 78: 159–71. doi: 10.1016/j.jprot.2012.11.006.

Casaletti, L., P. Lima, L. Oliveira, C. Borges, S. Báo, A. Bailão and C. Soares. 2017. Analysis of *Paracoccidioides lutzii* mitochondria: a proteomic approach. *Yeast* 34 (4): 179–188.

Copley, S. D., J. Rokicki, P. Turner, H. Daligault, M. Nolan and M. Land. 2011. The whole genome sequence of *Sphingobium chlorophenolicum* L-1: insights into the evolution of the pentachlorophenol degradation pathway. *Genome Biology and Evolution* 4 (2): 184–198.

Czaplicka, M. 2004. Sources and transformations of chlorophenols in the natural environment. *Science of the Total Environment* 322 (1): 21–39.

Chapman, J. B. and P. Rbson. 1965. Pentachlorophenol poisoning from bath-water. *Lancet* 1: 1266.

Deising, H., P. R. Jungblut and K. Mendgen. 1991. Differentiation-related proteins of the broad bean rust fungus *Uromyces viciae-fabae*, as revealed by high-resolution 2-dimensional polyacrylamide-gel electrophoresis. *Archives of Microbiology* 155 (2): 191–198. doi: 10.1007/bf00248616.

Del Chierico, F., A. Masotti, M. Onori, E. Fiscarelli, L. Mancinelli, G. Ricciotti, F. Alghisi, L. Dimiziani, C. Manetti, A. Urbani, M. Muraca and L. Putignani. 2012. MALDI-TOF MS proteomic phenotyping of filamentous and other fungi from clinical origin. *Journal of Proteomics* 75 (11): 3314–3330. doi: 10.1016/j.jprot.2012.03.048.

Domon, B. and R. Aebersold. 2010. Options and considerations when selecting a quantitative proteomics strategy. *Nature Biotechnology* 28 (7): 710–721. doi: 10.1038/nbt.1661.

El-Naas, M. H., H. A. Mousa and M. El Gamal. 2017. Microbial degradation of chlorophenols. In: *Microbe-induced degradation of pesticides*, edited by Shree Nath Singh, 23–58. Springer.

Eppink, M., E. Cammaart, D. Van Wassenaar, W. Middelhoven and W. van Berkel. 2000. Purification and properties of hydroquinone hydroxylase, a FAD-dependent monooxygenase involved in the catabolism of 4-hydroxybenzoate in *Candida parapsilosis* CBS604. *European Journal of Biochemistry* 267 (23): 6832.

Erdmann, O. L. 1841. Untersuchungen über den indigo. *Journal Fuer Praktische Chemie* 22 (1): 257–299.

Fenger, J. 2009. Air pollution in the last 50 years—from local to global. *Atmospheric Environment* 43 (1): 13–22.

Ferreira, R., H. Garcia, A. F. Sousa, C. S. R. Freire, A. J. D. Silvestre, L. P. N. Rebelo and C. Silva Pereira. 2013. Isolation of suberin from birch outer bark and cork using ionic liquids: A new source of macromonomers. *Industrial Crops and Products* 44: 520–527. doi: doi 10.1016/j.indcrop.2012.10.002.

Floudas, D., M. Binder, R. Riley, K. Barry, R. A. Blanchette, B. Henrissat, A. T. Martinez, R. Otillar, J. W. Spatafora, J. S. Yadav, A. Aerts, I. Benoit, A. Boyd, A. Carlson, A. Copeland, P. M. Coutinho, R. P. de Vries, P. Ferreira, K. Findley, B. Foster, J. Gaskell, D. Glotzer, P. Gorecki, J. Heitman, C.

Hesse, C. Hori, K. Igarashi, J. A. Jurgens, N. Kallen, P. Kersten, A. Kohler, U. Kues, T. K. Kumar, A. Kuo, K. LaButti, L. F. Larrondo, E. Lindquist, A. Ling, V. Lombard, S. Lucas, T. Lundell, R. Martin, D. J. McLaughlin, I. Morgenstern, E. Morin, C. Murat, L. G. Nagy, M. Nolan, R. A. Ohm, A. Patyshakuliyeva, A. Rokas, F. J. Ruiz-Duenas, G. Sabat, A. Salamov, M. Samejima, J. Schmutz, J. C. Slot, F. St John, J. Stenlid, H. Sun, S. Sun, K. Syed, A. Tsang, A. Wiebenga, D. Young, A. Pisabarro, D. C. Eastwood, F. Martin, D. Cullen, I. V. Grigoriev and D. S. Hibbett. 2012. The paleozoic origin of enzymatic lignin decomposition reconstructed from 31 fungal genomes. *Science* 336 (6089): 1715–9. doi: 10.1126/science.1221748.

Fraser, J. A., M. A. Davis and M. J. Hynes. 2002. The genes *gmdA*, encoding an amidase, and *bzuA*, encoding a cytochrome P450, are required for benzamide utilization in *Aspergillus nidulans*. *Fungal Genetics and Biology* 35 (2): 135–146.

Frija, L. M., R. F. Frade and C. A. Afonso. 2011. Isolation, chemical, and biotransformation routes of labdane-type diterpenes. *Chemical Reviews* 111 (8): 4418–4452.

Frija, L. M., H. Garcia, C. Rodrigues, I. Martins, N. R. Candeias, V. André, M. T. Duarte, C. Silva Pereira and C. A. Afonso. 2013. Short synthesis of the natural product 3β-hydroxy-labd-8 (17)-en-15-oic acid via microbial transformation of labdanolic acid. *Phytochemistry Letters* 6 (2): 165–169.

Fujii, T., K. Nakamura, K. Shibuya, S. Tanase, O. Gotoh, T. Ogawa and H. Fukuda. 1997. Structural characterization of the gene and corresponding cDNA for the cytochrome P450 from *Rhodotorula minuta* which catalyzes formation of isobutene and 4-hydroxylation of benzoate. *Molecular and General Genetics* 256 (2): 115–120.

Fukasawa, Y., T. Osono and H. Takeda. 2011. Wood decomposing abilities of diverse lignicolous fungi on nondecayed and decayed beech wood. *Mycologia* 103 (3): 474–482.

Gadd, G. M. 2001. *Fungi in bioremediation*. Vol. 23: Cambridge University Press.

Gautam, P., D. Mushahary, W. Hassan, S. K. Upadhyay, T. Madan, R. Sirdeshmukh, C. S. Sundaram and P. U. Sarma. 2016. In-depth 2-DE reference map of *Aspergillus fumigatus* and its proteomic profiling on exposure to itraconazole. *Medical Mycology* 54 (5): 524–536. doi: 10.1093/mmy/myv122.

Gérecová, G., M. Nebohácová, I. Zeman, L. Pryszcz, L'. Tomáška, T. Gabaldón and J. Nosek. 2015. Metabolic gene clusters encoding the enzymes of two branches of the 3-oxoadipate pathway in the pathogenic yeast *Candida albicans*. *FEMS Yeast Research* 15 (3).

Gonzalez-Rodriguez, V. E., E. Lineiro, T. Colby, A. Harzen, C. Garrido, J. M. Cantoral, J. Schmidt and F. J. Fernandez-Acero. 2015. Proteomic profiling of *Botrytis cinerea* conidial germination. *Archives of Microbiology* 197 (2): 117–133. doi: 10.1007/s00203-014-1029-4.

Gotelli, N. J., A. M. Ellison and B. A. Ballif. 2012. Environmental proteomics, biodiversity statistics and food-web structure. *Trends in Ecology & Evolution* 27 (8): 436–442. doi: 10.1016/j.tree.2012.03.001.

Harms, H., D. Schlosser and L. Y. Wick. 2011. Untapped potential: Exploiting fungi in bioremediation of hazardous chemicals. *Nature Reviews Microbiology* 9 (3): 177–192.

Harwood, C. S. and R. E. Parales. 1996. The beta-ketoadipate pathway and the biology of self-identity. *Annual Review of Microbiology* 50 (1): 553–90. doi: 10.1146/annurev.micro.50.1.553.

Hattemer-Frey, H. A. and C. C. Travis. 1989. Pentachlorophenol: Environmental partitioning and human exposure. *Archives of Environmental Contamination and Toxicology* 18 (4): 482–489.

Hawksworth, D. and R. Lücking. 2017. Fungal diversity revisited: 2.2 to 3.8 million species. *Microbiology Spectrum* 5 (4).

Herbst, F. A., A. Bahr, M. Duarte, D. H. Pieper, H. H. Richnow, M. Bergen, J. Seifert and P. Bombach. 2013. Elucidation of *in situ* polycyclic aromatic hydrocarbon degradation by functional metaproteomics (protein-SIP). *Proteomics* 13 (18–19): 2910–2920.

Hoferkamp, L., M. H. Hermanson and D. C. Muir. 2010. Current use pesticides in Arctic media: 2000–2007. *Science of the Total Environment* 408 (15): 2985–2994.

Hofrichter, M., R. Ullrich, M. J. Pecyna, C. Liers and T. Lundell. 2010. New and classic families of secreted fungal heme peroxidases. *Applied Microbiology and Biotechnology* 87 (3): 871–897. doi: 10.1007/s00253-010-2633-0.

Holesova, Z., M. Jakubkova, I. Zavadiakova, I. Zeman, L. Tomaska and J. Nosek. 2011. Gentisate and 3-oxoadipate pathways in the yeast *Candida parapsilosis*: identification and functional analysis of the genes coding for 3-hydroxybenzoate 6-hydroxylase and 4-hydroxybenzoate 1-hydroxylase. *Microbiology* 157 (7): 2152–2163. doi: 10.1099/mic.0.048215-0.

Hu, B. B., M. Z. Luo, X. L. Ji, L. B. Lin, Y. L. Wei and Q. Zhang. 2016. Proteomic analysis of *Mortierella isabellina* M6-22 during cold stress. *Archives of Microbiology* 198 (9): 869–876. doi: 10.1007/s00203-016-1238-0.

Hultman, J., M. P. Waldrop, R. Mackelprang, M. M. David, J. McFarland, S. J. Blazewicz, J. Harden, M. R. Turetsky, A. D. McGuire and M. B. Shah. 2015. Multi-omics of permafrost, active layer and thermokarst bog soil microbiomes. *Nature* 521 (7551): 208.

Ilyas, S., A. Rehman, A. V. Coelho and D. Sheehan. 2016. Proteomic analysis of an environmental isolate of *Rhodotorula mucilaginosa* after arsenic and cadmium challenge: Identification of a protein expression signature for heavy metal exposure. *Journal of Proteomics* 141: 47–56. doi: 10.1016/j.jprot.2016.04.012.

Jami, M. S., C. Barreiro, C. Garcia-Estrada and J. F. Martin. 2010. Proteome analysis of the penicillin producer *Penicillium chrysogenum* characterization of protein changes during the industrial strain improvement. *Molecular & Cellular Proteomics* 9 (6): 1182–1198. doi: 10.1074/mcp.M900327-MCP200.

Jehmlich, N., S. Kleinsteuber, C. Vogt, D. Benndorf, H. Harms, F. Schmidt, M. von Bergen and J. Seifert. 2010. Phylogenetic and proteomic analysis of an anaerobic toluene-degrading community. *Journal of Applied Microbiology* 109 (6): 1937–1945. doi: 10.1111/j.1365-2672.2010.04823.x.

Kallenborn, R., G. Christensen, A. Evenset, M. Schlabach and A. Stohl. 2007. Atmospheric transport of persistent organic pollutants (POPs) to Bjørnøya (bear island). *Journal of Environmental Monitoring* 9 (10): 1082–1091.

Kehoe, R. A., W. Deichmann-Gruebler and K. V. Kitzmiller. 1939. Toxic effects upon rabbits of pentachlorophenol and sodium pentachlorophenate. *Journal of Industrial Hygiene and Toxicology* 21 (160).

Keiblinger, K. M., I. C. Wilhartitz, T. Schneider, B. Roschitzki, E. Schmid, L. Eberl, K. Riedel and S. Zechmeister-Boltenstern. 2012. Soil metaproteomics–Comparative evaluation of protein extraction protocols. *Soil Biology & Biochemistry* 54: 14–24. doi: 10.1016/j.soilbio.2012.05.014.

Khan, A., K. Williams, M. P. Molloy and H. Nevalainen. 2003. Purification and characterization of a serine protease and chitinases from *Paecilomyces lilacinus* and detection of chitinase activity on 2D gels. *Protein Expression and Purification* 32 (2): 210–220. doi: 10.1016/j.pep.2003.07.007.

Kim, S. G., Y. Wang, K. H. Lee, Z. Y. Park, J. Park, J. Wu, S. J. Kwon, Y. H. Lee, G. K. Agrawal, R. Rakwal, S. T. Kim and K. Y. Kang. 2013. In-depth insight into *in vivo* apoplastic secretome of rice - *Magnaporthe oryzae* interaction. *Journal of Proteomics* 78: 58–71. doi: 10.1016/j.jprot.2012.10.029.

Kovacevic, G. and A. Sabljic. 2016. Atmospheric oxidation of hexachlorobenzene: New global source of pentachlorophenol. *Chemosphere* 159: 488–495.

Kües, U. 2015. Fungal enzymes for environmental management. *Current Opinion in Biotechnology* 33: 268–278.

Kuswandi and C. F. Roberts. 1992. Genetic control of the protocatechuic acid pathway in *Aspergillus nidulans*. *Journal of General Microbiology* 138 (4): 817–823.

Lah, L., B. Podobnik, M. Novak, B. Korosec, S. Berne, M. Vogelsang, N. Krasevec, N. Zupanec, J. Stojan, J. Bohlmann and R. Komel. 2011. The versatility of the fungal cytochrome P450 monooxygenase system is instrumental in xenobiotic detoxification. *Molecular Microbiology* 81 (5): 1374–1389. doi: DOI 10.1111/j.1365-2958.2011.07772.x.

Lamar, R. T. and D. M. Dietrich. 1990. *In situ* depletion of pentachlorophenol from contaminated soil by *Phanerochaete* spp. *Applied and Environmental Microbiology* 56 (10): 3093–3100.

Lamar, R. T., M. W. Davis, D. M. Dietrich and J. A. Glaser. 1994. Treatment of a pentachlorophenol and creosote contaminated soil using the lignin-degrading fungus *Phanerochaete sordida*: A field demonstration. *Soil Biology & Biochemistry* 26 (12): 1603–1611.

Lau, M. C., B. Stackhouse, A. C. Layton, A. Chauhan, T. Vishnivetskaya, K. Chourey, J. Ronholm, N. Mykytczuk, P. Bennett and G. Lamarche-Gagnon. 2015. An active atmospheric methane sink in high arctic mineral cryosols. *ISME journal* 9 (8): 1880.

Li, E. F., J. Ling, G. Wang, J. L. Xiao, Y. H. Yang, Z. C. Mao, X. C. Wang and B. Y. Xie. 2015. Comparative proteomics analyses of two races of *Fusarium oxysporum f. Sp conglutinans* that differ in pathogenicity. *Scientific Reports* 5. doi: 10.1038/srep13663.

Li, Y. and R. Macdonald. 2005. Sources and pathways of selected organochlorine pesticides to the Arctic and the effect of pathway divergence on HCH trends in biota: A review. *Science of the Total Environment* 342 (1): 87–106.

Lindemann, C., N. Thomanek, F. Hundt, T. Lerari, H. E. Meyer, D. Wolters and K. Marcus. 2017. Strategies in relative and absolute quantitative mass spectrometry based proteomics. *Biological Chemistry* 398 (5–6): 687–699.

Liu, D., K. M. Keiblinger, A. Schindlbacher, U. Wegner, H. Sun, S. Fuchs, C. Lassek, K. Riedel and S. Zechmeister-Boltenstern. 2017. Microbial functionality as affected by experimental warming of a temperate mountain forest soil: A metaproteomics survey. *Applied Soil Ecology* 117: 196–202.

Lünsmann, V., U. Kappelmeyer, R. Benndorf, P. M. Martinez-Lavanchy, A. Taubert, L. Adrian, M. Duarte, D. H. Pieper, M. Bergen and J. A. Müller. 2016a. *In situ* protein-SIP highlights burkholderiaceae as key players degrading toluene by para ring hydroxylation in a constructed wetland model. *Environmental Microbiology* 18 (4): 1176–1186.

Lünsmann, V., U. Kappelmeyer, A. Taubert, I. Nijenhuis, M. von Bergen, H. J. Heipieper, J. A. Müller and N. Jehmlich. 2016b. Aerobic toluene degraders in the rhizosphere of a constructed wetland model show diurnal polyhydroxyalkanoate metabolism. *Applied and Environmental Microbiology* 82 (14): 4126–4132.

Marco-Urrea, E., I. García-Romera and E. Aranda. 2015. Potential of non-ligninolytic fungi in bioremediation of chlorinated and polycyclic aromatic hydrocarbons. *New Biotechnology* 32 (6): 620–628. doi: https://doi.org/10.1016/j.nbt.2015.01.005.

Marra, R., P. Ambrosino, V. Carbone, F. Vinale, S. L. Woo, M. Ruocco, R. Ciliento, S. Lanzuise, S. Ferraioli, I. Soriente, S. Gigante, D. Turra, V. Fogliano, F. Scala and M. Lorito. 2006. Study of the three-way interaction between *Trichoderma atroviride*, plant and fungal pathogens by using a proteomic approach. *Current Genetics* 50 (5): 307–321. doi: 10.1007/s00294-006-0091-0.

Martins, C., A. Varela, C. C. Leclercq, O. Núñez, T. Větrovský, J. Renaut, P. Baldrian and C. Silva Pereira. 2018. Specialisation events of fungal metacommunities exposed to a persistent organic pollutant are suggestive of augmented pathogenic potential. Microbiome 6:208.

Martins, I., D. O. Hartmann, P. C. Alves, S. Planchon, J. Renaut, M. C. Leitao, L. P. N. Rebelo and C. Silva Pereira. 2013. Proteomic alterations induced by ionic liquids in *Aspergillus nidulans* and *Neurospora crassa*. *Journal of Proteomics* 94: 262–278. doi: 10.1016/j.jprot.2013.09.015.

Martins, I., H. Garcia, A. Varela, O. Núñez, S. Planchon, M. T. Galceran, J. Renaut, L. P. Rebelo and C. Silva Pereira. 2014a. Investigating *Aspergillus nidulans* secretome during colonisation of cork cell walls. *Journal of Proteomics* 98: 175–188.

Martins, I., D. O. Hartmann, P. C. Alves, C. Martins, H. Garcia, C. C. Leclercq, R. Ferreira, J. He, J. Renaut and J. D. Becker. 2014b. Elucidating how the saprophytic fungus *Aspergillus nidulans* uses the plant polyester suberin as carbon source. *BMC Genomics* 15 (1): 613.

Martins, I., A. Varela, L. M. Frija, M. A. Estevão, S. Planchon, J. Renaut, C. A. Afonso and C. Silva Pereira. 2017. Proteomic insights on the metabolism of *Penicillium janczewskii* during the biotransformation of the plant terpenoid labdanolic acid. *Frontiers in Bioengineering and Biotechnology* 5.

Martins, T. M., D. O. Hartmann, S. Planchon, I. Martins, J. Renaut and C. Silva Pereira. 2015. The old 3-oxoadipate pathway revisited: New insights in the catabolism of aromatics in the saprophytic fungus *Aspergillus nidulans*. *Fungal Genetics and Biology* 74: 32–44. doi: 10.1016/j.fgb.2014.11.002.

Martins, T. M., O. Nunez, H. Gallart-Ayala, M. C. Leitao, M. T. Galceran and C. Silva Pereira. 2014. New branches in the degradation pathway of monochlorocatechols by *Aspergillus nidulans*: A metabolomics analysis. *Journal of Hazardous Materials* 268: 264–72. doi: 10.1016/j.jhazmat.2014.01.024.

Matsuzaki, F. and H. Wariishi. 2005. Molecular characterization of cytochrome P450 catalyzing hydroxylation of benzoates from the white-rot fungus *Phanerochaete chrysosporium*. *Biochemical and Biophysical Research Communications* 334 (4): 1184–1190.

Matsuzaki, F., M. Shimizu and H. Wariishi. 2008. Proteomic and metabolomic analyses of the white-rot fungus *Phanerochaete chrysosporium* exposed to exogenous benzoic acid. *Journal of Proteome Research* 7 (6): 2342–50. doi: 10.1021/pr700617s.

McAllister, K. A., H. Lee and J. T. Trevors. 1996. Microbial degradation of pentachlorophenol. *Biodegradation* 7 (1): 1–40.

McGrath, R. and I. Singleton. 2000. Pentachlorophenol transformation in soil: A toxicological assessment. *Soil Biology & Biochemistry* 32 (8):1311-1314.

McLellan, I., M. Carvalho, C. Silva Pereira, A. Hursthouse, C. Morrison, P. Tatner, I. Martins, M. V. San Romao and M. Leitao. 2007. The environmental behaviour of polychlorinated phenols and its relevance to cork forest ecosystems: A review. *Journal of Environmental Monitoring* 9 (10): 1055–63. doi: 10.1039/b701436h.

McLellan, I., A. Hursthouse, C. Morrison, A. Varela and C. Silva Pereira. 2014. Development of a robust chromatographic method for the detection of chlorophenols in cork oak forest soils. *Environmental Monitoring and Assessment* 186 (2): 1281–93. doi: 10.1007/s10661-013-3457-z.

Mikesell, M. D. and S. A. Boyd. 1986. Complete reductive dechlorination and mineralization of pentachlorophenol by anaerobic microorganisms. *Applied and Environmental Microbiology* 52 (4): 861–865.

Monk, B. C. and A. Goffeau. 2008. Outwitting multidrug resistance to antifungals. *Science* 321 (5887): 367–369.

Moretti, M., A. Grunau, D. Minerdi, P. Gehrig, B. Roschitzki, L. Eberl, A. Garibaldi, M. L. Gullino and K. Riedel. 2010. A proteomics approach to study synergistic and antagonistic interactions of the fungal-bacterial consortium *Fusarium oxysporum* wild-type MSA 35. *Proteomics* 10 (18): 3292–3320. doi: 10.1002/pmic.200900716.

Morgan, M. K. 2015. Predictors of urinary levels of 2,4-dichlorophenoxyacetic acid, 3,5,6-trichloro-2-pyridinol, 3-phenoxybenzoic acid, and pentachlorophenol in 121 adults in Ohio. *International Journal of Hygiene and Environmental Health* 218 (5): 479–488.

Mueller, R. S., B. D. Dill, C. Pan, C. P. Belnap, B. C. Thomas, N. C. VerBerkmoes, R. L. Hettich and J. F. Banfield. 2011. Proteome changes in the initial bacterial colonist during ecological succession in an acid mine drainage biofilm community. *Environmental Microbiology* 13 (8): 2279–2292.

Muir, D. C. and P. H. Howard. 2006. Are there other persistent organic pollutants? A challenge for environmental chemists. *Environmental Science & Technology* 40 (23): 7157–7166.

Nakamura, T., H. Ichinose and H. Wariishi. 2012. Flavin-containing monooxygenases from *Phanerochaete chrysosporium* responsible for fungal metabolism of phenolic compounds. *Biodegradation* 23 (3): 343–350.

Nomura, S. 1953. Studies on chlorophenol. A clinical examination of workers exposed to pentachlorophenol. *Journal of Science of Labour, Part 2* 29: 474–483.

Pan, C., C. R. Fischer, D. Hyatt, B. P. Bowen, R. L. Hettich and J. F. Banfield. 2011. Quantitative tracking of isotope flows in proteomes of microbial communities. *Molecular & Cellular Proteomics* 10 (4): M110. 006049.

Patterson, S. D. 2003. Data analysis—the achilles heel of proteomics. *Nature Biotechnology* 21 (3): 221–222. doi: 10.1038/nbt0303-221.

Pinedo-Rivilla, C., J. Aleu and I. G. Collado. 2009. Pollutants biodegradation by fungi. *Current Organic Chemistry* 13 (12): 1194–1214.

Piskorska-Pliszczynska, J., P. Strucinski, S. Mikolajczyk, S. Maszewski, J. Rachubik and M. Pajurek. 2016. Pentachlorophenol from an old henhouse as a dioxin source in eggs and related human exposure. *Environmental Pollution* 208: 404–412.

Reddy, G. V. B. and M. H. Gold. 2000. Degradation of pentachlorophenol by *phanerochaete chrysosporium*: Intermediates and reactions involved. *Microbiology (London, United Kingdom)* 146 (2): 405–413.

Reddy, G. V. B. and M. H. Gold. 2001. Purification and characterization of glutathione conjugate reductase: a component of the tetrachlorohydroquinone reductive dehalogenase system from *Phanerochaete chrysosporium*. *Archives of Biochemistry and Biophysics* 391 (2): 271–277.

Rieble, S., D. K. Joshi and M. H. Gold. 1994. Purification and characterization of a 1,2,4-trihydroxybenzene 1,2-dioxygenase from the basidiomycete *Phanerochaete chrysosporium*. *Journal of Bacteriology* 176 (16): 4838–44.

Sandau, C. D., P. Ayotte, E. Dewailly, J. Duffe and R. J. Norstrom. 2000. Analysis of hydroxylated metabolites of PCBs (OH-PCBs) and other chlorinated phenolic compounds in whole blood from Canadian Inuit. *Environmental Health Perspectives* 108 (7): 611.

Saratale, G., S. Kalme, S. Bhosale and S. Govindwar. 2007. Biodegradation of kerosene by *Aspergillus ochraceus* NCIM-1146. *Journal of Basic Microbiology* 47 (5): 400–405.

Schägger, H. and G. von Jagow. 1991. Blue native electrophoresis for isolation of membrane protein complexes in enzymatically active form. *Analytical Biochemistry* 199 (2): 223–231.

Schneider, T., K. M. Keiblinger, E. Schmid, K. Sterflinger-Gleixner, G. Ellersdorfer, B. Roschitzki, A. Richter, L. Eberl, S. Zechmeister-Boltenstern and K. Riedel. 2012. Who is who in litter decomposition? Metaproteomics reveals major microbial players and their biogeochemical functions. *ISME Journal* 6 (9): 1749–1762. doi: 10.1038/ismej.2012.11.

Schöning, I., G. Morgenroth and I. Kögel-Knabner. 2005. O/N-alkyl and alkyl C are stabilised in fine particle size fractions of forest soils. *Biogeochemistry* 73 (3): 475–497.

Semeena, V. and G. Lammel. 2005. The significance of the grasshopper effect on the atmospheric distribution of persistent organic substances. *Geophysical Research Letters* 32 (7).

Shimizu, M., N. Yuda, T. Nakamura, H. Tanaka and H. Wariishi. 2005. Metabolic regulation at the tricarboxylic acid and glyoxylate cycles of the lignin-degrading basidiomycete *Phanerochaete chrysosporium* against exogenous addition of vanillin. *Proteomics* 5 (15): 3919–3931. doi: 10.1002/pmic.200401251.

Silva, A. J., D. P. Gomez-Mendoza, M. Junqueira, G. B. Domont, E. X. Ferreira, M. V. de Sousa and C. A. O. Ricart. 2012. Blue native-PAGE analysis of *Trichoderma harzianum* secretome reveals cellulases and hemicellulases working as multienzymatic complexes. *Proteomics* 12 (17): 2729–2738. doi: 10.1002/pmic.201200048.

Silva Pereira, C., J. J. F. Marques and M. San Romao. 2000a. Cork taint in wine: Scientific knowledge and public perception—a critical review. *Critical Reviews in Microbiology* 26 (3): 147–162.

Silva Pereira, C., A. Pires, M. Valle, L. Vilas Boas, J. Figueiredo Marques and M. San Romão. 2000b. Role of *Chrysonilia sitophila* in the quality of cork stoppers for sealing wine bottles. *Journal of Industrial Microbiology & Biotechnology* 24 (4): 256–261. doi: 10.1038/sj.jim.2900815.

Silva Pereira, C., G. A. Soares, A. C. Oliveira, M. E. Rosa, H. Pereira, N. Moreno and M. V. San Romão. 2006. Effect of fungal colonization on mechanical performance of cork. *International biodeterioration & biodegradation* 57 (4): 244–250.

Sobon, A., R. Szewczyk and J. Dlugonski. 2016. Tributyltin (TBT) biodegradation induces oxidative stress of *Cunninghamella echinulata*. *International Biodeterioration & Biodegradation* 107: 92–101. doi: 10.1016/j.ibiod.2015.11.013.

Speda, J., M. A. Johansson, U. Carlsson and M. Karlsson. 2017. Assessment of sample preparation methods for metaproteomics of extracellular proteins. *Analytical Biochemistry* 516: 23–36. doi: 10.1016/j.ab.2016.10.008.

Starke, R., R. Kermer, L. Ullmann-Zeunert, I. T. Baldwin, J. Seifert, F. Bastida, M. von Bergen and N. Jehmlich. 2016. Bacteria dominate the short-term assimilation of plant-derived N in soil. *Soil Biology & Biochemistry* 96: 30–38.

Stursova, M., J. Barta, H. Santruckova and P. Baldrian. 2016. Small-scale spatial heterogeneity of ecosystem properties, microbial community composition and microbial activities in a temperate mountain forest soil. *FEMS Microbiology Ecology* 92 (12). doi: 10.1093/femsec/fiw185.

Sugiura, S., H. Oota and S. Sato. 1956. A fatal case of acute poisoning caused by sodium pentachlorophenate. *Tokyo Iji Shinshi* 73: 161–162.

Szewczyk, R. and J. Długoński. 2009. Pentachlorophenol and spent engine oil degradation by *Mucor ramosissimus*. *International Biodeterioration & Biodegradation* 63 (2): 123–129.

Szewczyk, R., A. Sobon, R. Sylwia, K. Dzitko, D. Waidelich and J. Dlugonski. 2014. Intracellular proteome expression during 4-n-nonylphenol biodegradation by the filamentous fungus *Metarhizium robertsii*. *International Biodeterioration & Biodegradation* 93: 44–53. doi: DOI 10.1016/j.ibiod.2014.04.026.

Szewczyk, R., A. Soboń, M. Słaba and J. Długoński. 2015. Mechanism study of alachlor biodegradation by *Paecilomyces marquandii* with proteomic and metabolomic methods. *Journal of Hazardous Materials* 291: 52–64.

Tang, J., L. X. Liu, X. L. Huang, Y. Y. Li, Y. P. Chen and J. Chen. 2010. Proteomic analysis of *Trichoderma atroviride* mycelia stressed by organophosphate pesticide dichlorvos. *Canadian Journal of Microbiology* 56 (2): 121–127. doi: 10.1139/w09-110.

Taseli, B. K. and C. F. Gokcay. 2005. Degradation of chlorinated compounds by *Penicillium camemberti* in batch and up-flow column reactors. *Process Biochemistry* 40 (2): 917–923.

United Nations. 1972. Declaration of the United Nations Conference on the Human Environment. In: A/CONF.48/14, edited by United Nations. Stockholm.

United Nations. 1992. The Earth Summit agreements: a guide and assessment: an analysis of the Rio'92 UN Conference on Environment and Development. United Nations Conference on Environment and Development (1992: Rio de Janeiro, Brazil).

United Nations. 2001. Stockholm Convention on Persistent Organic Pollutants. In: *C.N.1017.2007. TREATIES-14*, edited by United Nations. Stockholm.

United Nations. 2015. Stockholm Convention on Persistent Organic Pollutants: Amendments to Annexes A and C. In: *C.N.681.2015.TREATIES-XXVII.15*, edited by United Nations. Stockholm.

Van Der Heijden, M. G., R. D. Bardgett and N. M. Van Straalen. 2008. The unseen majority: soil microbes as drivers of plant diversity and productivity in terrestrial ecosystems. *Ecology Letters* 11 (3): 296–310.

van Gorcom, R. F., J. G. Boschloo, A. Kuijvenhoven, J. Lange, A. J. van Vark, C. J. Bos, J. A. van Balken, P. H. Pouwels and C. A. van den Hondel. 1990. Isolation and molecular characterisation of the benzoate-para-hydroxylase gene (*bphA*) of *Aspergillus niger*: A member of a new gene family of the cytochrome P450 superfamily. *Molecular and General Genetics* 223 (2): 192–7.

Van Leeuwen, J., B. Nicholson, K. Hayes and D. Mulcahy. 1997. Degradation of chlorophenolic compounds by *Trichoderma harzianum* isolated from lake bonney, south-eastern South Australia. *Environmental Toxicology and Water Quality* 12 (4): 335–342.

Varela, A., C. Martins, O. Núñez, I. Martins, J. A. Houbraken, T. M. Martins, M. C. Leitão, I. McLellan, W. Vetter, M. T. Galceran and C. Silva Pereira. 2015. Understanding fungal functional biodiversity during the mitigation of environmentally dispersed pentachlorophenol in cork oak forest soils. *Environmental Microbiology* 17 (8): 2922–2934.

Varela, A., C. Martins and C. Silva Pereira. 2017. A three-act play: pentachlorophenol threats to the cork oak forest soils mycobiome. *Current Opinion in Microbiology* 37: 142–149.

Veiga, T., D. Solis-Escalante, G. Romagnoli, A. ten Pierick, M. Hanemaaijer, A. Deshmukh, A. Wahl, J. T. Pronk and J. M. Daran. 2011. Resolving phenylalanine metabolism sheds light on natural synthesis of penicilling in *Penicillium chrysogenum*. *Eukaryotic Cell* 11 (2): 238–249. doi: 10.1128/ec.05285-11.

Voříšková, J. and P. Baldrian. 2013. Fungal community on decomposing leaf litter undergoes rapid successional changes. *ISME Journal* 7 (3): 477–486. doi: http://www.nature.com/ismej/journal/v7/n3/suppinfo/ismej2012116s1.html.

Voříšková, J., V. Brabcová, T. Cajthaml and P. Baldrian. 2014. Seasonal dynamics of fungal communities in a temperate oak forest soil. *New Phytologist* 201 (1): 269–278.

Wang, D. Z., L. F. Kong, Y. Y. Li and Z. X. Xie. 2016. Environmental microbial community proteomics: Status, challenges and perspectives. *International Journal of Molecular Sciences* 17 (8). doi: 10.3390/ijms17081275.

Wang, H.-B., Z.-X. Zhang, H. Li, H.-B. He, C.-X. Fang, A.-J. Zhang, Q.-S. Li, R.-S. Chen, X.-K. Guo, H.-F. Lin, L.-K. Wu, S. Lin, T. Chen, R.-Y. Lin, X.-X. Peng and W.-X. Lin. 2011. Characterization of metaproteomics in crop rhizospheric soil. *Journal of Proteome Research* 10 (3): 932–940. doi: 10.1021/pr100981r.

Web of Knowledge, Clarivate Analytics. https://apps.webofknowledge.com, accessed in July 2017.

Weinbach, E. C. 1957. Biochemical basis for the toxicity of pentachlorophenol. *Proceedings of the National Academy of Sciences of the United States of America* 43 (5): 393–397.

Westermeier, R. 2016. 2D gel-based Proteomics: there's life in the old dog yet. *Archives of Physiology and Biochemistry* 122 (5): 236–237. doi: 10.1080/13813455.2016.1179766.

Wilkins, M. R., J. C. Sanchez, A. A. Gooley, R. D. Appel, I. HumpherySmith, D. F. Hochstrasser and K. L. Williams. 1996. Progress with proteome projects: Why all proteins expressed by a genome should be identified and how to do it. *Biotechnology and Genetic Engineering Reviews* 13: 19–50.

Wilmes, P. and P. L. Bond. 2004. The application of two-dimensional polyacrylamide gel electrophoresis and downstream analyses to a mixed community of prokaryotic microorganisms. *Environmental Microbiology* 6 (9): 911–920. doi: 10.1111/j.1462-2920.2004.00687.x.

Wilmes, P., A. Heintz-Buschart and P. L. Bond. 2015. A decade of metaproteomics: Where we stand and what the future holds. *Proteomics* 15 (20): 3409–3417. doi: 10.1002/pmic.201500183.

Williams, M. A., E. B. Taylor and H. P. Mula. 2010. Metaproteomic characterization of a soil microbial community following carbon amendment. *Soil Biology & Biochemistry* 42 (7): 1148–1156. doi: 10.1016/j.soilbio.2010.03.021.

Wu, L., H. Wang, Z. Zhang, R. Lin, Z. Zhang and W. Lin. 2011. Comparative metaproteomic analysis on consecutively *Rehmannia glutinosa* monocultured rhizosphere soil. *Plos One* 6 (5). doi: e20611 10.1371/journal.pone.0020611.

Wu, Y. R., Z. H. Luo, R. K. K. Chow and L. L. P. Vrijmoed. 2010. Purification and characterization of an extracellular laccase from the anthracene-degrading fungus *Fusarium solani* MAS2. *Bioresource Technology* 101 (24): 9772–9777. doi: 10.1016/j.biortech.2010.07.091.

Xie, C. L., L. Yan, W. B. Gong, Z. H. Zhu, S. W. Tan, D. Chen, Z. X. Hu and Y. D. Peng. 2016. Effects of different substrates on lignocellulosic enzyme expression, enzyme activity, substrate utilization and biological efficiency of *Pleurotus eryngii*. *Cellular Physiology and Biochemistry* 39 (4): 1479–1494. doi: 10.1159/000447851.

Yadav, J. S., M. B. Soellner, J. C. Loper and P. K. Mishra. 2003. Tandem cytochrome P450 monooxygenase genes and splice variants in the white rot fungus *Phanerochaete chrysosporium*: Cloning, sequence analysis, and regulation of differential expression. *Fungal genetics and biology* 38 (1): 10–21.

Yadid, I., J. Rudolph, K. Hlouchova and S. D. Copley. 2013. Sequestration of a highly reactive intermediate in an evolving pathway for degradation of pentachlorophenol. *Proceedings of the National Academy of Sciences of the United States of America* 110 (24): 2182–2190.

Yang, X., J.-Y. Sun, J.-L. Guo and X.-Y. Weng. 2012. Identification and proteomic analysis of a novel gossypol-degrading fungal strain. *Journal of the Science of Food and Agriculture* 92 (4): 943–951. doi: 10.1002/jsfa.4675.

Zeng, G. M., Z. Yu, Y. N. Chen, J. C. Zhang, H. Li, M. Yu and M. J. Zhao. 2011. Response of compost maturity and microbial community composition to pentachlorophenol (PCP)-contaminated soil during composting. *Bioresource Technology* 102 (10): 5905–5911. doi: 10.1016/j.biortech.2011.02.088.

Zhang, C., D. Suzuki, Z. Li, L. Ye and A. Katayama. 2012. Polyphasic characterization of two microbial consortia with wide dechlorination spectra for chlorophenols. *Journal of Bioscience and Bioengineering* 114 (5): 512–7. doi: 10.1016/j.jbiosc.2012.05.025.

Zhao, X., Y. Xie and J. Liu. 2017. Evaluating exosome protein content changes induced by virus activity using SILAC labeling and LC-MS/MS. In: *Methods in Enzymology*, edited by Barbara Imperiali, 193–209. Elsevier.

Zhao, Y., L. Yang and Q. Wang. 2008. Modeling persistent organic pollutant (POP) partitioning between tree bark and air and its application to spatial monitoring of atmospheric POPs in mainland China. *Environmental Science & Technology* 42 (16): 6046–6051.

Zheng, W., X. Wang, H. Yu, X. Tao, Y. Zhou and W. Qu. 2011. Global trends and diversity in pentachlorophenol levels in the environment and in humans: A meta-analysis. *Environmental Science & Technology* 45 (11): 4668–4675.

Zheng, W., H. Yu, X. Wang and W. Qu. 2012. Systematic review of pentachlorophenol occurrence in the environment and in humans in China: Not a negligible health risk due to the re-emergence of schistosomiasis. *Environment International* 42: 105–116.

Žifčáková, L., T. Větrovský, A. Howe and P. Baldrian. 2016. Microbial activity in forest soil reflects the changes in ecosystem properties between summer and winter. *Environmental Microbiology* 18 (1): 288–301.

CHAPTER-11

Bioreactors and other Technologies used in Fungal Bioprocesses

Avinash V. Karpe,[1,3,]* *Rohan M. Shah,*[3]
Vijay Dhamale,[3] *Snehal Jadhav,*[2] *David J. Beale*[1]
and *Enzo A. Palombo*[3]

1. Introduction

Fermented products have been used by humans since antiquity. Although some of these products have bacterial origins, fungi have been the major input sources of food, beverages, and animal feed. European history includes records of milk-based products, such as cheese, and grain-based products, such as beer/ale, whose manufacture involves the application of various fungi and yeasts, respectively. Similarly, Oriental-based fermentation products are predominately plant-based, utilising soybean (*miso* and *soy sauce*), rice (*sake*), and tubers (*schochu*), pulses (*idli*), palm sap (*toddy*), and coconut (*tempe*), among others.

More recently, the popularity of fermented foods has increased in an attempt to combat artificially flavored, high calorie foods (Dufosse et al. 2014) besides numerous other novel applications. For example, fungal food sources such as mushrooms are known to contain high levels of proteins, vitamins, and are low in lipids and saturated fatty acids (Ghorai et al. 2009). Additionally, fungi are also widely used in production of novel probiotics, such as modern versions of kombucha (Jayabalan et al. 2014), animal feed and feed-additives (Ghorai et al. 2009), agricultural applications, such as

[1] CSIRO Land & Water, Ecosciences Precinct, Dutton Park, QLD 4102. Australia.
[2] Centre for Advanced Sensory Science, School of Exercise and Nutrition Sciences, Deakin University, Burwood, VIC 3125. Australia.
[3] Department of Chemistry and Biotechnology, Swinburne University of Technology, Hawthorn, VIC 3122. Australia.
* Corresponding author: avinash.karpe@csiro.au

fertilisers and pesticides (Igiehon and Babalola 2017), biofuels (Aguieiras et al. 2015, Cardona and Sánchez 2007, Wilson 2009), nutraceuticals (Cheng et al. 2008, Dufosse et al. 2014, Jayakumar et al. 2006), antibiotics (Rasmussen et al. 2005) and in waste management (Aggelopoulos et al. 2014, Das et al. 2016, Karpe 2015). Although the variety of fungi used for a number of applications is very wide, the fungi used most often in fermentation processes belong to higher ascomycetes and basidiomycetes, such as white and brown rot fungi (Baldrian and Valášková 2008, Mäkelä et al. 2013, Karpe 2015).

This chapter presents an overview of research and commercial applications of fungal bioprocessing. It will also focus on the fungal bioreactor technologies and the techno-economic aspects of scaling up for various commercial applications. The review will focus on recent developments, especially those of the last decade (2010–2018).

2. Bioreactors and Bioprocessing: Developments and Issues

In recent decades, bioreactors have been used to commercialise a number of foods, beverages, and medicinal and industrial metabolites (See Sections 5.2 and 5.3). While submerged fermentation (SmF) is still widely used in bacterial and yeast based fermentations, the late 20th century re-emergence of solid state fermentation (SSF) from its oriental origin to a wider and global scale has played a significant role in the advancement of fungal bioprocessing. As previously mentioned, numerous commercial outputs have been reported regarding fungal systems. However, it is equally important that these developed processes are able to be translated to industrial scaled-up versions of their laboratory batch tests.

2.1 Issues and considerations of fungal bioreactor designs

Due to the differential sizes and nature of bioreactors, a number of parameters are required to be optimised to allow scaled-up versions. While earlier bioprocessing technologies were based on indigenous/cultural knowledge, many scale-ups were more arbitrary in nature. In recent decades, especially since 2000, a greater emphasis on mathematical or statistical modelling of these processes at laboratory (< 1 litre) or pilot scale (10–1000 litres) has resulted in technologies that are more fluently translated to larger industrial volumes.

Some of the important parameters considered to design mathematical/statistical models are thermodynamics, water content, metabolic rate and behavior, acidity/ alkalinity, and oxygen requirements. Simple yeast growth in a small reactor (1 litre), without temperature control, can increase the temperature of bioreactor system from an ideal 25°C to a lethal 35°C within 48 hours of fermentation. Similar tendencies, but with even higher temperatures (up to 40°C or more), have been observed in enclosed fungal biomass degradation in SSF conditions with no thermal control (Karpe 2015). Control of such parameters is easier and more economical at the laboratory stage with respect to larger post-pilot scale levels (≥ 1000 litres). Similarly, water activity (K_w) plays an important role in fungal bioprocessing, affecting substrate and product solubilities (Karpe et al. 2014).

Several mathematical/statistical models are used to efficiently predict scale-up of the fungal bioprocessing. Kinetic models aid growth-time dynamics and have been given attention in ethanol fermentation extensively. SmF primarily depends on variables, such as yeast cell load, product concentration, substrate concentration and metabolic limitation (substrate) or inhibition (product). Phisalaphong et al. (2006) described the model output in terms of specific growth rate (μ, Equation 11.1) and ethanol production rate (ν, Equation 11.2) as:

$$\mu = \mu_m \left(\frac{C_s}{K_s + C_s + \dfrac{C_s^2}{K_{ss}}} \right) \left(1 - \frac{C_p}{P_m} \right) \qquad \text{(Eq. 11.1)}$$

$$\nu = \nu_m \left(\frac{C_s}{K_{sp} + C_s + \dfrac{C_s^2}{K_{ssp}}} \right) \left(1 - \frac{C_p}{P'_m} \right) \qquad \text{(Eq. 11.2)}$$

Where:

C_s = substrate concentration (g/L); C_p = ethanol concentration (g/L); K_s = saturation growth constant (g/L); K_{sp} = saturation production constant (g/L); K_{ss} = substrate growth inhibition term (g/L); K_{ssp} = substrate production inhibition term (g/L); P_m = ethanol inhibition term-growth (g/L); P'_m = ethanol inhibition term-production (g/L).

The applied model (implemented on MATLAB statistical platform) using Arrhenius functions was able to provide the production rate activation energy of ethanol at 3.5×10^4 kJ/mol. The model predicted a temperature-based inverse relationship between cell growth/ethanol production and product inhibition/substrate load.

In SSF based biomass degradation, Karpe et al. (2015c) determined the fungal enzyme activity and degradation efficiency by full factorial (2^k design) statistics using Minitab® statistical software. The optimal activities obtained by factorial modelling resulted in considerable increase in mixed-fungal enzyme activities of cellulase, xylanase, and β-glucosidase, causing enhanced grape pomace degradation in a scaled-up process. Further models include Reaction–diffusion (Rahardjo et al. 2006), mechanistic (Mears et al. 2017) and logistic equation (Montoya et al. 2014). Detailed information on these models can be obtained from Desobgo et al. (2017).

Based on the mathematical/statistical modelling, there have been a number of scaled-up fungal bioreactors used in research and industrial scales. However, they can be broadly categorised into three types of submerged, semi-solid or solid and, biofilm-based fermentations. Depending on the type of product required, a particular strategy using a single type or a mixed version is generally adopted. Musoni et al. (2015) indicated several strategies and examples, primarily using *Trichoderma* spp., *Penicillium* spp. and *Aspergillus* spp. The authors indicate that in addition to a number of physico-chemical factors, fungal morphology also has an effect on the type of fermentation. For example, SmF primarily favors mycelial growth and, therefore, is unable to facilitate conidiation or sporulation. This physiological alteration is in contrast to SSF, where the fungi sporulate vigorously, which alters the output products. The reader is advised to read the work of Musoni et al. (2015) for further detailed information on bioreactors and their implementation approaches.

2.2 Submerged fermentation (SmF)

SmF is the classical fermentation method used for ethanol production and the manufacture of fermented foods. Also known as 'shake flask fermentation', SmF is an aqueous phase fermentation process, where the medium-to-substrate ratio is generally high (in excess of 10:1). Although the technique is well-established for wine and related alcohol production, its industrial applications flourished in Western countries during late 1940s for enhanced penicillin production from *P. chrysogenum*. The basic methods have been greatly optimized due to the requirement for scaling up of SmF (> 3000–5000 litres). Such optimizations include supplementation of oxygen, heat exchange, agitation, prevention of foaming, and culture loads. SmF remains the most important standardized fermentation process, largely because it has been thoroughly investigated and optimised (Humphrey 1998).

Fungal pellet reactors are some of the very recent types of SmFs that have been evaluated at research and development (pilot) scales. In its early days, these reactors were shown to ferment an *Aspergillus niger* mediated ferrocyanide treated beet-molasses to obtain citric acid (Clark 1962). Although this technology has been used at laboratory scale since the 1960s, scaled-up versions (> 15 kL) have been a relatively recent development (Pollard et al. 2007). The property of fungi to form large hyphae due to a considerable mycelial growth causes immobilised fungal bodies referred to as 'fungal pellets' (Fig. 11.1), thus eliminating the need for artificial immobilisation techniques.

Recent studies have demonstrated the ability for this technology to produce various commercial biomolecules, including phytase (*Aspergillus ficuum*) as an animal feed enhancer (Coban et al. 2015), fumaric acid produced by *Rhizopus oryzae* using pulp and paper waste as carbon source (Das et al. 2016), and the antimicrobial agent curvulamine (*Curvularia* spp.) (Yang et al. 2016). Darah et al. (2011) demonstrated the importance of agitation and related nutrient/fungal mixing to improve fungal pellet yields. Agitating *A. niger* at 200 rpm showed an increase in tannase production, while lower (50 rpm) and higher (> 200 rpm) resulted in depleted yields (Darah et al. 2011). Tannase is an important enzyme to degrade lignin/tannin originating metabolites that have previously been observed to impede fungal bioprocessing (Karpe et al. 2016, Karpe et al. 2017). This property of fungal enzymatic activity in pellet form has also been used in enhanced wastewater treatment. The pellet studies, at least on smaller scales, have elucidated their benefits in treating wastewater in batch and continuous flow process. The interested reader is suggested to follow the review by Espinosa-Ortiz et al. (2016). The authors compared various reactors, such as stirred tanks, airlift column, bubble column, and fluidised bed reactors. While all have some limitations, overall handling of most of these reactors is advantageous in terms of uniform energy dissipation, nutrient/oxygen distribution, and low operational inputs.

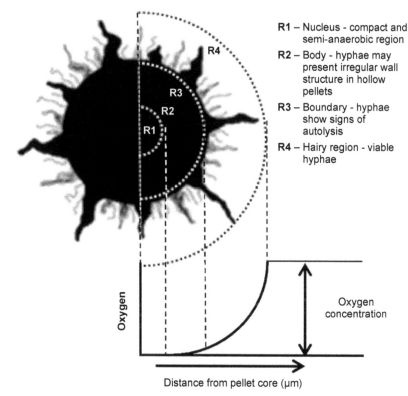

R1 – Nucleus - compact and semi-anaerobic region

R2 – Body - hyphae may present irregular wall structure in hollow pellets

R3 – Boundary - hyphae show signs of autolysis

R4 – Hairy region - viable hyphae

Figure 11.1 A general morphology of fungal pellet and oxygen availability in each layer (Picture courtesy of: (Espinosa-Ortiz et al. 2016); Copyright © 2015 Elsevier B.V.).

It has been long debated that SmF conditions create a stress environment for fungal growth after their spore stage. It has been observed in recent studies that SmF conditions may result in mycelial rupture, especially due to agitation in excess water. The secretome (molecules secreted by living cells) of *Neurospora sitophila* grown under SmF conditions displayed the presence of five intracellular proteins (formate dehydrogenase, ubiquitin-60S ribosomal protein L40, superoxide dismutase, enolase, and fructose biphosphatase) in the extracellular filtrate (Li et al. 2013). These proteins are known to be involved in the metabolism of glyoxylate and dicarboxylate, glycolysis, gluconeogenesis, and protection against oxidative stress. The presence of these enzymes in the secretome indicated possible fungal cell lysis.

Another issue related to SmF is the need for sterilization of the growth medium. Due to the aqueous conditions, SmF is more favorable for bacterial growth as compared to fungi, thus, SmF is highly prone to bacterial contamination. To prevent this, the medium is generally sterilized, primarily by hydrothermal methods, such as autoclaving or steaming. However, the application of these treatments generally results in the generation of furfurals and low molecular weight organic and fatty acids, which inhibit enzyme activity and overall fungal growth (Merino and Cherry 2007, Sanchez and Bautista 1988).

2.3 Solid state fermentation (SSF)

Solid state fermentation (SSF) is one of the second-generation bioconversion processes for the production of industrially important molecules, such as bio-hydrogen and bio-ethanol. The method employs a mixture of organisms and/or derived enzymes from them in a single step to generate a considerably higher bioconversion of biomass compared to SmF (Brijwani et al. 2010, Kausar et al. 2010, Lee 1997, Sarkar et al. 2012). As the water content in SSF is equal to or slightly greater than the substrate, the method allows greater aeration and surface attachment of filamentous fungi. This approach facilitates production of a vast array of lignocellulolytic enzymes. It also increases the amount of substrate to be degraded in a single step from 1–2% in SmF to more than 10% in SSF (Lee 1997, Ng et al. 2010), and reduces the need for separate enzyme production, isolation, and application in biomass degradation. Thus, SSF fulfils all the requirements for bioconversion in a single step, which makes it more cost-effective than SmF (Betini et al. 2009, Brijwani et al. 2010, Dashtban et al. 2010, Sarkar et al. 2012).

SSF provides the natural growth conditions for fungi, especially the filamentous fungi. Fungal spores and ultimately, the enzymes generated in SSF, exhibit higher tolerance to the environmental conditions. Thus, higher quantities of enzymes are produced under SSF conditions than SmF conditions. This pattern becomes more evident in mixed fungal cultures where a number of different enzymes are generated and utilized for various purposes. A number of studies have shown that the fungal and enzymatic behaviors in SSF conditions are generally synergistic or symbiotic, and competitive inhibition is not a widespread phenomenon (Gruntjes 2013, Singhania et al. 2010).

Similar to SmF-based biomass degradation, fungal bioprocessing under SSF is also dependent on the nutrient content of the biomass. The carbon-nitrogen (C/N) ratio of the substrate also plays an important role in SSF biodegradation with higher nitrogen content resulting in greater fungal degradation ability (Karpe et al. 2017). One of the other important factors affecting SSF is the temperature distribution. Due to limited water availability (80–90% moisture content, typically yielding < 1:1 solid liquid ratio) and poor heat conductivity of most biological materials (food, biomass, etc.), large system temperature fluctuations are observed. Some of these can be deleterious to the fungal processing. At small scales, fluctuations of 12–15°C have been observed, but in scaled up versions, such as composting, these differences can be much larger and temperature ranges of 20–70°C have been observed (Yong et al. 2011, Maeda et al. 2010). While small changes in temperature up to 35°C have been observed to make no significant differences in fungal output, a further increase causes considerable losses in enzyme activity (Singhania et al. 2009).

In the last 30 years, SSF has been widely used for a number of industrial bioprocesses. Bioconversion of the lignocellulose complex from various biomass sources has increased the production of industrial metabolites, biofuels, and secondary metabolites which can be used for medicinal purposes. The use of biomass addresses the issues related to managing the considerable amounts of waste material generated during numerous agricultural and allied processes.

SSF has been extensively applied in the production of enzymes. Singhania et al. (2009) have discussed the production of phytase, amylase, inulinase, cellulase, protease, α-galactosidase, lipase, tannase, laccase, chitinase, L-glutaminase, lipase in their review. The current focus of the technology appears to be on the production of fertilizers and pharmaceuticals, mainly antibiotics, as observed in recent patents and inventions. Other recent innovations focus on improved controlling systems, stirring apparatus, and adaptations for continuous cultures. For example, in their patent, Alitalo et al. (2017) describe an SSF process using a bioreactor with a solid support to produce hydrogen gas.

Considering the number of parameters mentioned above, the diversity of reactors used for SSF is typically less than that used in SmF. Although a number of reactors are used for research and development applications, overall, three types of reactors are widely used, with slight variations, as per commercial requirements. These consist of tray-type, packed-bed, and rotating drum. More information on these bioreactors can be obtained from Mitchell et al. (2006). This reference provides details about SSF reactors, their modelling, kinetics, thermodynamics, and scale-up.

2.4 Consolidated bioprocessing (CBP)

Consolidated Bio-processing (CBP) is one of the emerging techniques which employ a number of different strategies to improve the biomass degradation process. Briefly, it involves a single step process to degrade the substrate and convert it into the product of interest. This not only saves considerable time, but also minimises other resources, such as raw materials, thereby making it an economical technique. CBP involves numerous strategies, such as genetic strain improvements of the microbial populations involved in biomass degradation to yield the products of interest, metabolic engineering, and utilization of compatible organisms, including thermophilic microbes, to enhance the one-step degradation and successive bioconversion processes (Demain et al. 2005, Lynd et al. 2005).

As CBP was developed to address difficult and complex processing of biomass, it involves utilisation of numerous parameters, such as high enzyme load, simultaneous fermentation of a large variety of saccharides, and preventing or minimising the product/substrate inhibition. As discussed in Section 3.3, besides lignins and tannins, biomass consists of pentoses (e.g., xylose, rhamnose, and β-glucans), hexoses (e.g., glucose, fructose and galactose), heptoses (e.g., sedoheptulose), disaccharides (e.g., sucrose and maltose), trisaccharides (e.g., trehalose, raffinose, and maltotriose), and higher oligosaccharides (e.g., dextrins) (Karpe 2015). Although many ascomycetes, basidiomycetes, and thermophilic bacteria produce a battery of biomass degrading enzymes, no single organism has been reported to fulfil all necessary parameters (den Haan et al. 2015). Therefore, a general practice has been towards developing genetically enhanced microbial system using metabolic engineering (See Section 5.2), among other options. CBP combines either multi-enzymatic consortium (CBP I) or metabolic engineering of well-characterised bioprocessing organisms (CBP II), such as *Saccharomyces cerevisiae* to convert substrate to product in a one-step fermentation via SmF or SSF.

Traditionally, due to their aerobic nature, most moulds have been either unable to produce ethanol, or generate very limited quantities, mostly as a collateral by-product. However, recent CBP developments have been used to provide an add-on ethanol production function to these eukaryotes. Compared to bacterial CBP, developments in fungal CBP have been limited and not many fungi have been tested in CBP until recently (Olson et al. 2012). In terms of biofuel, and especially ethanol production, utilization of fungi, such as *Aspergillus*, *Rhizopus*, *Fusarium*, and *Trichoderma* has been reported (Hasunuma et al. 2013). Among these, *Trichoderma reesei* has been observed to generate moderate amounts of ethanol, while simultaneously degrading cellulosic biomass. One of the major limitations of this fungus, however, is its strict aerobic growth which prevents the production of commercially-viable levels of alcohols, such as ethanol. The utilization of facultative ethanologenic organisms, such as *Pichia stipites*, *Candida* spp., and *S. cerevisiae*, can address this limitation. Bacteria such as *Zymomonas mobilis* also have considerable potential to improve this process. This bacterium removes certain deficiencies of yeast-based fermentation, such as heterologous enzyme secretion and mass accumulation, resulting in recent developments in CBP (Linger and Darzins 2013).

A recent report on CBP for ethanol production involved construction of a laboratory-scale multi-species biofilm membrane reactor (MBM reactor) (Fig. 11.2).

Experimental analyses of microcrystalline cellulose (brand names such as Avicel®, FMC Corp., Philadelphia, PA, USA) based CBP using the MBM reactor showed that a co-culture of *S. cerevisiae* and *T. reesei* RUT C30 generated 7.2 g/L ethanol in the gas phase upon exogenous addition of β-glucosidase. When acid-treated wheat straw replaced Avicel® as the carbon source, addition of *Scheffersomyces stipites* (formerly *Pichia stipites*) generated about 9 g/L ethanol with complete utilization of xylose. The authors proposed that either the addition of a high β-glucosidase-producing organism or an ethnologenic strain, such as *Dekkera bruxxelensis*, would further improve ethanol production (Brethauer and Studer 2014). An earlier study on this aspect was reported by Zerva et al. (2014), in which *Paecilomyces variotii* was

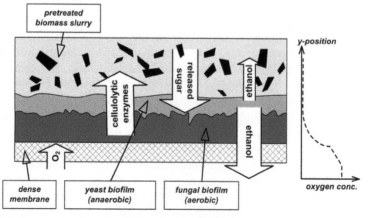

Figure 11.2 Conceptual design of the MBM reactor for CBP to generate ethanol from pre-treated wheat straw (Courtesy of (Brethauer and Studer 2014) . Published by The Royal Society of Chemistry (RSC) on behalf of the Centre National de la Recherche Scientifique (CNRS) and the RSC).

used to generate ethanol from corn cob and wheat bran. Higher nitrogen addition in the form of nitrate resulted in xylose and similar pentose metabolism, thereby creating noticeable deviation of the pentose phosphate pathway. The consumption of xylose generated up to 14 g/L ethanol in about 15 days of fermentation. Other studies conducted on similar lines are indicated in Table 11.1.

2.4.1 Multi-fungal consortium

One of the simpler methods of CBP-derived biomass degradation is the establishment of microbial symbiotic systems in order to improve the degradation. To overcome numerous limitations associated with first and second generation biomass degradation processes, mixed fungal culture degradation has been suggested (Brijwani et al. 2010, Chu et al. 2011). It is known that, apart from satisfactory cellulase production, *Aspergillus* spp. generate highly efficient xylanases and β-glucosidases in exceptional quantities (Betini et al. 2009, Singhania 2012). Additionally, *Penicillium* spp. can be used for lignin mineralization to enhance the overall degradation process (Rodríguez et al. 1994, Rodriguez et al. 1996).

A notable study of the multi-organism commensal mechanism was reported by Benoit et al. (2015). In this study, *Aspergillus niger* was grown in SmF conditions for 12 hours, followed by inoculation of *Bacillus subtilis*, resulting in a bacterial-fungal interaction (BFI) in which bacteria grew on fungal hyphae; this was reflected in considerable alterations of individual transcriptomes. A similar approach, but for direct production of ethanol from cellulose, utilised BFI with a combination of yeasts (genetically engineered *S. cerevisiae* and *Candida molischiana*) and the cellulolytic bacterium *Clostridium phytofermentans* under microaerophilic conditions. The co-culture not only resulted in bacterial cell protection by the yeasts, but also resulted in almost 3.5 times ethanol yield with respect to individual cultures (33 g/L vs 6–9 g/L) (Zuroff et al. 2013).

In a recent experiment, individual fungi (*Trichoderma harzianum*, *Aspergillus niger*, *Penicillium citrinum*, *Phanerochaete chrysosporium*, and *Trametes versicolor*) were cultured and crude enzymes were extracted. The enzyme mix was added to pre-treated grape pomace for degradation. About 13% lignin (*w/w*) degradation was observed and considerably high enzyme activities of cellulases and lignin peroxidases were seen, causing a release of substantial sugars and secondary metabolites, such as gallic acid and lithocholic acid, over a 16 day fermentation (Karpe 2015). More recently, the process was further streamlined and refined by the authors resulting in similar outputs within a 36 hour process (including pre-treatment) (Karpe et al. 2017).

However, one of the most popular and significant commercial outputs of multi-fungal consortium is associated with the 'Kombucha' beverage. Kombucha is a symbiotic consortium of several osmophilic yeasts and acetic acid bacteria (among others), which ferment substrates, such as tea-leaf extract and sugars. While the bacteria mostly consist of *Acetobacter* spp. and *Gluconobacter* spp., the yeast contribution comes from various species, such as *Saccharomyces* spp., *Brettanomyces* spp., *Candida* spp., *Torulospora* spp., *Pichia* spp., *Mycotorula* spp., and *Mycoderma* spp. The metabolic composition includes organic acids (C_2–C_6), sugars (C_6 and C_{12}), amino acids, nucleotides, hydrophilic vitamins (B and C), lipids, ethanol, CO_2 and

Table 11.1 Fungal CBP based production of metabolites.

Fungus	Substrate	Enzyme activity	Product yield*	Reference
Phlebia sp. MG-60	sugarcane bagasse	Cellulase (0.24 U/mL) Xylanase (1 U/mL)	21% (w/w) 4.5 g/L	(Khuong et al. 2014)
Trametes hirsuta	Sugars, wheat bran, rice straw	–	~ 0.5 g/g (sugar) Upto 9.1 g/L	(Okamoto et al. 2011)
Kluyveromyces marxianus *S. cerevisiae*	Jerusalem artichoke tubers	Inulinase (30.4 U/mL)	Upto 73.6 g/L	(Hu et al. 2012)
F. oxysporum	Rice straw	Cellulase (1.3 U/mL)	9.3 g/L 0.125 g/g	(Xu et al. 2015)
Endophytes	switchgrass, corn stover, eucalyptus mill powder	Various (20–450 U/g mycelia)	VOC (upto 13%)	(Wu et al. 2017)
Penicillium oxalicum	corn cob residue + wheat bran	Various (0.1–0.9 U/mL)	sodium gluconate (13.54 g/L)	(Han et al. 2018)
T. reesei *P. decumbens* *S. cerevisiae*	alkali-treated bagasse + wheat bran	β-glucosidase (~ 0.5 U/g)	5.8 g/L	(Liu et al. 2015)
Cryptococcus terricola	starch	–	Biodiesel $_{lipids}$ (Upto 62%)	(Tanimura et al. 2014)

* Product is ethanol unless indicated otherwise.

polyphenol, amongst others. Although several studies have reported the probiotic effects of Kombucha, the landmark study is that of Jayabalan et al. (2014).

2.4.2 *Genetically engineered CBP (GE-CBP)*

Genetically engineered CBP remains a popular processing method since its inception in early 2000s. Consumer preferences and religious biases still limit the use of genetically modified organisms in the food and feed sectors (McFadden and Lusk 2015), resulting in the use of much simpler and economical multi-fungal or BFI consortia (Section 3.4.1). However, from a chemical industry perspective, genetic modification and engineering are not considered problematic due to limited direct consumer consumption. The necessity of GE-CBP was driven by the obstacles faced in biomass treatment processes, as indicated above.

Saccharomyces cerevisiae has been preferably used in a majority of CBP processes, possibly due to a far better understanding of the yeast's genomic, transcriptomic, and proteomic systems, accumulated during the landmark Human Genome Project. Some of the early experiments were designed to induce fungal xylose reductase and dehydrogenase genes in *S. cerevisiae* to enhance the pentose phosphate pathway for one-step ethanol production from biomass-derived sugars (van Zyl et al. 2007). Manipulating more complex systems, such as the fungal lignocellulolytic machinery, has been more challenging and, progress has been slower, a number of genetically engineered basidiomycete and ascomycete systems have been expressed successfully in *S. cerevisiae* (van Zyl et al. 2007). Since early days, GE-CBP has been applied to a number of fungi to enhance their own cellulolytic activity. Additionally, a recent study demonstrated a symbiotic approach using enzymes from a genetically modified *Trichoderma reesei* using xylanolytic gene (A824) overexpression and genetically modified *Aspergillus niger* ($\Delta creA/ xlnR_c/araR_c$). The combination of both systems showed increased saccharification of wheat straw (Jiang et al. 2016).

Besides the above-mentioned symbiotic or synergistic approaches, recombinant DNA technologies have been widely used in developing CBP for biofuel production. However, their application to filamentous fungi remains limited due to the complex functional and expression systems. Hence, CBP has been more successful with simpler microorganisms, namely bacteria and yeasts. Indeed, *S. cerevisiae* has been the subject of numerous recombinant applications for CBP. Although various technologies have evolved to the point of developing a commercial-scale CBP process, most of the fundamental research is still in progress. Considerable developments have been made in bacterial-based CBP processes by introducing recombinant expression systems. However, significant research on integrating major biomass degraders (namely filamentous fungi) into CBP remains to be done.

3. Commercial Processes (Fungal Outputs)

3.1 *Food and beverage industry*

Food production (commercial and non-commercial) has been a part of human activity since antiquity. Various fermentation products, both foods and drinks, have been widely reported in historical documents of numerous early cultures throughout history. Such products include vinegar, cheeses, fermented vegetables (sauerkraut, cucumber, and olives), breads, alcoholic beverages, fermented milk, and protein foods (soy and miso sauces). Interestingly, these products are still widely used and form a significant part of the modern diet today. Further details about the individual food and beverages can be obtained from the published work of Wood (2012). According to the recent analysis by the Michigan State University-based GlobalEDGE, global exports of food and beverage in 2016 was valued at about US$720.7 billion, of which about 22% was either from fermented products or from fungal-derived food exports (globalEDGE 2017).

3.1.1 Fermentation products

Numerous fungi are known to cause food and beverage spoilage, both in human and animal feed. However, in recent years, fungi have been seen as the important part of the food and beverage industry due to the production and release of enzymes into the extracellular environment (Karpe et al. 2016). Due to this property, the extraction of fungal enzymes for industrial applications has been viewed economically, and, therefore, has been widely used. Table 11.2 provides a list of fungal enzymes used in industrial food and beverage applications.

3.1.2 Fungi as food

Similar to fermented products discussed above, fungal biomass, especially mushrooms and mushroom-type fungi, has been consumed by humans for a long time. Historically, mushrooms were predominantly commercially produced via SSF of agricultural biomass wastes, mostly soybean waste (Asia) or horticultural wastes (Europe). These edible fungi are rich in proteins and other nutritional secondary metabolites (Ghorai et al. 2009), which are presented in Table 11.3.

Besides mushrooms, numerous fungi are also used to make single cell proteins (SCPs) as food. These are considered valuable due to their high nutritional content, but lower saturated fat content, especially of cholesterol (Ghorai et al. 2009). Some yeast species, such as *Saccharomyces* spp., *Candida* spp., *Pichia* spp. and *Kluyveromyces* spp. and moulds, such as *Aspergillus* spp., *Rhizopus* spp., *Penicillium* spp. and, above mentioned basidiomycetes (Table 11.3) are widely used to produce commercial SCPs. Aggelopoulos et al. (2014) demonstrated that scaled-up SSP resulted in volatiles and unsaturated fatty-acid production, with a positive cost-benefit evaluation. SSF was performed with cultures of *S. cerevisiae*, *Kluyveromyces marxianus*, and kefir grain-derived species cultured on mix wastes (whey, molasses, and orange, potato and brewery residues) and generated high protein (> 33%), unsaturated fat (13–25.5%, primarily consisting of oleate and linoleate) and volatiles (e.g., pinene). Another study by Yunus et al. (2015) utilising wheat bran substrate reported a maximum SCP yield at 10% inoculum size, with *Candida utilis* and *Rhizopus oligosporus*. The fungal fermentation of wheat bran yielded protein content enhancement of up to 36.8% in 48 hours for the purpose of high nutrient supplementation in animal feed. Other generally known fungal SCPs include Marmite®, Vegemite®, Cenovis®, Vitam-R®

Table 11.2 Major fungal enzymes and their uses in food and beverage sector.

Enzyme	Property	Application	Source	References
α-amylase	α (1, 4) glycosidic bond hydrolysis	Corn liquefaction (Starch → glucose), Glucose and fructose syrups	*Aspergillus* spp., *Mucor* spp. *Penicillium* spp., *Pycnoporus sanguineus*	(Gopinath et al. 2017, Sundarram and Murthy 2014)
β-amylase	α (1, 4) glycosidic bond hydrolysis	(Starch → glucose), Brewing, maltose syrup	*Aspergillus* spp., *Penicillium* spp.	(Serna-Saldivar and Rubio-Flores 2017, Adeniran et al. 2010, do Prado et al. 2016)
Catalase	H_2O_2 mediated catalysis (breakdown)	Antioxidants, nutraceuticals	*Agaricus brasiliensis*, *Fusarium* spp., *Cryptococcus* spp.	(Mokochinski et al. 2015, Teder et al. 2017, Hur et al. 2014, Ray and Rosell 2017)
Isomerase	Isomerisation (aldose → ketose)	Glucose → Fructose, alcohol, juice and syrup	*Aspergillus oryzae*, *Candida* spp., *Piromyces* spp., *Saccharomyces* spp.	(Harhangi et al. 2003, Verhoeven et al. 2017, Ray and Rosell 2017)
Laccase	Monolignols/Cu^{2+}-based single electron oxidation	stabilizing agents, juice clarification	*Pleurotus* spp., *Phlebia* spp., *Trametes* spp., *Agaricus* spp.	(Lettera et al. 2016, Carabajal et al. 2013, Makela et al. 2013, Ray and Rosell 2017)
Lactase (β-Galactosidase)	β-d-galactoside hydrolysis	Lactose hydrolysis, dairy and nutraceuticals	*Kluyveromyceslactis* spp., *Guehomyce spullulans*, *Aspergillus oryzae*, *Kluyveromyces* spp.	(Saqib et al. 2017, Pandey 1992, Plou et al. 2017)
Lipase	Hydrolysis (Triglycerides → fatty acids + glycerol	Transesterfication, flavor modification (wine, cheese)	*Candida* spp., *Saccharomyces* spp., *Aspergillus* spp., *Penicillium* spp.	(Gupta et al. 2015, Ray and Rosell 2017, Aguieiras et al. 2015)
Pectinase	Cleavage: Methyl ester bond, α-1,4-glycosidic bond	Pectin hydrolysis in juices, alcohols (liquor clarification)	*Aspergillus* spp., *Penicillium* spp., *Fomes* spp., *Trichoderma* spp.	(Garg et al. 2016, Ray and Rosell 2017)
Protease	Peptide bond hydrolysis	Hydrolysing high protein food, e.g., soy sauce production	*Penicillium* spp., *Aspergillus* spp., *Beauveria* spp., *Fusarium* spp., *Mucor* spp., *Trichoderma* spp.	(Novelli et al. 2016, Souza et al. 2015, Ray and Rosell 2017)
Tannase	Ester bond hydrolysis	Tannin hydrolysis, tea processing, whitening agent in beverages	*Aspergillus* spp., *Pichia* spp., *Candida* spp.	(Liu et al. 2017, Dhiman et al. 2017, Ray and Rosell 2017)

Table 11.3 Selected major commercial fungi and their primary contents (Ghorai et al. 2009).

Fungus	Major components (%)				Major secondary metabolites	References
	Proteins	Polysaccharides	Chitin	Carbohydrates		
Lentimula edodes (Shitake mushroom)	29.7	14	9.2	–	Lentinam, vitamin-D; adenine, choline	(Jiang et al. 2010, Savci 2012)
Agaricus bisporus (Button mushroom)	32	18.7	14.7	–	Riboflavin	(Nitschke et al. 2011, Ghorai et al. 2009)
Pleurotus ostreaus (Oyster mushroom)	40	11.5	14.8	–	Levostatin, 1-octen-3-ol	(Vetter 2007, Jayakumar et al. 2006)
Tuber melanosporum (Truffle)	14	–	14.8	74	Phenols, flavonoids, carotenoids, 3,4-Dihydroxy-benzaldehyde	(Yan et al. 2017, Ghorai et al. 2009)

(*S. cerevisiae*), Pekilo® (*Paecilomyces varioti*), and, *Yarrowia* SCP (*Yarrowia lipolytica*).

For further information about recent SCP commercial developments and intellectual property rights, including fungal derivatives, the interested reader is referred to Gopinath et al. (2017) and Ritala et al. (2017).

3.2 *Agriculture*

Agricultural activities, especially harvesting and post-harvest processes, generate significant amounts of biomass wastes globally. It has been estimated that the major crops (corn, barley, oat, rice, sorghum, and sugarcane) generate about 7.39×10^{11} metric tonnes of biomass waste per year (Kim and Dale 2004). As of 2016, reported global agricultural production was about 1.07×10^{10} metric tonnes, of which only 1.1×10^8 metric tonnes were processed into final products. In addition, the Food and Agriculture Organization (FAO) of the United Nations indicated the generation of about 5.23×10^8 metric tonnes of agricultural wastes during 2011, of which 4.81×10^8 metric tonnes (CO_2 equivalent) was of lignocellulosic nature (FAO 2017). Most of this biomass waste remains underutilized and is either burned or landfilled, adding to overall production costs (Devesa-Rey et al. 2011).

3.2.1 *Biomass processing*

Biomass is composed of cellulose, hemicelluloses, and lignins (Carpita 1996, Singh et al. 2009) and other commercially valuable components, such as organic acids, oil, phenolics, colloids, and dietary fibres (Arvanitoyannis et al. 2006). Various secondary metabolites, such as flavanols, cathechins and proanthocyanidins, lignols and antioxidants, have been reported in horticultural and other food sources (González-Paramás et al. 2003, Jadhav et al. 2015). Cellulose is the chief source of carbon in lignocellulose complexes, as it is the richest source of glucose, the chief source of carbon for the microorganisms. Cellulose, however, exists in the core of the complex, surrounded by layers of hemicellulose and lignins (Fig. 11.3) and is often inaccessible (Karpe et al. 2017, Karpe et al. 2014).

Ascomycetes (*Trichoderma* spp., *Aspergillus* spp. and *Penicillium* spp.) and basidiomycetes have been well studied for their high cellulase and hemicellulase enzyme production (Brijwani et al. 2010, Klyosov 1987, Karpe et al. 2017, Karpe et al. 2014). Fungal enzymes have been used to generate important molecules, such as alcohols, flavonoids, organic acids and phenolics (Arvanitoyannis et al. 2006, Sánchez 2009, Strong and Burgess 2008). Basidiomycetes (white and brown-rot fungi), such as *Phanerochaete* spp., *Trametes* spp., *Ganoderma* spp., *Fomitopsis* spp., and *Postia* spp. are known for their ability to completely mineralize lignin component of biomass to CO_2 and H_2O (Sánchez 2009, Singh Arora and Kumar Sharma 2010, Karpe et al. 2017).

Fungal enzymes degrade a wide variety of substrates to make the utilisable carbon source available (Table 11.4). The systems differ in composition to hemicellulases whose secretions, even under similar conditions (Baldrian and Valášková 2008, Dashtban et al. 2009, Sánchez 2009, Sipos et al. 2010), depend on substrate variability (Liu et al. 2012).

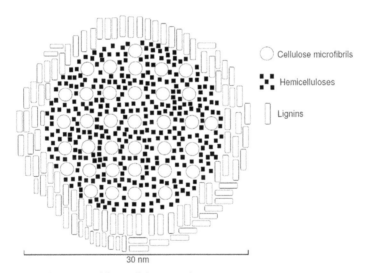

Figure 11.3 General structure of lignocellulose complex.

3.2.2 Bio-fertilizers and bio-pesticides

Another agricultural sector where fungal bioprocessing has grown in prominence is bio-fertiliser and pesticide production. Excessive salting of soils due to the deposition of nitrates (NO_3^-), phosphates (PO_4^{3-}), sulphates (SO_4^{2-}), and potassium (K), and ammonium (NH_4^+) salts is a growing concern for ecology and (sustainable) agriculture (Savci 2012). Additionally, the leeching of these salts and associated heavy metals in water bodies has been observed to cause significant changes in aquatic microflora and microfauna (Beale et al. 2017a, Beale et al. 2017b). The significant increase in

Table 11.4 A general classification of major fungal lignocellulolytic enzymes.

Enzyme	Subclasses	Target molecule/bond	Fungal source	References
Cellulases	Endoglucanases (EC 3.2.1.4)	Cellulose/β-1, 4-glycosidic	*Trichoderma* spp., *Aspergillus* spp., *Penicillium* spp., *Phanerochaete* spp., *Trametes* spp., *Sclerotium* spp.	Colussi et al. 2012, Dashtban et al. 2009, Francieli Colussi et al. 2011, Sweeney and Xu 2012, Karpe et al. 2014, Karpe et al. 2016
	Exoglucanases (EC 3.3.1.91)			
	β-glucosidases (EC 3.2.1.21)	Cellobiose/β-1, 4-glycosidic		
Hemicellulases	Xylanases (EC 3.2.1.8)	Xylans/β-(1, 4) xylo-pyranoside	*Trichoderma* spp., *Aspergillus* spp., *Penicillium* spp., *Malbranchea flava*, *Phanerochaete* spp.	Dashtban et al. 2009, Sweeney and Xu 2012, Cantarel et al. 2009, Karpe et al. 2015c, Karpe et al. 2017, Sharma et al. 2016, Tsai et al. 2017
	Xyloglucanases (3.2.1.151)	Xyloglucans/β-(1, 4)-D-glucosidic		
	β-xylosidases (3.2.1.37)	Xylans/β-(1, 4) xylopyrano-side		
	β-mannases (3.2.1.25)	β-mannosides/Hydrolyse terminal, non-reducing mannose residues		
	Arabinofuranosidases (3.2.1.55)	Arabino-xylans and galactans/(1, 3)- ; (1, 5)-arabino-furanosidic		
	α-L-arabinases (EC 3.2.1.99)	Arabinans/(1, 5)-α-arabino-furanosidic		
Lignases	Laccases (1.10.3.2)	Monolignols/Cu^{2+}-based single electron oxidation	*Postia* spp., *Fomitopsis* spp., *Pycnoporus* spp., *Phanerochaete* spp., *Trametes* spp., *Agaricus* spp., *Plebia* spp., *Pleurotus* spp.	Grinhut et al. 2011, Valentín et al. 2010, Takano et al. 2013, Saroj et al. 2013, Mäkelä et al. 2013, Carabajal et al. 2013, Fernández-Fueyo et al. 2012, Manole et al. 2008, Knop et al. 2014, Fernández-Fueyo et al. 2014, Choi et al. 2014, Floudas et al. 2012, Siddiqui et al. 2014, Gupta et al. 2016
	lignin peroxidases (1.11.1.14)	non-phenolic and β-O-4 non-phenolic lignin/H_2O_2-based oxidation depolymerization		
	manganese peroxidases (1.11.1.13)	phenolic lignin/Mn^{3+}-based peroxidase dependent oxidation		
	versatile peroxidases (1.11.1.16)	Both non-phenolic and phenolic lignin/H_2O_2- and Mn^{3+}- based oxidation		

these nutrients has often led to eutrophication in aquatic environments (Vassilev et al. 2015). Recent research has demonstrated that the use of fungal biofertilizers in combination with chemical fertilizers considerably decreases the negative effects of chemical fertilizers, but also improves crop output. A recent genomic study suggested that the combined use of NPK fertilizers with organic inputs increased the expression of genes related to nitrogen recycling in soil (Sun et al. 2015). Similar studies on the use of algal salvaged waste aquatic nutrients on tomato cropping showed an improved sugar and carotenoid content, resulting from slow released effects of these bio-fertilizers (Coppens et al. 2016). Fungal bio-fertilizers comprise a dominant position in the global bio-fertilizer market and have been reported to supplement the plants with both macronutrients (phosphorous, potassium, sulphate) and micronutrients (zinc, copper, iron). A number of ascomycetes, such as *Aspergillus* spp., *Penicillium* spp., *Fusarium* spp., and *Mucor* spp. function to stabilise soil phosphorous, whereas, the mycorrhizal species mobilize the phosphates. Similarly, a number of ascomycetes have been commercially used for enriching compost based bio-fertilizers (Pal et al. 2015, Owen et al. 2015).

The mycorrhizal fungi, often referred to as arbuscular mycorrhizal fungi (Wu et al. 2017), especially Glomeromycocetes, such as *Glomus intraradices* and *Rhizophagus irregularis*, have been widely reported to establish a symbiotic relationship within the rhizospheres of most plants (Igiehon and Babalola 2017). Another type of mycorrhiza, ectomycorrizal fungi (EcM), mostly Basidiomycetes and Ascomycetes, have been observed in numerous studies to form a dense, underground multi-species (incorporating Arbuscular-mycorrhizal fungi, AMFs) symbiotic fungi-plant-fungi networks (Fig. 11.4) (Tedersoo et al. 2010, Owen et

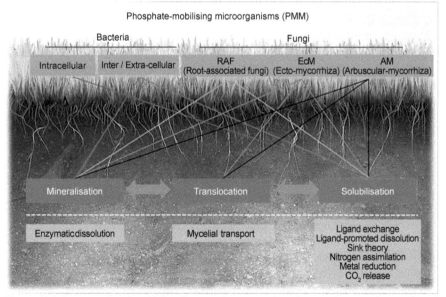

Figure 11.4 An example of multi-fungal consortia representing plant nutrient supplementation of phosphates through the rhizosphere (Picture courtesy of: (Owen et al. 2015), Copyright © 2014 Elsevier B.V.).

al. 2015). The mycorrhizal networks (MNs) not only fix and mobilise nutrients, but also aid in plant establishment, survival, biotic/abiotic stress metabolite exchange, and physiology. A recent review on this interaction by Gorzelak et al. (2015) will provide the reader with further details.

Due to these functions, the commercial demand for integrating AMF and EMF in crop fertigation has increased in recent decades. Although a number of laboratory observations have been unable to be replicated in the fields, novel biotechnological approaches have steadily improved the quality of commercial mycorrhizal products. Vassilev et al. (2015) in their extensive review of commercial production approaches have indicated two options of developing either 'compound inocula' or 'complex inocula'. The first approach involves co-inoculating two organisms with complementary properties. This not only has been observed to enhance the individual microbial functions, but considerably decreases the limitations of each member of multi-functional consortia. Other 'compound inocula' experiments consist of co-inoculating white rot fungi with rhizobium bacteria and was observed to provide nitrogen fixing, phosphate and micronutrient mobilisation, and bio-control of phytopathogenic fungi (Lemanceau et al. 2009, Tripathi et al. 2017). The second, 'complex inocula' approach uses an already established microbial consortia with differential bacterial or fungal species. An example would be the co-inoculation of ascomycetes/white rot fungi (individual) + AMF (consortium) + *Rhizobium* spp. (bacteria) to enhance plant growth and nutrition in legumes and cereals (Vassilev et al. 2015).

The novel use of next generation sequencing (Jiang et al. 2016) in combination with downstream techniques such as SSF have been shown to create considerable improvements in developing these consortia. Briefly, the SSF technique uses a lignocellulolytic fungal consortium to provide a pre-treatment, which is followed by the periodic addition of EMF and AMF fungi and, if applicable, nitrogen fixing bacteria (Igiehon and Babalola 2017, Vassilev et al. 2015). SSF and its applications are further discussed in Section 3.3. Additionally, the reader can obtain detailed information regarding fungal bio-fertilizers, bio-pesticides, and their commercial production from the recent works of Mukherjee (2017), Hatami and Ahangarani (2016), and Suthar et al. (2017).

3.3 *Biofuel production*

A significant rise in technological development has made energy an ever-decreasing commodity. Numerous environmental issues and high demand-to-supply ratio in recent years (2.1% annual increase in 2017) has necessitated an increased requirement for renewable energy (up to 25% of total energy produced) (www.iea. org) and biofuels (2.6% annual increase in 2016) (www.bp.com). The FAO/OECD 2016 outlook indicated ethanol and biodiesel contributions at about 1.2×10^{11} L and 3.3×10^{10} L, respectively (OECD/FAO 2016). As of 2016, maize (6.3×10^{10} L) and sugarcane (2.9×10^{10} L) have been the major contributors to ethanol production, while biomass contributed 6×10^8 L. Similarly, vegetable oil has been the largest contributor (2.6×10^{10} L) for biodiesel production (OECD/FAO 2016).

Traditionally, ethanol production has been a *S. cerevisiae* domain and contribution from moulds (ascomycetes or basidiomycetes) is considered relatively novel. Similarly, major biodiesel production technologies have been focused towards algal rather than fungal bioprocessing, due to the nature of the underlying biochemistry. Although most fungi (except yeasts) are exceptional in biomass processing, sugars represent major bioprocess outputs, with fuel-grade alcohols and lipids being the by-products. This comes from limited and slow utilisation of wide range of saccharides (produced from lignocellulolytic activities) (Karpe et al. 2017), limited growth inhibition, and possible autolysis caused on medium chained fatty acid (MCFAs) rich substrates (Karpe et al. 2015b).

However, some filamentous fungi have the potential to generate fuel-metabolites without requiring genetic engineering. While *Penicillium* spp. has potential for ethanol production on hemicellulose (especially, xylose) supplementation (Karpe et al. 2015b), others such as *Aspergillus* spp. and *Neurospora* spp. have indicated noticeable production (15–19 g/L ethanol) in a higher protein supplemented starch thin-stillage substrate. One of the issues that requires to be addressing to improve fungal ethanol is redox balancing without oxygen access. The aerobic nature of most fungi limits ethanol production, which requires a low or no oxygen tension. This has resulted in limited success of 1st and 2nd generation ethanol production by fungal bioprocessing. However, this can be considerably improved by a two-stage (lignocellulolysis and fermentation) process, as indicated by Lennartsson et al. 2014. Similarly, *Mortierella* spp. have been estimated to yield 19–23% (*w/w*) lipids, mostly MCFAs, from SSF of glucose-xylose rich hydrolysed wheat straw (Zheng et al. 2012) and individual monosaccharides (Gardeli et al. 2017). Additionally, *Mucor fragilis* fermentation has a lipid yield of about 13 g/L (Huang et al. 2016). Lipases are considered important contributors to biofuel-lipid generation and have been suggested to be used in combination with filamentous fungal bioprocessing to obtain higher transesterification/hydroesterification for better lipid yields. The interested reader is referred to the work of Aguieiras et al. (2015).

Hydrogen (H_2) is considered as an effective alternative fuel with almost 2.8-fold energy density to that of petrol/gasoline. Although the majority of H_2 production comes from petroleum and natural gas, newer bio-H_2 production technologies are continuously developed. Most bio-H_2 is based on either fermentative (anaerobic fungi and/or thermophilic bacteria/archaea) or photo-biological (blue-green algae) platforms (www.nrel.gov/bioenergy/biohydrogen). Although the latter platform is beyond this review's scope, the former platform is discussed here. While most of the bio-H_2 production is attributed to thermophilic *Clostridium* spp., numerous filamentous fungi are used to process biomass and food/feed residues. For example, SSF with *Phanerochaete chrysosporium* pre-treated wheat straw fermented with *Trichoderma atroviride* cellulases and *Clostridium perfringens* resulted in the generation of about 79 mL/g-wheat straw (Zhi and Wang 2014). An interesting study of bio-H_2 production was reported by Han et al. (2016). In this techno-economic pilot project, H_2 production was obtained by processing 2 tonnes/day waste bread. *Aspergillus awamori* and *Aspergillus oryzae* based protease-glucoamylase SSF pre-treated substrate was processed further to generate about 354 mL/h/L H_2, equating to about 499 m³ H$_2$/day @ $US 1.34/m³ H$_2$ ($US 14.89/kg H$_2$). Further information about bio-H$_2$ production, its process and economic considerations can be found in the review of Kumar et al. (2015).

4. Novel Industry Innovations and Future Prospects

The combination of 'multi-omics' platforms and increased intellectual property rights-based collaborative research and development strategies by industries and public-sector organisations have led to rapid development of novel processes and products from fungal bioprocessing. Recent innovations focus on improved controlling systems, stirring apparatus, and adaptations for continuous cultures. For example, in their patent, Alitalo et al. (2017) describe an SSF process using a bioreactor with a solid support to produce hydrogen gas. SSF has been extensively applied in the production of enzymes. Singhania et al. (2010) discussed the production of phytase, amylase, inulinase, cellulase, protease, α-galactosidase, lipase, tannase, laccase, chitinase, L-glutaminase, and lipase. The current focus appears to be on the production of fertilizers and pharmaceuticals, mainly antibiotics, as observed in recent patents and inventions.

One approach to advance bioprocessing is via the application of omics-based tools and approaches. The field of omics has made significant inroads to fungal bioprocess research. Although genomics and transcriptomics have been established fields since at least late 1990s, their application has been mainly targeted to health and clinical research (drug discovery). In addition, the more expression-based omics technologies (such as proteomics and metabolomics) are increasingly being applied. Figure 11.5 illustrates the application of these tools to fungal bioprocessing and how they can be used to provide insights into fungal and bacterial diversity of substrates and process (genomics), the potential of organisms to metabolise

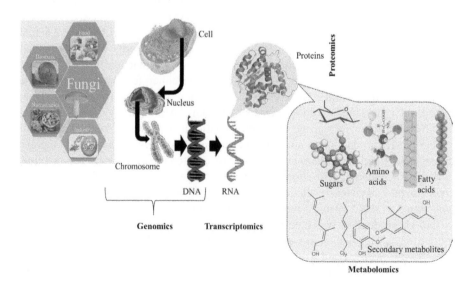

Figure 11.5 A general approach and pathway of multi-omics.

by-products (transcriptomics), and identify and measure metabolic outputs (proteomics and metabolomics) through a deeper fundamental understanding of the biochemical processes and the underlying mechanisms of biomass fermentation.

Omics-based applications are either implemented as an individual stream or as a consortium of omics-based tools. The advent of novel technologies, such as next generation sequencing and high sensitive instrumentation (gas and liquid chromatography) combined with mass spectral databases, such as NIST (National Institute of Standards and Technology), Wiley registry, Fiehn (Fiehnlab, UC Davis), and Golm (Max Planck Institute of Molecular Plant Physiology) have made significant developments in systems and functional biology. In their recent review, Beale et al. (2016a) provided an in-depth insight of applications of omics outputs in numerous systems biology and other general and commercial approaches. Various combinations of multi-omics have been applied to different matrices by the authors including dairy (proteomics + genomics) (Jadhav et al. 2015), wastewater (metagenomics + metabolomics) (Beale et al. 2016b), riverine (metagenomics + metabolomics + ionomics) (David Beale et al. 2017), and oceanic (metagenomics + metabolomics + polycyclic aromatic hydrocarbons (PAH) + emerging chemicals of concern (ECC)) (Beale et al. 2017).

4.1 Genomics, proteomics and metabolomics

Metagenomics has been the most prevalent of all omics in fungal bioprocessing, probably because of its wider scope and horizons. A more recent application of genomics/metagenomics in fungal bioprocessing has been in metabolic engineering to produce commercial yields of various products. Complementing conventional Sanger sequencing, newer techniques such as whole-genome sequencing, especially with Illumina sequencing, have rapidly found widespread application. Using QIIME (http://qiime.org/) and R (www.r-project.org) analysis software packages, Illumina sequencing was able to identify and evaluate petroleum degrading white-rot fungi based on community level expression of *alkB*, *phnAc*, and *nah* genes (Egan et al. 2018). The economic feasibility of these technologies was demonstrated by the novel Nextera tagmentation reaction using an automated workflow system (Shapland et al. 2015). This work showed that a process flow of > 4000 plasmids in a single run can be achieved with each sample run costing about $US 3.

Most metagenome processes focus either on developing targeted product expression in yeast machinery or making minor alterations in higher fungi to produce targeted products. For the yeast-based designs, cloning using plasmid vectors is often employed, either via non-homologous end joining or homologous recombination into the yeast genome auxotrophic regions (Fig. 11.6).

The application of these genome engineering techniques ensures a highly efficient metabolic engineered yeast production. In some very recent applications, the application of novel CRISPR-Cas9 editing has increased in metabolic engineering significantly and has been employed in addition to the homologous recombination techniques to the yeast cells (Löbs et al. 2017).

Considering the major role that proteins play in fungal physiology and biochemistry, proteomics has been a very important tool to quantify fungal

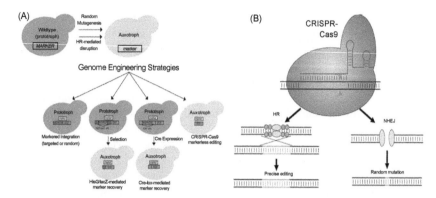

Figure 11.6 (A) Schematic diagram representing generation/utilization of auxotrophic markers for metabolic engineered yeast and, (B) CRISPR-Cas9-mediated genome editing (Courtesy of (Löbs et al. 2017). Under a creative commons attribution-non-commercial-no derivatives 4.0 international (CC BY-NC-ND 4.0)).

systems biology. Contrasting to genomics/transcriptomics, fungal proteomics and metabolomics represent a real-time fungi-environmental relationship and, as such, provide more rapid, efficient, and realistic approaches in terms of commercially valuable outputs. Additionally, structural proteomics and secretome analyses provide commercially important information that reflects differences in physiological and extracellular enzyme excretion (Bianco and Perrotta 2015). While the former (proteomics) focuses on the sutural characteristics of fungi (e.g., spores and mycelia), the latter (secretome) analyses the production of fungal enzymes.

The proteomic workflow involves various methods, such as gel-based approaches (SDS-PAGE, 1- and 2-dimensional electrophoresis, SDS-PAGE-MS) or mass spectral methods, such as matrix assisted laser desorption/ionisation time-of-flight (MALDI-TOF) MS and liquid chromatography mass spectrometry (LC-MS/MS) (Bianco and Perrotta 2015). The former technique uses an existing protein standard, such as bovine serum albumin; the latter techniques have increasingly focused on developed mass spectra, local selected reaction monitoring (SRM), or multiple reaction monitoring (MRM). The reader is referred to work of Bianco and Perrotta (2015) for further information on fungal bioprocess proteomics.

Metabolomics has the potential to provide biochemical information in order to understand and characterise various mechanisms of fungal bioprocessing. In particular, metabolic flux analysis provides a good understanding of the nature, time-dependence, and substrate-based limitations of fungal metabolism. Metabolomics also assists in extricating the correlation between cell phenotypes and their metabolic patterns and stoichiometry (Meijer et al. 2009). Numerous methods are available for metabolic flux analysis. The majority of these involve the use of heavy isotopes due to their considerably superior tracking abilities. In commonly utilized analyses, such as nuclear magnetic resonance (NMR) and, liquid and gas chromatography combined with mass spectrometry (LC- and GC-MS), metabolite tracking is performed using isotopes, such as ^1H, ^2H, ^{13}C, ^{14}N, ^{15}N and ^{18}O. While every isotope has its advantages and shortcomings, ^{13}C-labelled molecules have been used by most researchers

(Nelson et al. 2014, Karpe et al. 2015a, Karpe et al. 2015b), probably because of their wide-spread use in general chemistry. ^{2}H is a much more abundant element in nature, comprising about half of the peptide atomic population (Yang et al. 2010). It is much more economical as compared to other isotopes, and has a comparatively higher invasive rate and a slower decay rate (Kim et al. 2012). Complete ^{2}H labelling is not feasible, however, due to the limited tolerance of multicellular organisms, thereby resulting in partial labelling, generally from 8% (Kim et al. 2012) to 30% (Yang et al. 2010). Due to the complex nature of biomass composition and the allosteric nature of most lignocellulolytic enzymes, standard biochemical tests have proven less effective compared to metabolomic approaches for deciphering biomass conversion. Moreover, metabolic profiling methods indicate signature metabolites and critical pathway points which can be exploited to not only improve the biomass degradation process, but also to convert this biomass to products of industrial and medicinal value.

The reader is referred to works of Beale et al. (2016a) and Karpe et al. (2016) for further information on fungal metabolomics and techniques.

4.2 Secondary metabolites and nutraceuticals

Fungi produce a tremendous number of secondary metabolites which play important roles in the production of communication signals, virulence factors, and products which contribute to health and nutrition (Macheleidt et al. 2016). Diverse microbes co-existing in communities constantly interact with each other via secondary metabolites. Secondary metabolites may be used as signals for protection (i.e., competition) or serve as mediators for communication (i.e., co-operation) (Brakhage 2013). Abrudan et al. (2015) revealed that secondary metabolites function as signalling molecules and these molecules co-ordinate interspecies interaction between co-existing bacteria. They showed that the production of 2,4-diacetylphloroglucinol is repressed by fusaric acid produced by *Fusarium oxysporum*. 2,4-diacetylphloroglucinol is the most important component responsible for antimicrobial activity of *Pseudomonas fluorescens*.

Microorganisms have developed "quorum sensing" systems—a communication process that is dependent on cell densities. Extensively studied in bacteria, quorum sensing is often associated with several behaviors, such as secretion of virulence factors, biofilm formation, and bioluminescence (Grandclément et al. 2015). Albuquerque and Casadevall (2012) described quorum sensing in *Candida albicans* and identified the role of secondary metabolites, such as farnesol and tyrosol, in mycelium formation. The production of secondary metabolites could be a response to quorum sensing signals, for example, aldose reductase inhibitor sclerotiorin produced by *Penicillium sclerotiorum* seems to be related to quorum sensing molecules multicolic acid and its derivatives (Raina et al. 2010). Bjarnsholt et al. (2005) and Rasmussen et al. (2005) showed that quorum sensing molecules could be new drug targets; the quorum sensing response systems in *Pseudomonas aeruginosa* has been targeted with secondary metabolites, such as patulin and penicillic acid (Raina et al. 2010).

Many entomogenous fungi, such as *Metarhizium anisopliae* and *Beauveria bassiana*, produce secondary metabolites in submerged cultures that can act as insecticidal toxins. *M. anisopliae* produces cyclic depsipeptides, such as destruxins, while *B. bassiana* produces cyclic peptides, such as beauvericin, beauverolides, and bassianolide, that have insecticidal properties (Gillespie and Claydon 1989, Lozano-Tovar et al. 2015, Hu and Dong 2015).

4.3 Food additives and miscellaneous products

A mixed consortia of microbes is often used in preparation of many foods. This may include yeasts, filamentous fungi, acetic acid bacteria, and lactic acid bacteria. These co-cultures play an important role in the aroma, texture, flavor and, very often, the shelf life of fermented food items (Scherlach et al. 2013).

Filamentous fungi are non-conventional sources of natural pigments. Filamentous fungi can produce an extraordinary range of pigments of chemical classes, such as carotenoids, melanins, flavins, phenazines, quinones, monascins, violacein, or indigo (Dufosse et al. 2014). The polyketide pigments from the *Monascus* spp. are well-known, produced commercially, and have been used as food colorants for thousands of years (Feng et al. 2012). *Monascus* pigments are not approved food colorants in the US and the European Union due to the possible risk of being contaminated with citrinine, a metabolite of these pigments that is known to have nephrotoxic and hepatotoxic properties. Some strains among *Talaromyces* species produce *Monascus*-like polyketide azaphilone pigments without co-producing citrinine (Mapari et al. 2008, Mapari et al. 2009b). *Epicoccum nigrum* is another example of polyketide pigment orevactaene-producing fungi. Orevactaene exhibits anti-oxidant properties (Mapari et al. 2008, Mapari et al. 2009b). The red pigment extract from *T. aculeatus* and the yellow pigment extract from *E. nigrum* have enhanced photo-stability as compared to the red *Monascus* colorant and turmeric in liquid food systems (Mapari et al. 2009a).

Hydroxyanthraquinoids are major pigments found in lichens, plants, insects, and filamentous fungi, particularly *Pencillium* spp. and *Aspergillus* spp. Arpink red™ (now Natural Red™) manufactured by the Czech company Ascolor Biotech is produced by fermentation and bioprocess engineering using *Penicillium oxalicum* var. Armeniaca CCM 8242. This is the first commercial product from this chemical class and is a natural food colorant. *Aspergillus glaucus*, *A. cristatus*, and *A. repens* produce a range of red and yellow pigments, including emodin (yellow), physcion (yellow), questin (yellow to orange-brown), erythroglaucin (red), catenarin (red), and rubrocristin (red) (Caro et al. 2012).

Another secondary metabolite produced by *Monascus* and *Rhizopus* that has high demand is gamma amino butyric acid (GABA). GABA is an inhibitory neurotransmitter and has several health benefits including regulation of hormone secretion. Three species of *Monascus* viz., *M. purpureus*, *M. pilosus*, and *M. ruber* have been used in the production of GABA (Sun et al. 2008).

A well-known HMG-coA reductase inhibitor lovastatin (or Monocolin) is also a by-product of a fungal secondary metabolite. Lovastatin is the first naturally synthesised statin to be approved by the US FDA (Manzoni and Rollini 2002).

Researchers in late 1970s successfully isolated lovastatin from broths of *M. ruber* and *Aspergillus terreus* (Endo 1979, Alberts et al. 1980). Lovastatin has been isolated from filamentous fungi (predominantly from the *Monascus* and *Aspergillus* genera), yeasts (e.g., *Pichia labacensis* and *Candida cariosilignicola*), and basidiomycetes (*Pleurotus ostreatus*) since then (Alarcon et al. 2003, Cheng et al. 2008).

Conclusions

Rapid growth in industrial technologies and associated research and development has resulted in significant progress in fungal bioprocessing. The applications of novel platforms, such as multi-omics, as standalone techniques and as integrated workflows have helped in gaining deeper insights into fungal bioprocessing. This has, in turn, motivated considerable developments in bioreactor designs, process design, and scale-up for commercialisation. Further improvement in new technologies, especially synthetic biology aided by CRISPR-Cas9 gene editing, combined with bioreactor engineering and development, is expected to deliver further cutting-edge discoveries that will increase the efficiency and expand the potential applications of fungal bioprocessing. The Low-cost fungal bioprocessing is widely applied in bioremediation and thus useful for the environment.

Conflicts of Interest

The authors declare no conflict of interest.

References

Abrudan, M. I., F. Smakman, A. J. Grimbergen, S. Westhoff, E. L. Miller, G. P. van Wezel and D. E. Rozen. 2015. Socially mediated induction and suppression of antibiosis during bacterial coexistence. *Proceedings of the National Academy of Sciences* 112 (35): 11054–11059.

Adeniran, H., S. Abiose and A. Ogunsua. 2010. Production of fungal β-amylase and amyloglucosidase on some Nigerian agricultural residues. *Food and Bioprocess Technology* 3 (5): 693–698.

Aggelopoulos, T., K. Katsieris, A. Bekatorou, A. Pandey, I. M. Banat and A. A. Koutinas. 2014. Solid state fermentation of food waste mixtures for single cell protein, aroma volatiles and fat production. *Food Chemistry* 145: 710–716. doi: https://doi.org/10.1016/j.foodchem.2013.07.105.

Aguieiras, E. C., E. D. Cavalcanti-Oliveira and D. M. Freire. 2015. Current status and new developments of biodiesel production using fungal lipases. *Fuel* 159: 52–67.

Alarcon, J., S. Aguila, P. Arancibia-Avila, O. Fuentes, E. Zamorano-Ponce and M. Hernández. 2003. Production and purification of statins from *Pleurotus ostreatus* (Basidiomycetes) strains. *Zeitschrift für Naturforschung C* 58 (1–2): 62–64.

Alberts, A., J. Chen, G. Kuron, V. Hunt, J. Huff, C. Hoffman, J. Rothrock, M. Lopez, H. Joshua and E. Harris. 1980. Mevinolin: a highly potent competitive inhibitor of hydroxymethylglutaryl-coenzyme A reductase and a cholesterol-lowering agent. *Proceedings of the National Academy of Sciences* 77 (7): 3957–3961.

Albuquerque, P. and A. Casadevall. 2012. Quorum sensing in fungi–a review. *Medical mycology* 50 (4): 337–345.

Alitalo, A., E. Aura and M. Niskanen. 2017. Bioreactor and fermentation process for producing hydrogen. US Patent App. 15/536,002.

Arvanitoyannis, I. S., D. Ladas and A. Mavromatis. 2006. Potential uses and applications of treated wine waste: a review. *International Journal of Food Science and Technology* 41 (5): 475–487. doi: 10.1111/j.1365-2621.2005.01111.x.

Baldrian, P. and V. Valášková. 2008. Degradation of cellulose by basidiomycetous fungi. *FEMS Microbiology Reviews* 32 (3): 501–521. doi: 10.1111/j.1574-6976.2008.00106.x.

Beale, D., A. Karpe, W. Ahmed, S. Cook, P. Morrison, C. Staley, M. Sadowsky and E. Palombo. 2017a. A community multi-omics approach towards the assessment of surface water quality in an urban river system. *International Journal of Environmental Research and Public Health* 14 (3): 303.

Beale, D. J., A. V. Karpe and W. Ahmed. 2016a. Beyond Metabolomics: A review of multi-omics-based approaches. In: *Microbial Metabolomics: Applications in Clinical, Environmental, and Industrial Microbiology*, edited by David J. Beale, Konstantinos A. Kouremenos and Enzo A. Palombo, 289–312. Cham: Springer International Publishing.

Beale, D. J., A. V. Karpe, J. D. McLeod, S. V. Gondalia, T. H. Muster, M. Z. Othman, E. A. Palombo and D. Joshi. 2016b. An 'omics' approach towards the characterisation of laboratory scale anaerobic digesters treating municipal sewage sludge. *Water Research* 88: 346–357. doi: https://doi.org/10.1016/j.watres.2015.10.029.

Beale, D. J., J. Crosswell, A. V. Karpe, W. Ahmed, M. Williams, P. D. Morrison, S. Metcalfe, C. Staley, M. J. Sadowsky, E. A. Palombo and A. D. L. Steven. 2017b. A multi-omics based ecological analysis of coastal marine sediments from gladstone, in Australia's central queensland, and heron island, a nearby fringing platform reef. *Science of the Total Environment* 609: 842–853. doi: https://doi.org/10.1016/j.scitotenv.2017.07.184.

Benoit, I., M. H. van den Esker, A. Patyshakuliyeva, D. J. Mattern, F. Blei, M. Zhou, J. Dijksterhuis, A. A. Brakhage, O. P. Kuipers, R. P. de Vries and Á. T. Kovács. 2015. *Bacillus subtilis* attachment to *Aspergillus niger* hyphae results in mutually altered metabolism. *Environmental Microbiology* 17 (6): 2099–2113. doi: 10.1111/1462-2920.12564.

Betini, J. H. A., M. Michelin, S. C. Peixoto-Nogueira, J. A. Jorge, H. F. Terenzi and M. L. T. M. Polizeli. 2009. Xylanases from *Aspergillus niger*, *Aspergillus niveus* and *Aspergillus ochraceus* produced under solid-state fermentation and their application in cellulose pulp bleaching. *Bioprocess and Biosystems Engineering* 32 (6): 819–824.

Bianco, L. and G. Perrotta. 2015. Methodologies and perspectives of proteomics applied to filamentous fungi: from sample preparation to secretome analysis. *International Journal of Molecular Sciences* 16 (3): 5803–5829.

Bjarnsholt, T., P. Ø. Jensen, T. B. Rasmussen, L. Christophersen, H. Calum, M. Hentzer, H.-P. Hougen, J. Rygaard, C. Moser and L. Eberl. 2005. Garlic blocks quorum sensing and promotes rapid clearing of pulmonary *Pseudomonas aeruginosa* infections. *Microbiology* 151 (12): 3873–3880.

Brakhage, A. A. 2013. Regulation of fungal secondary metabolism. *Nature Reviews Microbiology* 11 (1):21.

Brethauer, S. and M. H. Studer. 2014. Consolidated bioprocessing of lignocellulose by a microbial consortium. *Energy & Environmental Science* 7 (4): 1446–1453.

Brijwani, K., H. S. Oberoi and P. V. Vadlani. 2010. Production of a cellulolytic enzyme system in mixed-culture solid-state fermentation of soybean hulls supplemented with wheat bran. *Process Biochemistry* 45 (1): 120–128.

Cantarel, B. L., P. M. Coutinho, C. Rancurel, T. Bernard, V. Lombard and B. Henrissat. 2009. The carbohydrate-active enzymes database (CAZy): an expert resource for glycogenomics. *Nucleic Acids Research* 37 (suppl 1): D233–D238. doi: 10.1093/nar/gkn663.

Carabajal, M., H. Kellner, L. Levin, N. Jehmlich, M. Hofrichter and R. Ullrich. 2013. The secretome of *Trametes versicolor* grown on tomato juice medium and purification of the secreted oxidoreductases including a versatile peroxidase. *Journal of Biotechnology* 168 (1): 15–23. doi: http://dx.doi.org/10.1016/j.jbiotec.2013.08.007.

Cardona, C. A. and Ó. J. Sánchez. 2007. Fuel ethanol production: Process design trends and integration opportunities. *Bioresource Technology* 98 (12): 2415–2457. doi: 10.1016/j.biortech.2007.01.002.

Caro, Y., L. Anamale, M. Fouillaud, P. Laurent, T. Petit and L. Dufosse. 2012. Natural hydroxyanthraquinoid pigments as potent food grade colorants: an overview. *Natural products and bioprospecting* 2 (5): 174–193.

Carpita, N. C. 1996. Structure and biogenesis of the cell walls of grasses. *Annual Review of Plant Physiology* 47 (1): 445–476. doi: doi:10.1146/annurev.arplant.47.1.445.

Cheng, M.-J., M.-D. Wu, I.-S. Chen and G.-F. Yuan. 2008. Secondary metabolites from the mycelia of the fungus *Monascus pilosus* BCRC 38072. *Chemical and Pharmaceutical Bulletin* 56 (3): 394–397.

Choi, J., N. Detry, K.-T. Kim, F. Asiegbu, J. Valkonen and Y.-H. Lee. 2014. fPoxDB: Fungal peroxidase database for comparative genomics. *BMC Microbiology* 14 (1): 117.

Chu, Y., Y. Wei, X. Yuan and X. Shi. 2011. Bioconversion of wheat stalk to hydrogen by dark fermentation: Effect of different mixed microflora on hydrogen yield and cellulose solubilisation. *Bioresource Technology* 102 (4): 3805–3809.

Clark, D. S. 1962. Submerged citric acid fermentation of ferrocyanide-treated beet molasses: morphology of pellets of *Aspergillus niger*. *Canadian Journal of Microbiology* 8 (1): 133–136. doi: 10.1139/m62-017.

Coban, H. B., A. Demirci and I. Turhan. 2015. Microparticle-enhanced *Aspergillus ficuum* phytase production and evaluation of fungal morphology in submerged fermentation. *Bioprocess and Biosystems Engineering* 38 (6): 1075–1080. doi: 10.1007/s00449-014-1349-4.

Colussi, F., W. Garcia, F. Rosseto, B. de Mello, M. de Oliveira Neto and I. Polikarpov. 2012. Effect of pH and temperature on the global compactness, structure, and activity of cellobiohydrolase Cel7A from *Trichoderma harzianum*. *European Biophysics Journal* 41 (1): 89–98. doi: 10.1007/s00249-011-0762-8.

Coppens, J., O. Grunert, S. Van Den Hende, I. Vanhoutte, N. Boon, G. Haesaert and L. De Gelder. 2016. The use of microalgae as a high-value organic slow-release fertilizer results in tomatoes with increased carotenoid and sugar levels. *Journal of Applied Phycology* 28 (4): 2367–2377. doi: 10.1007/s10811-015-0775-2.

Darah, I., K. Sumathi, K. Jain and S. H. Lim. 2011. Influence of agitation speed on tannase production and morphology of *Aspergillus niger* FETL FT3 in submerged fermentation. *Applied Biochemistry and Biotechnology* 165 (7): 1682–1690. doi: 10.1007/s12010-011-9387-8.

Das, R. K., S. K. Brar and M. Verma. 2016. Potential use of pulp and paper solid waste for the bio-production of fumaric acid through submerged and solid state fermentation. *Journal of Cleaner Production* 112: 4435–4444. doi: https://doi.org/10.1016/j.jclepro.2015.08.108.

Dashtban, M., H. Schraft and W. Qin. 2009. Fungal bioconversion of lignocellulosic residues: Opportunities and perspectives. *International Journal of Biological Sciences* 5: 578–595.

Dashtban, M., H. Schraft, T. A. Syed and W. Qin. 2010. Fungal biodegradation and enzymatic modification of lignin. *International Journal of Biochemistry and Molecular Biology* 1 (1): 36.

Demain, A. L., M. Newcomb and J. H. D. Wu. 2005. Cellulase, clostridia, and ethanol. *Microbiology and Molecular Biology Reviews* 69 (1): 124–154. doi: 10.1128/mmbr.69.1.124-154.2005.

den Haan, R., E. van Rensburg, S. H. Rose, J. F. Görgens and W. H. van Zyl. 2015. Progress and challenges in the engineering of non-cellulolytic microorganisms for consolidated bioprocessing. *Current Opinion in Biotechnology* 33: 32–38. doi: https://doi.org/10.1016/j.copbio.2014.10.003.

Desobgo, S. C., S. S. Mishra, S. K. Behera and S. K. Panda. 2017. Scaling-up and modelling applications of solid-state fermentation and demonstration in microbial enzyme production related to food industries. *Microbial Enzyme Technology in Food Applications*: 452.

Devesa-Rey, R., X. Vecino, J. L. Varela-Alende, M. T. Barral, J. M. Cruz and A. B. Moldes. 2011. Valorization of winery waste vs. the costs of not recycling. *Waste Management* 31 (11): 2327–2335. doi: https://doi.org/10.1016/j.wasman.2011.06.001.

Dhiman, S., G. Mukherjee, A. Kumar, P. Mukherjee, S. A. Verekar and S. K. Deshmukh. 2017. Fungal tannase: Recent advances and industrial applications. In: *Developments in Fungal Biology and Applied Mycology*, edited by Tulasi Satyanarayana, Sunil K. Deshmukh and B. N. Johri, 295–313. Singapore: Springer Singapore.

do Prado, H. F. A., A. A. dos Reis and E. A. de Oliveira. 2016. Fungal amylase: Production system. *Fungal Enzymes*: 423.

Dufosse, L., M. Fouillaud, Y. Caro, S. A. Mapari and N. Sutthiwong. 2014. Filamentous fungi are large-scale producers of pigments and colorants for the food industry. *Current Opinion in Biotechnology* 26: 56–61.

Egan, C. P., A. Rummel, V. Kokkoris, J. Klironomos, Y. Lekberg and M. Hart. 2018. Using mock communities of arbuscular mycorrhizal fungi to evaluate fidelity associated with Illumina sequencing. *Fungal Ecology* 33: 52–64.

Endo, A. 1979. Monacolin K, a new hypocholesterolemic agent produced by a *Monascus* species. *The Journal of antibiotics* 32 (8): 852–854.

Espinosa-Ortiz, E. J., E. R. Rene, K. Pakshirajan, E. D. van Hullebusch and P. N. L. Lens. 2016. Fungal pelleted reactors in wastewater treatment: Applications and perspectives. *Chemical Engineering Journal* 283: 553–571. doi: https://doi.org/10.1016/j.cej.2015.07.068.

FAO. 2017. FAOSTAT domain. [Electronic report]. The Food and Agriculture Organization of the United Nations, Last Modified 27 August 2014. http://faostat3.fao.org/faostat-gateway/go/to/download/Q/QD/E.

Feng, Y., Y. Shao and F. Chen. 2012. Monascus pigments. *Applied Microbiology and Biotechnology* 96 (6): 1421–1440.

Fernández-Fueyo, E., F. J. Ruiz-Dueñas, Y. Miki, M. J. Martínez, K. E. Hammel and A. T. Martínez. 2012. Lignin-degrading peroxidases from genome of selective ligninolytic fungus *Ceriporiopsis subvermispora*. *Journal of Biological Chemistry* 287 (20): 16903–16916. doi: 10.1074/jbc. M112.356378.

Fernández-Fueyo, E., F. J. Ruiz-Dueñas, M. J. Martínez, A. Romero, K. E. Hammel, F. J. Medrano and A. T. Martínez. 2014. Ligninolytic peroxidase genes in the oyster mushroom genome: heterologous expression, molecular structure, catalytic and stability properties, and lignin-degrading ability. *Biotechnology for Biofuels* 7 (1): 2.

Floudas, D., M. Binder, R. Riley, K. Barry, R. A. Blanchette, B. Henrissat, A. T. Martínez, R. Otillar, J. W. Spatafora, J. S. Yadav, A. Aerts, I. Benoit, A. Boyd, A. Carlson, A. Copeland, P. M. Coutinho, R. P. de Vries, P. Ferreira, K. Findley, B. Foster, J. Gaskell, D. Glotzer, P. Górecki, J. Heitman, C. Hesse, C. Hori, K. Igarashi, J. A. Jurgens, N. Kallen, P. Kersten, A. Kohler, U. Kües, T. K. A. Kumar, A. Kuo, K. LaButti, L. F. Larrondo, E. Lindquist, A. Ling, V. Lombard, S. Lucas, T. Lundell, R. Martin, D. J. McLaughlin, I. Morgenstern, E. Morin, C. Murat, L. G. Nagy, M. Nolan, R. A. Ohm, A. Patyshakuliyeva, A. Rokas, F. J. Ruiz-Dueñas, G. Sabat, A. Salamov, M. Samejima, J. Schmutz, J. C. Slot, F. St. John, J. Stenlid, H. Sun, S. Sun, K. Syed, A. Tsang, A. Wiebenga, D. Young, A. Pisabarro, D. C. Eastwood, F. Martin, D. Cullen, I. V. Grigoriev and D. S. Hibbett. 2012. The paleozoic origin of enzymatic lignin decomposition reconstructed from 31 fungal genomes. *Science* 336 (6089): 1715–1719. doi: 10.1126/science.1221748.

Francieli Colussi, Serpa Viviane, Da Silva Delabona, R. M. L. Priscila, Luiza Voltatodio Maria, Alves Renata, Luan Mello Bruno, Pereira Nei, Sanches Farinas Cristiane, Golubev Alexander M., Auxiliadora Morim Santos Maria and P. Igor. 2011. Purification, and biochemical and biophysical characterization of cellobiohydrolase I from *Trichoderma harzianum* IOC 3844. *Journal of Microbiology and Biotechnology* 21 (8): 10.

Gardeli, C., M. Athenaki, E. Xenopoulos, A. Mallouchos, A. A. Koutinas, G. Aggelis and S. Papanikolaou. 2017. Lipid production and characterization by *Mortierella* (Umbelopsis) *isabellina* cultivated on lignocellulosic sugars. *Journal of applied microbiology* 123 (6): 1461–1477. doi: doi:10.1111/jam.13587.

Garg, G., A. Singh, A. Kaur, R. Singh, J. Kaur and R. Mahajan. 2016. Microbial pectinases: an ecofriendly tool of nature for industries. *3 Biotech* 6 (1):47. doi: 10.1007/s13205-016-0371-4.

Ghorai, S. S. P. Banik, D. Verma, S. Chowdhury, S. Mukherjee and S. Khowala. 2009. Fungal biotechnology in food and feed processing. *Food Research International* 42 (5): 577–587. doi: https://doi.org/10.1016/j.foodres.2009.02.019.

Gillespie, A. T. and N. Claydon. 1989. The use of entomogenous fungi for pest control and the role of toxins in pathogenesis. *Pest Management Science* 27 (2): 203–215.

globalEDGE. 2017. Food and Beverage: Trade Statistics: UN Comtrade database. Michigan State University, accessed 01 December. https://globaledge.msu.edu/industries/food-and-beverage/resources.

González-Paramás, A. M., S. Esteban-Ruano, C. Santos-Buelga, S. de Pascual-Teresa and J. C. Rivas-Gonzalo. 2003. Flavanol content and antioxidant activity in winery byproducts. *Journal of Agricultural and Food Chemistry* 52 (2): 234–238. doi: 10.1021/jf0348727.

Gopinath, S. C. B., P. Anbu, M. K. M. Arshad, T. Lakshmipriya, C. H. Voon, U. Hashim and S. V. Chinni. 2017. Biotechnological processes in microbial amylase production. *BioMed Research International* 2017:9. doi: 10.1155/2017/1272193.

Gorzelak, M. A., A. K. Asay, B. J. Pickles and S. W. Simard. 2015. Inter-plant communication through mycorrhizal networks mediates complex adaptive behaviour in plant communities. *AoB PLANTS* 7:plv050-plv050. doi: 10.1093/aobpla/plv050.

Grandclément, C., M. Tannières, S. Moréra, Y. Dessaux and D. Faure. 2015. Quorum quenching: role in nature and applied developments. *FEMS Microbiology Reviews* 40 (1): 86–116.

Grinhut, T., T. Salame, Y. Chen and Y. Hadar. 2011. Involvement of ligninolytic enzymes and Fenton-like reaction in humic acid degradation by *Trametes* sp. *Applied Microbiology and Biotechnology* 91 (4): 1131–1140. doi: 10.1007/s00253-011-3300-9.

Gruntjes, T. 2013. Molecular mechanisms behind fungal-fungal and fungal-bacterial interactions. Master of Science, Universiteit Utrecht.

Gupta, R., A. Kumari, P. Syal and Y. Singh. 2015. Molecular and functional diversity of yeast and fungal lipases: Their role in biotechnology and cellular physiology. *Progress in Lipid Research* 57: 40–54. doi: https://doi.org/10.1016/j.plipres.2014.12.001.

Gupta, V. K., C. P. Kubicek, J.-G. Berrin, D. W. Wilson, M. Couturier, A. Berlin, E. X. F. Filho and T. Ezeji. 2016. Fungal enzymes for bio-products from sustainable and waste biomass. *Trends in Biochemical Sciences* 41 (7): 633–645. doi: https://doi.org/10.1016/j.tibs.2016.04.006.

Han, W., Y. Y. Hu, S. Y. Li, F. F. Li and J. H. Tang. 2016. Biohydrogen production from waste bread in a continuous stirred tank reactor: A techno-economic analysis. *Bioresource Technology* 221: 318–323. doi: https://doi.org/10.1016/j.biortech.2016.09.055.

Han, X., G. Liu, Y. Pan, W. Song and Y. Qu. 2018. Consolidated bioprocessing for sodium gluconate production from cellulose using *Penicillium oxalicum*. *Bioresource Technology* 251: 407–410. doi: https://doi.org/10.1016/j.biortech.2017.12.028.

Harhangi, H. R., A. S. Akhmanova, R. Emmens, C. van der Drift, W. T. A. M. de Laat, J. P. van Dijken, M. S. M. Jetten, J. T. Pronk and H. J. M. Op den Camp. 2003. Xylose metabolism in the anaerobic fungus *Piromyces* sp. strain E2 follows the bacterial pathway. *Archives of Microbiology* 180 (2): 134–141. doi: 10.1007/s00203-003-0565-0.

Hasunuma, T., F. Okazaki, N. Okai, K. Y. Hara, J. Ishii and A. Kondo. 2013. A review of enzymes and microbes for lignocellulosic biorefinery and the possibility of their application to consolidated bioprocessing technology. *Bioresource Technology* 135 (0): 513–522. doi: http://dx.doi.org/10.1016/j.biortech.2012.10.047.

Hatami, M. and F. Ahangarani. 2016. Role of beneficial fungi in sustainable agricultural systems. In: *Plant-Microbe Interaction: An Approach to Sustainable Agriculture*, edited by Devendra K. Choudhary, Ajit Varma and Narendra Tuteja, 397–416. Singapore: Springer Singapore.

Hu, N., B. Yuan, J. Sun, S.-A. Wang and F.-L. Li. 2012. Thermotolerant *Kluyveromyces marxianus* and *Saccharomyces cerevisiae* strains representing potentials for bioethanol production from jerusalem artichoke by consolidated bioprocessing. *Applied Microbiology and Biotechnology* 95 (5): 1359–1368. doi: 10.1007/s00253-012-4240-8.

Hu, Q. and T. Dong. 2015. Non-ribosomal peptides from entomogenous fungi. In: *Biocontrol of Lepidopteran Pests*, 169–206. Springer.

Huang, G., H. Zhou, Z. Tang, H. Liu, Y. Cao, D. Qiao and Y. Cao. 2016. Novel fungal lipids for the production of biodiesel resources by *Mucor fragilis* AFT7-4. *Environmental Progress & Sustainable Energy* 35 (6): 1784–1792. doi: doi:10.1002/ep.12395.

Humphrey, A. 1998. Shake flask to fermentor: What have we learned? *Biotechnology Progress* 14 (1): 3–7. doi: 10.1021/bp970130k.

Hur, S. J., S. Y. Lee, Y.-C. Kim, I. Choi and G.-B. Kim. 2014. Effect of fermentation on the antioxidant activity in plant-based foods. *Food Chemistry* 160: 346–356. doi: https://doi.org/10.1016/j.foodchem.2014.03.112.

Igiehon, N. O. and O. O. Babalola. 2017. Biofertilizers and sustainable agriculture: exploring arbuscular mycorrhizal fungi. *Applied Microbiology and Biotechnology* 101 (12): 4871–4881. doi: 10.1007/s00253-017-8344-z.

Jadhav, S., V. Gulati, E. M. Fox, A. Karpe, D. J. Beale, D. Sevior, M. Bhave and E. A. Palombo. 2015. Rapid identification and source-tracking of *Listeria monocytogenes* using MALDI-TOF mass spectrometry. *International Journal of Food Microbiology* 202: 1–9.

Jayabalan, R., R. V. Malbaša, E. S. Lončar, J. S. Vitas and M. Sathishkumar. 2014. A review on kombucha tea—Microbiology, composition, fermentation, beneficial effects, toxicity, and tea fungus. *Comprehensive Reviews in Food Science and Food Safety* 13 (4): 538–550. doi: 10.1111/1541-4337.12073.

Jayakumar, T., E. Ramesh and P. Geraldine. 2006. Antioxidant activity of the oyster mushroom, *Pleurotus ostreatus*, on CCl4-induced liver injury in rats. *Food and Chemical Toxicology* 44 (12): 1989–1996. doi: https://doi.org/10.1016/j.fct.2006.06.025.

Jiang, T., Q. Wang, S. Xu, M. M. Jahangir and T. Ying. 2010. Structure and composition changes in the cell wall in relation to texture of shiitake mushrooms (*Lentinula edodes*) stored in modified atmosphere packaging. *Journal of the Science of Food and Agriculture* 90 (5): 742–749. doi: 10.1002/jsfa.3876.

Jiang, Y., A. V. Duarte, J. van den Brink, A. Wiebenga, G. Zou, C. Wang, R. P. de Vries, Z. Zhou and I. Benoit. 2016. Enhancing saccharification of wheat straw by mixing enzymes from genetically-

348   *Fungal Bioremediation: Fundamentals and Applications*

modified *Trichoderma reesei* and *Aspergillus niger*. *Biotechnology Letters* 38 (1): 65–70. doi: 10.1007/s10529-015-1951-9.

Karpe, A. V., I. H. Harding and E. A. Palombo. 2014. Comparative degradation of hydrothermal pretreated winery grape wastes by various fungi. *Industrial Crops and Products* 59: 228–233.

Karpe, A. V. 2015. Biodegradation of winery biomass wastes by developing a symbiotic multi-fungal consortium. Swinburne University of Technology.

Karpe, A. V., D. J. Beale, N. Godhani, P. D. Morrison, I. H. Harding and E. A. Palombo. 2015a. Untargeted metabolic profiling of winery-derived biomass waste degradation by *Aspergillus niger*. *Journal of Chemical Technology and Biotechnology*: doi: 10.1002/jctb.4749. doi: 10.1002/jctb.4749.

Karpe, A. V., D. J. Beale, N. B. Godhani, P. D. Morrison, I. H. Harding and E. A. Palombo. 2015b. Untargeted metabolic profiling of winery-derived biomass waste degradation by *Penicillium chrysogenum*. *Journal of Agricultural and Food Chemistry* 63 (49): 10696–10704. doi: 10.1021/acs.jafc.5b04834.

Karpe, A. V., D. J. Beale, I. H. Harding and E. A. Palombo. 2015c. Optimization of degradation of winery-derived biomass waste by ascomycetes. *Journal of Chemical Technology & Biotechnology* 90 (10): 1793–1801. doi: 10.1002/jctb.4486.

Karpe, A. V., D. J. Beale, I. H. Harding and E. A. Palombo. 2016. Microbial metabolomics in biomass waste management. In: *Microbial Metabolomics: Applications in Clinical, Environmental, and Industrial Microbiology*, edited by David J. Beale, Konstantinos A. Kouremenos and Enzo A. Palombo, 261–288. Cham: Springer International Publishing.

Karpe, A. V., V. V. Dhamale, P. D. Morrison, D. J. Beale, I. H. Harding and E. A. Palombo. 2017. Winery biomass waste degradation by sequential sonication and mixed fungal enzyme treatments. *Fungal Genetics and Biology* 102: 22–30. doi: https://doi.org/10.1016/j.fgb.2016.08.008.

Kausar, H., M. Sariah, H. Mohd Saud, M. Zahangir Alam and M. Razi Ismail. 2010. Development of compatible lignocellulolytic fungal consortium for rapid composting of rice straw. *International Biodeterioration & Biodegradation* 64 (7): 594–600. doi: 10.1016/j.ibiod.2010.06.012.

Khuong, L. D., R. Kondo, R. De Leon, T. Kim Anh, K. Shimizu and I. Kamei. 2014. Bioethanol production from alkaline-pretreated sugarcane bagasse by consolidated bioprocessing using *Phlebia* sp. MG-60. *International Biodeterioration & Biodegradation* 88: 62–68. doi: https://doi.org/10.1016/j.ibiod.2013.12.008.

Kim, S. and B. E. Dale. 2004. Global potential bioethanol production from wasted crops and crop residues. *Biomass and Bioenergy* 26 (4): 361–375. doi: http://dx.doi.org/10.1016/j.biombioe.2003.08.002.

Kim, T.-Y., D. Wang, A. K. Kim, E. Lau, A. J. Lin, D. A. Liem, J. Zhang, N. C. Zong, M. P. Y. Lam and P. Ping. 2012. Metabolic labeling reveals proteome dynamics of mouse mitochondria. *Molecular and Cellular Proteomics* 11 (12): 1586–1594. doi: 10.1074/mcp.M112.021162.

Klyosov, A. A. 1987. Cellulases of the third generation. In *Biochemistry and genetics of cellulose degradation*, edited by J. -P. Aubert, P. Beguin and J. Millet, 71–99. Paris: Academic Press.

Knop, D., J. Ben-Ari, T. Salame, D. Levinson, O. Yarden and Y. Hadar. 2014. Mn^{2+}-deficiency reveals a key role for the *Pleurotus ostreatus* versatile peroxidase (VP4) in oxidation of aromatic compounds. *Applied Microbiology and Biotechnology* 98 (15): 6795–6804. doi: 10.1007/s00253-014-5689-4.

Kumar, G., P. Bakonyi, S. Periyasamy, S. H. Kim, N. Nemestóthy and K. Bélafi-Bakó. 2015. Lignocellulose biohydrogen: Practical challenges and recent progress. *Renewable and Sustainable Energy Reviews* 44: 728–737. doi: https://doi.org/10.1016/j.rser.2015.01.042.

Lee, J. 1997. Biological conversion of lignocellulosic biomass to ethanol. *Journal of Biotechnology* 56 (1): 1–24. doi: 10.1016/s0168-1656(97)00073-4.

Lemanceau, P. P. Bauer, S. Kraemer and J.-F. Briat. 2009. Iron dynamics in the rhizosphere as a case study for analyzing interactions between soils, plants and microbes. *Plant and Soil* 321 (1–2): 513–535.

Lennartsson, P. R., P. Erlandsson and M. J. Taherzadeh. 2014. Integration of the first and second generation bioethanol processes and the importance of by-products. *Bioresource Technology* 165: 3–8. doi: https://doi.org/10.1016/j.biortech.2014.01.127.

Lettera, V., C. Pezzella, P. Cicatiello, A. Piscitelli, V. G. Giacobelli, E. Galano, A. Amoresano and G. Sannia. 2016. Efficient immobilization of a fungal laccase and its exploitation in fruit juice clarification. *Food Chemistry* 196: 1272–1278. doi: https://doi.org/10.1016/j.foodchem.2015.10.074.

Li, Y., X. Peng and H. Chen. 2013. Comparative characterization of proteins secreted by *Neurospora sitophila* in solid-state and submerged fermentation. *Journal of Bioscience and Bioengineering* 116 (4): 493–498. doi: http://dx.doi.org/10.1016/j.jbiosc.2013.04.001.

Linger, J. G. and A. Darzins. 2013. Consolidated bioprocessing. In: *Advanced Biofuels and Bioproducts*, edited by James W. Lee, 267–280. Springer New York.

Liu, H.-Q., Y. Feng, D.-Q. Zhao and J.-X. Jiang. 2012. Evaluation of cellulases produced from four fungi cultured on furfural residues and microcrystalline cellulose. *Biodegradation* 23 (3): 465–472. doi: 10.1007/s10532-011-9525-6.

Liu, T. P. S. L., R. M. P. Brandão Costa, D. J. de Vasconcelos Freitas, C. Oliveira Nacimento, C. M. de Souza Motta, R. P. Bezerra, P. Nunes Herculano and A. L. F. Porto. 2017. Tannase from *Aspergillus melleus* improves the antioxidant activity of green tea: purification and biochemical characterisation. *International Journal of Food Science & Technology* 52 (3): 652–661. doi: 10.1111/ijfs.13318.

Liu, Y., Y. Zhang, J. Xu, Y. Sun, Z. Yuan and J. Xie. 2015. Consolidated bioprocess for bioethanol production with alkali-pretreated sugarcane bagasse. *Applied Energy* 157: 517–522. doi: https://doi.org/10.1016/j.apenergy.2015.05.004.

Löbs, A.-K., C. Schwartz and I. Wheeldon. 2017. Genome and metabolic engineering in non-conventional yeasts: Current advances and applications. *Synthetic and Systems Biotechnology* 2 (3): 198–207. doi: https://doi.org/10.1016/j.synbio.2017.08.002.

Lozano-Tovar, M., I. Garrido-Jurado, F. Lafont and E. Quesada-Moraga. 2015. Insecticidal activity of a destruxin-containing extract of *Metarhizium brunneum* against *Ceratitis capitata* (Diptera: Tephritidae). *Journal of economic entomology* 108 (2): 462–472.

Lynd, L. R., W. H. v. Zyl, J. E. McBride and M. Laser. 2005. Consolidated bioprocessing of cellulosic biomass: an update. *Current Opinion in Biotechnology* 16 (5): 577–583. doi: 10.1016/j.copbio.2005.08.009.

Macheleidt, J., D. J. Mattern, J. Fischer, T. Netzker, J. Weber, V. Schroeckh, V. Valiante and A. A. Brakhage. 2016. Regulation and role of fungal secondary metabolites. *Annual Review of Genetics* 50: 371–392.

Maeda, K., D. Hanajima, R. Morioka and T. Osada. 2010. Characterization and spatial distribution of bacterial communities within passively aerated cattle manure composting piles. *Bioresource Technology* 101 (24): 9631–9637. doi: https://doi.org/10.1016/j.biortech.2010.07.057.

Mäkelä, M. R., T. Lundell, A. Hatakka and K. Hildén. 2013. Effect of copper, nutrient nitrogen, and wood-supplement on the production of lignin-modifying enzymes by the white-rot fungus *Phlebia radiata*. *Fungal Biology* 117 (1): 62–70. doi: http://dx.doi.org/10.1016/j.funbio.2012.11.006.

Manole, A., D. Herea, H. Chiriac and V. Melnig. 2008. Laccase activity determination. *Cuza University Scientific Annals Lasi* 4: 17–24.

Manzoni, M. and M. Rollini. 2002. Biosynthesis and biotechnological production of statins by filamentous fungi and application of these cholesterol-lowering drugs. *Applied Microbiology and Biotechnology* 58 (5): 555–564.

Mapari, S. A., M. E. Hansen, A. S. Meyer and U. Thrane. 2008. Computerized screening for novel producers of *Monascus*-like food pigments in *Penicillium* species. *Journal of Agricultural and Food Chemistry* 56 (21): 9981–9989.

Mapari, S. A., A. S. Meyer and U. Thrane. 2009a. Photostability of natural orange–red and yellow fungal pigments in liquid food model systems. *Journal of Agricultural and Food Chemistry* 57 (14): 6253–6261.

Mapari, S. A., A. S. Meyer, U. Thrane and J. C. Frisvad. 2009b. Identification of potentially safe promising fungal cell factories for the production of polyketide natural food colorants using chemotaxonomic rationale. *Microbial Cell Factories* 8 (1):24.

McFadden, B. R. and J. L. Lusk. 2015. Cognitive biases in the assimilation of scientific information on global warming and genetically modified food. *Food Policy* 54: 35–43. doi: https://doi.org/10.1016/j.foodpol.2015.04.010.

Mears, L., S. M. Stocks, M. O. Albaek, G. Sin and K. V. Gernaey. 2017. Mechanistic fermentation models for process design, monitoring, and control. *Trends in Biotechnology* 35 (10): 914–924. doi: https://doi.org/10.1016/j.tibtech.2017.07.002.

Meijer, S., J. Otero, R. Olivares, M. R. Andersen, L. Olsson and J. Nielsen. 2009. Overexpression of isocitrate lyase—glyoxylate bypass influence on metabolism in *Aspergillus niger*. *Metabolic Engineering* 11 (2): 107–116. doi: http://dx.doi.org/10.1016/j.ymben.2008.12.002.

Merino, S. and J. Cherry. 2007. Progress and challenges in enzyme development for biomass utilization. In: *Biofuels*, edited by Lisbeth Olsson, 95–120. Springer Berlin Heidelberg.

Mitchell, D., N. Kriger and M. Berovic. 2006. Solid-State Fermentation Bioreactors-Fundamentals of Design and Operation. Mitchell DA. Berlin-Heidelberg-New York: Springer.

Mokochinski, J. B., B. G. C. López, V. Sovrani, H. S. Dalla Santa, P. P. González-Borrero, A. C. H. F. Sawaya, E. M. Schmidt, M. N. Eberlin and Y. R. Torres. 2015. Production of *Agaricus brasiliensis* mycelium from food industry residues as a source of antioxidants and essential fatty acids. *International Journal of Food Science & Technology* 50 (9): 2052–2058. doi: 10.1111/ijfs.12861.

Montoya, S., Ó. J. Sánchez and L. Levin. 2014. Mathematical modeling of lignocellulolytic enzyme production from three species of white rot fungi by solid-state fermentation. Cham.

Mukherjee, D. 2017. Microorganisms: Role for crop production and its interface with soil agroecosystem. In: *Plant-microbe interactions in agro-ecological perspectives: Volume 1: Fundamental mechanisms, methods and functions*, edited by Dhananjaya Pratap Singh, Harikesh Bahadur Singh and Ratna Prabha, 333–359. Singapore: Springer Singapore.

Musoni, M., J. Destain, P. Thonart, J.-B. Bahama and F. Delvigne. 2015. Bioreactor design and implementation strategies for the cultivation of filamentous fungi and the production of fungal metabolites: from traditional methods to engineered systems/conception de bioréacteurs et mise en oeuvre de stratégies pour la culture de champignons filamenteux et la production de métabolites d'origine fongique: des méthodes traditionnelles aux technologies actuelles. *Biotechnologie, Agronomie, Société et Environnement* 19 (4):430.

Nelson, C. J., L. Li and A. H. Millar. 2014. Quantitative analysis of protein turnover in plants. *Proteomics* 14 (4–5): 579–592. doi: www.doi.org/10.1002/pmic.201300240.

Ng, I. S., C.-W. Li, S.-P. Chan, J.-L. Chir, P. T. Chen, C.-G. Tong, S.-M. Yu and T.-H. D. Ho. 2010. High-level production of a thermoacidophilic β-glucosidase from *Penicillium citrinum* YS40-5 by solid-state fermentation with rice bran. *Bioresource Technology* 101 (4):1310-1317. doi: 10.1016/j.biortech.2009.08.049.

Nitschke, J., H.-J. Altenbach, T. Malolepszy and H. Mölleken. 2011. A new method for the quantification of chitin and chitosan in edible mushrooms. *Carbohydrate Research* 346 (11): 1307–1310. doi: https://doi.org/10.1016/j.carres.2011.03.040.

Novelli, P. K., M. M. Barros and L. F. Fleuri. 2016. Novel inexpensive fungi proteases: Production by solid state fermentation and characterization. *Food Chemistry* 198: 119–124. doi: https://doi.org/10.1016/j.foodchem.2015.11.089.

OECD/FAO. 2016. Biofuels. The Food and Agriculture Organization of the United Nations, accessed 5 April. http://www.fao.org/3/a-BO103e.pdf.

Okamoto, K., Y. Nitta, N. Maekawa and H. Yanase. 2011. Direct ethanol production from starch, wheat bran and rice straw by the white rot fungus *Trametes hirsuta*. *Enzyme and Microbial Technology* 48 (3): 273–277. doi: https://doi.org/10.1016/j.enzmictec.2010.12.001.

Olson, D. G., J. E. McBride, A. Joe Shaw and L. R. Lynd. 2012. Recent progress in consolidated bioprocessing. *Current Opinion in Biotechnology* 23 (3): 396–405.

Owen, D., A. P. Williams, G. W. Griffith and P. J. A. Withers. 2015. Use of commercial bio-inoculants to increase agricultural production through improved phosphrous acquisition. *Applied Soil Ecology* 86: 41–54. doi: https://doi.org/10.1016/j.apsoil.2014.09.012.

Pal, S., H. Singh, A. Farooqui and A. Rakshit. 2015. Fungal biofertilisers in indian agriculture: Perception, demand and promotion. *Journal of Eco-friendly Agriculture* 10 (2): 101–113.

Pandey, A. 1992. Recent process developments in solid-state fermentation. *Process Biochemistry* 27 (2): 109–117. doi: https://doi.org/10.1016/0032-9592(92)80017-W.

Phisalaphong, M., N. Srirattana and W. Tanthapanichakoon. 2006. Mathematical modeling to investigate temperature effect on kinetic parameters of ethanol fermentation. *Biochemical Engineering Journal* 28 (1): 36–43. doi: https://doi.org/10.1016/j.bej.2005.08.039.

Plou, F. J., J. Polaina, J. Sanz-Aparicio and M. Fernández-Lobato. 2017. β-Galactosidases for lactose hydrolysis and galactooligosaccharide synthesis. *Microbial Enzyme Technology in Food Applications*:121.

Pollard, D., T. Kirschner, G. Hunt, I. T. Tong, R. Stieber and P. Salmon. 2007. Scale up of a viscous fungal fermentation: Application of scale-up criteria with regime analysis and operating boundary conditions. *Biotechnology and Bioengineering* 96 (2): 307–317.

Rahardjo, Y. S. P., J. Tramper and A. Rinzema. 2006. Modeling conversion and transport phenomena in solid-state fermentation: A review and perspectives. *Biotechnology Advances* 24 (2): 161–179. doi: https://doi.org/10.1016/j.biotechadv.2005.09.002.

Raina, S., M. Odell and T. Keshavarz. 2010. Quorum sensing as a method for improving sclerotiorin production in *Penicillium sclerotiorum*. *Journal of biotechnology* 148 (2–3): 91–98.

Rasmussen, T. B., M. E. Skindersoe, T. Bjarnsholt, R. K. Phipps, K. B. Christensen, P. O. Jensen, J. B. Andersen, B. Koch, T. O. Larsen and M. Hentzer. 2005. Identity and effects of quorum-sensing inhibitors produced by *Penicillium* species. *Microbiology* 151 (5): 1325–1340.

Ray, R. C. and C. M. Rosell. 2017. *Microbial enzyme technology in food applications, Food Biology Series*: Boca Raton, Florida; London, England; New York : CRC Press.

Ritala, A., S. T. Häkkinen, M. Toivari and M. G. Wiebe. 2017. Single cell protein—State-of-the-Art, industrial landscape and patents 2001–2016. *Frontiers in Microbiology* 8 (2009). doi: 10.3389/fmicb.2017.02009.

Rodríguez, A., A. Carnicero, F. Perestelo, G. de la Fuente, O. Milstein and M. A. Falcón. 1994. Effect of *Penicillium chrysogenum* on lignin transformation. *Applied and Environmental Microbiology* 60 (8): 2971–2976.

Rodriguez, A., F. Perestelo, A. Carnicero, V. Regalado, R. Perez, G. de la Fuente and M. A. Falcon. 1996. Degradation of natural lignins and lignocellulosic substrates by soil-inhabiting fungi imperfecti. *FEMS Microbiology Ecology* 21 (3): 213–219. doi: 10.1111/j.1574-6941.1996.tb00348.x.

Sanchez, B. and J. Bautista. 1988. Effects of furfural and 5-hydroxymethylfurfural on the fermentation of *Saccharomyces cerevisiae* and biomass production from *Candida guilliermondii*. *Enzyme and Microbial Technology* 10 (5): 315–318. doi: http://dx.doi.org/10.1016/0141-0229(88)90135-4.

Sánchez, C. 2009. Lignocellulosic residues: Biodegradation and bioconversion by fungi. *Biotechnology Advances* 27 (2): 185–194. doi: 10.1016/j.biotechadv.2008.11.001.

Saqib, S., A. Akram, S. A. Halim and R. Tassaduq. 2017. Sources of β-galactosidase and its applications in food industry. *3 Biotech* 7 (1): 79. doi: 10.1007/s13205-017-0645-5.

Sarkar, N., S. K. Ghosh, S. Bannerjee and K. Aikat. 2012. Bioethanol production from agricultural wastes: An overview. *Renewable Energy* 37 (1): 19–27.

Saroj, S., P. Agarwal, S. Dubey and R. Singh. 2013. Manganese peroxidases: Molecular diversity, heterologous expression, and applications. In: *Adv. Enzyme Biotechnol.*, 67–87. Springer.

Savci, S. 2012. Investigation of effect of chemical fertilizers on environment. *APCBEE Procedia* 1: 287–292. doi: https://doi.org/10.1016/j.apcbee.2012.03.047.

Scherlach, K. K. Graupner and C. Hertweck. 2013. Molecular bacteria-fungi interactions: effects on environment, food, and medicine. *Annual Review of Microbiology* 67: 375–397.

Serna-Saldivar, S. O. and M. Rubio-Flores. 2017. Role of intrinsic and supplemented enzymes in brewing and beer properties. *Microbial Enzyme Technology in Food Applications*: 271.

Shapland, E. B., V. Holmes, C. D. Reeves, E. Sorokin, M. Durot, D. Platt, C. Allen, J. Dean, Z. Serber and J. Newman. 2015. Low-cost, high-throughput sequencing of DNA assemblies using a highly multiplexed nextera process. *ACS Synthetic Biology* 4 (7): 860–866.

Sharma, M., C. Mahajan, M. S. Bhatti and B. S. Chadha. 2016. Profiling and production of hemicellulases by thermophilic fungus *Malbranchea flava* and the role of xylanases in improved bioconversion of pretreated lignocellulosics to ethanol. *3 Biotech* 6 (1): 30. doi: 10.1007/s13205-015-0325-2.

Siddiqui, K. S., H. Ertan, T. Charlton, A. Poljak, A. K. Daud Khaled, X. Yang, G. Marshall and R. Cavicchioli. 2014. Versatile peroxidase degradation of humic substances: Use of isothermal titration calorimetry to assess kinetics, and applications to industrial wastes. *Journal of Biotechnology* 178 (0): 1–11. doi: http://dx.doi.org/10.1016/j.jbiotec.2014.03.002.

Singh Arora, D. and R. Kumar Sharma. 2010. Ligninolytic fungal laccases and their biotechnological applications. *Applied Biochemistry and Biotechnology* 160 (6): 1760–1788. doi: 10.1007/s12010-009-8676-y.

Singh, B., U. Avci, S. E. Eichler Inwood, M. J. Grimson, J. Landgraf, D. Mohnen, I. Sørensen, C. G. Wilkerson, W. G. T. Willats and C. H. Haigler. 2009. A specialized outer layer of the primary cell wall joins elongating cotton fibers into tissue-like bundles. *Plant Physiology* 150 (2): 684–699. doi: 10.1104/pp.109.135459.

Singhania, R. R., A. K. Patel, C. R. Soccol and A. Pandey. 2009. Recent advances in solid-state fermentation. *Biochemical Engineering Journal* 44 (1): 13–18.

Singhania, R. R., R. K. Sukumaran, A. K. Patel, C. Larroche and A. Pandey. 2010. Advancement and comparative profiles in the production technologies using solid-state and submerged fermentation for microbial cellulases. *Enzyme and Microbial Technology* 46 (7): 541–549. doi: http://dx.doi.org/10.1016/j.enzmictec.2010.03.010.

Singhania, R. R. 2012. β-glucosidase from *Aspergillus niger* NII 08121: Molecular characterization and applications in bioethanol production. Doctor of Science, National Institute for Interdisciplinary Science and Technology, Cochin University of Science and Technology.

Sipos, B., Z. Benkő, D. Dienes, K. Réczey, L. Viikari and M. Siika-aho. 2010. Characterisation of specific activities and hydrolytic properties of cell-wall-degrading enzymes produced by *Trichoderma reesei* Rut C30 on different carbon sources. *Applied Biochemistry and Biotechnology* 161 (1): 347–364. doi: 10.1007/s12010-009-8824-4.

Souza, P. M. d., M. L. d. A. Bittencourt, C. C. Caprara, M. d. Freitas, R. P. C. d. Almeida, D. Silveira, Y. M. Fonseca, E. X. Ferreira Filho, A. Pessoa Junior and P. O. Magalhães. 2015. A biotechnology perspective of fungal proteases. *Brazilian Journal of Microbiology* 46: 337–346.

Strong, P. J. and J. E. Burgess. 2008. Treatment methods for wine-related and distillery wastewaters: A review. *Bioremediation Journal* 12 (2): 70–87. doi: 10.1080/10889860802060063.

Sun, B., L. Zhou, X. Jia and C. Sung. 2008. Response surface modeling for y-aminobutyric acid production by *Monascus pilosus* GM100 under solid-state fermentation. *African Journal of Biotechnology* 7 (24).

Sun, R., X. Guo, D. Wang and H. Chu. 2015. Effects of long-term application of chemical and organic fertilizers on the abundance of microbial communities involved in the nitrogen cycle. *Applied Soil Ecology* 95: 171–178. doi: https://doi.org/10.1016/j.apsoil.2015.06.010.

Sundarram, A. and T. P. K. Murthy. 2014. α-amylase production and applications: a review. *Journal of Applied & Environmental Microbiology* 2 (4): 166–175.

Suthar, H., K. Hingurao, J. Vaghashiya and J. Parmar. 2017. Fermentation: A process for biofertilizer production. In: *Microorganisms for Green Revolution: Volume 1: Microbes for Sustainable Crop Production*, edited by Deepak G. Panpatte, Yogeshvari K. Jhala, Rajababu V. Vyas and Harsha N. Shelat, 229–252. Singapore: Springer Singapore.

Sweeney, M. D. and F. Xu. 2012. Biomass converting enzymes as industrial biocatalysts for fuels and chemicals: Recent developments. *Catalysts* 2 (2): 244–263.

Takano, M., M. Yamaguchi, H. Sano, M. Nakamura, H. Shibuya and Y. Miyazaki. 2013. Genomic gene encoding manganese peroxidase from a white-rot fungus *Phanerochaete crassa* WD1694. *Journal of Wood Science* 59 (2): 141–148. doi: 10.1007/s10086-012-1309-z.

Tanimura, A., M. Takashima, T. Sugita, R. Endoh, M. Kikukawa, S. Yamaguchi, E. Sakuradani, J. Ogawa, M. Ohkuma and J. Shima. 2014. *Cryptococcus terricola* is a promising oleaginous yeast for biodiesel production from starch through consolidated bioprocessing. *Scientific Reports* 4: 4776.

Teder, T., W. E. Boeglin, C. Schneider and A. R. Brash. 2017. A fungal catalase reacts selectively with the 13S fatty acid hydroperoxide products of the adjacent lipoxygenase gene and exhibits 13S-hydroperoxide-dependent peroxidase activity. *Biochimica et Biophysica Acta (BBA)—Molecular and Cell Biology of Lipids* 1862 (7): 706–715. doi: https://doi.org/10.1016/j.bbalip.2017.03.011.

Tedersoo, L., T. W. May and M. E. Smith. 2010. Ectomycorrhizal lifestyle in fungi: global diversity, distribution, and evolution of phylogenetic lineages. *Mycorrhiza* 20 (4): 217–263.

Tripathi, S., S. K. Mishra and A. Varma. 2017. Mycorrhizal fungi as control agents against plant pathogens. In: *Mycorrhiza—Nutrient Uptake, Biocontrol, Ecorestoration*, edited by Ajit Varma, Ram Prasad and Narendra Tuteja, 161–178. Cham: Springer International Publishing.

Tsai, A. Y.-L., K. Chan, C.-Y. Ho, T. Canam, R. Capron, E. R. Master and K. Bräutigam. 2017. Transgenic expression of fungal accessory hemicellulases in *Arabidopsis thaliana* triggers transcriptional patterns related to biotic stress and defense response. *PLoS ONE* 12 (3): e0173094. doi: 10.1371/journal.pone.0173094.

Valentín, L., B. Kluczek-Turpeinen, S. Willför, J. Hemming, A. Hatakka, K. Steffen and M. Tuomela. 2010. Scots pine (*Pinus sylvestris*) bark composition and degradation by fungi: Potential substrate for bioremediation. *Bioresource Technology* 101 (7): 2203–2209. doi: http://dx.doi.org/10.1016/j.biortech.2009.11.052.

van Zyl, W. H., L. R. Lynd, R. den Haan and J. E. McBride. 2007. Consolidated bioprocessing for bioethanol production using *Saccharomyces cerevisiae*. In: *Biofuels*, edited by Lisbeth Olsson, 205–235. Berlin, Heidelberg: Springer Berlin Heidelberg.

Vassilev, N., M. Vassileva, A. Lopez, V. Martos, A. Reyes, I. Maksimovic, B. Eichler-Löbermann and E. Malusà. 2015. Unexploited potential of some biotechnological techniques for biofertilizer production and formulation. *Applied Microbiology and Biotechnology* 99 (12): 4983–4996. doi: 10.1007/s00253-015-6656-4.

Verhoeven, M. D., M. Lee, L. Kamoen, M. Van Den Broek, D. B. Janssen, J.-M. G. Daran, A. J. Van Maris and J. T. Pronk. 2017. Mutations in PMR1 stimulate xylose isomerase activity and anaerobic growth on xylose of engineered *Saccharomyces cerevisiae* by influencing manganese homeostasis. *Scientific Reports* 7:46155.

Vetter, J. 2007. Chitin content of cultivated mushrooms *Agaricus bisporus*, *Pleurotus ostreatus* and *Lentinula edodes*. *Food Chemistry* 102 (1): 6–9. doi: https://doi.org/10.1016/j.foodchem.2006.01.037.

Wilson, D. B. 2009. Cellulases and biofuels. *Current Opinion in Biotechnology* 20 (3): 295–299. doi: 10.1016/j.copbio.2009.05.007.

Wood, B. J. 2012. *Microbiology of fermented foods*: Springer Science & Business Media.

Wu, W., R. W. Davis, M. B. Tran-Gyamfi, A. Kuo, K. LaButti, S. Mihaltcheva, H. Hundley, M. Chovatia, E. Lindquist, K. Barry, I. V. Grigoriev, B. Henrissat and J. M. Gladden. 2017. Characterization of four endophytic fungi as potential consolidated bioprocessing hosts for conversion of lignocellulose into advanced biofuels. *Applied Microbiology and Biotechnology* 101 (6): 2603–2618. doi: 10.1007/s00253-017-8091-1.

Xu, J., X. Wang, L. Hu, J. Xia, Z. Wu, N. Xu, B. Dai and B. Wu. 2015. A novel ionic liquid-tolerant *Fusarium oxysporum* BN secreting ionic liquid-stable cellulase: Consolidated bioprocessing of pretreated lignocellulose containing residual ionic liquid. *Bioresource Technology* 181: 18–25. doi: https://doi.org/10.1016/j.biortech.2014.12.080.

Yan, X., Y. Wang, X. Sang and L. Fan. 2017. Nutritional value, chemical composition and antioxidant activity of three Tuber species from China. *AMB Express* 7: 136. doi: 10.1186/s13568-017-0431-0.

Yang, J., R.-H. Jiao, L.-Y. Yao, W.-B. Han, Y.-H. Lu and R.-X. Tan. 2016. Control of fungal morphology for improved production of a novel antimicrobial alkaloid by marine-derived fungus *Curvularia* sp. IFB-Z10 under submerged fermentation. *Process Biochemistry* 51 (2): 185–194. doi: https://doi.org/10.1016/j.procbio.2015.11.025.

Yang, X.-Y., W.-P. Chen, A. K. Rendahl, A. D. Hegeman, W. M. Gray and J. D. Cohen. 2010. Measuring the turnover rates of arabidopsis proteins using deuterium oxide: an auxin signaling case study. *The Plant Journal* 63 (4): 680–695. doi: 10.1111/j.1365-313X.2010.04266.x.

Yong, X., Y. Cui, L. Chen, W. Ran, Q. Shen and X. Yang. 2011. Dynamics of bacterial communities during solid-state fermentation using agro-industrial wastes to produce poly-γ-glutamic acid, revealed by real-time PCR and denaturing gradient gel electrophoresis (DGGE). *Applied Microbiology and Biotechnology* 92 (4): 717–725. doi: 10.1007/s00253-011-3375-3.

Yunus, F.-u.-N., M. Nadeem and F. Rashid. 2015. Single-cell protein production through microbial conversion of lignocellulosic residue (wheat bran) for animal feed. *Journal of the Institute of Brewing* 121 (4): 553–557. doi: 10.1002/jib.251.

Zerva, A., A. L. Savvides, E. A. Katsifas, A. D. Karagouni and D. G. Hatzinikolaou. 2014. Evaluation of *Paecilomyces variotii* potential in bioethanol production from lignocellulose through consolidated bioprocessing. *Bioresource Technology* 162: 294–299. doi: https://doi.org/10.1016/j.biortech.2014.03.137.

Zheng, Y., X. Yu, J. Zeng and S. Chen. 2012. Feasibility of filamentous fungi for biofuel production using hydrolysate from dilute sulfuric acid pretreatment of wheat straw. *Biotechnology for Biofuels* 5: 50–50. doi: 10.1186/1754-6834-5-50.

Zhi, Z. and H. Wang. 2014. White-rot fungal pretreatment of wheat straw with *Phanerochaete chrysosporium* for biohydrogen production: simultaneous saccharification and fermentation. *Bioprocess and Biosystems Engineering* 37 (7): 1447–1458. doi: 10.1007/s00449-013-1117-x.

Zuroff, T. R., S. B. Xiques and W. R. Curtis. 2013. Consortia-mediated bioprocessing of cellulose to ethanol with a symbiotic *Clostridium phytofermentans*/yeast co-culture. *Biotechnology for Biofuels* 6 (1): 59. doi: 10.1186/1754-6834-6-59.

Electron Microscopy Techniques Applied to Bioremediation and Biodeterioration Studies with Moulds

State of the Art and Future Perspectives

Antoni Solé,[1,]* Maria Àngels Calvo,[2] María José Lora,[3] and Alejandro Sánchez-Chardi[4]

1. Introduction

Fungi are a widespread and diverse group of eukaryotic organisms with about 70,000 species described and 1,5 million estimated (Hawksworth 1991, Speranza et al. 2010), covering macroscopic mushrooms, unicellular fungi (yeasts), and filamentous fungi (moulds or molds). They mainly inhabit terrestrial ecosystems and grow on a variety of organic and inorganic substrates, even dead organic materials, being considered saprophytic microorganisms. Although most fungi are free-living, some of them form symbiotic associations with phototrophic organisms, such as mycorrhizae, with plants (Rhodes 2012, Wu et al. 2016, Sánchez-Castro et al. 2017) and lichens, with

[1] Departament de Genètica i Microbiologia, Facultat de Biociències, Universitat Autònoma de Barcelona, Bellaterra, Cerdanyola del Vallès, 08193, Barcelona, Spain.
[2] Departament de Sanitat i Anatomia Animals, Facultat de Veterinària, Universitat Autònoma de Barcelona, Bellaterra, Cerdanyola del Vallès, 08193, Barcelona, Spain.
[3] Departament d'Educació, Generalitat de Catalunya, Institut Montserrat Roig, Terrassa, 08221, Barcelona, Spain.
[4] Servei de Microscòpia, Universitat Autònoma de Barcelona, Bellaterra, Cerdanyola del Vallès, 08193, Barcelona, Spain.
* Corresponding author: antoni.sole@uab.cat

cyanobacteria and/or microalgae (Gadd et al. 2012, Gadd et al. 2014), using organic matter as electron donor and energy and carbon sources.

As part of the metabolic activity of the moulds, the mycelial networks, the vegetative fragment of these pluricellular organisms, produce extracellular oxidative enzymes, and organic acids, which allow them to participate actively in the decomposition of wood (lignin and cellulose), paper, cloth, and other products derived from these natural sources, used as heterotrophic substrates, to carbon dioxide (Rhodes 2013). They are able to solubilize (releasing phosphates and other ligands make them available to the plants, and metal cations liberation) and to immobilize (metal biomineralization, mainly extracellularly, as oxalates, phosphates, and/or carbonates) metal compounds (Rhodes 2012, Gadd et al. 2012, Gadd et al. 2014, Rhee et al. 2014).

The high tolerance of some fungi to natural and anthropogenic pollutants and their metabolic pathways, that reduce the effect of such potentially toxic compounds, make them suitable for bioremediation studies. This tolerance may be exemplified by environments polluted with heavy metals or polycyclic aromatic hydrocarbons that are often characterized by a high fungal diversity despite to these constraining conditions (Gadd et al. 2012, Cébron et al. 2015, Gutiérrez-Corona et al. 2016, Sánchez-Castro et al. 2017). Among them, moulds also play an important role in the remediation (mycoremediation) of toxic polluted environments not only with heavy metals but also oil, chemical toxic compounds, petroleum products, and/or xenobiotics, these last are the synthetic chemicals not produced by organisms in nature (Singh 2006, Reyes-César et al. 2014, Rhodes 2014, Marco-Urrea et al. 2015, Kadri et al. 2017, Varjani 2017). Due to their ability to remove contaminants with harmful concentrations through several interaction mechanisms, they are increasingly considered to be useful bioremediation agents (Harms et al. 2011, Gadd et al. 2012, Abd-El-Aziz Abeer and Sayed Shaban 2016, Gutiérrez-Corona et al. 2016), and often used to clean up wastes and wastewaters in contaminated environments (Binupriya et al. 2006, Singh 2006, Pramila and Ramesh 2011, Coreño-Alonso et al. 2014, Asemoloye et al. 2017, Gola et al. 2018).

In addition, they are also involved in the biodeterioration of exposed cultural heritage, including buildings composed of natural (rock and mineral) or concrete substrates, and paper, wood, textiles and metallic materials, among others. Hence, these materials are subject to fungal colonization that may contribute to a slow gain or loss of structural integrity through their metabolic activities that can cause the mineral surface dissolution or lignin/cellulose decomposition (Bastian et al. 2010, Pinzari et al. 2010, Pinzari et al. 2013, George et al. 2013, Rhodes 2013, Dussud and Ghiglione 2014, Gadd et al. 2014, Pottier et al. 2014, Mensah-Attipoe et al. 2016, Oetari et al. 2016, Salem 2016, Mansour et al. 2017, Obidi and Okekunjo 2017).

Studies referent to the interaction between mould biomass and the surface of different type of materials and compounds, and the produced effect of those interactions on both, have been analyzed through a large number of methodologies, including the generation of high resolution images. Thus, the electron microscopy (EM) imaging has become a crucial tool to understand such interactions and their subsequent effects, and can be applied to achieve nanoscale imaging in environmental samples, where free-living or symbiotic moulds are present.

2. Electron Microscopy and High Resolution Imaging

Since the construction of the first electron microscope by Ernst Ruska in 1931 and the expansion of their applications around 1950, EM revolutionized existent microscopy (light microscopy, LM) and high resolution imaging until new levels of resolution, such as molecular and atomic levels (Hayat 1972, Glauert 1974, Bozzola and Russell 1999, Das Murtey 2016, Graham and Orenstein 2007) were reached. Briefly, electron microscopes are based on the use of electrons as source of information, in contrast with photons used in LM. For that, a coherent electron beam is generated by a gun in the form of filament or single crystal cathode of tungsten or lanthanum hexaboride, respectively. An anode is used to accelerate these electrons (from less than 1 kV to more than 400 kV) in a high vacuum environment and focused by electrostatic and electromagnetic lenses (Veneklasen 1972, Garcia and Rohrer 1989, Reimer 1998, Bozzola and Russell 1999, Egerton 2005, Reichelt et al. 2007, Carter and Williams 2016). In contrast with conventional EM that uses thermionic electron guns working by heating filaments, field emission microscopes use crystals.

Interestingly, the field emission microscopes produce electron beams of small diameters and high coherence, offering higher spatial resolution imaging and sensitivity levels of detectors (Veneklasen 1972, Reichelt et al. 2007, Williams and Carter 2009, Carter and Williams 2016). Images are generated by the detection of interactions between samples and the electron beam, offering a wide diversity of high resolution images from material and biological samples (Julián et al. 2010, Maldonado et al. 2011, Rodríguez-Cariño et al. 2011, Kuo 2014, Seras-Franzoso et al. 2016, Gola et al. 2018). However, particularly for life sciences, cells and tissues often need to be processed with long protocols prior to be useful for the observation by EM. As a general overview of the procedures, biological samples need to be fixed to avoid degradation, dehydrated to work at high vacuum conditions, and covered by a layer of high conductive metal to dissipate electron beam energy or embedded in a hard material to be cut (Kuo 2014, Graham and Orenstein 2007). Specific protocols for each type of sample or approach and improving protocols reducing time or reagents (McDonald 2014, Prakash and Nawani 2014) are of interest to optimize high resolution imaging and to reduce artifacts, false positive results, and costs (Skepper 2000, McDonald and Webb 2011, Esteve et al. 2013, Kuo 2014, McDonald 2014, Melo et al. 2014, Noguera-Ortega et al. 2016), the two aspects that may be explored in bioremediation/biodeterioration studies with moulds. Table 12.1 summarizes the main characteristics of LM and EM, focusing on the most interesting aspects for moulds and other biological samples.

Today, EM has become an important tool to visualize the ultrastructure of organisms as well as to understand the relationships organism/environment, relating form (size and shape) and function in qualitative or quantitative approaches. In this sense, EM offers a wide plethora of options to visualize intimate details of these

Table 12.1 Characteristics from light and electron microscopies: Differences and similarities.

Feature	Light microscopy	Electron microscopy
Illuminating source	photons from lamp or laser	electrons (filament or crystal); minimum electron acceleration: 1–400 kV
Wavelength	UV, visible, IR	electron beam (0.5 Å)
Type of Lens	glass or quartz	electromagnetic
Medium of illuminating particles	air at atmospheric pressure	high vacuum
Magnification	1,000–1,500 (conventional LM)	< 1,000,000
Resolution	200 nm (LM); few nm (SRLM)	> 0.5 nm
Type of samples	live (cell cultures, organisms), dead	died (dried, frozen)
Sample preparation time	direct to hours	direct to days
Visualization/Localization	coloured dyes (staining of conventional LM)	heavy metals (U, Pb, Os, Rt, Pt, Ta, Cr) to increase the conductivity and contrast of samples
	autofluorescence, fluorescent dyes (FM, CM)	electrodense trackers (gold and other metallic nanoparticles, quantum dots)
Support of samples	glass slide (conventional LM, FM, CM)	metal stub (SEM)
	glass or plastic dish (FM, CM)	metal grid (TEM)
Samples thickness	5–10 μm (paraffin or other embedding media)	70 nm (ultrathin sections for morphometric studies)
	1–100 μm (frozen); 1–400 μm (CM)	till 200–300 nm (ultrathin sections for analytical studies)
Image dimension	2D (sections)	2D (ultrathin sections for TEM)
	3D (stereology (LM), reconstruction (CM))	3D (SEM and TEM reconstruction (tomography))

UV—Ultraviolet; IR—Infrared; LM—Light microscopy; FM—Fluorescence microscopy; CM—Confocal microscopy; SRLM—Superresolution light microscopy; SEM—Scanning electron microscopy; TEM—Transmission electron microscopy.

organisms at nanoscale using the interaction between electrons and samples as source of information. However, when compared to other fields of knowledge, the potential applications of conventional and new EM techniques with moulds have been understudied in bioremediation/biodeterioration studies.

There is a great number of high resolution imaging devices coupled to a wide variety of detectors, which allows 2D or 3D more or less realistic images of a wide spectrum of liquid and solid samples to be obtained. Then, common nanoscale imaging has been used to characterize cell surface's roughness, geometry, and ultrastructural details at more or less native state. For instance, information about cells morphology and composition may be obtained with atomic force microscopy (AFM), tunneling force microscopy (TFM), near-field scanning optical microscopy (NSOM), and other types of scanning probe microscopy (SPM) have been used to study surfaces (Cech et al. 2013) and EM to study surfaces and inner details of mould samples (Cámara et al. 2017). For each specific type of samples or experimental approaches, these devices have different configurations, detectors, or tools to obtain better (rapid, easy, reproducible) specific results (Engel et al. 1997, Pakulska et al. 2016, Klein et al. 2017), such as focused ion beam (FIB), electron loss spectroscopy (EELS), or tip-enhanced Raman scattering (TERS), among others. Moreover, super resolution techniques, such as termed photo-activated localization microscopy (PALM), structured illumination microscopy (SIM), and stochastic optical reconstruction microscopy (STORM) are based in LM reaching a resolution of few nanometers (Betzig et al. 2006, Britton et al. 2013, Green et al. 2017). However, at present, EM is probably the most used high resolution imaging to study material and biological samples due to its versatility to reveal different ultrastructural and compositional information at nanoscale level. EM is used as a routine tool in other fields of knowledge, such as pathology or cell biology (Graham and Orenstein 2007, Melo et al. 2014), but ultrastructural studies related to morphometry of the moulds and their interactions with environment are still understudied.

The way we indicated above, there is a high number of electron microscopes with different configurations, characteristics, and detectors, however, we will focus on the most interesting aspects of the EM techniques applied to bioremediation or biodeterioration studies with moulds, taking into account the available information published until now (state of the art) and exploring future perspectives. Therefore, taking into account the foregoing, the main aims of this chapter are: (i) to summarize the available information on EM principles focusing on studies with moulds, (ii) to describe a general view of the potential uses of EM in bioremediation and biodeterioration processes in which moulds are involved, and (iii) to explore the EM future perspectives with moulds.

3. Electron Microscopy Equipment

Although other types of electron microscopes are available (Venables et al. 1994, Egerton 2005), scanning and transmission electron microscopies (SEM and TEM, respectively) are probably the most common and versatile EM techniques used. They have an important number of applications on life sciences, including in mould studies.

3.1 Scanning electron microscopy (SEM)

SEM is a type of electron microscopy mainly used to study the morphology of sample surfaces. Images are produced by scanning of the sample surface with a focused beam of electrons, usually working from less than 1 kV to 30 kV. The electrons from the beam interact with atoms of the sample and each of these interactions may be visualized with specific detectors to explore surface morphology and elemental composition. However, during the last decades, the SEM development, in techniques and resolution until less than 1 nm, has been possible to establish new and effective applications to obtain valuable information at nanoscale level. Different configurations and modes of use of SEM allow the imaging of surfaces until native state and study composition and distribution of elements. Dried samples may be observed at conventional high vacuum after conventional protocol of preparation, obtaining typical 3D images of SEM. In contrast, hydrated samples at room temperature (wet samples) may be observed at nearly native or native state using low vacuum of variable pressure or environmental (ESEM). Also, native frozen samples may be observed at low temperatures (LTSEM or cryoSEM, de los Ríos et al. (2009)). Interestingly, field emission scanning electron microscopes (FESEM, Rueda et al. (2015), Seras-Franzoso et al. (2016), Serna et al. 2016, 2017a, 2017b, Wu et al. (2016), Mukherjee et al. (2017)) or high resolution scanning electron microscopes (HRSEM) have a more stable and coherent electron gun that reach high resolution images at low voltages (LVSEM, Reimer (1998)).

3.2 Transmission electron microscopy (TEM)

TEM, the original EM, is based on highly accelerated electron beam transmitted through the sample (entire samples or sections) placed in a metal grid (copper, gold, titanium, nickel, among others) alone or coated with relatively conductive materials, such as plastic polymers (formvar, butvar) or carbon. The specimen or ultrathin section is often less than 100 nm in thickness (although specific studies needs thick sections of about 200 nm) that allows the beam to be transmitted through the sample obtaining the typical 2D images. Hence, TEM imaging is formed by the transmitted electrons that interact with and cross the samples (ultrathin sections or entire samples embedded in resins or frozen). This equipment is routinely used in several research fields as the best complementary technique to the light microscopy (LM), but reaching higher resolution images. TEM, and specially the high resolution TEM (HRTEM), is the most powerful electron microscope, working at high acceleration voltages and achieving a level of magnification and resolution capable of imaging virus and atoms (Wierzchos and Ascaso 1998). On the other hand, cryoTEM is another robust TEM technique based on the observation of frozen samples at nearly or native state (de los Ríos et al. 2009, Céspedes et al. 2014). With a rapid preparation of samples (plunged in a cryogen), cryoTEM has resolved multiple molecular difficulties due to formation of 2D and 3D high resolution images or reconstructions (Vinson et al. 1989, Frank 1996, Trejo et al. 2011, Frankl et al. 2015). Nowadays, as an interesting remark, the essential importance of cryoEM in different scientific fields, such as structural biology, has been recognized with the 2017 Nobel Prize of Chemistry to Dr. Ch. Dubochet, Dr. J. Frank, and Dr. R. Henderson for the development of this robust technique. Finally, energy filtered TEM (EFTEM) using inelastically scattered electrons or high resolution TEM (HRTEM) are a set of powerful imaging and analytical techniques using properties of the energy loss spectrum to increase contrast and, then, remove chromatic aberrations (Egerton 2009, 2011, Povedano-Priego et al. 2017).

Interestingly, scanning transmission electron microscopy (STEM) is a hybrid mode used in both SEM and TEM by incorporating appropriate detectors (X-rays detectors). In this equipment the electron beam is focused to a fine spot, which is then scanned over the sample. Images are formed by combining information of both scattered and transmitted electrons passing through a thin specimen or section, and obtaining 2D or 3D images (de Jonge and van Druten 2003, Pennycook and Colliex 2012, Gola et al. 2018).

4. Electron Microscopy Detectors

High resolution EM images are formed using the information of a variety of elastic and inelastic collisions between electrons of electron beam and atoms within the sample. Although virtually all phenomena of such interaction of an accelerated electron beam with a target sample may be detected, here we summarize the most significant image forming detectors for the study of ultrastructure of moulds and their interactions with the environment.

4.1 Secondary electron (SE) detectors

Images are produced by the emitted electrons from sample surface atoms by the collision of electrons from beam. The areas with the highest number of SE will appear as a "brighter" SE intensity and form the typical 3D images. Due to the high resolution until less than 1 nm reached by these detectors and the high variability of samples than can be checked, this type of detectors is the most used in SEM studies and become a crucial tool to explore ultrastructural morphometry of moulds samples, among others (Julián et al. 2010, Sánchez-Chardi et al. 2011, Cano-Garrido et al. 2016, Seras-Franzoso et al. 2016, Morando et al. 2017, Serna et al. 2017a). SE detectors have been used to study nanoscale details of surfaces (membranes, morphometry, etc). As a relevant detail, a new generation of high resolution detectors working efficiently at low voltage, low vacuum, or low temperatures allow the imaging of non-coated samples in a more native state (Céspedes et al. 2014, Rueda et al. 2015, Rueda et al. 2016, Cano-Garrido et al. 2016, Seras-Franzoso et al. 2016, Serna et al. 2016, Serna et al. 2017a, Serna et al. 2017b, Pesarrodona et al. 2017). These detectors are used in SEM and STEM (Bartoli et al. 2014). SEM images of the mycelium (typical hyphae structure) of *Aspergillus* sp., and the typical asexual reproductive structures (conidia) of the most common moulds, *Aspergillus* sp., *Penicillium* sp., and *Rhizopus* sp., are shown in Fig. 12.1.

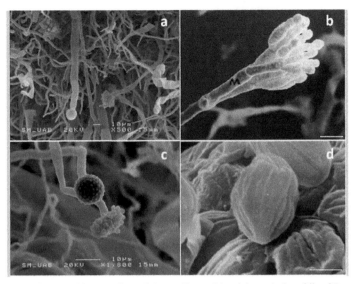

Figure 12.1 SEM images of the mycelium of *Aspergillus* sp. (a) and the typical conidia of *Penicillium* sp. Scale bar represents 1 μm (b), *Aspergillus* sp. (c), and *Rhizopus* sp. Scale bar represents 1 μm (d).

4.2 Back-scattered electron (BSE) detectors

There is a higher probability of producing an elastic collision of electrons and atoms with greater atomic number (for example: heavy metals) because of their greater cross-sectional area. Consequently, the number of backscattered electrons (BSE) reaching a BSE detector will be higher in areas with heavy elements and will appear as "brighter" BSE intensity. The energy of fluorescence may be also detected with BSE, obtaining similar information than confocal microscopy, but at nanoscale (Cámara et al. 2011, Cano-Garrido et al. 2016, Seras-Franzoso et al. 2016). With lower resolution than SE detectors, BSE images are often used for analytical ultrastructural studies in conjunction with other detectors, such as SE for morphological studies or X-rays detectors for compositional studies. These detectors are used in SEM and STEM (Cámara et al. 2017, Sohrabi et al. 2017) to detect extra and intracellular localization of metals, gold particles of secondary antibodies in immunolabeling studies linked to specific proteins, or fluorescence in cells and tissues (Seras-Franzoso et al. 2016). A SEM-BSE image of moulds is shown in Fig. 12.2.

4.3 Characteristic X-Rays (EDS, EDX, EDAX) detectors

Characteristic X-rays are emitted when the electron beam removes an inner shell electron from the sample, causing a higher-energy electron to fill the shell and release energy. These characteristic X-rays are an effective analytical technique based on two step signals used to identify the elemental composition, to measure the abundance, and to map the distribution of elements in the samples, as reported in Fig. 12.3. In available literature, this measurement is referred to as energy dispersive X-ray analysis (EDX or EDAX) or as energy dispersive spectroscopy (EDS) and

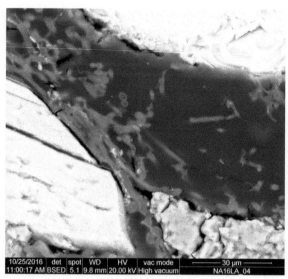

Figure 12.2 SEM-BSE image of fungal hyphae within fissures of a rock from the Namib deserts. Image courtesy of Asunción de los Ríos (MNCN-CSIC).

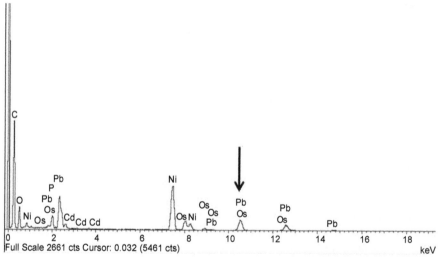

Figure 12.3 EDX spectrum. Ultrathin section from a culture polluted with lead and placed on a nickel grid. Arrow indicated the main Pb peak at 10.5 keV. Spectrum courtesy of Eduard Villagrasa (Dept. Genètica i Microbiologia UAB).

depends on the detection of characteristic X-rays produced by the interaction of electrons from beam and from atoms of sample. Peculiarly, electron microscopes with field emission gun increase the sensitivity of X-rays detection techniques in relation to thermionic ones by a high reduction of minimum detectable mass of atoms. As an example, a thickness of few atoms of iron may be detected with field emission gun, in contrast to a minimum thickness of 100–1000 atoms in thermionic ones (Michael and Williams 1987, Reimer 1997). The X-rays detectors have been widely used for qualitative or quantitative microanalyses of elements (mainly heavy metals), especially in bioremediation studies in metal polluted sites (Goldstein et al. 1992, Bozzola and Russell 1999, Baratelli et al. 2010, Maldonado et al. 2011, Carter and Williams 2016). These detectors may be used coupled to SEM and TEM microscopes to test the capacity of microorganisms for extra- and intracellular uptake of metals and have been widely used in experimental and field studies with different organisms (Burgos et al. 2013, Esteve et al. 2013) including moulds (Coreño-Alonso et al. 2014, Wu et al. 2016, Salem 2016, Mukherjee et al. 2017, Mansour et al. 2017, Povedano-Priego et al. 2017).

On the other hand, electron energy-loss spectrometry (EELS) is another TEM analytical technique less used than characteristic X-rays detectors, which offers chemical and structural information at high spatial resolution. This TEM technique is a one-step signal based on the detection of inelastic scattering of fast electrons after crossing a thin specimen (Egerton 2009, 2011, Hofer et al. 2016) only available in STEM. An EELS spectrum is shown in Fig. 12.4.

Finally, synchrotron radiation is produced by the acceleration of charged particles, including electrons to nearly the speed of light (Mobilio et al. 2015), and may offer suitable complementary analytical and morphological information to EM detectors. This approach was poorly studied in bioremediation and biodeterioration studies with moulds, but offered interesting results in heavy metals detection by the use of both synchrotron radiation based on X-ray absorption fine structure (XAFS) or X-ray absorption near edge structure (XANES) spectroscopies and scanning transmission soft-X-ray microscopy (STXM) (Muscente et al. 2015, Wu et al. 2016). The emerging full field soft X-ray microscopy enables obtaining images at resolutions

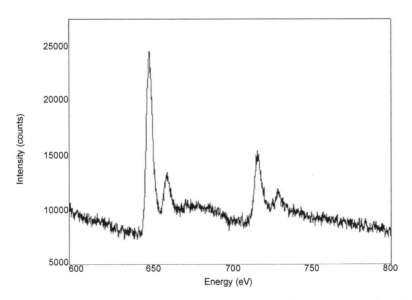

Figure 12.4 EELS spectrum. Sample containing manganese, peak at 640 eV, and iron, peak at 710 eV. Spectrum courtesy of Belén Ballesteros (ICN 2).

in the order of 50 nm and even lower and a quantitative estimation of the 3D structure of a whole cell by tomographic reconstruction techniques (Otón et al. 2016).

5. Electron Microscopy Protocols

Samples for SEM have to be prepared to withstand the vacuum conditions and high energy beam of electrons. Here, we summarize two examples of standard protocols (a rapid protocol and a standard methodology) for the imaging of ultrastructure for SEM and TEM observations. Although there are a high variety of protocols, including cryomethods that substitute chemical reagents for physical agents, the conventional chemical protocols for SEM and TEM continue to be the set of most used methodologies for EM in life sciences and the most adequate approach for a high variety of studies of morphology at ultrastructural level.

5.1 SEM protocols

A rapid protocol is a useful way to obtain near native state imaging of small size samples (from nanometers to few micrometers), preferably with low water content or in liquid suspension. The method consists in the deposition of a drop with sample in a silicon wafer (Ted Pella) or adhesive carbon film (Julián et al. 2010, Seras-Franzoso et al. 2016, Serna et al. 2016, Serna et al. 2017a, Serna et al. 2017b) from few seconds to few minutes, and dry excess with Whatman paper. As other rapid methods, samples are not fixed and need to be immediately observed (with or without coating) in a SEM at normal (15–30 kV) or low voltage (0.6–5 kV). A variety of samples (protein nanoparticles, lipid structures, viruses, bacteria, hard tissues) were observed with this rapid method (Bozzola and Russell 1999, Kuo 2014, Céspedes et al. 2014, Cano-Garrido et al. 2016, Noguera-Ortega et al. 2016, Rueda et al. 2016, Seras-Franzoso et al. 2016, Serna et al. 2016, Serna et al. 2017a, Serna et al. 2017b, Genís et al. 2017), but at present was not checked in several fields of science including moulds studies.

A conventional electron microscopy protocol for SEM involves: a chemical fixation of samples to preserve the original ultrastructure in general with aldehydes that mainly fixate proteins; a post-fixation usually with osmium or ruthenium tetraoxide to fixate unsaturated lipids when required to minimize the lipid extraction; dehydration usually with ethanol; critical point drying and coating with metals to increase SE signal (Julián et al. 2010, Burgos et al. 2013, Prakash and Nawani 2014, Cano-Garrido et al. 2016, Seras-Franzoso et al. 2016, Serna et al. 2017b). In detail, an example of conventional SEM protocol initiates with samples fixed in 2.5% (or more %) glutaraldehyde in Millonig buffer phosphate (Millonig 1961) for 2 hours and washed four times in the same buffer 10 minutes each, dehydrated in a graded series (30, 50, 70, 90, and 100%) of ethanol and dried by critical-point drying (namely the particular conditions of pressure and temperature in which liquid and vapor phases co-exist and have the same density, allowing the transition from liquid to gaseous form without abrupt change of volume) with CO_2 at high pressure or with hexamethyldisilazane (HMDS) at atmospheric pressure. Dried samples are mounted on metal stubs and coated with gold, platinum, or carbon, among other conductive metals, for better image contrast, and are observed in a SEM usually working at 15–30 kV. These processed samples are adequate for morphological and compositional studies, such as reported by Diestra et al. (2005), Esteve et al. (2013), Coreño-Alonso et al. (2014).

5.2 *TEM protocols*

A rapid protocol for TEM is negative staining, a widely used method to observe a variety of biological structures, including fungi in liquid suspension (Brenner and Horne 1959, Gregersen 1978, Larone 2014, Barreto-Vieira and Barth 2015, Rueda et al. 2015, Serna et al. 2016). A drop of few μl of a sample is deposited in a carbon or formvar (or other polymer)-coated TEM grid for few seconds to few minutes, liquid excess eliminated with a Whatman paper, contrasted with a drop of few μl of a water solution of 2% uranyl acetate for 1–2 minutes, excess dried with paper, air dried, and observed with a TEM working a normal voltage (75–200 kV) (Serna et al. 2016, Seras-Franzoso et al. 2016, Pesarrodona et al. 2017).

Conventional EM protocols involve a chemical fixation and the dehydration of samples, and moreover for TEM the subsequent embedding in acrylic or epoxy resins of samples, and sectioning at room temperature (ultramicrotomy) is also needed. During all this process, many cellular components, such as, structural or storage lipids, and cell wall proteins may be extracted, delocalized, lost, or form artifacts. In TEM, a primary chemical fixation with aldehydes and a secondary chemical fixation with osmium tetroxide (OsO_4) or ruthenium (RtO_4) is often required to minimize the protein and lipid extraction, respectively (Hayat 2000, Graham and Orenstein 2007, Burgos et al. 2013, Kuo 2014). The embedding media are not water miscible, and the use of intermediary organic solvents, during the dehydration process, is necessary. In detail, TEM samples are fixed in 2.5% (or more %) glutaraldehyde in Millonig buffer phosphate (Millonig 1961) for 2 hours, washed in the same buffer 4 times, post-fixed in 1% OsO_4 at 4°C for 2 hours, and washed again. Then, samples are dehydrated in a graded series (30, 50, 70, 90, and 100%) of acetone and embedded in Spurr resin. Ultrathin sections of 50–200 nm (usually 70 nm) are placed on metal TEM grids (directly or coated with carbon or a plastic film), air dried, and contrasted or post-stained with conventional uranyl acetate and lead citrate solutions to get a better sample contrast. As for negative staining samples, the samples may be observed at a wide variety of conditions in TEM (Diestra et al. 2005, Maldonado et al. 2011, Coreño-Alonso et al. 2014, Serna et al. 2017b). As a representative example with moulds, a TEM image of the hyphae ultrastructure of *Aspergillus* sp., is shown in Fig. 12.5.

As mentioned above, an EDX detector can be coupled with SEM and TEM, but some specifications about the sample treatment have to be taken into consideration in specific metal sorption studies in order to avoid obtaining positive results, when in reality they are not there. Pb citrate is one of the most common solutions to contrast the samples used in TEM protocols. The use of it should be ruled out for Pb sorption studies to prevent interference with the Pb really captured by the cells. On the other hand, to obtain a better contrast, the metallization is carried out with a noble metal (SEM), but if this metal has an energy peak in the EDX spectrum that overlaps with

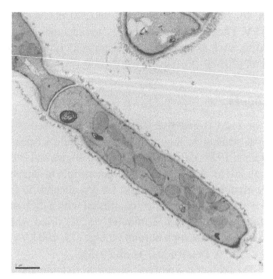

Figure 12.5 TEM image of the ultrastructure of an *Aspergillus* sp. hyphae. Scale bar represents 2 μm.

the metal being evaluated it should not be used, with the consequent lost in the image resolution.

Moreover, the presence of metals in some components of the electron microscopes, and mainly in the support grids, has to be considered prior to use in ecotoxicological studies with metals to avoid possible external sources of metal pollution. For example, in the case of copper unreliable results were obtained by TEM-EDX, since there were interferences with the copper material of the microscope itself (grid holder), and authors therefore considered the results to be not conclusive (Burgos et al. 2013). In addition, copper and gold were also detected in the EDX spectrum of the unpolluted cells because the gold grids, usually used in TEM studies, are made of Cu and recovered with Au (Baratelli et al. 2010, Esteve et al. 2013). Therefore, in studies with copper, other types of grids are recommended, such as titanium or nickel grids.

6. Applications

The ability of moulds to form extended mycelia networks, the low specificity of their catabolic enzymes, and their independence from using pollutants or different kinds of materials as a growth substrate make these moulds well-suited for bioremediation and biodeterioration processes (Singh 2006, Gadd et al. 2014, Harms et al. 2011, Gola et al. 2018, Obidi and Okekunjo 2017).

Although the light microscopy is the methodology mostly used to quickly and easily observe the structure of the fungal cells, up till now, the EM imaging has been applied to different kind of studies related to bioremediation and/or biodeterioration fungal abilities, providing successful results. SEM is used to observe the external morphology of microorganisms and their structural distribution in any natural/ polluted environmental sample, and TEM to analyze their ultrastructure. Thus, in order to assess any change in the fungal ultrastructure or in the structure of the material of interest, a combination of both LM and EM images of the material and fungus previous to their interaction are often required as controls of unaltered material and normal fungus ultrastructure.

6.1 Biorremediation

Environmental chemicals of particular concern include petroleum hydrocarbons, halogenated solvents from industrial sources, endocrine disrupting agents and drugs, explosives, agricultural chemicals, heavy metals, metalloids, and radionuclides (Harms et al. 2011). Moulds have the capacity to remove environmental organic chemicals and to decrease the risk associated with metals, metalloids, and radionuclides.

Up till now, most of the bioremediation studies in which fungi are involved have been focused in laboratory experiments with cultures of microorganisms isolated and characterized from polluted environments where there is high selective pressure. Moulds interact with metal and other pollutants using different ways, among them processes of: mobilization or immobilization, sorption to cells externally (exopolysaccharide envelopes) or internally bind to intracellular components (cytoplasm), and chemical transformation among metal species (Say et al. 2003, Binupriya et al. 2006, Chen et al. 2011, Coreño-Alonso et al. 2014, Urík et al. 2014, Gutiérrez-Corona et al. 2016, Mukherjee et al. 2017). In this sense, although the use of EM techniques in bioremediation studies is undervalued, they constitute a set of methodologies that has been applied, mainly in samples polluted with heavy metals, to get new information about the metal-fungal mycelium interactions and to allow a quick diagnosis about the capacity of fungi to adsorb and/or accumulate metals.

Most of this research has been carried out in axenic cultures. However, it is important to remark that in natural environments these microorganisms are constantly interacting with others and are influenced by environmental parameters changeable over time. Therefore, the effects of the interaction between moulds and pollutants could be different according to environmental or laboratory samples.

6.1.1 Structural changes on the surface of materials as a consequence of mould growth

Povedano-Priego et al. (2017) studied the lead biomineralization by the fungus *Penicillium chrysogenum* A15 growing on metallic lead (Pb shot pellets) by ESEM-EDX. In addition to abiotic corrosion, moulds cause corrosion of the metallic materials. They form crystals of different size and abundance, usually called secondary minerals, surrounding their own mycelia, although hyphae were not always detected. Biomineralization of lead phosphates (pyromorphite) by moulds is a well-known process (Leitão 2009, Rhee et al. 2014), which has been described to be mainly mediated by acidic phosphatase activity from organic phosphate substrates (Liang et al. 2015). Accumulated uranium was also biomineralized as uranyl phosphate minerals with abundant uranium precipitates encrusting the hyphae (Fomina et al. 2007, Fomina et al. 2008, Gadd and Fomina 2011). On the other hand, Gadd et al. (2014) showed SEM and BSE images of mycogenic metal oxalate precipitates formed by *Beauveria caledonica* and crystalline aggregates of copper oxalate (moolooite) in association with fungal hyphae within the lichen medulla of *Lecanora polytropa*. It is known that the formation of toxic metal oxalates may contribute to fungal metal tolerance (Gadd et al. 2014).

The moulds not only grow on or degrade metal surfaces; for example, Pramila and Ramesh (2011) studied the changes over time on the surface of low density polyethylene (LDPE, synthetic plastic) by *Aspergillus flavus* and *Mucor circinelloides* isolated from municipal landfill area. These changes observed from SEM images included cracks, formation of pits, and the presence of sporangia and spores grown throughout the LDPE film. Morphological changes on plastic surfaces were also described by Krueger et al. (2015) as a microbiological solution to environmental pollution with plastics.

Moulds are also effective in removing color from textile wastewater containing organic contaminants. Mahmoud et al. (2017) studied the bioremediation of red azo dye (organic nitrogen compounds) from aqueous solutions using *Aspergillus niger* strain isolated from textile wastewater. SEM images showed the high affinity of this mould to the direct red azo dye, confirming the discoloration process. Agnihotri and Agnihotri (2015) evaluated the effluent decolorizing ability of *Aspergillus oryzae* JSA-1 as a system of bioremediation of biomethanated distillery effluent (BME). TEM images proved that the mould decolorized the BME mainly by adsorption of colorants on their surface cells. Mycelia accumulated electron dense material majority around the cell wall and minimally in the cytoplasm. These results indicated that the discoloration process occurred in two steps, an important initial adsorption at mycelia cell wall level, followed by a minimal intracellular adsorption.

6.1.2 *Morphological and structural changes into the moulds as a consequence of the interaction with pollutants*

Prior to the analysis of the possible ultrastructural changes due to the interaction between the mould assayed and the material of interest, it is necessary to obtain LM and EM images from both parts separately. In addition, in the case of contamination by heavy metals and with the aim of proving whether the selected microorganism have the capacity to adsorb (externally) or accumulate (internally) metals, cells from laboratory cultures growing with and without the tested metal (control samples) generally are assessed by EDX coupled to SEM and TEM. In control samples, the metal should not be detected (negative result) either internally or externally in any of the microorganisms assayed. On the other hand, if a positive result is detected in polluted samples, we recommend to analyze the different parts of the filter/grid to ensure that the metal tested was retained only in cells, which corroborate their capacity to immobilize it. As an example, SEM and TEM images and the corresponding EDX spectra from a mould grown in absence or presence of chromium are shown in Fig. 12.6 (control or not polluted samples) and Fig. 12.7 (samples polluted with chromium).

Usually conventional EM images exhibited discernible changes (distortion of the cells, spores or mycelia) after exposure to pollutants, mainly at the highest concentrations. In the study of Povedano-Priego et al. (2017), changes in the morphology and the size of mould colonies growing in increasing Pb concentrations,

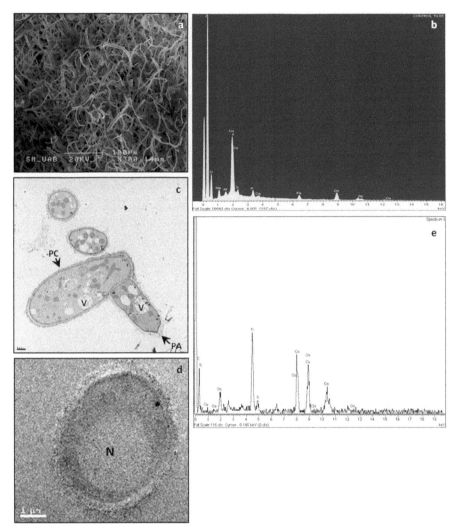

Figure 12.6 Unpolluted culture of *Aspergillus tubingensis*. SEM image (a). EDX spectrum coupled to SEM (b). TEM image of ultrathin section (70 nm). Arrows show apical pole (PA) and cell wall (PC). Vacuoles (V). Scale bar represents 2 μm (c). TEM image of ultrathin section (200 nm). Nucleus (N). Scale bar represent 1 μm (d). EDX spectrum coupled to TEM, from cellular cytoplasm (e).

the inhibition of spore formation, and the disappearance of the mycelium were detected by ESEM over time. In addition, the lead was localized as electron-dense precipitates on the fungal cell walls, as indicated by the HRTEM-EDX spectra. Furthermore, Wu et al. (2016) used the FESEM-EDX to analyze the effect of Cr (VI) in the structure of arbuscular mycorrhizal fungi (fungal-plant symbiosis). They observed, mainly at higher metal concentrations, the formation of large quantities of metal particles immobilized on the surface of hyphae and spores (probably in the extracellular polymeric substances (EPS) of the moulds), which tended to get bigger with time. Chitin and pigments (such as melanin) and their functional groups (e.g.,

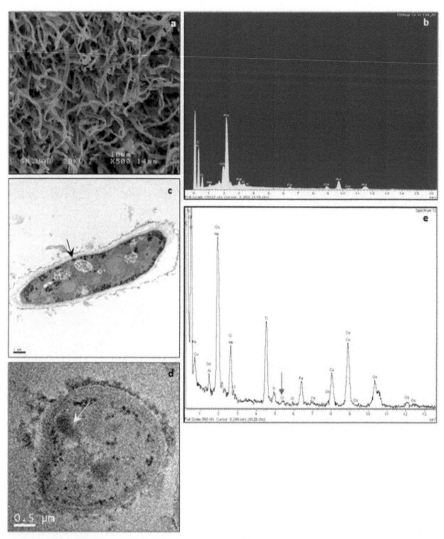

Figure 12.7 *Aspergillus tubingensis* culture treated with Cr (VI). SEM image (a). EDX spectrum coupled to SEM (b). TEM image of ultrathin section (70 nm). Arrow indicates the distribution of high electrondense inclusions. Scale bar represents 1 µm (c). TEM image of ultrathin section (200 nm). Arrow indicates low electron-dense inclusions. Scale bar represents 0.5 µm (d). EDX spectrum coupled to TEM, from a low electron-dense inclusion. Arrow indicated the main Cr peak at 5.4 keV (e).

phosphate and carboxyl groups) located in the EPS play an important role in this biosorption process (Gadd 2007).

Results obtained by Coreño-Alonso et al. (2014) showed that although the interaction of *Aspergillus niger* var *tubingensis* strain Ed8 with Cr (VI) is based mainly in a reduction process, the sorption process, determined by EM-EDX, is also carried out by this mould due to its ability to incorporate Cr intracellularly into low electrodense inclusions, but not extracellularly. The ability of *Paecilomyces lilacinus* XLA to capture chromium at extra- and intra-cellular levels and the possible structure alteration on fungus cells due to the metal presence was assessed by Xu et al. (2017). A large number of bigger vacuoles were observed in cells exposed to Cr (VI) and EDX analysis, and elemental mapping evidenced the intracellular distribution of chromium, in particular around the vacuoles, inner vacuoles, and periplasmic space, indicating that these intracellular components might be involved in the metal immobilization.

On the other hand, although the mycelium of *Penicillium chrysogenum* strain F1 was broken into fragments during the bioleaching of heavy metals from a contaminated soil process, the damage in the living cells analyzed by EM was minimal: a thinner cell wall, the presence of smaller vacuoles, and the cytoplasm concentration (Deng et al. 2012). Gola et al. (2016) applied EM for assessing the separate and combined effect of multiple metals in the morphological characteristics and surface properties of *Beauveria bassiana*. Tightly packed deformed hyphae with shortened length forming aggregates, a higher ultrastructural cell damage, and the metals accumulation both in membrane and inside the cell (dark dense area) occurred in metal mixture samples, and all metals peaks were detected in the same EDX spectrum. Thus, the heavy metal removal via sorption and accumulation mechanisms was proved by means of SEM-EDX and TEM-EDX. Moreover, Gola et al. (2018) using several EM techniques, designate this entomopathogenic fungus as a potential candidate for the remediation not only of samples polluted with heavy metals but also dyes.

All these filamentous eukaryotic microorganisms had strong resistance to heavy metals and had a great potential to be used to remove metals from polluted environments.

6.1.3 Intracellular biosynthesis of metal nanoparticles

The intracellular biosynthesis of metal nanoparticles from metal ions by different kind of microorganisms and their ability to remove metals from polluted environments is of great importance as an eco-friendly alternative to classical remediation methods. The most part of studies published are related to the synthesis of these nanoparticles by mainly bacteria and also by fungi.

The average size, morphology, and cellular location of these nanoparticles is determined by EM techniques. Images obtained by Salvadori et al. (2014) show, as an example, the intracellular localization of copper nanoparticles in *Rhodotorula mucilaginosa* (read yeast) with a mainly spherical shape and an average size of 10.5 nm. On the other hand, Mukherjee et al. (2017) studied the synthesis and adhesion of silver nanoparticles to the cell surface of *Aspergillus foetidus* MTCC8876 and their utility as an absorbent of arsenic. EM techniques were applied to evaluate the growth in size and diameter of these nanoparticles in presence of arsenic.

6.1.4 Localization and Chemical speciation of most chemical elements in nanoscale

These studies have been realized by means of Synchrotron radiation based X-ray absorption fine structure (XAFS) spectroscopy, used to determine the metal speciation, and Scanning transmission soft-X-ray microscopy (STXM) to determine localization and chemical speciation of elements at nanoscale. For example, according to Wu et al. (2016), the elemental distribution maps of Cr (III) and Cr (VI) in the hyphae contaminated with Cr (VI) corroborated the Cr surface accumulation obtained by SEM-EDS, previously indicated above, but in addition, the presence of Cr (III) indicates a reduction (transformation) of this Cr (VI) to Cr (III) as a strategy of bioremediation by the fungus.

6.2 Biodeterioration of cultural heritage

Any process arising from biological activity that affects cultural heritage is called biodeterioration. However, many environmental factors may positively or negatively contribute to the deteriorating actions of microorganisms (Scheerer et al. 2009, van Laarhoven et al. 2016). Thus, moulds play a considerable role for the deterioration of cultural heritage; they are able to destroy or modify paintings, textiles, paper, parchment, leather, oil, casein, glue, cork, and other materials (Allsopp et al. 2004, Mensah-Attipoe et al. 2016, Oetari et al. 2016, Cámara et al. 2017, Obidi and Okekunjo 2017). Biological deterioration takes place when microorganisms adhere to material surfaces, through the segregation of a complex matrix made of polysaccharides and proteins. The growth of filamentous microorganisms can also weaken polymers through the penetration of mycelia, which causes swelling and bursting of the material (Rhodes 2013, Dussud and Ghiglione 2014, Gadd et al. 2014, Salem 2016, Morando et al. 2017).

In museums and their storage rooms, climate control, regular cleaning, and microbiological monitoring are essential in order to prevent fungal contamination. The prevention of mould growth in museums as well as the development of appropriate treatment measures for contaminated objects is a challenge for restorers, museum curators, and architects. This has implications for the techniques of cleaning and conservation of objects, but also consequences for the occupational safety and health of restorers and other museum personnel. In outdoor environments moulds and lichens are the most important agents of biodeterioration of historic monuments and sculptures made of stone, mortar, and plaster (Barton and Wellheiser 1985, Manoharachary et al. 1997, Camuffo 1998, Gysels et al. 2004, Wainwright 2008, Geneste and Mauriac 2014, Mensah-Attipoe et al. 2016, Morando et al. 2017).

6.2.1 Fungi as deteriorative agents of museums

Contamination of pieces of art presented in exhibition rooms or stored in depots and their spoilage by moulds is not exceptional, but rather frequent in old and newly built museums. Moulds are able to grow, to alter, and to degrade all types of organic and inorganic materials. Pieces of art were made of all types of organic materials and these materials are again used for an authentic restoration or conservation of the objects in recent times: Paint was made of mineral pigments bound with organic binders, such as egg yolk, casein, linseed oil (poppy seed), hempseed oil, chinese wood oil, or different resins. Linen canvas clamped on wooden frames serves as painting ground and was often primed with rabbit skin glue before painting. Gold leaf on precious wooden or stucco frames was applied using organic glues, linseed, or turpentine oil. Glues were based on cellulose or rabbit skin. Sculptures and other art objects were made of textiles, leather, straw, clay, natural hair, or feathers (Barton and Wellheiser 1985, Bech-Andersen and Elborne 2004, Bastian et al. 2010, Bastian and Alabouvette 2009, George et al. 2013, Gadd et al. 2014, Oetari et al. 2016, Mansour et al. 2017, Obidi and Okekunjo 2017).

The majority of documents are books and scrolls made of paper, papyrus, and parchment. Because of the diversity of exoenzymes produced by moulds (cellulases, glucanases, laccases, phenolases, keratinases, mono-oxygenases, and many more) and their remarkable ability to grow at low water activity values, the preservation of museum objects is connected with prevention, monitoring, and treatment of mould on contaminated objects. A compilation of moulds frequently occurring on paper, paintings, and other materials in museums is given in Table 12.2.

Table 12.2 List of moulds frequently occurring on paper, paintings, and other materials.

Material	Mould species
Paintings: oil, water color, acrylic	*Alternaria* sp.,*Aspergillus flavus, Aspergillus sect. niger, A. sydowii, A. versicolor, Aureobasidium pullulans, Chaetomium funicola, Cladosporium herbarum, C. cladosporioides, Eurotium chevalieri, E. rubrum, Fusarium* sp., *Mucor* sp., *Penicillium chrysogenum, P. citrinum, P. decumbens* and many other species of the genus
Paper (laid-paper, wood pulp paper) and cellulose textiles (cotton, linen)	*Alternaria* sp., *Aspergillus clavatus, A. flavus, A. glaucus, A. terreus, A. repens, A. ruber, A. fumigatus, A. ochraceus, A. nidulans, Aspergillus sect. niger, Botrytis cinerea, Chaetomium globosum, C. elatum, C. indicum, Eurotium amstelodami, Fusarium* sp., *Mucor* sp., *Paecilomyces variotii, Penicillium chrysogenum, P. funiculosum, P. pupurogenum, P. rubrum, P. variabile, P. spinulosum, P. fellutatum, P. frequentans, P. citrinum, Pichia guilliermondi, Rhizopus oryzae, Stachybotrys chartarum, Trichoderma harzianum, T. viride, Stemphilium* sp., *Ulocladium* sp.
Parchment	*Cladosporium cladosporioides, Epicoccum nigrum, Penicillium chrysogenum*
Keratinous substrates (leather, wool, feathers, fur, hair)	*Absidia glauca, A. cylindrospora, A. spinosa, Acremonium* sp., *Alternaria alternate, Aspergillus sydowii, A. candidus, A. clavatus, A. carneus, A. foetidus, A. flavus, A. fumigatus, and many other species of the genus Arthroderma* sp., *Aureobasidium pullulans, Chaetomium globosum, Chrysosporium* sp., *Coniosporium* sp., *Cladosporium cladosporioides, Cunninghamella echinulata, C. elegans, Epicoccum nigrum, Emericella* sp., *Geotrichum candidum, Mucor* sp., *Penicillium brevicompactum, Penicillium chrysogenum* and many other species of the genus *Phoma medicaginis. Scopulariopsis* sp., *Stachybotrys chartarum, Trichophyton* sp. and *Rhizopus* sp.
Archeological findings: bones, ceramics. Archaeological findings often carry a large load of spores and mycelium	*Alternaria* spp., *Aspergillus flavus, A. glaucus, A. terreus, A. repens, A. ruber, A. fumigatus, A. ochraceus, A. nidulans, Aspergillus sect. niger, Botrytis cinerea, Chaetomium globosum, C. elatum, C. indicum, Eurotium amstelodami, Fusarium* spp., *Mucor* spp., *Paecilomyces variotii, Penicillium chrysogenum, P. funiculosum, P. pupurogenum, P. rubrum, P. variabile, P. spinulosum, P. fellutatum, P. frequentans, P. citrinum, Pichia guilliermondi, Rhizopus oryzae, Stachybotrys chartarum, Trichoderma harzianum, T. viride, Stemphilium* spp., *Ulocladium* spp., *Aureobasidium pullulans, Artrhinium* spp., *Epicoccum* spp., *Phoma* spp.

Fungal growth on objects of cultural heritage often causes a serious aesthetical spoiling due to colony formation and fungal pigments. Moreover, moulds degrade materials and thus affect objects substantially: the enzymatic degradation of organic binders causes reduction or even loss of paint layers. Moulds penetrate and migrate underneath paint layers, thus causing detachment. In paper conservation moulds are a special problem due to their ability to excrete cellulases. Lignin degrading moulds are rarely observed in indoor environments, but considerable damage can be caused by the cellulose degraders *Serpula lacrymans* or *Conophora puteana* in churches and other objects of cultural heritage if wooden altars or the roof structures are attacked. Also, originals and museum reconstructions of historical buildings are considerably damaged by *S. lacrymans* (Camuffo 1998, Nittérus 2000, Meier and Petersen 2006, Błyskal 2009, Mesquita et al. 2009, Pottier et al. 2014).

The development of moulds in museums is due to a large extent determined by several factors and parameters, including the indoor climate, the amount of available nutrients (from the atmosphere and from the materials themselves), and also by the cleaning intervals in the museum. The indoor climate as indicated by temperature, relative humidity, and by specific humidity is the most important factor for fungal growth (van Laarhoven et al. 2016). Then, because humidity up to 70% during weeks or months in storage rooms is a suitable condition to establish high fungal diversity, a range of 55% of humidity is generally regarded as the border line for fungal growth, and thus climate control is adjusted below this value in museums. It is also closely related to the buildings physical properties, especially the thermal insulation and the tendency to generate condensation from warm indoor air on cold walls of the building envelope. Depending on the climate in the museum or storage rooms, the fungal diversity is restricted to few xerophilic and xerotolerant species of genera *Eurotium*, *Aspergillus*, and *Wallemia*.

The chemical composition and potential biodegradability of objects also influence the development of fungal communities with species with different exo-enzymes. Also, the influence of air stream through doors, warming by sunlight, and daily changes of temperature gradients, as well as the isolation and exposition of the building envelope have to be considered as important factors. In fact, fungal growth mostly happens between shelves with little aeration or near walls with temperatures below the dew point. The atmospheric particular matter carrying carbonates, minerals, and others also influence the mycobiota in museums. Finally, it is also important to remark that micro-habitats are often also created by wrapping of single objects into plastic foils or putting them into extremely tight boxes, not allowing an exchange of air. In these conditions, vapor may produce fungal communities.

6.2.2 *Moulds as deteriorative agents of stone monuments*

From the biological point of view, a stone is an extreme environment, poor in nutrients, with enormous changes of humidity, with mechanical erosion due to wind and rain, and high doses of UV radiation. Nevertheless, stone is inhabited by moulds and other microorganisms in all climate regions of the earth. Epilitic moulds (living on the rock) and endolitic moulds (living inside of pores and fissures) play a major role in the weathering of monuments made of rock (de los Ríos et al. 2012, Miller et al. 2012, Morando et al. 2017, Sohrabi et al. 2017). Moulds might be the most important endolits on building stone, on mortar, and on plaster because their activity is high and they are extremely erosive.

On monuments there are two major morphological and ecological groups of moulds that are adapted to different environmental conditions. In moderate or humid climates, the fungal communities are dominated by hyphomycetes, including species of *Alternaria, Arthrinium, Cladosporium, Epicoccum, Aureobasidium*, and *Phoma*. In arid and semi-arid environments, the fungal community shifts towards black yeasts and microcolonial moulds. Black moulds belonging to the genera *Hortaea, Sarcinomyces, Coniosporium, Capnobotryella, Exophiala*, and *Trimmatostroma* form small black colonies on and inside the stone and often occur in close association with lichen. Famous monuments covered and deteriorated by moulds are the Acropolis of Athens and the antique temples of Delos (Gorbushina et al. 1993, Chatzidakis et al. 1997). Due to their thick, melanized cell walls, moulds also resist chemical attack and cannot easily be killed by biocides or other anti-microbial treatments. Black moulds dwell deep inside granite, calcareous limestone, and marble, and deteriorate those stones. The phenomenon of *biopitting*, namely the formation of lesions in a size range of up to 2 cm in diameter and depth on stone, is caused by the black moulds, among other microorganisms (de la Rosa et al. 2012, Lombardozzi et al. 2012). Due to the strong melanization of the cell walls, the stones inhabited by these moulds appear spotty or are even completely covered by black layers. In addition to outdoor environments, the black moulds are also found on rock surfaces of caves. The oldest and most precious objects suffering from serious fungal invasions are rock art caves, especially the Lascaux cave (Bastian et al. 2010, Geneste and Mauriac 2014, de la Rosa et al. 2017).

6.2.3 EM Methodologies on cultural heritage

Mycological research on biodeterioration in museums or monuments is still mainly based on classical cultivation methods using a variety of standardized media as MEA, DG18, and dichlorane rose bengale (DRBC) medium. Interestingly, the recovery rate concerning fungi in cultivation methods is assumed to be more than 70% of the total present in environmental samples, in contrast to recover less than 1% in bacteria. For this reason, culture-based approaches are still extremely useful in mycology. Nevertheless, the use of modern and sophisticated molecular techniques for the detection and identification of moulds on and in materials of cultural heritage will provide a deeper insight and understanding of fungal community structures and their consequences for the material (Sterflinger and Prillinger 2001, Urzì et al. 2003, Teertstra et al. 2004, Martin and Rygiewicz 2005, Michaelsen et al. 2006, de los Ríos et al. 2009, Portillo et al. 2008, Trejo et al. 2011, Obidi and Okekunjo 2017). However, concerning the practice of conservation, consolidation, and prevention of biodeterioration, the phenomenological analysis of the objects using a combination of both LM and EM techniques is the first and most important step (de los Ríos and Ascaso 2005, Cámara et al. 2017, Sohrabi et al. 2017).

Ultrastructural changes on the material surface as a consequence of mould growth have been analyzed by EM. As an example, the calcium oxalate crystals formation during fungal biodeterioration of paper by *Aspergillus terreus* was studied by Gadd et al. (2014) through high and low magnification EM images. This organic/inorganic interaction results in replacement of the original minerals with calcium oxalate with both the fungal hyphae and cellulose fibrils, providing preferential sites for precipitation of calcium oxalate dihydrate and monohydrate (Pinzari et al. 2010). Moreover, SEM is utilized to characterize the microbial communities attached to the surface of different sort of materials (Harshvardhan and Jha 2013, Zettler et al. 2013, Bartoli et al. 2014, Krueger et al. 2015, Salem 2016). For instance, Verma et al. (2011) used SEM micrographs to determine morphological changes based on the growth of *Aspergillus niger* in liquid media depending on the presence and type of solid support. The authors concluded that the growth on polyester sheets can be an interesting promise as a suitable solid support in industrial and environmental processes. Because of this, mould develops thick biofilms on its surface with a more channeled structure. SEM micrographs of various materials interfaces colonized by moulds are shown in Fig. 12.8.

The microorganisms can penetrate actively into the stones for protection or to obtain the carbon source for their metabolism. EM has been also used to study the role of the interface between microorganisms and rocks as deteriorative agents of stone monuments (de los Ríos et al. 2009, Cámara et al. 2011, Miller et al. 2012, Aloise et al. 2014, Bartoli et al. 2014, Pottier et al. 2014, Morando et al. 2017, Sohrabi et al. 2017). The application of several microscopy techniques, known as *in situ* microscopy, allows study of these processes of biodeterioration without the need to separate the biological components, epilithic or endolithic microorganisms,

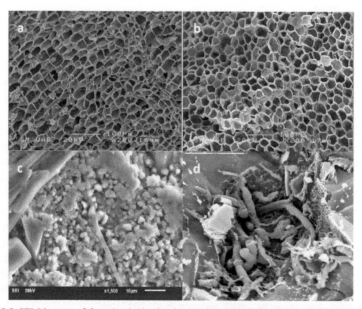

Figure 12.8 SEM images of fungal colonization in a cork sample before (a) and after an antifungical treatment (b). SEM image of a mould on the surface of a cement sample (c). LTSEM image of fungal hyphae within fissures of a rock from the Namib deserts. Scale bar represents 20 μm (d). Image (d) courtesy of Asunción de los Ríos (MNCN-CSIC).

from the lithic material, and thus observe the samples in their native state. SEM-BSE has been applied, for example, to investigate the endolithic colonization from stone (de los Ríos et al. 2009, Cámara et al. 2017, Maldonado et al. 2011), and to analyze the efficacy of different biocides against the microorganisms inhabiting in it (Cámara et al. 2011). In both cases, the pieces of rock were fixed, dehydrated, embedded in resin, and then polished previously to be visualized and analyzed by SEM (Wierzchos and Ascaso 1994). EDX in conjunction with both SEM and TEM is a very useful technique to study the chemical effects of moulds growing on different materials, such as rocks, observing mineral deposits associated with microbial cells, and the spatial distribution of the most abundant elements detected on its surface (Salem 2016, Cámara et al. 2017). On the other hand, the application of LTSEM in biodeterioration studies, in which moulds are involved, allows observing, in a shorter time, hydrated and non-conducting rock samples with a good image resolution (de los Ríos and Ascaso 2005, de los Ríos et al. 2009). An endolithic fungal image obtained by LTSEM is shown in Fig. 12.8D.

Future Perspectives

As summarized in this study, there is a plethora of available conventional and new EM techniques and applications, often routinely used in other fields of knowledge but understudied in bioremediation and biodeterioration studies with moulds. However, these techniques allow visualizing at nanoscale the morphology, composition, and interactions of moulds with their substrates in normal conditions or exposed to different types and levels of pollutants for bioremediation studies or to different environmental conditions (humidity, temperature, etc.) for biodeterioration studies.

Moreover, the improvement of these techniques for mould studies, such as adapting specific protocols for each type of sample or approach, improving methods by the reduction of time and reagents of long protocols, and exploring techniques, such as immunolocalization of specific proteins or different imaging with available detectors, are the three ways that may be considered for future experimental designs. Virtually in all the studies, the EM imaging may become the perfect complement of newer molecular or biochemical techniques, and routinely used in bioremediation/biodeterioration studies as occurs in other fields of knowledge, such as nanotechnology, pathology, or cell biology.

In conclusion, there is still a long way to explore the major benefits of the existing EM approaches in bioremediation and biodeterioration studies in which moulds are involved. The use of this powerful set of techniques will allow us to understand the morphometry, composition, and relationships of these organisms at a nanoscale level.

References

Abd-El-Aziz Abeer, R. M. and R. M. Sayed Shaban. 2016. Impact of heavy metals on genotoxic by ISSR, growth and biosorption by some fungi. *Research Journal Biotechnology* 11: 72–79.

Agnihotri, S. and S. Agnihotri. 2015. Study of bioremediation of biomethanated distillery effluent by *Aspergillus oryzae* JSA-1, using electron micrography and column chromatography techniques. *International Journal of Current Microbiology and Applied Sciences* 4 (7): 308–314.

Aloise, P., M. Ricca, M. F. La Russa, S. A. Ruffolo, C. M. Belfiore, G. Padeletti and G. M. Crisci. 2014. Diagnostic analysis of stone materials from underwater excavations: the case study of the roman archaeological site of baia (Nepal, Italy). *Applied Physics A* 114 (3): 655–662. doi: 10.1007/s00339-013-7890-1.

Allsopp, D., K. J. Seal and C. C. Gaylarde. 2004. *Introduction to Biodeterioration*. 2 ed. Cambridge: Cambridge University Press.

Asemoloye, M. D., R. Ahmad and J. Segun Gbolagade. 2017. Synergistic rhizosphere degradation of γ-hexachlorocyclohexane (lindane) through the combinatorial plant-fungal action. *PLOS ONE* 12 (8): e0183373. doi: 10.1371/journal.pone.0183373.

Baratelli, M., J. Maldonado, I. Esteve, A. Solé and E. Diestra. 2010. Electron microscopy techniques and energy dispersive X-ray applied to determine the sorption of lead in *Paracoccus* sp. DE2007 In: *Current Research Technology and Education topics in Applied Microbiology and Microbial Biotechnology. Microbial Book Series*, edited by A. Méndez-Vilas, 1601–1608. Formatex Research Center.

Barreto-Vieira, D. F. and O. M. Barth. 2015. Negative and positive staining in transmission electron microscopy for virus diagnosis. In: *Microbiology in Agriculture and Human Health*, edited by Mohammad Manjur Shah, 45–56. IntechOpen.

Bartoli, F., A. C. Municchia, Y. Futagami, H. Kashiwadani, K. H. Moon and G. Caneva. 2014. Biological colonization patterns on the ruins of angkor temples (cambodia) in the biodeterioration vs bioprotection debate. *International Biodeterioration & Biodegradation* 96: 157–165. doi: https://doi.org/10.1016/j.ibiod.2014.09.015.

Barton, J. P. and J. G. Wellheiser. 1985. *An ounce of prevention: a handbook on disaster contingency planning for archives, libraries and record centers*. Toronto, Canáda: Toronto Area Archivists Group Education Foundation.

Bastian, F. and C. Alabouvette. 2009. Lights and shadows on the conservation of a rock art cave: The case of *lascaux cave*. *International Journal of Speleology* 38 (1): 55–60.

Bastian, F., V. Jurado, A. Nováková, C. Alabouvette and C. Saiz-Jimenez. 2010. The microbiology of lascaux cave. *Microbiology* 156 (3): 644–652. doi: 10.1099/mic.0.036160-0.

Bech-Andersen, J. and S. A. Elborne. 2004. The true dry rot fungus (*Serpula lacrymans*) from nature to houses. In: *European research on cultural heritage: state-of-the-art studies*, edited by Miloš Drdácký, 445–448. Prague: Institute for Theoretical and Applied Mechanics, Academy of Sciences of the Czech Republic.

Betzig, E., G. H. Patterson, R. Sougrat, O. W. Lindwasser, S. Olenych, J. S. Bonifacino, M. W. Davidson, J. Lippincott-Schwartz and H. F. Hess. 2006. Imaging intracellular fluorescent proteins at nanometer resolution. *Science* 313 (5793): 1642.

Binupriya, A. R., M. Sathishkumar, K. Swaminathan, E. S. Jeong, S. E. Yun and S. Pattabi. 2006. Biosorption of metal ions from aqueous solution and electroplating industry wastewater by *Aspergillus japonicus*: phytotoxicity studies. *Bulletin of Environmental Contamination and Toxicology* 77 (2): 219–227. doi: 10.1007/s00128-006-1053-4.

Błyskal, B. 2009. Fungi utilizing keratinous substrates. *International Biodeterioration & Biodegradation* 63 (6): 631–653. doi: https://doi.org/10.1016/j.ibiod.2009.02.006.

Bozzola, J. J. and L. D. Russell. 1999. *Electron microscopy principles and techniques for biologists, The Jones and Bartlett series in biology*. Sudbury, Mass: Jones and Bartlett.

Brenner, S. and R. W. Horne. 1959. A negative staining method for high resolution electron microscopy of viruses. *Biochimica et Biophysica Acta* 34: 103–110. doi: https://doi.org/10.1016/0006-3002(59)90237-9.

Britton, S., J. Coates and S. P. Jackson. 2013. A new method for high-resolution imaging of Ku foci to decipher mechanisms of DNA double-strand break repair. *The Journal of Cell Biology* 202 (3): 579–595. doi: 10.1083/jcb.201303073.

Burgos, A., J. Maldonado, A. De los Rios, A. Solé and I. Esteve. 2013. Effect of copper and lead on two consortia of phototrophic microorganisms and their capacity to sequester metals. *Aquatic Toxicology* 140–141: 324–336. doi: https://doi.org/10.1016/j.aquatox.2013.06.022.

Cámara, B., A. de los Ríos, M. Urizal, M. Álvarez de Buergo, M. J. Varas, R. Fort and C. Ascaso. 2011. Characterizing the microbial colonization of a dolostone quarry: Implications for stone biodeterioration and response to biocide treatments. *Microbial Ecology* 62 (2): 299–313. doi: 10.1007/s00248-011-9815-x.

Cámara, B., M. Álvarez de Buergo, M. Bethencourt, T. Fernández-Montblanc, M. F. La Russa, M. Ricca and R. Fort. 2017. Biodeterioration of marble in an underwater environment. *Science of the Total Environment* 609: 109–122. doi: https://doi.org/10.1016/j.scitotenv.2017.07.103.

Camuffo, D. 1998. *Microclimate for Cultural Heritage*. 1 ed. Amsterdam: Elsevier Science

Cano-Garrido, O., A. Sánchez-Chardi, S. Parés, I. Giró, W. I. Tatkiewicz, N. Ferrer-Miralles, I. Ratera, A. Natalello, R. Cubarsi, J. Veciana, À. Bach, A. Villaverde, A. Arís and E. Garcia-Fruitós. 2016. Functional protein-based nanomaterial produced in microorganisms recognized as safe: A new platform for biotechnology. *Acta Biomaterialia* 43: 230–239. doi: https://doi.org/10.1016/j.actbio.2016.07.038.

Carter, B. and D. B. Williams. 2016. *Transmission Electron Microscopy: Diffraction, Imaging, and Spectrometry*. Switzerland Springer International Publishing.

Cébron, A., T. Beguiristain, J. Bongoua-Devisme, J. Denonfoux, P. Faure, C. Lorgeoux, S. Ouvrard, N. Parisot, P. Peyret and C. Leyval. 2015. Impact of clay mineral, wood sawdust or root organic matter on the bacterial and fungal community structures in two aged PAH-contaminated soils. *Environmental Science and Pollution Research* 22 (18): 13724–13738. doi: 10.1007/s11356-015-4117-3.

Cech, V., E. Palesch and J. Lukes. 2013. The glass fiber–polymer matrix interface/interphase characterized by nanoscale imaging techniques. *Composites Science and Technology* 83: 22–26. doi: https://doi.org/10.1016/j.compscitech.2013.04.014.

Céspedes, M. V., U. Unzueta, W. Tatkiewicz, A. Sánchez-Chardi, O. Conchillo-Solé, P. Álamo, Z. Xu, I. Casanova, J. L. Corchero, M. Pesarrodona, J. Cedano, X. Daura, I. Ratera, J. Veciana, N. Ferrer-Miralles, E. Vazquez, A. Villaverde and R. Mangues. 2014. *In vivo* architectonic stability of fully de *novo* designed protein-only nanoparticles. *ACS Nano* 8 (5): 4166–4176. doi: 10.1021/nn4055732.

Chatzidakis, P., G. Katsikis and M. Vartis-Matarangas. 1997. Delos sacred greece: characterization of the building stones, their origin and decay factors. In: *Engineering geology and the environment*, edited by P. G. Marinos, G. C. Koukis, G. C. Tsiambaos and G. C. Stournaras, 3089–3094. Rotterdam: A. A. Balkema.

Chen, S., Q. Hu, M. Hu, J. Luo, Q. Weng and K. Lai. 2011. Isolation and characterization of a fungus able to degrade pyrethroids and 3-phenoxybenzaldehyde. *Bioresource Technology* 102 (17): 8110–8116. doi: https://doi.org/10.1016/j.biortech.2011.06.055.

Coreño-Alonso, A., A. Solé, E. Diestra, I. Esteve, J. F. Gutiérrez-Corona, G. E. Reyna López, F. J. Fernández, and A. Tomasini. 2014. Mechanisms of interaction of chromium with *Aspergillus niger* var *tubingensis* strain Ed 8. *Bioresource Technology* 158: 188–192. doi: https://doi.org/10.1016/j.biortech.2014.02.036.

Das Murtey, M. 2016. Immunogold techniques in electron microscopy. In: *Modern Electron Microscopy in Physical and Life Sciences*, edited by Milos Janecek and Robert Kral, 143–160. IntechOpen.

de Jonge, N. and N. J. van Druten. 2003. Field emission from individual multiwalled carbon nanotubes prepared in an electron microscope. *Ultramicroscopy* 95 (1–4): 85–91. doi: https://doi.org/10.1016/S0304-3991(02)00301-7.

de la Rosa, J. M., P. M. Martin-Sanchez, S. Sanchez-Cortes, B. Hermosin, H. Knicker and C. Saiz-Jimenez. 2017. Structure of melanins from the fungi *Ochroconis lascauxensis* and *Ochroconis anomala* contaminating rock art in the lascaux cave. *Scientific Reports* 7 (1): 13441. doi: 10.1038/s41598-017-13862-7.

de la Rosa, J. P. M., P. A. Warke and B. J. Smith. 2012. Microscale biopitting by the endolithic lichen *Verrucaria baldensis* and its proposed role in mesoscale solution basin development on limestone. *Earth Surface Processes and Landforms* 37 (4): 374–384. doi: doi:10.1002/esp.2244.

de los Ríos, A. and C. Ascaso. 2005. Contributions of *in situ* microscopy to the current understanding of stone biodeterioration. *International Microbiology* 8: 181–188.

de los Ríos, A., B. Cámara, M. Á. García del Cura, V. J. Rico, V. Galván and C. Ascaso. 2009. Deteriorating effects of lichen and microbial colonization of carbonate building rocks in the romanesque churches of segovia (spain). *Science of the Total Environment* 407 (3): 1123–1134. doi: https://doi.org/10.1016/j.scitotenv.2008.09.042.

de los Ríos, A., S. Pérez-Ortega, J. Wierzchos and C. Ascaso. 2012. Differential effects of biocide treatments on saxicolous communities: Case study of the segovia cathedral cloister (spain). *International Biodeterioration & Biodegradation* 67: 64–72. doi: https://doi.org/10.1016/j.ibiod.2011.10.010.

Deng, X., L. Chai, Z. Yang, C. Tang, H. Tong and P. Yuan. 2012. Bioleaching of heavy metals from a contaminated soil using indigenous *Penicillium chrysogenum* strain F1. *Journal of Hazardous Materials* 233–234: 25–32. doi: https://doi.org/10.1016/j.jhazmat.2012.06.054.

Diestra, E., A. Solé, M. Martí, T. G. de Oteyza, J. O. Grimalt and I. Esteve. 2005. Characterization of an oil-degrading microcoleus consortium by means of confocal scanning microscopy, scanning electron microscopy and transmission electron microscopy. *Scanning* 27 (4): 176–180. doi: doi:10.1002/sca.4950270404.

Dussud, C. and J. F. Ghiglione. 2014. Bacterial degradation of synthetic plastics. In: *CIESM Workshop Monograph (No. 46). http://oceans.taraexpeditions.org/en/m/science/news/bacterial-degradation-of-synthetic-plastics/.*

Egerton, R. 2005. *Physical principles of electron microscopy: An introduction to TEM, SEM, and AEM*: Springer.

Egerton, R. F. 2009. Electron energy-loss spectroscopy in the TEM. *Reports on Progress in Physics* 72 (1):016502.

Egerton, R. F. 2011. *Electron Energy-Loss Spectroscopy In: The Electron Microscope.* 3 ed: Springer US.

Engel, A., C. A. Schoenenberger and D. J. Müller. 1997. High resolution imaging of native biological sample surfaces using scanning probe microscopy. *Current Opinion in Structural Biology* 7 (2): 279–284. doi: https://doi.org/10.1016/S0959-440X(97)80037-1.

Esteve, I., J. Maldonado, A. Burgos, E. Diestra, M. Burnat and A. Solé. 2013. Confocal laser scanning and electron microscopy techniques as powerful tools for determining the *in vivo* effect and sequestration capacity of lead in cyanobacteria. In: *Cyanobacteria: Ecology, Toxicology and Management*, edited by Aloysio Da S. Ferrão-Filho. Nova Publishers.

Fomina, M., J. Charnock, S. Hillier, R. Alvarez and G. M Gadd. 2007. Fungal transformations of uranium oxides. *Environmental Microbiology* 9 (7): 1696–710. doi: 10.1111/j.1462-2920.2007.01288.x.

Fomina, M., J. M. Charnock, S. Hillier, R. Alvarez, F. Livens and G. M. Gadd. 2008. Role of fungi in the biogeochemical fate of depleted uranium. *Current Biology* 18 (9): R375–R377. doi: https://doi.org/10.1016/j.cub.2008.03.011.

Frank, J. 1996. Three-dimensional electron microscopy of macromolecular assemblies. San Diego, Academic Press.

Frankl, A., M. Mari and F. Reggiori. 2015. Electron microscopy for ultrastructural analysis and protein localization in *Saccharomyces cerevisiae*. *Microbial Cell* 2 (11): 412–428. doi: 10.15698/mic2015.11.237.

Gadd, G. M. 2007. Geomycology: biogeochemical transformations of rocks, minerals, metals and radionuclides by fungi, bioweathering and bioremediation. *Mycological Research* 111 (1): 3–49. doi: https://doi.org/10.1016/j.mycres.2006.12.001.

Gadd, G. M., J. Bahri-Esfahani, Q. Li, Y. J. Rhee, Z. Wei, M. Fomina and X. Liang. 2014. Oxalate production by fungi: significance in geomycology, biodeterioration and bioremediation. *Fungal Biology Reviews* 28 (2): 36–55. doi: https://doi.org/10.1016/j.fbr.2014.05.001.

Gadd, G. M. and M. Fomina. 2011. Uranium and fungi. *Geomicrobiology Journal* 28 (5–6): 471–482. doi: 10.1080/01490451.2010.508019.

Gadd, G. M., Y. J. Rhee, K. Stephenson and Z. Wei. 2012. Geomycology: metals, actinides and biominerals. *Environmental Microbiology Reports* 4 (3): 270–296. doi: doi:10.1111/j.1758-2229.2011.00283.x.

Garcia, N. and H. Rohrer. 1989. Coherent electron beams and sources. *Journal of Physics: Condensed Matter* 1 (23): 3737.

Geneste, J. M. and M. Mauriac. 2014. The conservation of lascaux cave, france. In: *The Conservation of Subterranean Cultural Heritage*, edited by C. Saiz-Jimenez. CRC Press-Taylor and Francis Group.

Genís, S., A. Sánchez-Chardi, À. Bach, F. Fàbregas and A. Arís. 2017. A combination of lactic acid bacteria regulates *Escherichia coli* infection and inflammation of the bovine endometrium. *Journal of Dairy Science* 100 (1): 479–492. doi: https://doi.org/10.3168/jds.2016-11671.

George, R. P., S. Ramya, D. Ramachandran and U. Kamachi Mudali. 2013. Studies on biodegradation of normal concrete surfaces by fungus *Fusarium* sp. *Cement and Concrete Research* 47: 8–13. doi: https://doi.org/10.1016/j.cemconres.2013.01.010.

Glauert, A. M. 1974. The high voltage electron microscope in biology. *The Journal of Cell Biology* 63 (3): 717–748.

Gola, D., P. Dey, A. Bhattacharya, A. Mishra, A. Malik, M. Namburath and S. Z. Ahammad. 2016. Multiple heavy metal removal using an entomopathogenic fungi *Beauveria bassiana*. *Bioresource Technology* 218: 388–396. doi: https://doi.org/10.1016/j.biortech.2016.06.096.

Gola, D., A. Malik, M. Namburath and S. Z. Ahammad. 2018. Removal of industrial dyes and heavy metals by beauveria bassiana: FTIR, SEM, TEM and AFM investigations with Pb(II). *Environmental Science and Pollution Research* 25 (21): 20486–20496. doi: 10.1007/s11356-017-0246-1.

Goldstein, J., D. E. Newbury, P. Echlin, D. Joy, C. Fiori and E. Lifshin. 1992. *Scanning Electron Microscopy and X-Ray Microanalysis. A Text for Biologists, Materials Scientists, and Geologists*. 2 ed. Boston, MA: Springer.

Gorbushina, A. A., W. E. Krumbein, C. H. Hamman, L. Panina, S. Soukharjevski, and U. Wollenzien. 1993. Role of black fungi in color change and biodeterioration of antique marbles. *Geomicrobiology Journal* 11 (3–4): 205–221. doi: 10.1080/01490459309377952.

Graham, L. and J. M. Orenstein. 2007. Processing tissue and cells for transmission electron microscopy in diagnostic pathology and research. *Nature Protocols* 2: 2439. doi: 10.1038/nprot.2007.304.

Green, C. M., K. Schutt, N. Morris, R. M. Zadegan, W. L. Hughes, W. Kuang and E. Graugnard. 2017. Metrology of DNA arrays by super-resolution microscopy. *Nanoscale* 9 (29): 10205–10211. doi: 10.1039/c7nr00928c.

Gregersen, T. 1978. Rapid method for distinction of gram-negative from gram-positive bacteria. *European Journal of Applied Microbiology and Biotechnology* 5 (2):123-127. doi: 10.1007/bf00498806.

Gutiérrez-Corona, J. F., P. Romo-Rodríguez, F. Santos-Escobar, A. E. Espino-Saldaña and H. Hernández-Escoto. 2016. Microbial interactions with chromium: basic biological processes and applications in environmental biotechnology. *World Journal of Microbiology and Biotechnology* 32 (12): 191. doi: 10.1007/s11274-016-2150-0.

Gysels, K., F. Delalieux, F. Deutsch, R. Van Grieken, D. Camuffo, A. Bernardi, G. Sturaro, H.-J. Busse and M. Wieser. 2004. Indoor environment and conservation in the royal museum of fine arts, antwerp, belgium. *Journal of Cultural Heritage* 5 (2): 221–230. doi: https://doi.org/10.1016/j.culher.2004.02.002.

Harms, H., D. Schlosser and L. Y. Wick. 2011. Untapped potential: exploiting fungi in bioremediation of hazardous chemicals. *Nature Reviews Microbiology* 9:177. doi: 10.1038/nrmicro2519.

Harshvardhan, K. and B. Jha. 2013. Biodegradation of low-density polyethylene by marine bacteria from pelagic waters, Arabian Sea, India. *Marine Pollution Bulletin* 77 (1): 100–106. doi: https://doi.org/10.1016/j.marpolbul.2013.10.025.

Hawksworth, D. L. 1991. The fungal dimension of biodiversity: magnitude, significance, and conservation. *Mycological Research* 95 (6): 641–655. doi: https://doi.org/10.1016/S0953-7562(09)80810-1.

Hayat, M. A. 1972. *Basic electron microscopy techniques*. New York: Van Nostrand Reinhold Co.

Hayat, M. A. 2000. *Principles and Techniques of Electron Microscopy: Biological Applications*. 4 ed. London, UK: Cambridge University Press.

Hofer, F., F. P. Schmidt, W. Grogger and G. Kothleitner. 2016. Fundamentals of electron energy-loss spectroscopy. *IOP Conference Series: Materials Science and Engineering* 109 (1): 012007.

Julián, E., M. Roldán, A. Sánchez-Chardi, O. Astola, G. Agustí and M. Luquin. 2010. Microscopic cords, a virulence-related characteristic of *Mycobacterium tuberculosis*, are also present in nonpathogenic mycobacteria. *Journal of Bacteriology* 192 (7): 1751–1760. doi: 10.1128/JB.01485-09.

Kadri, T., T. Rouissi, S. Kaur Brar, M. Cledon, S. Sarma and M. Verma. 2017. Biodegradation of polycyclic aromatic hydrocarbons (PAHs) by fungal enzymes: A review. *Journal of Environmental Sciences* 51:52-74. doi: https://doi.org/10.1016/j.jes.2016.08.023.

Klein, A. E., N. Janunts, S. Schmidt, S. Bin Hasan, C. Etrich, S. Fasold, T. Kaiser, C. Rockstuhl and T. Pertsch. 2017. Dual-SNOM investigations of multimode interference in plasmonic strip wave guides. *Nanoscale* 9 (20): 6695–6702. doi: 10.1039/C6NR06561A.

Krueger, M. C., H. Harms and D. Schlosser. 2015. Prospects for microbiological solutions to environmental pollution with plastics. *Applied Microbiology and Biotechnology* 99 (21): 8857–8874. doi: 10.1007/s00253-015-6879-4.

Kuo, J. 2014. *Electron Microscopy: Methods and Protocols. Methods in Molecular Biology (Book 1117)*. Third ed. Totowa, NJ: Humana Press Inc.

Larone, D. H. 2014. *Medically Important Fungi: A Guide to Identification*. 5 ed. Vol. 45, *Laboratory Medicine*. Washington, DC: ASM Press.

Leitão, A. L. 2009. Potential of *Penicillium* species in the bioremediation field. *International Journal of Environmental Research and Public Health* 6 (4): 1393–1417. doi: 10.3390/ijerph6041393.

Liang, X., S. Hillier, H. Pendlowski, N. Gray, A. Ceci and G. M. Gadd. 2015. Uranium phosphate biomineralization by fungi. *Environmental Microbiology* 17 (6): 2064–2075. doi: doi:10.1111/1462-2920.12771.

Lombardozzi, V., T. Castrignanò, M. D'Antonio, A. C. Municchia and G. Caneva. 2012. An interactive database for an ecological analysis of stone biopitting. *International Biodeterioration & Biodegradation* 73: 8–15. doi: https://doi.org/10.1016/j.ibiod.2012.04.016.

Mahmoud, M. S., M. K. Mostafa, S. A. Mohamed, N. A. Sobhy and M. Nasr. 2017. Bioremediation of red azo dye from aqueous solutions by *Aspergillus niger* strain isolated from textile wastewater. *Journal of Environmental Chemical Engineering* 5 (1): 547–554. doi: https://doi.org/10.1016/j.jece.2016.12.030.

Maldonado, J., A. Solé, Z. M. Puyen and I. Esteve. 2011. Selection of bioindicators to detect lead pollution in ebro delta microbial mats, using high-resolution microscopic techniques. *Aquatic Toxicology* 104 (1): 135–144. doi: https://doi.org/10.1016/j.aquatox.2011.04.009.

Manoharachary, C., P. J. M. Reddy, B. Prabhakar and K. C. Mohan. 1997. Fungal spora and biodeterioration in some museums and libraries of Hyderabad, India. *Journal of Environmental Biology* 18: 37–42.

Mansour, M., R. Hassan and R. Salem. 2017. Characterization of historical bookbinding leather by FTIR, SEM-EDX and investigation of fungal species isolated from the leather. *Egyptian Journal of Archaeological and Restoration Studies* 7 (1): 1–10. doi: 10.21608/ejars.2017.6823.

Marco-Urrea, E., I. García-Romera and E. Aranda. 2015. Potential of non-ligninolytic fungi in bioremediation of chlorinated and polycyclic aromatic hydrocarbons. *New Biotechnology* 32 (6): 620–628. doi: https://doi.org/10.1016/j.nbt.2015.01.005.

Martin, K. J. and P. T. Rygiewicz. 2005. Fungal-specific PCR primers developed for analysis of the ITS region of environmental DNA extracts. *BMC Microbiology* 5: 28–28. doi: 10.1186/1471-2180-5-28.

McDonald, K. L. and R. I. Webb. 2011. Freeze substitution in 3 hours or less. *Journal of Microscopy* 243 (3): 227–233. doi: doi:10.1111/j.1365-2818.2011.03526.x.

McDonald, K. L. 2014. Rapid embedding methods into epoxy and LR white resins for morphological and immunological analysis of cryofixed biological specimens. *Microscopy and Microanalysis* 20 (1): 152–163. doi: 10.1017/S1431927613013846.

Meier, C. and K. Petersen. 2006. *Schimmelpilze auf Papier. Ein Handbuch für Restauratoren: Biologische Grundlagen, Erkennung, Behandlung und Prävention* Der Andere Verlag (Tönning).

Melo, R. C. N., E. Morgan, R. Monahan-Earley, A. M. Dvorak and P. F. Weller. 2014. Pre-embedding immunogold labeling to optimize protein localization at subcellular compartments and membrane microdomains of leukocytes. *Nature Protocols* 9: 2382. doi: 10.1038/nprot.2014.163.

Mensah-Attipoe, J., T. Reponen, A. M. Veijalainen, H. Rintala, M. Täubel, P. Rantakokko, J. Ying, A. Hyvärinen and P. Pasanen. 2016. Comparison of methods for assessing temporal variation of growth of fungi on building materials. *Microbiology* 162 (11): 1895–1903. doi: doi:10.1099/mic.0.000372.

Mesquita, N., A. Portugal, S. Videira, S. Rodríguez-Echeverría, A. M. L. Bandeira, M. J. A. Santos and H. Freitas. 2009. Fungal diversity in ancient documents. A case study on the archive of the university of coimbra. *International Biodeterioration & Biodegradation* 63 (5): 626–629. doi: https://doi.org/10.1016/j.ibiod.2009.03.010.

Michael, J. R. and D. B. Williams. 1987. A consistent definition of probe size and spatial resolution in the analytical electron microscope. *Journal of Microscopy* 147 (3): 289–303. doi: doi:10.1111/j.1365-2818.1987.tb02840.x.

Michaelsen, A., F. Pinzari, K. Ripka, W. Lubitz and G. Piñar. 2006. Application of molecular techniques for identification of fungal communities colonising paper material. *International Biodeterioration & Biodegradation* 58 (3): 133–141. doi: https://doi.org/10.1016/j.ibiod.2006.06.019.

Miller, A. Z., P. Sanmartín, L. Pereira-Pardo, A. Dionísio, C. Saiz-Jimenez, M. F. Macedo and B. Prieto. 2012. Bioreceptivity of building stones: A review. *Science of the Total Environment* 426: 1–12. doi: https://doi.org/10.1016/j.scitotenv.2012.03.026.

Millonig, G. 1961. A modified procedure for lead staining of thin sections. *The Journal of Biophysical and Biochemical Cytology* 11 (3): 736–739.

Mobilio, S., F. Boscherini and C. Meneghini. 2015. *Synchrotron radiation: Basics, methods and applications*: Springer-Verlag Berlin Heidelberg.

Morando, M., K. Wilhelm, E. Matteucci, L. Martire, R. Piervittori, H. A. Viles and S. E. Favero-Longo. 2017. The influence of structural organization of epilithic and endolithic lichens on limestone weathering. *Earth Surface Processes and Landforms* 42 (11): 1666–1679. doi: doi:10.1002/esp.4118.

Mukherjee, T., S. Chakraborty, A. A. Biswas and T. Das Kumar. 2017. Bioremediation potential of arsenic by non-enzymatically biofabricated silver nanoparticles adhered to the mesoporous carbonized fungal cell surface of *Aspergillus foetidus* MTCC8876. *Journal of Environmental Management* 201: 435–446. doi: https://doi.org/10.1016/j.jenvman.2017.06.030.

Muscente, A. D., M. F. Marc, J. G. Dale and S. Xiao. 2015. Assessing the veracity of precambrian 'sponge' fossils using *in situ* nanoscale analytical techniques. *Precambrian Research* 263: 142–156. doi: https://doi.org/10.1016/j.precamres.2015.03.010.

Nittérus, M. 2000. Moulds in archives and libraries, a literary survey. *Restaurator* 21 (1): 25–40.

Noguera-Ortega, E., N. Blanco-Cabra, R. M. Rabanal, A. Sánchez-Chardi, M. Roldán, S. Guallar-Garrido, E. Torrents, M. Luquin and E. Julián. 2016. Mycobacteria emulsified in olive oil-in-water trigger a robust immune response in bladder cancer treatment. *Scientific Reports* 6: 27232. doi: 10.1038/srep27232.

Obidi, O. and F. Okekunjo. 2017. Bacterial and fungal biodeterioration of discolored building paints in lagos, nigeria. *World Journal of Microbiology and Biotechnology* 33 (11): 196. doi: 10.1007/s11274-017-2362-y.

Oetari, A., T. Susetyo-Salim, W. Sjamsuridzal, E. A. Suherman, M. Monica, R. Wongso, R. Fitri, D. G. Nurlaili, D. C. Ayu and T. P. Teja. 2016. Occurrence of fungi on deteriorated old dluwang manuscripts from Indonesia. *International Biodeterioration & Biodegradation* 114: 94–103. doi: https://doi.org/10.1016/j.ibiod.2016.05.025.

Otón, J., E. Pereiro, A. J. Pérez-Berná, L. Millach, C. O. S. Sorzano, R. Marabini and J. M. Carazo. 2016. Characterization of transfer function, resolution and depth of field of a soft X-ray microscope applied to tomography enhancement by wiener deconvolution. *Biomedical Optics Express* 7 (12): 5092–5103. doi: 10.1364/BOE.7.005092.

Pakulska, M. M., S. Miersch and M. S. Shoichet. 2016. Designer protein delivery: From natural to engineered affinity-controlled release systems. *Science* 351 (6279).

Pennycook, S. J. and C. Colliex. 2012. Spectroscopic imaging in electron microscopy. *MRS Bulletin* 37 (1): 13–18. doi: 10.1557/mrs.2011.332.

Pesarrodona, M., E. Crosas, R. Cubarsi, A. Sánchez-Chardi, P. Saccardo, U. Unzueta, F. Rueda, L. Sanchez-García, N. Serna, R. Mangues, N. Ferrer-Miralles, E. Vázquez and A. Villaverde. 2017. Intrinsic functional and architectonic heterogeneity of tumor-targeted protein nanoparticles. *Nanoscale* 9 (19): 6427–6435. doi: 10.1039/C6NR09182B.

Pinzari, F., M. Zotti, A. De Mico and P. Calvini. 2010. Biodegradation of inorganic components in paper documents: Formation of calcium oxalate crystals as a consequence of *Aspergillus terreus* thom growth. *International Biodeterioration & Biodegradation* 64 (6): 499–505. doi: https://doi.org/10.1016/j.ibiod.2010.06.001.

Pinzari, F., J. Tate, M. Bicchieri, Y. J. Rhee and G. M. Gadd. 2013. Biodegradation of ivory (natural apatite): possible involvement of fungal activity in biodeterioration of the lewis chessmen. *Environmental Microbiology* 15 (4): 1050–1062. doi: doi:10.1111/1462-2920.12027.

Portillo, M. d. C., J. M. Gonzalez and C. Saiz-Jimenez. 2008. Metabolically active microbial communities of yellow and grey colonizations on the walls of altamira cave, spain. *Journal of Applied Microbiology* 104: 681–91. doi: 10.1111/j.1365-2672.2007.03594.x.

Pottier, D., V. Andre, J. P. Rioult, A. Bourreau, C. Duhamel, V. K. Bouchart, E. Richard, M. Guibert, P. Verite and D. Garon. 2014. Airborne molds and mycotoxins in *Serpula lacrymans*–damaged homes. *Atmospheric Pollution Research* 5 (2): 325–334. doi: https://doi.org/10.5094/APR.2014.038.

Povedano-Priego, C., I. Martín-Sánchez, F. Jroundi, I. Sánchez-Castro and M. L. Merroun. 2017. Fungal biomineralization of lead phosphates on the surface of lead metal. *Minerals Engineering* 106: 46–54. doi: https://doi.org/10.1016/j.mineng.2016.11.007.

Prakash, D. and N. N. Nawani. 2014. A rapid and improved technique for scanning electron microscopy of actinomycetes. *Journal of Microbiological Methods* 99: 54–57. doi: https://doi.org/10.1016/j.mimet.2014.02.005.

Pramila, R. and K. V. Ramesh. 2011. Biodegradation of low density polyethylene (LDPE) by fungi isolated from municipal landfill area. *Journal of Microbiology and Biotechnology Research* 1 (4): 6.

Reichelt, H., C. A. Faunce and H. H. Paradies. 2007. Elusive forms and structures of N-hydroxyphthalimide: the colorless and yellow crystal forms of N-hydroxyphthalimide. *The Journal of Physical Chemistry A* 111 (13): 2587–2601. doi: 10.1021/jp068599y.

Reimer, L. 1997. Elements of a transmission electron microscope. In: *Transmission Electron Microscopy: Physics of Image Formation and Microanalysis*, edited by Ludwig Reimer, 86–135. Berlin, Heidelberg: Springer Berlin Heidelberg.

Reimer, L. 1998. *Scanning Electron Microscopy: Physics of Image Formation and Microanalysis, Springer Series in Optical Sciences*: Springer-Verlag Berlin Heidelberg.

Reyes-César, A., Á. E. Absalón, F. J. Fernández, J. M. González and D. V. Cortés-Espinosa. 2014. Biodegradation of a mixture of PAHs by non-ligninolytic fungal strains isolated from crude oil-contaminated soil. *World Journal of Microbiology and Biotechnology* 30 (3): 999–1009. doi: 10.1007/s11274-013-1518-7.

Rhee, Y., S. Hillier, H. Pendlowski and G. M. Gadd. 2014. Pyromorphite formation in a fungal biofilm community growing on lead metal. *Environmental Microbiology* 16: 1441–1451. doi: 10.1111/1462-2920.12416.

Rhodes, C. J. 2012. Feeding and healing the world: through regenerative agriculture and permaculture. *Science Progress* 95 (4): 345–446. doi: 10.3184/003685012X13504990668392.

Rhodes, C. J. 2013. Applications of bioremediation and phytoremediation. *Science progress* 96 (Pt 4): 417–427. doi: 10.3184/003685013x13818570960538.

Rhodes, C. J. 2014. Mycoremediation (bioremediation with fungi)—growing mushrooms to clean the earth. *Chemical Speciation & Bioavailability* 26 (3): 196–198. doi: 10.3184/095422914X140474 07349335.

Rodríguez-Cariño, C., C. Duffy, A. Sánchez-Chardi, F. McNeilly, G. M. Allan and J. Segalés. 2011. Porcine circovirus type 2 morphogenesis in a clone derived from the L35 lymphoblastoid cell line. *Journal of Comparative Pathology* 144 (2): 91–102. doi: https://doi.org/10.1016/j.jcpa.2010.07.001.

Rueda, F., M. V. Céspedes, O. Conchillo-Solé, A. Sánchez-Chardi, J. Seras-Franzoso, R. Cubarsi, A. Gallardo, M. Pesarrodona, N. Ferrer-Miralles, X. Daura, E. Vázquez, E. García-Fruitós, R. Mangues, U. Unzueta and A. Villaverde. 2015. Bottom-up instructive quality control in the biofabrication of smart protein materials. *Advanced Materials* 27 (47): 7816–7822. doi: 10.1002/adma.201503676.

Rueda, F., B. Gasser, A. Sánchez-Chardi, M. Roldán, S. Villegas, V. Puxbaum, N. Ferrer-Miralles, U. Unzueta, E. Vázquez, E. Garcia-Fruitós, D. Mattanovich and A. Villaverde. 2016. Functional inclusion bodies produced in the yeast *Pichia pastoris*. *Microbial Cell Factories* 15 (1): 166. doi: 10.1186/s12934-016-0565-9.

Salem, M. Z. M. 2016. EDX measurements and SEM examination of surface of some imported woods inoculated by three mold fungi. *Measurement* 86: 301–309. doi: https://doi.org/10.1016/j.measurement.2016.03.008.

Salvadori, M. R., R. A. Ando, C. A. Oller do Nascimento and B. Corrêa. 2014. Intracellular biosynthesis and removal of copper nanoparticles by dead biomass of yeast isolated from the wastewater of a mine in the brazilian amazonia. *PLOS ONE* 9 (1): e87968. doi: 10.1371/journal.pone.0087968.

Sánchez-Castro, I., V. Gianinazzi-Pearson, J. C. Cleyet-Marel, E. Baudoin and D. van Tuinen. 2017. Glomeromycota communities survive extreme levels of metal toxicity in an orphan mining site. *Science of the Total Environment* 598: 121–128. doi: https://doi.org/10.1016/j.scitotenv.2017.04.084.

Sánchez-Chardi, A., F. Olivares, T. F. Byrd, E. Julián, C. Brambilla and M. Luquin. 2011. Demonstration of cord formation by rough *Mycobacterium abscessus* variants: Implications for the clinical microbiology laboratory. *Journal of Clinical Microbiology* 49 (6): 2293–2295. doi: 10.1128/JCM.02322-10.

Say, R., N. Yilmaz and A. Denizli. 2003. Removal of heavy metal ions using the fungus *Penicillium Canescens*. *Adsorption Science & Technology* 21 (7): 643–650. doi: 10.1260/026361703772776420.

Scheerer, S., O. Ortega-Morales and C. Gaylarde. 2009. Microbial deterioration of stone monuments—An updated overview. In: *Advances in Applied Microbiology*, 97–139. Academic Press.

Seras-Franzoso, J., A. Sánchez-Chardi, E. Garcia-Fruitós, E. Vázquez and A. Villaverde. 2016. Cellular uptake and intracellular fate of protein releasing bacterial amyloids in mammalian cells. *Soft Matter* 12 (14): 3451–3460. doi: 10.1039/C5SM02930A.

Serna, N., M. V. Céspedes, P. Saccardo, Z. Xu, U. Unzueta, P. Álamo, M. Pesarrodona, A. Sánchez-Chardi, M. Roldán, R. Mangues, E. Vázquez, A. Villaverde and N. Ferrer-Miralles. 2016. Rational engineering of single-chain polypeptides into protein-only, BBB-targeted nanoparticles.

Nanomedicine: Nanotechnology, Biology and Medicine 12 (5): 1241–1251. doi: https://doi. org/10.1016/j.nano.2016.01.004.

Serna, N., M. V. Céspedes, L. Sánchez-García, U. Unzueta, R. Sala, A. Sánchez-Chardi, F. Cortés, N. Ferrer-Miralles, R. Mangues, E. Vázquez and A. Villaverde. 2017a. Peptide-based nanostructured materials with intrinsic proapoptotic activities in CXCR4+ solid tumors. *Advanced Functional Materials* 27 (32): 1700919. doi: doi:10.1002/adfm.201700919.

Serna, N., L. Sánchez-García, A. Sánchez-Chardi, U. Unzueta, M. Roldán, R. Mangues, E. Vázquez and A. Villaverde. 2017b. Protein-only, antimicrobial peptide-containing recombinant nanoparticles with inherent built-in antibacterial activity. *Acta Biomaterialia* 60: 256–263. doi: https://doi. org/10.1016/j.actbio.2017.07.027.

Singh, H. 2006. *Mycoremediation: fungal Bioremediation. Wiley-Interscience. John Wiley & Sons, Inc. Publication.*

Skepper, J. N. 2000. Immunocytochemical strategies for electron microscopy: choice or compromise. *Journal of Microscopy* 199 (1): 1–36. doi: doi:10.1046/j.1365-2818.2000.00704.x.

Sohrabi, M., S. E. Favero-Longo, S. Pérez-Ortega, C. Ascaso, Z. Haghighat, M. H. Talebian, H. Fadaei and A. de los Ríos. 2017. Lichen colonization and associated deterioration processes in pasargadae, UNESCO world heritage site, Iran. *International Biodeterioration & Biodegradation* 117: 171–182. doi: https://doi.org/10.1016/j.ibiod.2016.12.012.

Speranza, M., J. Wierchos, J. Alonso, L. Bettucci, A. Martín-González and C. Ascaso. 2010. Traditional and new microscopy techniques applied to the study of microscopic fungi included in amber In: *Microscopy: Science, Technology, Applications and Education*, edited by A. Mendez-Vilas and J. Diaz, 1135–1145. Formatex Research Center.

Sterflinger, K. and H. Prillinger. 2001. Molecular taxonomy and biodiversity of rock fungal communities in an urban environment (Vienna, Austria). *Antonie van Leeuwenhoek* 80 (3): 275–286. doi: 10.1023/a:1013060308809.

Teertstra, W. R., L. G. Lugones and H. A. B. Wösten. 2004. *In situ* hybridisation in filamentous fungi using peptide nucleic acid probes. *Fungal Genetics and Biology* 41 (12): 1099–1103. doi: https:// doi.org/10.1016/j.fgb.2004.08.010.

Trejo, R., T. Dokland, J. Jurat-Fuentes and F. Harte. 2011. Cryo-transmission electron tomography of native casein micelles from bovine milk. *Journal of Dairy Science* 94 (12): 5770–5775. doi: https:// doi.org/10.3168/jds.2011-4368.

Urík, M., M. Hlodák, P. Mikušová and P. Matúš. 2014. Potential of microscopic fungi isolated from mercury contaminated soils to accumulate and volatilize Mercury(II). *Water, Air, & Soil Pollution* 225 (12): 2219. doi: 10.1007/s11270-014-2219-z.

Urzì, C., V. La Cono, F. De Leo and P. Donato. 2003. Fluorescent *in situ* hybridization (FISH) to study biodeterioration. International Congress on Molecular Biology and Cultural Heritage, Seville, Spain.

van Laarhoven, K. A., H. P. Huinink and O. C. G. Adan. 2016. A microscopy study of hyphal growth of *Penicillium rubens* on gypsum under dynamic humidity conditions. *Microbial Biotechnology* 9 (3): 408–418. doi: 10.1111/1751-7915.12357.

Varjani, S. J. 2017. Microbial degradation of petroleum hydrocarbons. *Bioresource Technology* 223: 277–286. doi: https://doi.org/10.1016/j.biortech.2016.10.037.

Venables, J. A., C. J. Harland, P. A. Bennett and T. E. A. Zerrouk. 1994. Electron diffraction in UHV SEM, REM, and TEM. Proceedings—Annual Meeting, Microscopy Society of America, New Orleans, LA, USA.

Veneklasen, L. H. 1972. Some general considerations concerning the optics of the field emission illumination system. *Optik* 36: 410–433.

Verma, N., M. C. Bansal and V. Kumar. 2011. Scanning electron microscopic analysis of *Aspergillus niger* pellets and biofilms under various process conditions. *International Journal of Microbiological Research* 2 (1): 08–11.

Vinson, P. K., Y. Talmon and A. Walter. 1989. Vesicle-micelle transition of phosphatidylcholine and octyl glucoside elucidated by cryo-transmission electron microscopy. *Biophysical Journal* 56 (4): 669–681.

Wainwright, M. 2008. Some highlights in the history of fungi in medicine—A personal journey. *Fungal Biology Reviews* 22 (3): 97–102. doi: https://doi.org/10.1016/j.fbr.2008.11.001.

Wierzchos, J. and C. Ascaso. 1994. Application of back-scattered electron imaging to the study of the lichen-rock interface. *Journal of Microscopy* 175 (1): 54–59. doi: doi:10.1111/j.1365-2818.1994. tb04787.x.

Wierzchos, J. and C. Ascaso. 1998. Mineralogical transformation of bioweathered granitic biotite, studied by HRTEM; evidence for a new pathway in lichen activity. *Clays and Clay Minerals* 46 (4): 446–452.

Williams, D. B. and C. B. Carter. 2009. Electron sources. In: *Transmission Electron Microscopy: A Textbook for Materials Science*, edited by David B. Williams and C. Barry Carter, 73–89. Boston, MA: Springer US.

Wu, S., X. Zhang, Y. Sun, Z. Wu, T. Li, Y. Hu, J. Lv, G. Li, Z. Zhang, J. Zhang, L. Zheng, X. Zhen and B. Chen. 2016. Chromium immobilization by extra- and intraradical fungal structures of arbuscular mycorrhizal symbioses. *Journal of Hazardous Materials* 316: 34–42. doi: https://doi.org/10.1016/j.jhazmat.2016.05.017.

Xu, X., L. Xia, W. Chen and Q. Huang. 2017. Detoxification of hexavalent chromate by growing paecilomyces lilacinus XLA. *Environmental Pollution* 225: 47–54. doi: https://doi.org/10.1016/j.envpol.2017.03.039.

Zettler, E. R., T. J. Mincer and L. A. Amaral-Zettler. 2013. Life in the "plastisphere": Microbial communities on plastic marine debris. *Environmental Science & Technology* 47 (13): 7137–7146. doi: 10.1021/es401288x.

Index

A

aeration in decoloration 145
agricultural practices 239
Antibiotics 186–196, 198, 199, 212–215, 218–220, 226, 227
Applications 356, 358, 359, 366, 377
Arbuscular mycorrhizae 244–247, 252, 253, 255
Aromatic peroxygenases 25, 34
Arthromyces ramosus peroxidase 25, 31
available transfer factors (ATF) 243
azo dye 125–127, 129, 132, 141, 144–147, 149–152, 157, 159–162
azo dye structure 149

B

basidiomycetes 5, 10, 11, 13
Bioaccumulation 12, 176
Biodegradation 4, 10, 11, 13
Biodeterioration 354–356, 358, 363, 366, 372, 375–377
Biofertilizers 335
Biofilters 265–268, 271–280
Biofuels 320, 324, 336
biomass 4, 9, 10
Bioreactor 192, 193, 198–211
bioremediation 3–7, 9, 12, 14, 22, 24, 26, 28, 46, 49, 62, 63, 65, 70, 72–75, 77, 298, 300, 354–356, 358, 363, 366–368, 372, 377
biosorption 9, 10
biosorption and bioreduction of Cr (VI) 178
Biosorption of metal 9
biotechnological applications 175
Biotrickling filters 265, 266

C

Caldariomyces fumago chloroperoxidase 25, 33
carbon source and decoloration 144
Catabolism of aromatics 288, 304
Chernobyl 238, 239, 241–244, 247, 248, 250, 251, 255
Chromium biotransformation 177, 367
Co-metabolism 96, 113, 117
coprinus cinereus peroxidase 25, 31

D

Dechlorination 94, 96, 98, 101, 105–108, 116, 117
decoloration 125–130, 132–136, 138–154, 156–162
degradation 4, 6–8, 10–14
Degradation of polycyclic aromatic hydrocarbons 11
Degradation Pathway 94, 97, 100–103, 106–110, 113, 114, 117
Dichlorophenols 94, 98, 103, 104
Dye decolorization peroxidases 25, 31
Dyes 22, 25, 26, 30–32, 37–41, 46–49, 147

E

Ecotoxicity 188, 226–228
ectomycorrhizae 242, 247, 250
Electron microscopy 354–360, 364
Elimination capacity 266, 267, 269–271, 273, 277–279
Endocrine disrupting compounds 41
Environmental pollution 288, 290, 307
enzyme and decoloration 147, 150
Enzymes 323–325, 327, 329–331, 333, 334, 338, 340, 341

F

Filamentous fungi 287, 288, 294, 298, 303, 306, 309
filamentous fungi 3–6, 9, 11
food chain 239, 245, 247
Fukushima 238, 239
Fungal Peroxidases 22, 24–26, 28, 30–32, 35, 36, 38, 39, 41–46, 48, 49
fungi 125–128, 132, 133, 144–152, 156, 160–162, 174–178, 181, 186–188, 190–195, 212, 214

G

Gaseous pollutants 274
Gene 161

H

Heme-thiolate peroxidases 25, 32
hot particles 239, 240, 251
Hydrophobic volatile compounds 268
Hydrophobins 268, 269, 275
hyperaccumulators 244

I

Immobilized fungi 151
Immobilization of enzyme 13, 46
Immobilization of radionuclides 242

L

Laccases 63–65, 67–70, 73–75, 77, 96, 97,
        99–102, 104, 106, 107, 109, 111, 115
lichens 241, 255
Lignin peroxidase 25, 28, 30
Ligninolytic peroxidases 24, 25, 27, 45, 48
Lignocellulose 324, 332, 333
LiP 100, 101, 103, 104, 106, 107, 111

M

Manganese peroxidase 25, 29
Mass transfer 268, 269, 273, 274, 276–278, 280
mechanisms 10–13
Metabolomics 302, 303
Metal bioaccumulation 9
Methylation 100, 101, 103, 109, 112, 117
microscopy detectors 360
microscopy protocols 364
mineralization 10
MnP 99–104, 106, 107, 109, 111
Monochlorophenols 94–99
Moulds 354–356, 358, 360, 361, 363–369,
        372–377
Multi-Omics 338, 339, 343
myco-remediation 240

N

nitrogen source and decoloration 145
Non-ligninolytic peroxidases 25, 45, 48

O

Oxidation 96, 97, 100, 101, 103, 106, 107, 109

P

Pentachlorophenol 94, 109, 110, 112, 113, 115,
        287–299, 306, 308
Peroxidases 22, 24–39, 41–46, 48, 49

Pesticides 22, 25, 44
Pharmaceuticals 22, 42, 43
Phenol oxidases 62, 67
Phenols 24–26, 29–33, 45, 46, 48, 49
phytoremediation 240, 244–246, 249, 250, 253
Polycyclic aromatic hydrocarbons 22, 35
Polymerization 94, 97, 102, 105, 109, 117
Pressure drop 269, 277, 279, 280
Proteomics 287, 288, 292–294, 298, 303

Q

Quorum Sensing 341

R

Radionuclides 238–255
Removal 186, 188–196, 198–200, 202, 204, 206,
        208, 210–216, 218–220, 222, 224, 226, 227
Removal efficiency 266, 272, 273

S

saprotrophic fungi 244
saprotrophs 5
Single Cell Proteins 330
Sorption 4, 6–10

T

Tetrachlorophenols 94, 109, 112
Transfer ratios 248, 249
translocation 241, 245–249, 252, 254, 255
transport and toxicity of chromium 181
treatment 10
Trichlorophenols 94, 96, 105, 109, 111, 112
Tyrosinases 63, 65–70, 73, 75–77

U

Ultrastructure 356, 360, 364–367
Uranium 239, 244, 252–254

V

Versatile peroxidase 25, 30
Volatile organic compounds 267–269

W

White-rot fungi 10, 188–193

Y

yeast 126, 128–130, 132, 133, 135, 137,
        139–145, 161

Milton Keynes UK
Ingram Content Group UK Ltd.
UKHW020319111024
449327UK00040B/1397